WITHDRAWN FROM
KENT STATE UNIVERSITY LIBRARIES

ETHYLENE AND PLANT DEVELOPMENT

Proceedings of Previous Easter Schools in Agricultural Science, published by Butterworths, London

*SOIL ZOOLOGY   Edited by D. K. McL. Kevan (1955)
*THE GROWTH OF LEAVES   Edited by F. L. Milthorpe (1956)
*CONTROL OF THE PLANT ENVIRONMENT   Edited by J. P. Hudson (1957)
*NUTRITION OF THE LEGUMES   Edited by E. G. Hallsworth (1958)
*THE MEASUREMENT OF GRASSLAND PRODUCTIVITY   Edited by J. D. Ivins (1959)
*DIGESTIVE PHYSIOLOGY AND NUTRITION OF THE RUMINANT   Edited by D. Lewis (1960)
*NUTRITION OF PIGS AND POULTRY   Edited by J. T. Morgan and D. Lewis (1961)
*ANTIBIOTICS IN AGRICULTURE   Edited by M. Woodbine (1962)
*THE GROWTH OF THE POTATO   Edited by J. D. Ivins and F. L. Milthorpe (1963)
*EXPERIMENTAL PEDOLOGY   Edited by E. G. Hallsworth and D. V. Crawford (1964)
*THE GROWTH OF CEREALS AND GRASSES   Edited by F. L. Milthorpe and J. D. Ivins (1965)
*REPRODUCTION IN THE FEMALE MAMMAL   Edited by G. E. Lamming and E. C. Amoroso (1967)
*GROWTH AND DEVELOPMENT OF MAMMALS   Edited by G. A. Lodge and G. E. Lamming (1968)
*ROOT GROWTH   Edited by W. J. Whittington (1968)
*PROTEINS AS HUMAN FOOD   Edited by R. A. Lawrie (1970)
*LACTATION   Edited by I. R. Falconer (1971)
*PIG PRODUCTION   Edited by D. J. A. Cole (1972)
*SEED ECOLOGY   Edited by W. Heydecker (1973)
HEAT LOSS FROM ANIMALS AND MAN: ASSESSMENT AND CONTROL   Edited by J. L. Monteith and L. E. Mount (1974)
*MEAT   Edited by D. J. A. Cole and R. A. Lawrie (1975)
*PRINCIPLES OF CATTLE PRODUCTION   Edited by Henry Swan and W. H. Broster (1976)
*LIGHT AND PLANT DEVELOPMENT   Edited by H. Smith (1976)
PLANT PROTEINS   Edited by G. Norton (1977)
ANTIBIOTICS AND ANTIBIOSIS IN AGRICULTURE   Edited by M. Woodbine (1977)
CONTROL OF OVULATION   Edited by D. B. Crighton, N. B. Haynes, G. R. Foxcroft and G. E. Lamming (1978)
POLYSACCHARIDES IN FOOD   Edited by J. M. V. Blanshard and J. R. Mitchell (1979)
SEED PRODUCTION   Edited by P. D. Hebblethwaite (1980)
PROTEIN DEPOSITION IN ANIMALS   Edited by P. J. Buttery and D. B. Lindsay (1981)
PHYSIOLOGICAL PROCESSES LIMITING PLANT PRODUCTIVITY   Edited by C. Johnson (1981)
ENVIRONMENTAL ASPECTS OF HOUSING FOR ANIMAL PRODUCTION   Edited by J. A. Clark (1981)
EFFECTS OF GASEOUS AIR POLLUTION IN AGRICULTURE AND HORTICULTURE   Edited by M. H. Unsworth and D. P. Ormrod (1982)
CHEMICAL MANIPULATION OF CROP GROWTH AND DEVELOPMENT   Edited by J. S. McLaren (1982)
CONTROL OF PIG REPRODUCTION   Edited by D. J. A. Cole and G. R. Foxcroft (1982)
SHEEP PRODUCTION   Edited by W. Haresign (1983)
UPGRADING WASTE FOR FEEDS AND FOOD   Edited by D. A. Ledward, A. J. Taylor and R. A. Lawrie (1983)
FATS IN ANIMAL NUTRITION   Edited by J. Wiseman (1984)
IMMUNOLOGICAL ASPECTS OF REPRODUCTION IN MAMMALS   Edited by D.B. Crighton (1984)

The titles are now out of print but are available in microfiche editions

Ethylene and Plant Development

J. A. ROBERTS, PhD
G. A. TUCKER, PhD
University of Nottingham School of Agriculture

BUTTERWORTHS
London Boston Durban Singapore Sydney Toronto Wellington

All rights reserved. No part of this publication may be reproduced or transmitted in any form or by any means, including photocopying and recording, without the written permission of the copyright holder, application for which should be addressed to the Publishers. Such written permission must also be obtained before any part of this publication is stored in a retrieval system of any nature.

This book is sold subject to the Standard Conditions of Sale of Net Books and may not be re-sold in the UK below the net price given by the Publishers in their current price list.

First published 1985

© The several contributors named in the list of contents 1985

British Library Cataloguing in Publication Data

Ethylene and plant development.
  1. Plants, Effect of ethylene on
  I. Roberts, J.A.   II. Tucker, G.A.
  581.19′27   QK753.E8

ISBN 0–407–00920–5

Library of Congress Cataloging in Publication Data
Main entry under title:

Ethylene and plant development.

   Proceedings of the 39th University of Nottingham Easter School in Agricultural Science, held March 26–30, 1984, in Sutton Bonington, England.
   Includes index.
   1. Plants, Effect of ethylene on—Congresses.
2. Ethylene—Synthesis—Congresses.   3. Plants—Development—Congresses.   I. Tucker, G.A. (Gregory A.)
II. Roberts, J.A. (Jeremy A.)   III. Easter School in Agricultural Science (39th : 1984 : Sutton Bonington, Nottinghamshire)
QK753.E8E84   1984   581.3   84-23115

Typeset by Scribe Design, Gillingham, Kent
Printed and bound in England by Robert Hartnoll Ltd, Bodmin, Cornwall

PREFACE

This volume contains the Proceedings of the Thirty-ninth University of Nottingham Easter School in Agricultural Science which was held at Sutton Bonington from 26th–30th March 1984. The conference was entitled 'Ethylene and Plant Development' and included a workshop, organized in conjunction with the Association of Applied Biologists, on the 'Practical control of ethylene in fruit, vegetables and flowers'. The contents are a mixture of review and research papers thus giving a thorough and up-to-date presentation of the subject. Ethylene is of great agricultural and horticultural significance by virtue of its role in such developmental processes as growth, ripening, abscission and senescence. The workshop reviewed the practical methods and advantages of either applying ethylene to, or removing ethylene from, various commercial products. The rest of the conference dealt with the more fundamental aspects of ethylene synthesis and action during the developmental processes in which the gas is active. Emphasis was particularly placed on the effects of ethylene on gene expression and cell development since advances in these areas may eventually lead to a more scientifically-based control of ethylene levels and action within the plant. The organizers gratefully acknowledge the financial support of ICI, Bayer AG, Monsanto and Shell. The success of the conference was largely due to the administrative skills and patience of Mrs E. Wyss and Mrs S. Bruce.

Thanks also go to Dr M. Knee for his assistance in the organization of the workshop, and to Dr F.B. Abeles, Professor S.F. Yang, Professor J. Bruinsma, Dr M.B. Jackson, Dr R.O. Sharpes and Dr D.J. Osborne for their confident chairing of the various sessions. Finally we would like to thank all the participants, both delegates and helpers, for contributing to such an enjoyable conference.

CONTENTS

1 **ETHYLENE AND PLANT DEVELOPMENT: AN INTRODUCTION** 1
   Fred B Abeles, *US Department of Agriculture, Agricultural Research Service, Appalachian Fruit Research Station, Kearneysville, West Virginia, USA*

2 **METABOLISM OF 1-AMINOCYCLOPROPANE-1-CARBOXYLIC ACID** 9
   S F Yang, Y Liu, L Su, G D Peiser, N E Hoffman *and* T McKeon, *Department of Vegetable Crops, University of California, Davis, California, USA*

3 **STUDIES ON THE ENZYMES OF ETHYLENE BIOSYNTHESIS** 23
   H Kende, M A Acaster *and* M Guy, *MSU-DOE Plant Research Laboratory, Michigan State University, East Lansing, Michigan, USA*

4 **THE OXYGEN AFFINITY OF 1-AMINOCYCLOPROPANE-1-CARBOXYLIC ACID OXIDATION IN SLICES OF BANANA FRUIT TISSUE** 29
   N H Banks, *Cambridge University*

5 **CARBON DIOXIDE FLUX AND ETHYLENE PRODUCTION IN LEAVES** 37
   Roger F Horton, *Department of Botany, University of Guelph, Ontario, Canada*

6 **THE EFFECT OF TEMPERATURE ON ETHYLENE PRODUCTION BY PLANT TISSUES** 47
   Roger J Field, *Plant Science Department, Lincoln College, Canterbury, New Zealand*

7 **THE RELATIONSHIP BETWEEN POLLINATION, ETHYLENE PRODUCTION AND FLOWER SENESCENCE** 71
   A D Stead, *Department of Botany, Royal Holloway & Bedford Colleges, Egham, Surrey, UK*

8 THE ETHYLENE FORMING ENZYME SYSTEM IN CARNATION
   FLOWERS   83
   K Manning, *Glasshouse Crops Research Institute, Littlehampton, West Sussex, UK*

9 ETHYLENE BIOSYNTHESIS IN *PENICILLIUM DIGITATUM*
   INFECTED CITRUS FRUIT   93
   Oded Achilea, Edo Chalutz, Yoram Fuchs *and* Ilana Rot, *Department of Fruit and Vegetable Storage, Agricultural Research Organization, The Volcani Center, Bet Dagan 50250, Israel*

10 ETHYLENE BINDING   101
   A R Smith *and* M A Hall, *Department of Botany and Microbiology, University College of Wales, Aberystwyth, Dyfed, UK*

11 ETHYLENE BINDING IN *PHASEOLUS VULGARIS* L COTYLEDONS   117
   C J Howarth, A R Smith *and* M A Hall, *Department of Botany and Microbiology, University College of Wales, Aberystwyth, Dyfed, UK*

12 ETHYLENE METABOLISM   125
   E M Beyer Jr, *E I du Pont de Nemours & Co Inc, Agricultural Chemicals Department, Building 402, Experimental Station, Wilmington, Delaware, USA*

13 ETHYLENE METABOLISM IN *PISUM SATIVUM* L AND *VICIA FABA* L   139
   A R Smith, D E Evans, P G Smith *and* M A Hall, *Department of Botany and Microbiology, University College of Wales, Aberystwyth, Dyfed, UK*

14 REGULATION OF THE EXPRESSION OF TOMATO FRUIT
   RIPENING GENES: THE INVOLVEMENT OF ETHYLENE   147
   D Grierson, A Slater, M Maunders, P Crookes, *Department of Physiology and Environmental Science, University of Nottingham Faculty of Agricultural Science,* G A Tucker, *Department of Applied Biochemistry and Food Science, University of Nottingham Faculty of Agricultural Science,* W Schuch *and* K Edwards, *ICI Corporate Bioscience and Colloid Laboratory, Runcorn, Cheshire*

15 INDUCTION OF CELLULASE BY ETHYLENE IN AVOCADO FRUIT   163
   Mark L Tucker, *Department of Molecular Plant Biology, University of California, Berkeley, CA 94720, USA,* Rolf E Christoffersen, *Mann Laboratory, University of California, Davis, CA 95616, USA and* Lisa Woll *and* George G Laties, *Department of Biology and Molecular Biology Institute, University of California, Los Angeles, USA*

16 ETHYLENE AND ABSCISSION   173
   R Sexton, *Department of Biological Science, Stirling University, Stirling, UK,* L N Lewis, *Molecular Plant Biology, University of California, Berkeley, California, USA and* A J Trewavas *and* P Kelly, *Department of Botany, Edinburgh University, Edinburgh, UK*

17 **TARGET CELLS FOR ETHYLENE ACTION** 197
    Daphne J Osborne, *Developmental Botany, Weed Research Organization, Oxford,* Michael T McManus, *Department of Biochemistry, University of Oxford and* Jill Webb, *Electron Microscopy, Weed Research Organization, Oxford*

18 **ETHYLENE, LATERAL BUD GROWTH AND INDOLE-3-ACETIC ACID TRANSPORT** 213
    J R Hillman, *Botany Department, The University, Glasgow, UK,* H Y Yeang, *Rubber Research Institute of Malaysia, Kuala Lumpur, Malaysia and* V J Fairhurst, *Botany Department, The University, Glasgow, UK*

19 **ETHYLENE AND PETIOLE DEVELOPMENT IN AMPHIBIOUS PLANTS** 229
    Irene Ridge, *Biology Department, Open University, Walton Hill, Milton Keynes, UK*

20 **ETHYLENE AND THE RESPONSES OF PLANTS TO EXCESS WATER IN THEIR ENVIRONMENT—A REVIEW** 241
    Michael B Jackson, *Agricultural and Food Research Council, Letcombe Laboratory, Letcombe Regis, Wantage, Oxfordshire, UK*

21 **ETHYLENE AND FOLIAR SENESCENCE** 267
    Jeremy A Roberts, *Physiology and Environmental Science,* Gregory A Tucker, *Applied Biochemistry and Food Science and* Martin J Maunders, *Physiology and Environmental Science, University of Nottingham Faculty of Agricultural Science*

22 **ETHYLENE AS AN AIR POLLUTANT** 277
    David M Reid *and* Kevin Watson, *Plant Physiology Research Group, Biology Department, University of Calgary, Alberta, Canada*

23 **SOURCES OF ETHYLENE OF HORTICULTURAL SIGNIFICANCE** 287
    Fred B Abeles, *US Department of Agriculture, Agricultural Research Service, Appalachian Fruit Research Station, Kearneysville, West Virginia, USA*

24 **EVALUATING THE PRACTICAL SIGNIFICANCE OF ETHYLENE IN FRUIT STORAGE** 297
    Michael Knee, *East Malling Research Station, East Malling, Maidstone, Kent, UK*

25 **ETHYLENE IN COMMERCIAL POST-HARVEST HANDLING OF TROPICAL FRUIT** 317
    F J Proctor *and* J C Caygill, *Tropical Development and Research Institute, Gray's Inn Road, London, UK*

26 **RESPIRATION AND ETHYLENE PRODUCTION IN POST-HARVEST SOURSOP FRUIT (*ANNONA MURICATA* L)** 333
    J Bruinsma, *Department of Plant Physiology, Agricultural University,*

Wageningen, *The Netherlands* and R E Paull, *Department of Botany, University of Hawaii at Manoa, Honolulu, Hawaii, USA*

27 **THE EFFECT OF HEAVY METAL IONS ON TOMATO RIPENING** 339
Graeme E Hobson, Royston Nichols and Carol E Frost, *Glasshouse Crops Research Institute, Littlehampton, West Sussex, UK*

28 **POST-HARVEST EFFECTS OF ETHYLENE ON ORNAMENTAL PLANTS** 343
R Nichols and Carol E Frost, *Glasshouse Crops Research Institute, Littlehampton, West Sussex, UK*

29 **SIGNIFICANCE OF ETHYLENE IN POST-HARVEST HANDLING OF VEGETABLES** 353
S P Schouten, *Sprenger Instituut, Wageningen, The Netherlands*

30 **RELATIONSHIP BETWEEN ETHYLENE PRODUCTION AND PLANT GROWTH AFTER APPLICATION OF ETHYLENE RELEASING PLANT GROWTH REGULATORS** 363
K Lürssen and J Konze, *Bayer AG, Pflanzenschutz, Anwendungstechnik, Biologische Forschung, Leverkusen, FRG*

31 **COMMERCIAL SCALE CATALYTIC OXIDATION OF ETHYLENE AS APPLIED TO FRUIT STORES** 373
C J Dover, *East Malling Research Station, East Malling, Maidstone, Kent, UK*

32 **LOW ETHYLENE CONTROLLED-ATMOSPHERE STORAGE OF McINTOSH APPLES** 385
F W Liu, *Cornell University, Ithaca, New York, USA*

33 **A COMMERCIAL DEVELOPMENT PROGRAMME FOR LOW ETHYLENE CONTROLLED-ATMOSPHERE STORAGE OF APPLES** 393
G D Blanpied, James A Bartsch and J R Turk, *Cornell University, Ithaca, New York, USA*

APPENDIX 405

LIST OF PARTICIPANTS 407

INDEX 413

1
ETHYLENE AND PLANT DEVELOPMENT: AN INTRODUCTION

FRED B. ABELES
US Department of Agriculture, Agricultural Research Service, Appalachian Fruit Research Station, Kearneysville, West Virginia, USA

Introduction

Early work on ethylene dealt with reports on the effects of leaking illuminating gas on plants (Girardin, 1864). In 1901, Neljubow (1901) demonstrated that ethylene was the physiologically active ingredient of illuminating gas. Later, Crocker and Knight (1908) reported that ethylene was a potent plant growth regulator and that it was capable of causing floral senescence, epinasty, abscission, intumescences and inhibition of growth. They also noted that as little as $0.1\,\mu l\,l^{-1}$ was capable of causing these various effects. The ability of gases produced by oranges to ripen bananas was noted by Cousins (1910), and later Denny (1924) reported that ethylene accelerated the ripening and respiration of lemons. The work by Kidd and West (1925) on the climacteric was seminal in focusing attention on the varied metabolic changes in fruit ripening and set the stage for contemporary research on fruit storage physiology.

Research on phytohormones was accelerated with the discovery of auxin in 1925 (Went and Thimann, 1937). However, the greater appeal of working with indole-acetic acid (IAA), which was known to be produced by plants as opposed to ethylene whose presence was difficult to quantitatively identify, captured the energies and attention of most plant hormone physiologists. Gane (1934) provided chemical evidence for the production of ethylene by plants, but it was not until Burg and Stolwijk (1959) showed that gas chromatography could be used to quantitatively measure physiologically significant levels of ethylene that the field attracted significant numbers of workers.

The concept that ethylene might be an important second messenger in plant development was initiated with the observation by Zimmerman and Wilcoxon (1935) that auxin increased ethylene production and that the ethylene so produced might play a role in auxin action. They suggested that auxin-induced ethylene production might play a role in the ability of auxin to induce epinasty, swelling, root initiation and inhibition of growth. We now know that auxin-induced ethylene production plays a role in many processes (*Table 1.1*).

As mentioned above, ethylene enhances fruit ripening. Regeimbal and Harvey (1927) reported that both invertase and protease activity of pineapples increased after they were treated with ethylene suggesting that the control of enzyme synthesis might play a role in ethylene action. Since that time, ethylene has been

Table 1.1 DEVELOPMENTAL PROCESSES WHERE AUXIN-INDUCED ETHYLENE PRODUCTION IS THOUGHT TO MEDIATE AUXIN ACTION

Abscission (Abeles and Rubinstein, 1964)
Apical dominance (Blake, Reid and Rood, 1983)
Branch angle (Blake, Pharis and Reid, 1980)
Bud growth, inhibition (Burg and Burg, 1968)
Callus, shoot initiation and growth (Huxter, Thorpe and Reid, 1981)
Epinasty (Amrhein and Schneebeck, 1980)
Flowering inhibition (Abeles, 1967)
Flowering, promotion in bromeliads (Burg and Burg, 1966b)
Flowering, senescence (Burg and Dijkman, 1967)
Flowering, sex expression in cucurbits (Shannon and de le Guardia, 1969)
Hypertrophy of hypocotyls (Wample and Reid, 1979)
Hypocotyl hook opening (Kang *et al.*, 1967)
Isocoumarin formation in carrots (Chalutz, De Vay and Maxie, 1969)
Latex flow, promotion (D'Auzac and Ribaillier, 1969)
Phenylalanine ammonia lyase (Rhodes and Wooltorton, 1971)
Root elongation, inhibition (Chadwick and Burg, 1970)
Root initiation (Fabijan, Taylor and Reid, 1981)
Stem elongation, inhibition (Burg and Burg, 1966a)
Swelling, onion leaf bases (Levy and Kedar, 1970)

Table 1.2 ENZYMES REGULATED BY ETHYLENE

Abscission
　Cellulase (Horton and Osborne, 1967)
　Polygalacturonase (Hashinaga *et al.*, 1981)
Aerenchyma
　Cellulase (Kawase, 1981)
Ripening
　Cellulase (Pesis, Fuchs and Zauberman, 1978)
　Chlorophyllase (Looney and Patterson, 1967)
　Invertase (Jeffery *et al.*, 1984)
　Laccase (Mayer and Harel, 1981)
　Malate dehydrogenase (Rhodes *et al.*, 1968)
　Polygalacturonase (Grierson, Tucker and Robertson, 1981)
Senescence
　Ribonuclease (Sacher, Engstrom and Broomfield, 1979)
Stress
　Beta-1,3-glucanase (Abeles *et al.*, 1971)
　Chitinase (Boller *et al.*, 1983)
　Cinnamate 4-hydroxylase (Rhodes, Wooltorton and Hill, 1981)
　Hydroxycinnamate CoA ligase (Rhodes, Wooltorton and Hill, 1981)
　Hydroxyproline rich glycoprotein (Toppan, Roby and Esquerre-Tugaye, 1982)
　Phenylalanine ammonia lyase (Rhodes, Wooltorton and Hill, 1981)
Function not known
　Ethylene mono-oxygenase (Abeles and Dunn, 1984)
　Peroxidase (Gahagan, Holm and Abeles, 1968)

Table 1.3 INHIBITORS OF PYRIDOXAL PHOSPHATE DEPENDENT ENZYMES

Rhizobitoxin: alpha-amino-gamma-(2'-amino-3'-hydroxypropoxy)-trans-beta-butenoic acid.
　$HOCH_2-CHNH_2-CH-O-CH=CH-CHNH_2-COOH$
L-Canaline: alpha-amino-gamma-amino-oxybutyric acid
　$H_2N-O-CH_2-CH_2-CHNH_2-COOH$
AVG: L-alpha-amino-gamma-(2'amino-ethoxy)-trans-beta-butenoic acid
　$H_2N-CH_2-CH_2-O-CH=CH-CHNH_2-COOH$
AOA: Amino-oxyacetic acid
　$H_2N-O-CH_2-COOH$

shown to increase the activity of a number of enzymes associated with ripening, abscission, senescence and stress. A partial list of the enzymes associated with these processes is given in *Table 1.2*.

Lieberman and Mapson (1964) were the first to show that methionine was a precursor of ethylene. Later, they demonstrated that rhizobitoxin was an effective inhibitor of ethylene production (Owens, Lieberman and Kunishi, 1971). Since that time, other inhibitors of pyridoxal phosphate dependent enzymes such as canaline, aminoethoxyvinylglycine (AVG), and amino-oxyacetic acid (AOA) have been used to unravel the pathway from methionine to ethylene (Amrhein and Wenker, 1979). The structures of these inhibitors are shown in *Table 1.3*.

Ethylene biosynthesis and action

While some of the details of the ethylene pathway are still being determined, the general outline is that methionine is converted to S-adenosylmethionine (SAM), then to 1-aminocyclo-propane-1-carboxylic acid (ACC), and finally to ethylene (Adams and Yang, 1979). The ethylene forming enzyme (EFE) has been difficult to study because it may be localized for instance on the vacuolar membrane (Guy and Kende, 1984) and activity of this enzyme is lost when membranes are destroyed (*see* Kende, Acaster and Guy, Chapter 3).

Some progress is being made on the mechanism of ethylene action. Valuable tools in these studies are the observations that CO_2 (Burg and Burg, 1967), silver ions (Beyer, 1976), hypobaric atmospheres (Burg and Burg, 1965) and the chemical TH6241 (Thompson Hayward Chemical Co.) (1,5-methyl-4-ethoxycarbonylmethoxy-1,2,3-benzothiodiazole) (Parups, 1973; Daalen and Daams, 1970) can block ethylene action. Additional aids in such studies are the concepts that a similarity exists between ethylene effects in terms of dose response curves and the effect of hydrocarbon gas analogues (Burg and Burg, 1967).

As far as we know, all plant cells make ethylene all the time. Because of this, its ability to act as a regulator is dependent on one of the following mechanisms. The first mechanism involves a change in the sensitivity of the cell to the ethylene that is already there while the second involves a response caused by a change in the level of ethylene produced by the tissue. In the first case, the rate of ethylene production remains constant during the physiological process. An example of this would be abscission. During abscission, it appears as if the ability of ethylene to act depends on the amount of auxin (acting as a juvenility factor) that is in the tissue (Abeles and Rubinstein, 1964). An example of a process which is controlled by an increase in the rate of ethylene production is wound-induced protein synthesis. Tissue damage results in an increase in ethylene production which in turn stimulates the synthesis of enzymes such as beta-1,3-glucanase and chitinase (Abeles *et al.*, 1971).

For ethylene to act it must bind to some part of the cell. As far as we know, one binding site seems to be used for most, if not all, ethylene effects. This interpretation is based on the results obtained from studies with ethylene action inhibitors, dose response curves and ethylene analogues. Though exceptions exist, many physiological effects of ethylene that are blocked with CO_2 and silver ions show similar dose response curves and respond similarly to various hydrocarbon analogues such as propylene and acetylene.

The problem of hormone binding, and the initial effects of the hormone binding site complex, has been a major research challenge. Ethylene has a special appeal for these studies because it is a simple molecule. It is also relatively easy to add and monitor ethylene in an experimental system. At the present time, investigators are evaluating the possibility that ethylene undergoes chemical modification at its site of action (*see* Beyer, 1981, and Chapter 12 for a review on ethylene action and metabolism). The binding (covalent, coordinate, or van der Waal's) between ethylene and its site of action can be reversible and relatively rapid. For example, the response time for epinasty (Funke *et al.*, 1938) and the inhibition of root elongation (Chadwick and Burg, 1970) is 1 h. In the case of ethylene induced inhibition of elongation of seedlings, about 30 min are required for ethylene action (Eisinger, 1983). In the case of epinasty, recovery from the effect of ethylene requires some hours, while in the root elongation system, removal of ethylene results in a rapid, almost instantaneous resumption of normal growth (Chadwick and Burg, 1970). In a similar fashion, Biale and Young (1981) indicated that ethylene caused a rapid and reversible increase in the rate of lemon fruit respiration.

Direct and indirect approaches have been used in hormone action studies. The direct approach has been to examine the physical binding of ethylene to the plant or plant parts. The indirect approach has been to study ethylene mediated processes such as ripening and work backwards, learning more about the details of earlier elements in the system. For example, ripening is a softening process involving the induction of polygalacturonase. The increase in polygalacturonase is preceded by the synthesis of its mRNA (Grierson, Tucker and Robertson, 1981 and Chapter 14). The same may also be true of abscission, another cell wall degrading process, involving cellulase (Abeles and Holm, 1966). The logical conclusion then is the belief that ethylene can activate a particular part of the nucleus. For these processes and others like them involving protein synthesis, ethylene alone or in conjunction with another substance, activates the genome.

The reports that ethylene promotes seed germination (Ketring, 1977; Taylorson, 1979) and bud break (Morgan, Meyer and Merkle, 1969) suggest that ethylene can also act by regulating the translation of preformed mRNA. Others have shown that such preformed or stored mRNA plays an important role in seed germination (Payne, 1976; Suzuki and Minamikawa, 1983). Since leaves may also contain stored mRNA (Giles, Grierson and Smith, 1977), processes involving the rapid production and response to ethylene such as wounding and stress may also involve stored mRNA.

The direct approach for ethylene action studies has been to look for binding sites by using labelled ethylene (Sisler, 1979, 1980). This approach faces a number of obstacles. The amount of ethylene needed for a physiological effect is small (about $0.1\,\mu l\,l^{-1}$, or 10^{-9} M in the liquid phase) so even with highly labelled ethylene, only small amounts are bound. The binding forces are weak. For example, as discussed above, the removal of ethylene from the gas phase surrounding stem or root tissue can result in a rapid return to normal growth rates. Finally, some plants have the ability to oxidize ethylene to ethylene oxide, carbon dioxide and other derivatives (Beyer, 1981), and the presence of these metabolic products complicates the interpretation of data.

We have used a modification of affinity chromatography to estimate the binding of ethylene to plants. In this technique, a pulse of ethylene or other hydrocarbon gas and methane acting as an internal standard are flushed through a glass column filled with ethylene sensitive tissue such as germinating seeds. The length of time required for the gases to appear in the effluent of this 'plant chromatograph' was used as an indication of their relative affinity for the tissue. For example, if the transit time of ethylene through the column was greater than that for methane, this would indicate binding. However, we have observed that the relative affinity of ethylene and other hydrocarbon gases for plant and fungal tissue was more closely associated with their solubility in water than any physiological activity. In addition, competitive inhibitors of ethylene action such as silver ions and CO_2 did not decrease binding (Abeles, 1984a).

The concept that ethylene acts without undergoing any dissociation has been tested a number of times. Earlier studies indicated that ethylene itself was not rearranged as a result of its contact with the cell (Beyer, 1981). While some early reports suggested that plants metabolize ethylene, it wasn't until the work of Jerie and Hall (1978), Dodds *et al.* (1979) and Beyer (1981) that ethylene oxidation by plants was conclusively demonstrated. The significance of this phenomenon is not fully understood and various explanations have been advanced. It is conceivable that oxidation of ethylene is a side reaction or effect of the ethylene binding site complex. For example, an increase in ethylene oxidation is associated with ripening, abscission and floral senescence (Beyer, 1981). Carbon disulphide (CS_2) has been a useful probe to test the role of ethylene oxidation in ethylene action. As Beyer originally observed, CS_2 was an effective inhibitor of ethylene oxidation. The action appears to be specific because CS_2 at the levels used in our experiments had no effect on growth, respiration and photosynthesis (Abeles, 1984b). We have observed that CS_2 totally blocked ethylene oxidation without inhibiting ethylene action. The reports cited above, that ethylene oxidation increases during ripening, abscission and floral senescence, suggested that ethylene may control the enzyme which oxidizes it. We have obtained evidence in favour of that view and have observed that ethylene can cause a manifold increase in oxidase activity and that the effect is blocked by cycloheximide (Abeles and Dunn, 1984).

Conclusion

We have learned much about ethylene and plant biology in the 83 years since a Russian graduate student showed that the ethylene in illuminating gas caused pea seedlings to grow horizontally (Neljubow, 1901). Through the efforts of many workers we currently possess a good deal of information on what ethylene does and how the cell synthesizes it. We have also developed a finer appreciation of the normal role of ethylene in plant growth and development. The need to learn more about the initial binding sites and the effects of the bound ethylene still remains a major challenge. Even though this goal remains elusive, we are learning to ask better questions, use better tools, and design more sophisticated experiments.

Agriculture has been quick to exploit the lessons learned thus far in this field of phytohormones. For example, most fruit and vegetable storage strategies include attempts to remove or eliminate ethylene action. Ethrel (2-chloroethyphosphonic acid), an ethylene releasing compound, is being incorporated in a variety of horticultural practices. It is safe to assume that the work presented here, and to be performed in the future, will also benefit agriculture and basic plant biology.

References

ABELES, F.B. (1967), *Plant Physiology*, **42**, 608–609
ABELES, F.B. (1984a). *Plant Physiology*, **74**, 525–528
ABELES, F.B. (1984b). *Journal of Plant Growth Regulation*, **3**, 85–89

ABELES, F.B., BOSSHART, R.P., FORRENCE, L.E., and HABIG, W.H. (1971). *Plant Physiology*, **47**, 129–134
ABELES, F.B. and DUNN, L.S. (1984). *Plant Physiology* (submitted for publication)
ABELES, F.B. and HOLM, R.E. (1966). *Plant Physiology*, **41**, 1337–1342
ABELES, F.B. and RUBINSTEIN, B. (1964). *Plant Physiology*, **39**, 963–969
ADAMS, D.O. and YANG, S.F. (1979). *Proceedings of the National Academy of Sciences*, **76**, 170–174
AMRHEIN, N. and SCHNEEBECK, D. (1980). *Physiologia Plantarum*, **49**, 62–64
AMRHEIN, N. and WENKER, D. (1979). *Plant and Cell Physiology*, **20**, 1635–1642
BEYER, E.M. (1976). *Plant Physiology*, **58**, 268–271
BEYER, E.M. (1981). In *Recent Advances in the Biochemistry of Fruits and Vegetables*, pp. 107–121. Ed. by Friend, J. and Rhodes, M.J.C. Phytochemical Society of Europe. Symposium Series No. 19. Academic Press, New York
BIALE, J.B. and YOUNG, R.E. (1981). In *Recent Advances in the Biochemistry of Fruits and Vegetables*, pp. 1–9. Ed. by Friend, J. and Rhodes, M.J.C. Phytochemical Society of Europe. Symposium Series No. 19. Academic Press, New York
BLAKE, T.J., PHARIS, R.P. and REID, D.M. (1980). *Planta*, **148**, 64–68
BLAKE, T.J., REID, D.M. and ROOD, S.B. (1983). *Physiologia Plantarum*, **59**, 481–487
BOLLER, T., GEHRI, A., MAUCH, F. and VÖGELI, U. (1983). *Planta*, **157**, 22–31
BURG, S.P. and BURG, E.A. (1965). *Science*, **148**, 1190–1196
BURG, S.P. and BURG, E.A. (1966a). *Proceedings of the National Academy of Sciences*, **55**, 262–269
BURG, S.P. and BURG, E.A. (1966b). *Science*, **152**, 1269
BURG, S.P. and BURG, E.A. (1967). *Plant Physiology*, **42**, 144–152
BURG, S.P. and BURG, E.A. (1968). *Plant Physiology*, **43**, 1069–1074
BURG, S.P. and DIJKMAN, M.J. (1967). *Plant Physiology*, **42**, 1648–1650
BURG, S.P. and STOLWIJK, J.A.J. (1959). *Journal of Biochemistry and Microbiological Technology and Engineering*, **1**, 245–259
CHADWICK, A.V. and BURG, S.P. (1970). *Plant Physiology*, **45**, 192–200
CHALUTZ, E., DE VAY, J.E. and MAXIE, E.C. (1969). *Plant Physiology*, **44**, 235–241
CROCKER, W. and KNIGHT, L.I. (1908). *Botanical Gazette*, **46**, 259–276
COUSINS, H.H. (1910). *Annual Report of the Department of Agriculture, Jamaica*
DAALEN, J.J. VAN and DAAMS, J. (1970). *Naturwissenschaften*, **8**, 395
D'AUZAC, J. and RIBAILLIER, D. (1969). *Comptes Rendus de L'Academie des Sciences, Paris*, **268**, 3046–3050
DENNY, F.E. (1924). *Journal of Agricultural Research*, **27**, 757–769
DODDS, J.H., MUSA, S.K., JERIE, P.H. and HALL, M.A. (1979). *Plant Science Letters*, **17**, 109–114
EISINGER, W. (1983). *Annual Reviews of Plant Physiology*, **34**, 225–240
FABIJAN, D., TAYLOR, J.S. and REID, D.M. (1981). *Physiologia Plantarum*, **53**, 589–597
FUNKE, G.L., DE COEYER, F., DE DECKER, A. and MATON, J. (1938). *Biologisch Jaarboek*, **5**, 335–381
GAHAGAN, H.E., HOLM, R.E. and ABELES, F.B. (1968). *Physiologia Plantarum*, **21**, 1270–1279
GANE, R. (1934). *Nature (London)*, **134**, 1008
GILES, A.B., GRIERSON, D. and SMITH, H. (1977). *Planta*, **136**, 31–36
GIRARDIN, J.P.L. (1864). *Jahrerbauch über die Agrikulture-Chemie*, **7**, 199–200
GRIERSON, D., TUCKER, G.A. and ROBERTSON, N.G. (1981). In *Recent Advances in*

the Biochemistry of Fruits and Vegetables, pp. 149–160. Ed. by Friend, J. and Rhodes, M.J.C. Phytochemical Society of Europe. Symposium Series No. 19. Academic Press, New York

GUY, M. and KENDE, H. (1984). *Planta*, **160**, 281–287

HASHINAGA, F., IWAHORI, S., NISHI, Y. and ITOO, S. (1981). *Nippon Nogeikagaku Kaishi*, **55**, 1217–1223

HORTON, R.F. and OSBORNE, D.J. (1967). *Nature (London)*, **214**, 1086–1088

HUXTER, T.J., THORPE, T.A. and REID, D.M. (1981). *Physiologia Plantarum*, **53**, 319–326

JEFFERY, D., SMITH, C., GOODENOUGH, P., PROSSER, I. and GRIERSON, D. (1984). *Plant Physiology*, **74**, 32–38

JERIE, P.H. and HALL, M.A. (1978). *Proceedings of the Royal Society of London B*, **200**, 87–94

KANG, B.G., YOCUM, C.S., BURG, S.P. and RAY, P.M. (1967). *Science*, **156**, 958–959

KAWASE, M. (1981). *American Journal of Botany*, **68**, 651–658

KETRING, D.L. (1977). In *Physiology and Biochemistry of Seed Dormancy and Germination*, pp. 136–178. North Holland Publ. Co., Amsterdam

KIDD, F. and WEST, C. (1925). *Great Britain Department of Science and Industry Research. Food Investigation Board Report. 1924*, 27–33

LEVY, D. and KEDAR, N. (1970). *HortScience*, **5**, 80–82

LIEBERMAN, M. and MAPSON, L.W. (1964). *Nature (London)*, **204**, 343–345

LOONEY, N.E. and PATTERSON, M.E. (1967). *Nature (London)*, **214**, 1245–1246

MAYER, A.M. and HAREL, E. (1981). In *Recent Advances in the Biochemistry of Fruits and Vegetables*, pp. 161–180. Ed. by Friend, J. and Rhodes, M.J.C. Phytochemical Society of Europe. Symposium Series No. 19. Academic Press, New York

MORGAN, P.W., MEYER, R.E. and MERKLE, M.G. (1969). *Weed Science*, **17**, 353–355

NELJUBOW, D. (1901). *Beihefte zur Botanisches Centralblatt*, **10**, 128–129

OWENS, L.D., LIEBERMAN, M. and KUNISHI, A. (1971). *Plant Physiology*, **48**, 1–4

PARUPS, E.V. (1973). *Physiologia Plantarum*, **29**, 365–370

PAYNE, P.I. (1976). *Biological Review*, **51**, 329–363

PESIS, E., FUCHS, Y. and ZAUBERMAN, G. (1978). *Plant Physiology*, **61**, 416–419

REGEIMBAL, L.O. and HARVEY, R.B. (1927). *Journal of the American Chemical Society*, **49**, 1117–1118

RHODES, M.J.C., GALLIARD, T., WOOLTORTON, L.S.C. and HULME, A.C. (1968). *Phytochemistry*, **7**, 405–408

RHODES, M.J.C. and WOOLTORTON, L.S.C. (1971). *Phytochemistry*, **10**, 1989–1997

RHODES, M.J.C., WOOLTORTON, L.S.C. and HILL, A.C. (1981). In *Recent Advances in the Biochemistry of Fruits and Vegetables*, pp. 191–218. Ed. by Friend, J. and Rhodes, M.J.C. Phytochemical Society of Europe. Symposium Series No. 19. Academic Press, New York

SACHER, J.A., ENGSTROM, D. and BROOMFIELD, D. (1979). *Planta*, **144**, 413–418

SHANNON, S. and DE LE GUARDIA, M.D. (1969). *Nature (London)*, **223**, 186

SISLER, E.C. (1979). *Plant Physiology*, **64**, 538–542

SISLER, E.C. (1980). *Plant Physiology*, **66**, 404–406

SUZUKI, Y. and MINAMIKAWA, T. (1983). *Plant and Cell Physiology*, **24**, 1371–1377

TAYLORSON, R.B. (1979). *Weed Science*, **27**, 7–10

TOPPAN, A., ROBY, D. and ESQUERRE-TUGAYE, M-T. (1982). *Plant Physiology*, **70**, 82–86

WAMPLE, R.L. and REID, D.M. (1979). *Physiologia Plantarum*, **45**, 219–226

WENT, F.W. and THIMANN, K.V. (1937). *Phytohormones*. Macmillan, New York

ZIMMERMAN, P.W. and WILCOXON, F. (1935). *Contributions of the Boyce Thompson Institute*, **7**, 209–229

2
METABOLISM OF 1-AMINOCYCLOPROPANE-1-CARBOXYLIC ACID

S.F. YANG, Y. LIU, L. SU, G.D. PEISER, N.E. HOFFMAN and T. McKEON
Department of Vegetable Crops, University of California, Davis, California, USA

Introduction

ACC (1-aminocyclopropane-1-carboxylic acid) was first isolated over 25 years ago from ripe cider apples and perry pears by Burroughs (1957) and from ripe cowberries by Vahatalo and Virtanen (1957). Although Burroughs (1960) could not detect ACC in other varieties of apple or pear he examined, he observed that the amount of ACC in perry pears increased during storage, and speculated that ACC might be related in some way to fruit ripening. However, ACC remained simply as one of the non-protein amino acids of plants, arousing little interest until its recognition as an ethylene precursor in 1979. Since Adams and Yang (1979) demonstrated that ethylene is biosynthesized in apple tissue via the following sequence: methionine → S-adenosylmethionine (SAM) → ACC → ethylene, this pathway has been shown to operate throughout the diversity of higher plant tissues (Yang and Hoffman, 1984). The formation of ACC from SAM is catalysed by ACC synthase, which is considered to control the rate-limiting step in ethylene biosynthesis. In addition to the metabolism of ACC to ethylene, ACC has recently been recognized to be conjugated into N-malonyl-ACC (MACC).

In this chapter we shall summarize the present status of knowledge pertaining to the metabolism of ACC to ethylene and to MACC in various plant systems.

Metabolism of ACC to ethylene

REGULATION OF THE ETHYLENE-FORMING ENZYME (EFE) SYSTEM *IN VIVO*

When ACC is applied to various plant organs (with the exception of preclimacteric fruit and flowers) from a number of plant species, a marked increase in ethylene production is observed (Cameron *et al.*, 1979; Lürssen, Naumann and Schröder, 1979). This suggests that the enzyme system which converts ACC to ethylene (EFE) is largely constitutive and the formation of ACC is the rate-limiting step in these plant tissues. In preclimacteric fruit and young petals of carnation flower, ethylene production is very low because they have a very limited ability not only to convert SAM to ACC, but also to convert ACC to ethylene. However, at the onset of ripening or senescence, their ability to convert SAM to ACC and ACC to ethylene increases dramatically resulting in a surge in ethylene production. Thus, EFE can be induced during certain developmental stages. Since preclimacteric (unripe) fruit lack both ACC synthase and EFE, a massive increase in ethylene production requires development of both enzymes. However, when green tomato fruit were treated with ethylene for a short period (18 h), there was no increase in ACC content or in ethylene production rate, however the tissue's ability to convert ACC to ethylene increased markedly (*Table 2.1*). These data indicate that when preclimacteric fruit tissues are exposed to ethylene, the increase in EFE precedes the increase in ACC synthase. Whether or not this is also true during the natural ripening of fruits remains to be clarified. Ethylene is also known to promote the development of EFE in other excised tissues (Hoffman and Yang, 1982; Riov and Yang, 1982; Chalutz *et al.*, 1984). EFE can also be promoted by various environmental factors such as water stress (McKeon, Hoffman and Yang, 1982).

Table 2.1 PROMOTION BY ETHYLENE OF THE CAPABILITIES FOR CONVERTING ACC TO ETHYLENE AND MACC IN PRECLIMACTERIC TOMATO FRUIT

Pretreatment	C_2H_4 production (nmol g^{-1})	ACC levels (nmol g^{-1})	ACC→C_2H_4 production (nmol g^{-1})	ACC→MACC levels (nmol g^{-1})
Air	1.4	1.6	8.7	2.2
Ethylene	1.3	0.4	48.4	25.6

Immature green tomato fruits were treated with air or 10 μl l^{-1} ethylene for 18 h. Discs (0.5 cm diameter) were then prepared from the pericarp tissue, and ethylene produced and ACC content at the end of a 6 h incubation period were determined. For measurement of the capabilities to convert ACC to ethylene or to MACC, the discs were incubated in a solution containing 2 mM ACC and the amount of ethylene produced during, and MACC accumulated at the end of, a 6 h incubation period were determined

Recently many investigators have observed that light markedly inhibited ethylene production by various green leaf tissues in enclosed systems. Gepstein and Thimann (1980) were the first to report that the conversion of ACC to ethylene was inhibited by light. Since CO_2 is known to promote ethylene production in leaf tissues (Dhawan, Bassi and Spencer, 1981), Grodzinski, Boesel and Horton (1982) and Kao and Yang (1982) reasoned that the inhibition of ethylene evolution by light might result from a decrease in internal CO_2 concentration. Indeed, when CO_2 was added into the incubation flask, the rate of ethylene production in the light increased markedly, to a level which was even higher than that produced in the dark; carbon dioxide, however, had no appreciable effect on leaf segments incubated in the dark. The concentration of CO_2 giving half-maximal activity was 0.06% and 0.18% for rice and tobacco leaves, respectively. Thus, it is the CO_2 metabolism rather than light *per se*, which regulates the conversion of ACC to ethylene. Unlike the development of EFE mentioned above, the modulation of ACC conversion to ethylene by CO_2 or light is rapid and reversible, indicating that CO_2 regulates the activity, but not the synthesis of EFE (Kao and Yang, 1982). The mechanism by which CO_2 modulates the conversion of ACC to ethylene is not understood.

SOME CHARACTERISTICS OF EFE

It has long been recognized that dinitrophenol, high temperature, various lipophilic compounds, and osmotic shock treatment, all of which could modify membrane structure and function, greatly reduce the rate of ethylene synthesis in plant tissues. Moreover, when those tissues which are producing ethylene actively are homogenized, the ethylene-forming capability is totally lost. These observations lead to the suggestion that the ethylene-forming system is highly structured and requires membrane integrity (Lieberman, 1979). Recently, John (1983) has suggested that the generation of ethylene from ACC is coupled to a transmembrane flow of protons from the outside to the inside of the plasma membrane. This model explains the strict dependence of ethylene biosynthesis on membrane integrity and the marked inhibition by the protonophore, dinitrophenol. While this is an interesting hypothesis, direct experimental evidence is lacking.

Although the ACC molecule possesses two enantiotopic methylene groups, they are not geometrically equivalent and can be distinguished by a regiospecific enzyme. Ethyl substitution at one of each of the four methylene hydrogens results in four stereoisomers of 1-amino-2-ethylcyclopropane-1-carboxylic acid (AEC). If ACC conversion to ethylene by plant tissues were to proceed in regiospecific fashion, Hoffman *et al.* (1982a) reasoned that these four stereoisomers of AEC might not be converted into 1-butene with equal efficiency. In apple and etiolated mungbean hypocotyls, (+)-allocoronamic acid (for structure, *see Figure 2.2*) was preferentially converted to 1-butene. By chemical oxidation using NaOCl, in contrast, all AEC isomers were converted with nearly equal efficiency to 1-butene. ACC and AEC appear to be degraded by the same enzyme since both reactions are inhibited to the same extent by nitrogen atmosphere or by Co^{2+}, and since, when both substrates are present simultaneously, each acts as an inhibitor with respect to the other. These observations indicate that the enzyme converting ACC to ethylene exhibits regiospecificity.

Soon after ACC was established as the immediate precursor of ethylene, Konze and Kende (1979) reported an enzyme extract from etiolated pea seedling capable of converting ACC to ethylene. Many similar systems have since been reported, including carnation microsomes (Mayak, Legge and Thompson, 1981), pea microsomes (McRae, Baker and Thompson, 1982), pea mitochondria (Vinkler and Apelbaum, 1983), IAA-oxidase and peroxidase (Vioque, Albi and Vioque, 1981). Although these systems are oxygen-dependent, heat-denaturable and inhibited by radical scavengers, there are many characteristics which do not resemble those of the natural *in vivo* system. McKeon and Yang (1984) have compared the regiospecificity, with regard to the conversion of AEC isomers to 1-butene, by pea epicotyls and by the pea epicotyl extract. While pea epicotyls displayed the same regiospecificity observed in mungbean hypocotyls and apple tissues, and exhibited high affinity for ACC with a K_m of about 66 µM with respect to the internal ACC concentration, the pea homogenate did not differentiate between AEC isomers (*Table 2.2*) and exhibited very low affinity for ACC (reported Kms for the pea enzyme range from 15–400 mM). Moreover, the pea enzyme required Mn^{2+}, and was very sensitive to inhibition by EDTA and mercaptoethanol, whereas the *in vivo* system was not. These data contradict the view that the reported enzymic systems function to catalyse the conversion of ACC to ethylene *in vivo*. It should be noted that ACC can be converted to ethylene chemically by various oxidants, including oxidative free radicals (Legge, Thompson and Baker, 1982). A simple explanation accounting for the low affinity and lack of regiospecificity in these enzyme systems is that the physiological enzyme (EFE) catalyses directly the oxidation of ACC yielding ethylene, whereas the reported enzymes catalyse the activation of molecular oxygen probably to free radicals, such as O_2^- and OH, which

Table 2.2 CONVERSION OF AEC ISOMERS (3 mM) TO 1-BUTENE BY PEA STEM SEGMENTS OR BY THE PEA STEM HOMOGENATE

AEC isomer	1-Butene production (nmol g^{-1} h^{-1})	
	Pea stem	Pea stem homogenate
(+)-Allocoronamic acid	1.26	0.56
(−)-Allocoronamic acid	0.03	0.57
(+)-Coronamic acid	0.01	0.20
(−)-Coronamic acid	0.02	0.24

From McKeon and Yang (1984)

in turn react with ACC nonenzymatically to form ethylene. Such reaction mechanisms are in agreement with the observation that the *in vitro* systems have high Km values but are saturable with ACC, because as the concentrations of ACC are increased, the rate-limiting reaction shifted to the reaction of oxygen activation, the rate of which is independent of ACC concentration. Recently Guy and Kende (1984) observed that the vacuole fraction isolated from pea protoplasts accounted for 80% of the protoplast ethylene production. Unlike the pea enzyme system, the vacuoles resemble the *in vivo* system in that they differentiate between AEC isomers, have an apparent K_m of 70 μM for ACC, are sensitive to Co^{2+} inhibition, and very sensitive to membrane disruption. Thus, intact vacuoles appear to be the smallest biological units which possess properties characteristic of EFE (*see* Kende, Acaster and Guy, Chapter 3).

MECHANISM OF THE REACTION

Although the enzyme system responsible for the conversion of ACC to ethylene remains to be characterized, some progress has been made with respect to the reaction mechanism and the other degradation products. Analogous to the chemical oxidation by hypochlorite, we have previously proposed that ACC can be oxidized by a hydroxylase to N-hydroxy-ACC, which is a nitrenium equivalent, or by a dehydrogenase to form the nitrenium intermediate, which is then fragmented into ethylene and cyanoformic acid (Yang, 1981). Cyanoformic acid is very labile and breaks down spontaneously to CO_2 (derived from the carboxyl group of ACC) and HCN (derived from C-1 of ACC). The overall reaction represents a two-electron oxidation.

The validity of this pathway is supported by the recent observation of Peiser *et al.* (1984). They showed that [1-^{14}C]ACC was primarily converted into [4-^{14}C]-asparagine, a hydrated product of β-cyanoalanine, in mungbean hypocotyls, and into a derivative of β-cyanoalanine, γ-glutamyl-β-[4-^{14}C]cyanoalanine, in *Vicia sativa* epicotyls, in amounts similar to the ethylene produced. When $Na^{14}CN$ was administered to mungbean hypocotyls or *Vicia sativa* epicotyls, it was similarly incorporated into asparagine in mungbean, and into the β-cyanoalanine conjugate in *Vicia sativa*. When [carboxyl-^{14}C]ACC was administered into these plant tissues,

CO_2 was produced in an amount equivalent to the amount of ethylene produced. Thus, in the conversion of ACC to ethylene, the carboxyl group yields CO_2, and C-1 gives off HCN. Since HCN is toxic to plants, it is pertinent to ask whether plants have ample capacity to detoxify the HCN thus formed. β-Cyanoalanine synthase, which catalyses the formation of β-cyanoalanine from cysteine and HCN, is widely distributed in higher plants. While ethylene production rates in higher plants range from 0 to 0.2 nmol g^{-1} min^{-1}, β-cyanoalanine synthase activity ranges from 4–1000 nmol g^{-1} min^{-1} (Miller and Conn, 1980). Recently, Adlington et al. (1983) have fed [cis-2,3-^{2}H$_2$]ACC to apple slices and observed that it yielded a mixture of equal amounts of cis- and trans-[1,2-^{2}H$_2$]ethylene. This indicates that the configuration of hydrogens is lost during the conversion of ACC to ethylene by apple tissue. In contrast to the biosynthetic results, the chemical oxidation of ACC to ethylene with NaOCl results in complete retention of ethylene configuration. These data indicate that NaOCl oxidation of ACC to ethylene may proceed by a concerted elimination mechanism, while biosynthesis proceeds by a stepwise mechanism involving an intermediate that allows scrambling of the hydrogens in the ring, resulting in loss of stereochemistry. Most recently, Pirrung (1983) has studied the electrochemical oxidation of [cis-2,3-^{2}H$_2$]ACC, which results in ethylene formation with loss of stereochemistry, as observed in plant tissue. He has therefore suggested that oxidation of ACC proceeds in vivo in two sequential one-electron oxidation reactions, and the initial oxidation yields free radical intermediates which undergo rapid ring-opening and loss of stereochemistry. However, it should be noted that the free radical mechanism is only one of the possible pathways, as an elimination reaction involving an ionic intermediate that allows scrambling of the hydrogens following ring-opening, cannot be ruled out.

Metabolism of ACC to malonyl-ACC

PROPERTIES OF MALONYL-ACC (MACC)

While ACC is known to be metabolized to 2-oxobutyrate and ammonia in some micro-organisms (Honma and Shimomura, 1978), Amrhein et al. (1981) found no such deamination reaction occurring in plant tissues, but rather discovered, initially in buckwheat seedlings, that ACC supplied exogenously was efficiently converted into a conjugate, which was identified as MACC. Working on the biosynthesis of stress ethylene in wilted wheat leaves, Apelbaum and Yang (1981) have previously observed that the loss of ACC during an incubation period was greater than the quantity of ethylene produced during the same period, suggesting that ACC must have been metabolized by some pathway other than ethylene production. Hoffman, Yang and McKeon (1982b) therefore examined the metabolism of exogenously supplied ACC in light grown wheat leaves and independently identified MACC as the major non-volatile metabolite of ACC in water-stressed as well as non-stressed tissues; the natural occurrence of MACC in the wilted wheat leaves was identified by GC-MS techniques. Amrhein et al. (1982) have examined a number of tissues from various plants for their ability to metabolize exogenous ACC and found that with the exception of ripe apple, all other tissues are capable of metabolizing ACC to MACC. It appears that the enzyme responsible for the MACC formation is mostly constitutive, like the enzyme converting ACC to ethylene. Thus, MACC would be formed in most plant tissues as long as ACC synthesis is induced. Since MACC is a poor ethylene producer and the conjugation of ACC to MACC is essentially irreversible, it is thought that MACC is a biologically inactive end-product of ACC rather than a storage form of ACC (Amrhein et al., 1982; Hoffman, Liu and Yang, 1983b). Support for such a conclusion is exemplified by the observation that germinating peanut seeds contained large amounts of MACC (50–100 nmol g^{-1}) and produced ethylene, which was not derived from MACC but was instead synthesized *de novo* from SAM (Hoffman, Fu and Yang, 1983a). In agreement with this view is the observation that MACC does not evoke ethylene-specific responses in various test systems (K. Lürssen, personal communication). Like the other plant hormones, whose level can be regulated via conjugation, malonylation of ACC to MACC results in regulation of endogenous ACC level and thereby of ethylene production rate.

MACC FORMATION IN WATER-STRESSED WHEAT LEAVES

When wheat leaves are stressed by wilting, both the ACC content and ethylene synthesis increase markedly, and both decline later (Apelbaum and Yang, 1981). The increase in ethylene production in response to wilting is caused primarily by an increase in ACC synthesis, although the capacity to convert ACC to ethylene also increases (McKeon, Hoffman and Yang, 1982). The capacity to conjugate ACC to MACC exists in both turgid and stressed wheat leaves and is not affected by stress status (Hoffman, Yang and McKeon, 1982b). The time courses of change in ACC and MACC levels and of total ethylene production in the water-stressed and non-stressed wheat leaves are illustrated in *Figure 2.1*. It should be noted that while there is a sharp rise and then a decline in the level of ACC and the production rate of ethylene in the stressed leaves, the level of MACC increased gradually until it reached a plateau. This increase in MACC levels is positively correlated with severity of water stress and the increased synthesis of ACC. Once formed, the MACC levels do not decrease even after the stressed tissues are rehydrated. Administration of labelled ACC and MACC shows that the conjugation of ACC to MACC is essentially irreversible. Repeated wilting treatments following the first wilting and rehydration cycle result in no further increase in ethylene production or in the levels of ACC and MACC. However, when benzyladenine is supplied during the rehydration process, a subsequent wilting treatment resulted in a rise in MACC level and a rapid rise followed by a decline in ethylene production rate and in the level of ACC. The magnitude of these increases is, however, smaller in these rewilted tissues than that observed in the first wilting treatment.

Since various forms of stress result in elevated ethylene levels, production of this gas has been suggested to be a potential indicator of stress. As shown in *Figure 2.1*, water stress results in a transient increase in the rate of ethylene production and the level of ACC. Thus, neither measurement of ethylene production rate nor of the level of ACC during a given period serves as an effective stress indicator. In contrast, MACC, which accumulates with water stress and is not appreciably metabolized, may be a better indicator of water-stress. A similar conclusion has

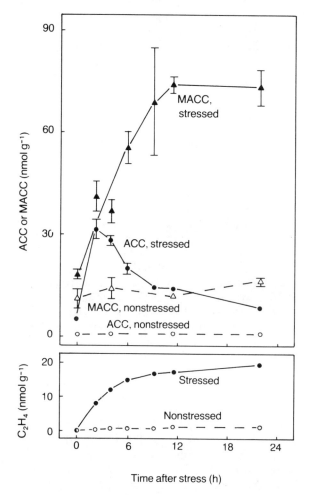

Figure 2.1 Time courses of total ethylene production and of changes in ACC and MACC content in excised wheat leaves which have been subjected to water-stress until they lost 9% of their initial fresh weight. From Hoffman, Liu and Yang (1983b)

been presented by Amrhein *et al.* (1984). The rapid decline in ACC content and consequently of ethylene production in these wilted wheat leaves can be attributed, in part, to the efficient conjugation of ACC to MACC. Thus, malonylation of ACC serves as a mechanism to dissipate excess ACC, and participates in the regulation of ethylene biosynthesis.

STIMULATION BY ETHYLENE OF THE CAPABILITY TO CONVERT ACC TO MACC IN GREEN TOMATO FRUIT

Conjugation of ACC to MACC has been examined in many vegetative tissues, and it is generally concluded that the enzyme responsible for this process is constitutive

Table 2.3 PROMOTION OF THE DEVELOPMENT OF ACC MALONYLTRANSFERASE IN GREEN TOMATO FRUIT BY ETHYLENE

Treatment	ACC → MACC (% conversion)	Malonyltransferase (nmol mg protein^{-1} h^{-1})
Air	8.4	4.7
Ethylene	35.7	28.8

After green tomato fruits were treated with ethylene (27 µl l^{-1}) for 18 h, ACC malonyltransferase was extracted from the pericarp tissue. For measurement of *in vivo* capability to convert ACC to MACC, [2,3-^{14}C]ACC (10 nmol, 17 nCi) was applied to the discs prepared from pericarp tissue. After incubation for 6 h, the metabolites were extracted and analysed for MACC.

(Amrhein *et al.*, 1982, 1984; Hoffman *et al.*, 1982b, 1983b). Recently Liu *et al.* (unpublished results) using preclimacteric tomato tissue have found that this tissue had little capability to form MACC from exogenously administered ACC, but that this capability, like EFE, was greatly promoted when the intact green tomato fruit were treated with ethylene for 18 h as shown in *Table 2.1*. They have therefore compared the activities of ACC malonyltransferase isolated from green tomato fruit, which had been treated with air or with 27 µl l^{-1} ethylene for 18 h. In parallel with the increase in their capability to convert ACC to MACC, ethylene treatment increased the development of malonyltransferase sixfold (*Table 2.3*). These results are in full agreement with the observations of Su *et al.* (1984) who reported that the level of ACC conjugate in green tomatoes increased progressively as their ethylene production rate increased during the maturation and ripening process.

RELATIONSHIP BETWEEN THE MALONYLATIONS OF ACC AND OF D-AMINO ACIDS

N-Malonylation of D-amino acids commonly occurs in higher plants (Kawasaki, Ogawa and Sasaoka, 1982). It is thought that the physiological significance of N-malonylation is to inactivate foreign and potentially toxic substances such as D-amino acids or herbicides which might otherwise be harmful to plants (Iwan, 1976; Lamoureux *et al.*, 1981). Since ACC has no asymmetric carbon, it can be recognized as a D- as well as an L-amino acid. Conceivably, the malonylations of ACC and of D-amino acids may be interrelated. Indeed, various D-amino acids (D-Phe, D-Met, D-Ala) inhibit the malonylation of exogenously administered ACC in mungbean hypocotyl segments, resulting in an increase in free ACC and ethylene production rate; L-enantiomers are, however, ineffective (Amrhein *et al.*, 1982; 1984; Liu, Hoffman and Yang, 1983). Reciprocally, ACC or D-Phe greatly inhibited the formation of N-malonyl-D-methionine from exogenously administered D-methionine (Liu, Hoffman and Yang, 1983). These results indicate an intimate relationship between the malonylation of ACC and of D-amino acids, and further suggest that both reactions may be catalysed by the same enzyme. Such a conclusion has been fully confirmed from the data obtained with the cell-free system, which will be discussed in the next section.

Previously Satoh and Esashi (1981, 1982) have reported that D-amino acids stimulated ethylene production and increased ACC content in cocklebur cotyledons. The above observations can now be explained on the basis that D-amino acids inhibit malonylation of ACC resulting in a higher ACC level, and thereby a higher ethylene production rate. These data clearly indicate that malonylation of ACC regulates the endogenous ACC level and hence the ethylene production rate.

ACC AND D-AMINO ACID MALONYLTRANSFERASE

A cell-free extract capable of catalysing the formation of MACC from ACC and malonyl-CoA has been recently reported by Amrhein and Kionka (1983). We have also isolated ACC malonyltransferase from mungbean hypocotyls and from tomato fruit using procedures similar to those reported by Amrhein and Kionka (1983). Malonyltransferase is assayed in a reaction mixture consisting of 0.1 M KCl, 0.1 M K-phosphate at pH 8.0, and 0.25 mM labelled (10 nCi) ACC, D-Met, or D-Phe, and 1 mM malonyl-CoA, in a total volume of 50 µl. After incubation for 1 h at 35°C, the labelled N-malonates produced are separated from unreacted amino acids with a small column of Dowex 50(H^+ form) ion-exchange resin, and their radioactivity assayed by scintillation counting. The products as N-malonyl conjugates were characterized by paper radiochromatography and radioelectrophoresis and by acid hydrolysis as previously described (Hoffman, Yang and McKeon, 1982b; Liu, Hoffman and Yang, 1983). Some characteristics of ACC malonyltransferase are listed in *Table 2.4*. In addition to ACC, the enzyme preparation also utilizes D-Phe, and D-Met as substrates and their K_m and V_{max} have been determined. ACC malonyltransferase is very sensitive to inhibition by a number of sulph-hydryl reagents including *p*-chloromercuribenzoate, Hg^{2+}, Cu^{2+}, Zn^{2+}, Co^{2+}, and N-ethylmaleimide, listed in decreasing order of their inhibitory activity. These results suggest the involvement of a sulph-hydryl function at the active site of the enzyme.

Malonyl-CoA + Enz-SH ⇌ Malonyl-S-Enz + CoA
Malonyl-S-Enz + ACC ⇌ N-Malonyl-ACC + Enz-SH

An intimate relationship between malonylation of ACC and of D-amino acids *in vivo* has been established (preceding section). Of particular interest is the characterization of this relationship in the cell-free enzyme system. As in mungbean hypocotyls, ACC malonyltransferase in the present preparation is competitively inhibited by D-Phe and D-Met, with K_i values of 1 and 3 mM, respectively. A comparison of these K_i values with their corresponding K_m values (*Table 2.4*) reveals that these two sets of values agree well. Such a relationship is expected if the malonylations of ACC and D-amino acids are catalysed by the same enzyme. The view that malonylations of ACC and D-amino acids are catalysed by the same enzyme is further supported by the observation that the ratio of ACC malonyltransferase activity to D-Phe malonyltransferase activity remains constant throughout

Table 2.4 CHARACTERISTICS OF ACC MALONYLTRANSFERASE FROM MUNGBEAN HYPOCOTYLS

pH optimum	8
Temperature optimum	35°C
K_m for malonyl-CoA	0.5 mM
K_m (V_{max} relative to ACC) for amino acids)	
ACC	0.2 mM (100)
D-Phe	1 mM (79)
D-Met	3 mM (160)

the different enzyme fractionation steps and both enzyme activities are inhibited similarly by various sulph-hydryl reagents, D-amino acids and cycloalkane amino acids. Those cycloalkane amino acids examined can be arranged in the following sequence with respect to their K_i values: 1-aminocyclobutanecarboxylic acid < 1-aminocyclopentanecarboxylic acid < 1-aminocyclohexanecarboxylic acid. Confirmation that a single enzyme carries out the malonylations awaits purification of the enzyme.

Although L-Met and L-Phe exert small but significant inhibition on the malonylation of ACC, these L-amino acids are not malonylated to a significant extent by the enzyme preparation. Thus, the enzyme is stereospecific. Although the ACC molecule possesses two methylene groups and hence can be recognized as a D- or L-amino acid, these two methylene groups are not geometrically identical, and can be distinguished by a stereospecific enzyme. As described earlier, ethyl substitution of each of the four methylene hydrogens of ACC results in four stereoisomers: (+)-allocoronamic acid, (−)-allocoronamic acid, (+)-coronamic acid and (−)-coronamic acid. These stereoisomers have been instrumental in the demonstration that ACC is stereoselectively converted to 2-oxobutyric acid by a microbial ACC deaminase (Honma and Shimomura, 1978) and to ethylene by plant tissues (see earlier section). We have therefore examined the efficiencies of these isomers serving as substrates of malonyltransferase and as inhibitors of ACC malonyltransferase.

Since radioactive AEC isomers are not available, their enzymic product, N-malonyl-AEC, has to be assayed chemically. After reaction N-malonyl-AEC was separated from non-reacted AEC by cation ion exchange resin, Dowex 50 (H^+ form), and then hydrolysed into AEC in 2N HCl. AEC was then chemically degraded into 1-butene, which was determined by a gas chromatograph. Because the conversion efficiency of AEC to 1-butene is low, the accuracy and sensitivity of this chemical assay are not high. *Figure 2.2* shows the K_m (AEC isomers were employed as substrates) and K_i (AEC isomers were employed as inhibitors of the conversion of radioactive ACC to MACC) values for each AEC isomer. It is to be noted that (+)-allocoronamic acid and (−)-coronamic acid have lower K_m and K_i values indicating that they have higher affinities for the enzyme. Since the malonyltransferase is capable of differentiating between D- and L-amino acids, we may assume that such a stereospecific catalysis is imparted by at least three points of attachment between enzyme and substrate. A diagram depicting such an interaction is shown in *Figure 2.2*. Those amino acids examined can be arranged in the following sequence according to their affinity for the enzyme: ACC > D-Phe > D-Met > D-Ala > 2-aminoisobutyric acid (AIB) > glycine and L-amino acids. An amino acid has a structure of

$$\begin{array}{c} R_R \\ \diagdown \\ C \\ \diagup \\ R_S \end{array} \begin{array}{c} NH_3^+ \\ \diagup \\ \\ \diagdown \\ COO^- \end{array}$$

where R_R and R_S are Pro-(R) and pro-(S) substituents, respectively. To be an effective substrate of malonyltransferase, R_S should be an H (as in a D-amino acid) or a methylene group of a cyclopropane (as in ACC), while R_R should be a large non-polar group (as in a D-amino acid) or a methylene group of a cyclopropane (as in ACC). As illustrated in *Figure 2.2*, when the pro-(R) methylene group of ACC is substituted by an ethyl group, it becomes (+)-allocoronamic acid or (−)-coronamic acid, both of which have an R-configuration as a D-amino acid, and are effective

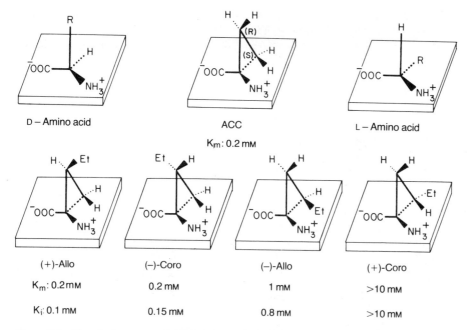

Figure 2.2 Chemical structure of AEC analogues in relation to D- and L-amino acids. K_m values of AEC analogues serving as substrates of malonyltransferase and K_i values of AEC analogues serving as competitive inhibitors of ACC malonyltransferase are also given

substrates of malonyltransferase as indicated by their low K_m and K_i values. Conversely, when the pro-(S) methylene group of ACC is substituted by an ethyl group, it yields (−)-allocoronamic acid or (+)-coronamic acid, both of which have an S-configuration as an L-amino acid, and are not as effective substrates as their enantiomers. These results are in agreement with the view that ACC is recognized by the enzyme as a D-amino acid. In parallel with such a view is the observation that (R)-isovaline ($R_R = C_2H_5-$, $R_S = CH_3-$) which has an R-configuration as a D-amino acid, inhibits more effectively ($K_i = 2.4$ mM) the conversion of ACC to MACC by malonyltransferase than does (S)-isovaline ($K_i = 15$ mM). These observations indicate that ACC is recognized by the enzyme as a D-amino acid and support the view that ACC and D-amino acid malonyltransferases are the same enzyme.

Acknowledgements

Our work reported here was supported by research grants from the National Science Foundation (PCM-8114933) and from the US–Israel Agricultural Research and Development Fund.

References

ADAMS, D.O. and YANG, S.F. (1979). *Proceedings of the National Academy of Sciences, USA*, **76**, 170–174

ADLINGTON, R.M., BALDWIN, J.E. and RAWLINGS, B.J. (1983). *Journal of the Chemistry Society and Chemical Communications*, **1983**, 290-292
AMRHEIN, N. and KIONKA, C. (1983). *Plant Physiology, Suppl.*, **72**, No. 207. (Abstr.)
AMRHEIN, N., SCHNEEBECK, D., SKORUPKA, H., TOPHOF, S. and STOCKIGT, J. (1981). *Naturwissenschaften*, **68**, 619-620
AMRHEIN, N., BREUING, F., EBERLE, J., SKORUPKA, H. and TOPHOF, S. (1982). In *Plant Growth Substances* 1982, pp. 249-258. Ed. by P.F. Wareing. Academic Press, London
AMRHEIN, N., DORZOK, U., KIONKA, C., KONDIZIOLKA, U., SKORUPKA, H. and TOPHOF, S. (1984). In *Biochemical, physiological and applied aspects of ethylene*, pp. 11-20. Ed. by Y. Fuchs and E. Chalutz. Martinus Nijhoff Publ., The Hague
APELBAUM, A. and YANG, S.F. (1981). *Plant Physiology*, **68**, 594-596
BURROUGHS, L.F. (1957). *Nature*, **179**, 360-361
BURROUGHS, L.F. (1960). *Journal of the Science of Food and Agriculture*, **11**, 14-18
CAMERON, A.C., FENTON, C.A.L., YU, Y.B., ADAMS, D.O. and YANG, S.F. (1979). *HortScience*, **14**, 178-180
CHALUTZ, E., MATTOO, A.K., SOLOMOS, T. and ANDERSON, J.D. (1984). *Plant Physiology*, **74**, 99-103
DHAWAN, K.R., BASSI, P.K. and SPENCER, M.S. (1981). *Plant Physiology*, **68**, 831-834
GEPSTEIN, S. and THIMANN, K.V. (1980). *Planta*, **149**, 196-199
GRODZINSKI, B., BOESEL, I. and HORTON, R.F. (1982). *Journal of Experimental Botany*, **33**, 1185-1193
GUY, M. and KENDE, H. (1984). *Planta*, **160**, 281-287
HOFFMAN, N.E. and YANG, S.F. (1982). *Plant Physiology*, **69**, 317-322
HOFFMAN, N.E., YANG, S.F., ICHIHARA, A. and SAKAMURA, S. (1982a). *Plant Physiology*, **70**, 195-199
HOFFMAN, N.E., YANG, S.F. and McKEON, T. (1982b). *Biochemistry and Biophysics Research Communications*, **104**, 765-770
HOFFMAN, N.E., FU, J.R. and YANG, S.F. (1983a). *Plant Physiology*, **71**, 197-199
HOFFMAN, N.E., LIU, Y. and YANG, S.F. (1983b). *Planta*, **157**, 518-523
HONMA, M. and SHIMOMURA, T. (1978). *Agricultural and Biological Chemistry*, **42**, 1825-1831
IWAN, J. (1976). *ACS Symposium Series*, **1976**, 132-152
JOHN, P. (1983). *FEBS Letters*, **152**, 141-143
KAO, C.H. and YANG, S.F. (1982). *Planta*, **155**, 251-266
KAWASAKI, Y., OGAWA, T. and SASAOKA, K. (1982). *Agricultural and Biological Chemistry*, **46**, 1-5
KONZE, J.R. and KENDE, H. (1979). *Planta*, **146**, 293-301
LAMOUREUX, G., GOUOT, J.M., DAVIS, D.G. and RUSNESS, D.G. (1981). *Journal of Agriculture and Food Chemistry*, **29**, 996-1002
LEGGE, R.L., THOMPSON, J.E. and BAKER, J.E. (1982). *Plant Cell Physiology*, **23**, 171-177
LIEBERMAN, M. (1979). *Annual Review of Plant Physiology*, **30**, 533-591
LIU, Y., HOFFMAN, N.E. and YANG, S.F. (1983). *Planta*, **158**, 437-441
LÜRSSEN, K., NAUMANN, K. and SCHRODER, R. (1979). *Z. Pflanzenphysiologie*, **92**, 285-294
MAYAK, S., LEGGE, R.L. and THOMPSON, J.E. (1981). *Planta*, **153**, 49-55
McKEON, T.A., HOFFMAN, N.E. and YANG, S.F. (1982). *Planta*, **155**, 437-443
McKEON, T.A. and YANG, S.F. (1984). *Planta*, **16**, 84-87
McRAE, D.G., BAKER, J.E. and THOMPSON, J.E. (1982). *Plant Cell Physiology*, **23**, 375-383

MILLER, J.M. and CONN, E.E. (1980). *Plant Physiology,* **65**, 1199–1202
PEISER, G.D., WANG, T-T., HOFFMAN, N.E., YANG, S.F., LIU, H. and WALSH, C.T. (1984). *Proceedings of the National Academy of Sciences, USA*, **81**, 3059–3063
PIRRUNG, M.C. (1983). *Journal of the American Chemical Society*, **105**, 7207–7209
RIOV, J. and YANG, S.F. (1982). *Plant Physiology*, **70**, 136–141
SATOH, S. and ESASHI, Y. (1981). *Phytochemistry*, **20**, 947–949
SATOH, S. and ESASHI, Y. (1982). *Physiologia Plantarum*, **54**, 147–152
SU, L., McKEON, T., GRIERSON, D., CANTWELL, M. and YANG, S.F. (1984). *HortScience*, **19**, 576–578
VAHATALO, M.L. and VIRTANEN, A.I. (1957). *Acta Chemica Scandinavica*, **11**, 741–743
VINKLER, C. and APELBAUM, A. (1983). *FEBS Letters*, **162**, 252–256
VIOQUE, A., ALBI, M.A. and VIOQUE, B. (1981). *Phytochemistry*, **20**, 1473–1475
YANG, S.F. (1981). In *Recent Advances in the Biochemistry of Fruits and Vegetables*, pp. 89–106. Ed. by J. Friend, M.J.C. Rhodes. Academic Press, London
YANG, S.F. and HOFFMAN, N.E. (1984). *Annual Review of Plant Physiology*, **35**, 155–189

STUDIES ON THE ENZYMES OF ETHYLENE BIOSYNTHESIS

H. KENDE, M.A. ACASTER and M. GUY
MSU-DOE Plant Research Laboratory, Michigan State University, East Lansing, Michigan, USA

Introduction

Since the discovery of 1-aminocyclopropane-1-carboxylic acid (ACC) as the immediate precursor of ethylene in plants (Adams and Yang, 1979; Lürssen, Naumann and Schröder, 1979) great progress has been made in elucidating the control of ethylene biosynthesis at the enzymatic level. Two enzymatic reactions appear to be unique for this pathway, the conversion of S-adenosylmethionine (SAM) to ACC and the oxidation of ACC to ethylene. Below, we shall discuss recent results concerning the regulation and properties of both of these enzymatic activities.

The conversion of SAM to ACC

PROPERTIES OF ACC SYNTHASE

The enzyme which converts SAM to ACC, ACC synthase, was first isolated from tomato pericarp tissue (Boller, Herner and Kende, 1979), and its basic properties have been described by these authors and by Yu, Adams and Yang (1979). ACC synthase requires pyridoxal phosphate as cofactor and utilizes SAM specifically as substrate. Its K_m with regard to SAM is 13–20 µM. It is competitively inhibited by aminoethoxyvinylglycine (AVG) and by amino-oxyacetic acid (AOA) with K_i values of 0.2 and 0.8 µM, respectively. This is in contrast to the inhibition of aspartate aminotransferase, another pyridoxal phosphate-requiring enzyme, which is irreversibly inhibited by analogues of vinylglycine (Rando, 1974).

REGULATION OF ACC-SYNTHASE ACTIVITY

In plant tissues that do not synthesize ethylene, the availability of ACC is usually the limiting factor. Therefore, the regulation of ACC-synthase activity has been of particular interest in studies concerning the control of ethylene synthesis. ACC-synthase activity increases during ripening of tomatoes (Boller, Herner and Kende, 1979), as a result of stress, e.g. wounding (Boller and Kende, 1980; Yu and Yang,

1980), and in response to indole-3-acetic acid (Jones and Kende, 1979; Yu and Yang, 1979). The most rapid and the largest increases in ACC-synthase activity have been observed following cutting of tomato pericarp tissue. Therefore, much of the work on the regulation of ACC-synthase activity has been performed with wounded pericarp tissue.

Discs excised from the pericarp of green tomatoes produce relatively little ethylene, but when they are further cut into 12 sectors (wounding), there is a substantial increase in ethylene production (Boller and Kende, 1980). The activity of ACC synthase is low in freshly cut sectors, triples within 20 min and increases tenfold within 40 min after wounding. The amount of ACC in the tissue rises approximately linearly after a lag of about 20 min, paralleling the enhancement of ACC-synthase activity. The rate of ethylene formation increases 30 min after wounding and is proportional to the amount of ACC in the tissue. These results indicate that ethylene is produced from ACC in tomato pericarp tissue in response to wounding. Such ethylene formation seems to be induced through rapid enhancement of ACC-synthase activity. The increase in ACC-synthase activity is substantially reduced by treatment of the tissue with cordycepin and is completely inhibited when cycloheximide is added to the incubation medium (Acaster and Kende, 1983). Following treatment with cycloheximide, ACC-synthase activity declines in both green and pink tomato tissue. The apparent half-life of ACC-synthase activity in green tissue is between 30 and 40 min, in pink pericarp it is in the order of 2 h (Kende and Boller, 1981). When pericarp tissue is cut and incubated in a solution containing [^{35}S]methionine, radioactivity is incorporated into protein. Following addition of cycloheximide, the radioactivity incorporated into protein of pink fruits declines in parallel with the activity of ACC synthase. Radioactivity incorporated into protein of green fruit remains essentially at the same level for at least 5 h after addition of cycloheximide while the activity of ACC synthase declines (Acaster and Kende, 1983). Therefore, ACC-synthase activity seems to turn over with the bulk protein in pink fruits but does not follow the general pattern of protein turnover in green fruit. Density labelling has been used to distinguish between activation of an ACC-synthase precursor and *de novo* synthesis of this enzyme. Wounded pericarp tissue has been incubated on D_2O or H_2O, and the buoyant density of ACC synthase has been determined by isopycnic equilibrium centrifugation in a CsCl gradient. The buoyant density of ACC synthase isolated from H_2O-treated pericarp is $1.316 \pm 0.001 \mathrm{\,g\,ml^{-1}}$ compared to $1.326 \pm 0.002 \mathrm{\,g\,ml^{-1}}$ for ACC synthase from D_2O-treated pericarp discs. The average density increase of deuterated ACC synthase has been 0.75%. These results are consistent with the hypothesis that the enhancement of ACC-synthase activity in wounded pericarp tissue is based on *de novo* synthesis of this enzyme.

PURIFICATION OF ACC SYNTHASE

Further characterization of ACC synthase and rigorous elucidation of its mode of regulation requires purification of this enzyme. This has proved to be quite difficult because the amount of ACC synthase in terms of protein is very low, even in wounded tomato pericarp tissue. In addition, ACC synthase appears to be quite labile. It loses activity during purification and is very sensitive to small pH changes. Unfortunately, no suitable affinity column has been found for the purification of ACC synthase. The most successful protocol for partial purification of ACC synthase includes precipitation with ammonium sulphate, chromatography on Sephadex G-100 and hydrophobic interaction chromatography using phenyl-Sepharose (Acaster and Kende, 1983). This latter technique yields approximately a 70-fold increase in the specific activity of ACC synthase over that in the crude extract. At this level of purification, ACC synthase is not apparent as a distinct band on an SDS-polyacrylamide gel. However, labelling of wounded pericarp tissue with [^{35}S]methionine reveals two new radioactive protein bands of 38 000 and 25 000 dalton molecular weight. Since the estimated molecular weight of ACC synthase is 57 000 ± 1500 dalton, it is not clear whether the newly labelled protein bands are related in any way to ACC synthase. More recent efforts to purify ACC synthase by High Performance Liquid Chromatography have been very promising because a substantial level of purification can be achieved in a short time (Bleecker and Kende, unpublished).

The conversion of ACC to ethylene

PROPERTIES OF THE ETHYLENE-FORMING ENZYME SYSTEM

A number of cell-free systems have been described that are capable of converting ACC to ethylene. Konze and Kende (1979), e.g. have reported that homogenates of etiolated peas contain an enzymatic activity which, in the presence of a low-molecular-weight cofactor, oxidizes ACC to ethylene. This reaction exhibits some of the features of the *in vivo* ethylene-forming system, such as requirement for molecular O_2 and inhibition by Co(II). However, the K_m of this enzyme with regard to ACC is much higher than that of the *in vivo* enzyme. Other *in vitro* ethylene-forming enzyme systems that have been described up to now suffer from similar discrepancies (Yang and Hoffman, 1984).

Hoffman *et al.* (1982) have provided an analytical tool to differentiate between cell-free ACC-dependent ethylene formation that is likely to be identical with the *in vivo* system and cell-free ethylene formation that is different from the *in vivo* system. They have reported that one of the four stereoisomers of 1-amino-2-ethylcyclopropane-1-carboxylic acid (AEC), namely (1R, 2S)-AEC, is preferentially converted to 1-butene by apple and mungbean hypocotyl tissue and have suggested that conversion of (1R, 2S)-AEC to 1-butene is mediated by the same enzyme as conversion of ACC to ethylene. Applying this test, McKeon and Yang (1984) and Guy and Kende (1984) have shown that the cell-free ethylene-forming system of Konze and Kende (1979) does not differentiate between stereoisomers of AEC as the intact tissue does; therefore, the ethylene-forming system of Konze and Kende is unlikely to represent the *in vivo* ethylene-forming enzyme.

CELLULAR LOCALIZATION OF THE ETHYLENE-FORMING ENZYME SYSTEM

Guy and Kende (1984) have studied the compartmentation of components of the ethylene-forming system in protoplasts isolated from pea leaves and have found that over 80% of the ACC is localized in the vacuole. Isolated protoplasts are capable of forming ACC and ethylene. Isolated vacuoles also evolve ethylene but cannot synthesize ACC. Over 80% of the ethylene produced by the protoplast can be accounted for as originating from the vacuole.

Vacuoles obtained from protoplasts that have been isolated in the presence of 200 μM AVG produce only trace amounts of ethylene from the low levels of endogenous ACC but convert exogenously supplied ACC very effectively to ethylene. Such vacuoles have been used to study the kinetic parameters of ACC-dependent ethylene synthesis. Ethylene synthesis in vacuoles is saturated at 1 mM ACC, as is ethylene synthesis in pea leaf discs incubated on increasing concentrations of ACC. The apparent K_m for ACC-dependent ethylene synthesis in isolated vacuoles is 61 μM; in pea epicotyls it is 66 μM (McKeon and Yang, 1984). Ethylene formation from endogenous ACC is greatly inhibited in isolated vacuoles by Co(II), n-propyl gallate and in an atmosphere of N_2. Ethylene synthesis also ceases when vacuoles are lysed by passage through a syringe and a hypodermic needle.

We have investigated whether isolated vacuoles convert AEC to 1-butene and whether the vacuolar system discriminates between the stereoisomers of AEC. We have not had the four stereoisomers of AEC at our disposal but the racemic mixtures of (1R, 2S)- and (1S, 2R)-AEC, called (±)-allocoronamic acid, and (1S, 2S)- and (1R, 2R)-AEC, called (±)-coronamic acid. It is important to note that (±)-allocoronamic acid contains the isomer that is converted preferentially to 1-butene while (±)-coronamic acid does not. Vacuoles produce 1-butene from (±)-allocoronamic acid but not in detectable amounts from (±)-coronamic acid. Therefore, isolated vacuoles exhibit the same stereospecificity as does the intact tissue with regard to 1-butene formation from AEC. Ethylene synthesis from 0.1 mM ACC is reduced by about 70% in the presence of 1 mM (±)-allocoronamic acid and, conversely, synthesis of 1-butene from 1 mM (±)-allocoronamic acid is inhibited by about 60% in the presence of 4 mM ACC. This indicates that synthesis of 1-butene from AEC and ethylene from ACC is catalysed by the same enzyme in isolated vacuoles. The difference in the conversion of (±)-allocoronamic acid and (±)-coronamic acid to 1-butene is not based on differences in the uptake of these compounds. Neither in leaf discs nor in isolated vacuoles is the high stereospecificity of the butene-forming enzyme reflected in a similar stereospecificity of the AEC uptake system.

Vacuoles prepared from *Vicia faba* protoplasts produce ethylene much as do pea vacuoles. We have also established that more than 75% of the ACC per protoplast is sequestered in the vacuole. Ethylene synthesis in isolated *V. faba* vacuoles is inhibited by Co(II).

Conclusion

We have shown that vacuoles isolated from pea and *V. faba* protoplasts produce ethylene from ACC and that the characteristics of the ethylene-forming system of isolated vacuoles resemble those of the intact tissue in every respect tested. The vacuole is probably one but perhaps not the sole organelle of the plant cell where ethylene is synthesized.

References

ACASTER, M.A. and KENDE, H. (1983). *Plant Physiology*, **72**, 139–145
ADAMS, D.O. and YANG, S.F. (1979). *Proceedings of the National Academy of Sciences of the USA*, **76**, 170–174

BOLLER, T. and KENDE, H. (1980). *Nature*, **286**, 259–260
BOLLER, T., HERNER, R.C. and KENDE, H. (1979). *Planta*, **145**, 293–303
GUY, M. and KENDE, H. (1984). *Planta*, **160**, 281–287
HOFFMAN, N.E., YANG, S.F., ICHIHARA, A. and SAKAMURA, S. (1982). *Plant Physiology*, **70**, 195–199
JONES, J.F. and KENDE, H. (1979). *Planta*, **146**, 649–656
KENDE, H. and BOLLER, T. (1981). *Planta*, **151**, 476–481
KONZE, J.R. and KENDE, H.(1979). *Planta*, **146**, 293–301
LÜRSSEN, K., NAUMANN, K. and SCHRÖDER, R. (1979). *Zeitschrift für Pflanzenphysiologie*, **92**, 285–294
McKEON, T.A. and YANG, S.F. (1984). *Planta*, **160**, 84–87
RANDO, R.R. (1974). *Science*, **185**, 320–324
YANG, S.F. and HOFFMAN, N.E. (1984). *Annual Review of Plant Physiology*, **35**, 155–189
YU, Y.-B. and YANG, S.F. (1979). *Plant Physiology*, **64**, 1074–1077
YU, Y.-B. and YANG, S.F. (1980). *Plant Physiology*, **66**, 281–285
YU, Y.-B., ADAMS, D.O. and YANG, S.F. (1979). *Archives of Biochemistry and Biophysics*, **198**, 280–286

4
THE OXYGEN AFFINITY OF 1-AMINOCYCLOPROPANE-1-CARBOXYLIC ACID OXIDATION IN SLICES OF BANANA FRUIT TISSUE

N.H. BANKS
Cambridge University

Introduction

Ethylene production in fruits has for a long time been known to be dependent upon oxygen (Abeles, 1973), and the inhibition of both ethylene production and respiration by low oxygen atmospheres is an important component in the success of controlled atmosphere storage of fruits. Mapson and Robinson (1966) reported that keeping banana fruit in low oxygen atmospheres (5–7.5% oxygen) prevented the onset of the climacteric, but that the inhibitory effects of such an atmosphere on ripening and ethylene production could be overcome by the inclusion of ethylene in physiological concentrations. This suggests that the oxygen mixture they employed was effective in preventing the onset of ethylene synthesis, but had little effect on the rate at which ethylene was produced after it had been initiated. Burg (1973) reported that the K_m O_2 of the ethylene forming system of apple cortex discs was equivalent to 0.2% oxygen in the gaseous phase. Since that time, the pathway for the biosynthesis of ethylene in a wide variety of fruit tissues has been elucidated, and can be summarized as follows (Yang, 1981):

methionine → S-adenosylmethionine → ACC → ethylene

Hoffman and Yang (1980) reported that the rate of ethylene synthesis in fruits is limited by the concentration of 1-aminocyclopropane-1-carboxylic acid (ACC) in the tissue. ACC, which is the precursor of ethylene in this pathway, accumulates in climacteric apple tissue under nitrogen (Yang, 1981). Thus, it appears that the oxygen affinity of the ethylene-producing system of apple tissue as measured by Burg must relate to that of the last step, namely the oxidation of ACC to ethylene.

Ethylene production by early climacteric banana fruit is reduced by coating them with TAL Pro-long, an aqueous dispersion of sucrose esters of fatty acids and the sodium salt of carboxymethylcellulose (Banks, 1984). Coating can inhibit the ethylene production of the fruit by 70% within 1 h and this effect can be mimicked by placing them in an atmosphere containing only 5% oxygen (Banks, unpublished data). Both of these treatments reduced the intercellular oxygen concentrations inside the fruit to approximately 1.4% and caused ACC to accumulate in the fruit pulp. If the affinity as measured by Burg for apple tissue discs applied to the banana, then this oxygen level should only have resulted in a 12% inhibition of ethylene production. This suggests that the affinity for oxygen of the enzyme which converts ACC to ethylene (the ethylene forming enzyme, EFE) in banana tissue must be considerably lower than that measured by Burg. If this were the case, the interaction of ACC and oxygen concentrations in controlling the rate of ethylene synthesis could then become an interesting area for further study because both could be limiting substrates within the ranges in which they are encountered in fruit held in modified atmospheres.

This work was undertaken in order to estimate the oxygen affinity of the EFE in slices of banana fruit tissue. The estimation was complicated by several factors. Firstly, the enzyme has two substrates, ACC and oxygen, and is therefore not strictly susceptible to Michaelis–Menten analysis. However, if the following assumptions are made about this enzyme system, it can be treated as though it is limited by a single substrate, oxygen:

(1) the enzyme binds ACC and then subsequently binds molecular oxygen;
(2) there is no accumulation of any product of ACC oxidation;
(3) the ACC concentration of the tissue is unaffected by the availability of oxygen over short periods in different oxygen concentrations.

Assumptions (1) and (2) made for this system await further information on the nature of the conversion of ACC to ethylene before they can be tested. Assumption (3) was tested experimentally as described below.

Another difficulty lies in the possibility of gradients in oxygen concentration which may be found in banana tissue. While the tissues of most fruits and vegetables are thought to be sufficiently porous for there to be no physiologically significant gradient of oxygen between any two parts of the tissue, the banana has been cited as an exception to this generalization because of its high rate of respiration (Burton, 1982). Thus, it was necessary to determine the magnitude of oxygen gradients within the pieces of tissue under study in order to be able to ascertain the effective oxygen concentration available for ACC oxidation in a given external oxygen atmosphere.

All fruit used in this study were obtained directly from the importers in the green, preclimacteric state and brought to the same stage of ripening by treating with ethylene at 1000 μl l^{-1} for 15 h (Banks, 1984) and then leaving at 20 °C for 24 h in air (relative humidity > 90%).

Estimation of oxygen gradients within banana tissue slices

MEASURED AVERAGE OXYGEN DEPRESSION

For the estimation of the oxygen gradient in whole fruit pulps, peels were removed from seven fruit and their pulps left to equilibrate in air for 1 h. The internal atmosphere of each pulp was sampled using a syringe while it was held under water (Banks, 1983). The sampling canula (16 gauge, 12 cm long) was inserted 7 cm longitudinally into the pulp centre and withdrawn 3 cm so that a 1.5 ml sample could be drawn into the syringe via a cavity in the centre of the pulp. Gas concentrations were measured using gas chromatography (Banks, 1983).

Each sample would have been drawn from the intercellular spaces at various depths within each pulp and thus represented an 'average' depression of oxygen concentration in the internal atmosphere relative to outside air. The mean depression was 0.71% (± 0.092%) oxygen and the mean radius was 1.35 cm. Thus, 99% confidence limits for the true 'average' concentration difference ($\overline{\Delta C}$) between the internal and external atmospheres of a banana fruit pulp were:

$$0.39\% < \overline{\Delta C} < 1.03\%$$

The maximum difference which might be expected inside the 4 mm thick pieces of tissue used in the affinity study was derived, using a mathematical model, assuming the upper of these two limits.

GRADIENT IN A MODEL SYSTEM

In the model employed, the pulp is considered to be a cylinder of tissue comprising concentric layers of uniform cells. Each cell absorbs the same quantity of oxygen in unit time and resistance to gaseous diffusion is uniform throughout the tissue. The ends of the cylinder contribute little to its total gaseous exchange so that a cross-sectional, two-dimensional model can be used to describe the gradient which exists between the centre and the surface of the model pulp.

For such a system, the oxygen depression at distance x from the tissue centre (ΔC_x) relative to the external atmosphere is given by an equation of the form (Goddard, 1945):

$$\Delta C_x = k(R^2 - x^2)$$

where R is the pulp radius and k is a constant. At the centre of the tissue ($x = 0$) the oxygen depression (ΔC_0) is given by:

$$\Delta C_0 = k.R^2$$

and therefore:

$$k = \Delta C_0/R^2$$

The average depression of oxygen in the intercellular system ($\overline{\Delta C}$) is given by:

$$\overline{\Delta C} = \text{SUM } p(x).\Delta C_x$$

where $p(x)$ is the proportion of the cross-sectional area at depth x,

$$= \int_0^R \frac{2\pi x}{\pi R^2} \, dx . \Delta C_x$$

$$= \int_0^R \frac{2x}{R^2} \, dx . \Delta \frac{C_0(R^2 - x^2)}{R^2}$$

$$= \frac{2\Delta C_0}{R^4} \int_0^R R^2 x - x^3 \, dx$$

$$= \frac{2\Delta C_0}{R^4} \left[\frac{R^2 x^2}{2} - \frac{x^4}{4} \right]_0^R$$

$$= \frac{2\Delta C_0}{R^4} \cdot \frac{R^4}{4}$$

$$= \frac{\Delta C_0}{2}$$

GRADIENT IN A SLICE OF TISSUE

Since the measured average depression of oxygen concentration in a real pulp ($\overline{\Delta C}$) lay between 0.39% and 1.03%, the mathematical model suggests that the depression at its centre (ΔC_0) could have been as high as 2.06%. With pulps of mean radius 1.35 cm, this corresponds to a maximum value for k of 1.13% oxygen cm^{-2}. If the resistances to diffusion in the radial and longitudinal directions of the fruit pulp were of similar magnitude and the majority of the gaseous exchange of a tissue slice t cm thick occurred across the cut surfaces, the oxygen depression at its centre ($\Delta C_{t/2}$) would be given by (Goddard, 1945):

$$\Delta C_{t/2} = 2k \left(\frac{t^2}{2} - \left(\frac{t}{2} \right)^2 \right)$$

$$= \frac{kt^2}{2}$$

Thus, the oxygen depression in a 4 mm slice would not be expected to affect estimation of the $K_m O_2$ significantly, since the mathematical model predicts that the oxygen depression at the centre of such a slice would not exceed 0.09%.

Estimation of the oxygen affinity of the ethylene-forming enzyme

Preliminary analyses using 12 mm diameter cores of banana pulp tissue showed that maximum ethylene production occurred in air, and that no further stimulation resulted from exposure to 36% oxygen (*Figure 4.1*). For the subsequent study, each fruit was peeled and its pulp cut into quadrants along the vertical axis and then into 4 mm thick wedges of tissue. Wedges were blotted with tissue paper under a vacuum to remove from the intercellular spaces any cell sap released by cutting. Seven to ten wedges (4–6 g) were placed on a bed of glass beads in each of five 65 ml, gas-tight containers and flushed with one of the following humidified gas mixtures: 0.7, 1.5, 6, 10 or 21% oxygen in nitrogen. The argon content of mixtures containing less oxygen than air was less than 40 µl l^{-1}. Samples (0.5 ml) were removed at 12 min intervals over a 48 min period for the analysis of oxygen and

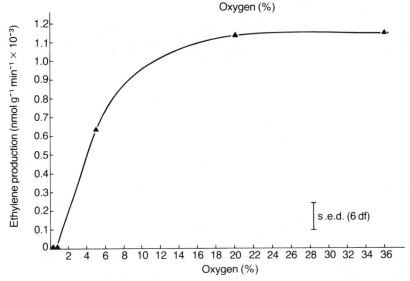

Figure 4.1 Ethylene production by 12 mm diameter cores of banana fruit pulp in different concentrations of oxygen (means of three replicates)

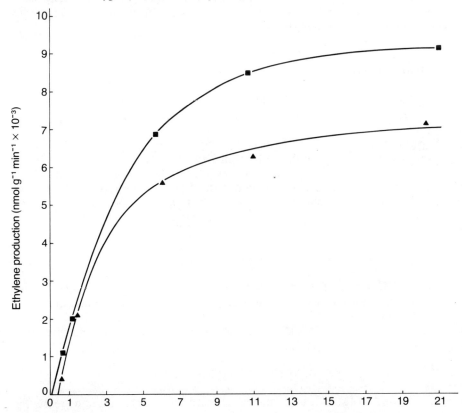

Figure 4.2 Ethylene production by 4 mm slices of banana fruit pulp in different concentrations of oxygen over two consecutive 24 min periods. Graph shows data from one of five replicates. (P_1 (▲) = 79–103 min and P_2 (■) = 103–127 min after fruit was peeled)

ethylene contents, and the same volume of an appropriate gas mixture was added to maintain the oxygen content of the jar at the same level. The oxygen content of each jar and the rate of ethylene production were determined from data obtained over two 24 min periods. Ethylene production was calculated from the difference between the initial and the final ethylene measurements for each period, and the oxygen content estimated as the mean of the measurements taken over that time.

The affinity was estimated as the K_mO_2 according to traditional Michaelis–Menten kinetics (i.e. as the substrate concentration at half the maximum velocity). Individual estimates of the K_mO_2 were made from separate curves fitted by eye to the data for the two periods from each of five replicates. The rate of ethylene production in air increased from the first to the second period for each replicate (*Figure 4.2*) but this had no effect on the estimates of the K_mO_2 (2.52% and 2.54% respectively, SED (4 df) = 0.139%). The overall mean estimate was 2.53 ± 0.362% oxygen. Ethylene production by an anaerobic piece of tissue would have been zero (Burg, 1973). Thus, extrapolating curves of the type shown in *Figure 4.2* to the x-axis provided further estimates of the average depression of oxygen concentration within the tissue. The mean value was 0.31 ± 0.089% oxygen, which is considerably higher than the value derived using the mathematical model. This suggests that the vacuum blotting had not entirely removed all the cell sap from the intercellular spaces. Each estimate of the K_mO_2 was corrected for its corresponding estimate of tissue oxygen gradient and the mean value after the adjustment was 2.22 ± 0.413% oxygen.

ACC content of the tissue

Samples of tissue were frozen in liquid nitrogen immediately after the fruit were peeled and at intervals until the end of the experiment, then freeze-dried and homogenized on a ball mill. Starch-free extracts were prepared by shaking 0.8 g of the dried powder with 5 ml absolute ethanol, and then filtering through glass wool and a Whatman GF/B glass microfibre filter in a Swynnex holder. Since ethanol can yield ethylene in the ACC assay, 1 ml aliquots in 25 ml bottles were vacuum-dried at 50 °C and then resuspended in distilled water. The assay employed mercuric chloride and alkaline hypochlorite reagents at 0 °C as described by Lizada and Yang (1979).

ACC ACCUMULATION OVER TIME IN AIR

The level of ACC in wedges kept in air increased in a linear fashion with time by about 60% from the point at which the fruit were peeled until the last measurements of ethylene production were made (*Figure 4.3*, $P<0.001$) and this corresponds with the increased rate of ethylene production between periods one and two.

ACC CONTENT AT DIFFERENT OXYGEN CONCENTRATIONS

Measurements of ACC at the end of the experiment showed that there were highly significant differences between replicate fruit ($P<0.001$). In addition the ACC contents of those wedges held in 0.7, 1.5 and 6% oxygen mixtures were higher than

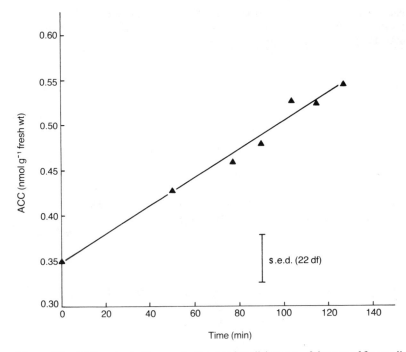

Figure 4.3 ACC content of banana fruit pulp after slicing at t = 0 (means of five replications with fitted polynomials)

Table 4.1 EFFECTS OF A 48 MIN INCUBATION IN DIFFERENT OXYGEN MIXTURES ON THE ACC CONTENT OF 4 mm THICK BANANA PULP WEDGES

O_2 (%)	ACC (nmol g^{-1} fresh wt)
21	0.55
10	0.54
6	0.72
1.5	0.71
0.7	0.68
	SED (14 df) 0.057

those of wedges kept in 10% oxygen or in air ($P<0.01$, Table 4.1). Under these circumstances, the proportion of EFE molecules with bound ACC at any given time in slices kept in low oxygen would be expected to be higher than that of wedges kept in air, which would result in them showing a correspondingly greater rate of ethylene production than would have been observed if measurements could have been made instantaneously. This may have decreased the estimates of the K_mO_2 made at the second period, but there was no evidence for this effect.

Conclusion

The K_mO_2 for the conversion of ACC to ethylene in banana fruit pulp tissue as estimated by this system is 2.2%, ten times greater than that estimated by previous workers for discs of apple cortex. This provides a mechanism whereby the ethylene production of early climacteric banana fruit could be directly inhibited by the oxygen tensions that are found in fruit coated with TAL Pro-long or kept in a 5% oxygen atmosphere. If the oxygen affinities of the EFE in other commodities are found to be of similar magnitude to that of this system, their rates of ethylene production could be substantially affected by means of controlled or modified atmosphere storage.

Acknowledgements

I am deeply indebted to the Wolfson Foundation for financial support for this work, to Marianne Hammat for technical assistance and to David Brown for advice on the mathematical model. I would also like to thank the Geest Organization for the provision of fruit and my wife for her criticism of this manuscript.

References

ABELES, F.B. (1973). *Ethylene in plant biology*. Academic Press, London
BANKS, N.H. (1983). *Journal of Experimental Botany*, **34**, 871–879
BANKS, N.H. (1984). *Journal of Experimental Botany*, **35**, 127–137
BURG, S.P. (1973). *Proceedings of the National Academy of Sciences, USA*, **70**, 591–597
BURTON, W.G. (1982). *Postharvest physiology of food crops*. Longman, London
GODDARD, D.R. (1945). In *Physical chemistry of cells and tissues*. Edited by R. Höber. Churchill, London
HOFFMAN, N.E. and YANG, S.F. (1980). *Journal of the American Society for Horticultural Science*, **105**, 492–495
LIZADA, M.C.C. and YANG, S.F. (1979). *Analytical Biochemistry*, **100**, 140–145
MAPSON, L.W. and ROBINSON, J.E. (1966). *Journal of Food Technology*, **1**, 215–225
YANG, S.F. (1981). In *Recent Advances in the Biochemistry of Fruits and Vegetables*. Edited by J. Friend and M.J.C. Rhodes. Academic Press, London

5
CARBON DIOXIDE FLUX AND ETHYLENE PRODUCTION IN LEAVES

ROGER F. HORTON
Department of Botany, University of Guelph, Ontario, Canada

Introduction

Since the discovery of the ethylene biosynthesis pathway in higher plants (Adams and Yang, 1979; Lürssen, Naumann and Schröder, 1979), there have been considerable efforts to develop our understanding of the physiological significance of the gas in terms of the regulation of this pathway (Yang, 1980). Many chemical treatments and environmental shifts had earlier been shown to alter ethylene evolution rates (Abeles, 1973), developmental changes had been correlated with alterations in ethylene production (Leiberman, 1979) and a wide variety of treatments, including those with other plant growth regulators, were known to alter the apparent sensitivity of certain plant tissues to applied ethylene (Osborne, 1976).

Much of the earlier work on ethylene production was carried out using dark-grown tissue or with green tissues held in the dark during the testing period. Although some experimenters have used airflow systems (Dhawan, Bassi and Spencer, 1981; Veroustraete *et al.*, 1982), the requirement to confine this gaseous regulator to allow accurate sampling induced many workers to employ 'sealed-flask' conditions in which changes in oxygen and carbon dioxide levels were likely to occur.

Recently, using both airflow (Bassi and Spencer, 1982) and sealed-flask conditions (Gepstein and Thimann, 1980; de Laat, Brandenburg and van Loon, 1981; Kao and Yang, 1982; Grodzinski, Boesel and Horton, 1982a,b, 1983) several groups have focused their attention on the effects of light and carbon dioxide on ethylene production (Bassi and Spencer, 1983), and, to a lesser extent, ethylene action, in photosynthetic tissues. In this chapter, I will describe some of our experiments with a variety of plant species, and discuss our results, not only in terms of the biochemistry of ethylene synthesis, but also in the larger context of the role of ethylene in the overall physiology of leaves.

Light and ethylene release

In the absence of an added carbon dioxide source, leaf tissues held in the light in a sealed flask released less ethylene than those in the dark. *Table 5.1* illustrates this

Table 5.1 EFFECT OF LIGHT AND BICARBONATE IONS ON ETHYLENE RELEASE FROM LEAF DISCS OF *RANUNCULUS SCELERATUS*

Treatment	Ethylene release (nmol g^{-1} fresh wt 4 h^{-1})
Light	0.40 ± 0.01
Light + 50 mM NaHCO$_3$	0.78 ± 0.07
Dark	0.77 ± 0.02
Dark + 50 mM NaHCO$_3$	0.86 ± 0.01

(Woodrow and Horton, unpublished)
Light at a photon flux density of 180 µmol m^{-2} s^{-1} (PAR). Fifteen 7 mm discs incubated on 2 ml 100 mM NaEPPS, pH 8.0 at 25 °C ± NaHCO$_3$ in 25 ml MicroFernbach flasks. Ethylene values are the means of three replicates ± standard error.

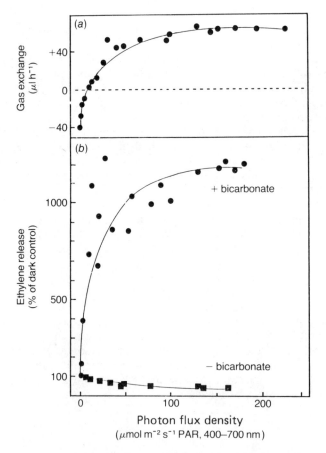

Figure 5.1 Effect of light intensity on photosynthesis (gas exchange) and ACC-dependent ethylene release from leaf discs of *Gomphrena globosa* in the presence or absence of bicarbonate (from Grodzinski, Boesel and Horton, 1983)

effect with leaf discs from the celery-leaved buttercup, *Ranunculus sceleratus*. The apparent inhibition of ethylene release by light was relieved when bicarbonate was included in the incubation medium in sealed flasks. We had earlier demonstrated this effect of light using leaf tissue of *Xanthium strumarium* and it has since been confirmed in a range of C3 species including sunflower, soybean, tomato (Bassi and Spencer, 1983), tobacco and rice (Kao and Yang, 1982). The effect of bicarbonate, and the effect of the photosynthetic inhibitor DCMU, in the light (Grodzinski, Boesel and Horton, 1982a), suggested that the prime factor in this regulation was, in fact, carbon dioxide (Grodzinski, Boesel and Horton, 1982a,b; Horton *et al.*, 1982). Changes in the flux of this gas, which occur when the balance between tissue carboxylations (photosynthesis) and decarboxylations (respiration, photorespiration) was altered during light/dark shift, correlated with changes in ethylene release (Grodzinski, Boesel and Horton, 1982a). Experiments on the effects of different light intensities on ethylene release also suggested that light was acting via photosynthesis, and that other possible effects of light acting via phytochrome or a blue-light receptor could be largely excluded for these systems (Grodzinski, Boesel and Horton, 1982b). Of particular interest were our demonstrations, initially with *Zea mays* (Grodzinski, Boesel and Horton, 1982a) and later with amaranth (*Gomphrena globosa*) (Grodzinski, Boesel and Horton, 1983) that the relationship between carbon dioxide, light and ethylene release was different in C4 tissues. The effect of light intensity on photosynthesis (measured in a Gilson apparatus with Warburg buffer) and on ethylene release from the amaranth leaves is shown in *Figure 5.1*. When the tissue is held without an added carbon dioxide source the ethylene remains at a level below that found in the dark. However, in this C4 tissue, there is a very dramatic enhancement of ethylene release with increased light intensity in the presence of 200 mM bicarbonate.

Carbon dioxide and ethylene release

Although carbon dioxide is established as a critical factor in regulating ethylene production from leaves, the biological mechanism of this effect needs further careful consideration. One possibility, apart from the obvious fact that carbon dioxide fixation is an essential prerequisite for the production of all intermediary metabolites including methionine, is that carbon dioxide is directly regulating the biosynthetic rate of ethylene. Because our studies, and those of others, show that regulation by carbon dioxide is still apparent when the tissue is incubated in 1-aminocyclopropane-1-carboxylic acid (ACC) it would seem reasonable to suggest that it is the step between ACC and ethylene which is controlled. Experiments on leaf discs (Grodzinski, Boesel and Horton, 1982a; Kao and Yang, 1982; McRae *et al.*, 1983) and on thylakoid and microsomal membrane systems (McRae *et al.*, 1983) have been interpreted in these terms. Because the precise nature of the ACC to ethylene step in the intact plant remains unclear, and because carbon dioxide can influence the physiological effects of ethylene (Abeles, 1973), it was also suggested (Grodzinski, Boesel and Horton, 1982a) that carbon dioxide might act to regulate the retention of ethylene within plant tissues and that part of this 'retention' could be at 'action' sites within the cells. Studies on the binding of ethylene to plant tissues has continued (Hall *et al.*, 1982 and Chapter 10) but, as with other plant growth regulators, the relationship between 'binding' sites and 'action' sites remains obscure. In this regard the demonstration by Beyer (1975) that ethylene,

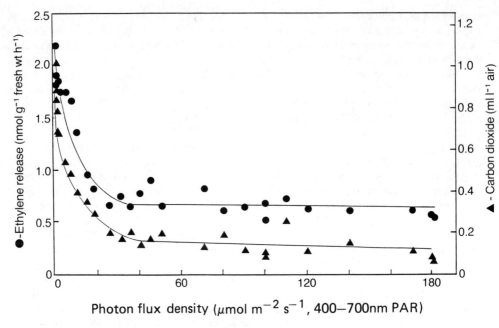

Figure 5.2 Ethylene release (●) and carbon dioxide levels (▲) from leaf discs of *Ranunculus sceleratus* in sealed flasks at a range of light intensities (Woodrow and Horton, unpublished)

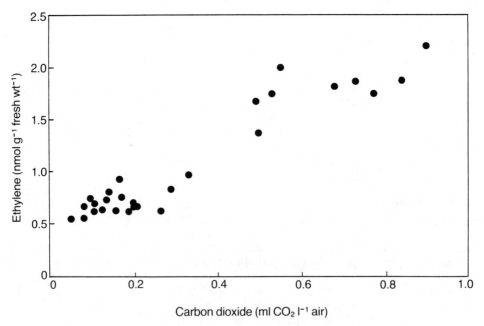

Figure 5.3 Ethylene release at various carbon dioxide levels in sealed flasks containing *Ranunculus sceleratus* leaf discs (Woodrow and Horton, unpublished)

albeit in small amounts under the conditions employed, could be metabolized to other compounds including carbon dioxide is of great interest. Beyer (1979) further suggested that this metabolism was related to the physiological action of ethylene (*see* Chapter 12).

The use of bicarbonate as a carbon dioxide source for some of our experiments could give rise to suggestions that we are looking at some other, indirect, effect of the ionic components of the incubation media on ethylene metabolism. It should be stressed that there is clear evidence for the involvement of carbon dioxide itself in these systems. *Figure 5.2* shows the parallelism between ethylene release and carbon dioxide levels in sealed flasks containing leaf discs of *R. sceleratus* at various light intensities in the absence of bicarbonate. Ethylene release is lower when light intensities above the light compensation point are able to cause the tissue to lower the carbon dioxide levels in the flask to the carbon dioxide compensation point. When the data are replotted to show ethylene release against the final carbon dioxide level, the relationship shown in *Figure 5.3* is obtained. It should be noted that the effect of carbon dioxide is seen at levels very close to the ambient concentrations of the gas in air; effects on ethylene release are not confined to high levels (1000 $\mu l\, l^{-1}$ and above) of carbon dioxide. The level of carbon dioxide in the flasks in these experiments will not necessarily be the same as that within the tissue, but will be related to it (Sharkey and Raschke, 1981).

Stomata and ethylene release

The demonstrations that ethylene release was lowered under conditions where carbon dioxide availability was limited would appear to argue against the direct involvement of stomatal apertures in regulating ethylene release. In many plants low carbon dioxide levels act to enhance stomatal opening (Raschke, 1975). We had earlier (Grodzinski, Boesel and Horton, 1982a; 1983) attempted to check stomatal opening by a microscopic examination of leaf discs at the end of

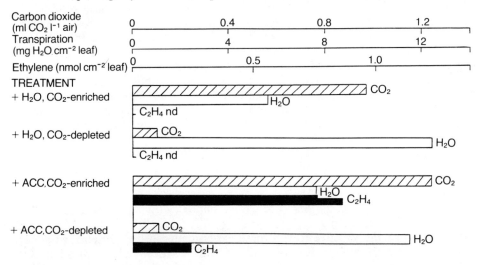

Figure 5.4 Transpiration and ethylene release from oat (*Avena sativa*) leaves under carbon dioxide enrichment or depletion (Horton and Saville, unpublished)

experiments and found little difference between treatments. In an attempt to assess stomatal behaviour more directly, we have compared transpiration and ethylene release from oat leaves (*Avena sativa* cv. Elgin) in the light under conditions of CO_2-depletion (to the carbon dioxide compensation point) and carbon dioxide enrichment (by including a buffered well of bicarbonate solution). To ensure continued transpiration within a sealed system, the relative humidity of the air in the 1 litre jars enclosing the leaves was maintained close to 30% by including a 50 ml saturated calcium chloride solution. Ethylene release from endogenous sources was not detectable under either carbon dioxide regime (C_2H_4 n.d., *Figure 5.4*) although the effect of carbon dioxide on water loss was readily determined. When 0.5 mM ACC was included in treatment vials, ethylene was released and this release was enhanced under carbon dioxide enrichment. However, ethylene release was low when transpiration was high. Similar experiments, both in sealed jars and on an open bench top, showed that ACC (to levels of 10 mM) and ethylene (to levels of $100 \mu l \, l^{-1}$) did not affect transpiration rates in these leaves. Although the absolute rates of transpiration determined in the largely unmoving air of these sealed jars will be very different from those occurring in an open situation, the data do suggest that the control of ethylene release by carbon dioxide is not simply related to effects on stomatal aperture.

The effects of chemical treatments

BUFFERS

The use of leaf discs allows the application of a wide range of precise chemical and environmental treatments to probe ethylene metabolism. In our studies we have employed a variety of buffers (Grodzinski, Boesel and Horton, 1983) but many of these were shown to be inhibitory to ethylene release. Although some of the work on leaf discs (Kao and Yang, 1982; Bassi and Spencer, 1983) has been run in unbuffered conditions, it is important, if we are to fully define these systems, to buffer incubation media. Tissues from *R. sceleratus* appear to be less prone to infiltration by the media than some other species we have used, and the ethylene release is less inhibited by buffer components. The effect of media pH on ethylene release from *R. sceleratus* has been shown to be directly related to the availability of carbon dioxide in the head space of the flasks, with the buffering regulating the release of carbon dioxide from bicarbonate in the medium (Woodrow and Horton, unpublished). Further, the controlled release of carbon dioxide from buffered wells of bicarbonate not in direct contact with the tissue, also shows that carbon dioxide itself causes changes in ethylene release.

ACC UPTAKE

Small amounts of leaf tissue may not evolve enough ethylene from endogenous sources to allow accurate measurement during short-term experiments. Accordingly it has become the practice to run these experiments in the presence of added ACC to ensure sufficient ethylene production. This practice begs the question as to whether some of the data can be explained in terms of differences in ACC uptake. The data in *Figure 5.4* indicate that in oat leaves when ACC uptake via the

Table 5.2 THE UPTAKE OF ^{14}C-ACC BY *RANUNCULUS SCELERATUS* LEAF DISCS

Treatment	^{14}C-ACC uptake[a] (nmol g^{-1} fresh wt)	
	1 h	2 h
Dark	120 ± 10	175 ± 11
Light (CO$_2$-depleted)	99 ± 8	185 ± 7
Light + 100 mM NaHCO$_3$	65 ± 3	125 ± 2

(Woodrow and Horton, unpublished)
[a]2,3-^{14}C-ACC equivalents in ethanol soluble fraction. Each value represents the mean of three determinations ± 1 standard error. Light at a photon flux density of 180 µmol m^{-2} s^{-1} (PAR).

transpiration stream is lower, by 33%, under high carbon dioxide levels, the release of ethylene from a given amount of ACC in the tissue is simultaneously increased by 400%. Further, the incubation of *R. sceleratus* leaf discs in ^{14}C-ACC under different light and carbon dioxide conditions which cause markedly different ethylene evolution rates (*see Table 5.1*) do not show very different patterns of uptake of radioactivity from the medium (*Table 5.2*).

PROMOTERS AND INHIBITORS OF ETHYLENE SYNTHESIS

The effects of a wide range of putative inhibitors and promoters of ethylene synthesis have been reported (Lieberman, 1979; Yang, 1980) and many of these have been ascribed specific sites of action within the ethylene biosynthetic pathway. Our work suggests that it is essential to ensure that these compounds are not acting indirectly by altering carbon dioxide fluxes in the tissue, and to ensure that they are acting on synthesis *per se*—in a system where the key steps from S-adenosylmethionine to ACC and from ACC to ethylene remain enzymically ill-defined—rather than on retention. Often the amount of ethylene finally released is the only factor that has been measured.

Ethylene production is stimulated by auxin treatment in many tissues (Abeles, 1973). Although it has been demonstrated that endogenous auxin levels and

Table 5.3 EFFECT OF LIGHT, ACC AND IAA ON ETHYLENE RELEASE FROM PEA LEAF TISSUE (*PISUM SATIVUM*) IN SEALED FLASKS

Treatment	Ethylene release (nmol g^{-1} fresh wt 4 h^{-1})
Light + H$_2$O	0.69 ± 0.07
Dark + H$_2$O	1.60 ± 0.25
Light + 0.5 mM ACC	2.96 ± 0.06
Dark + 0.5 mM ACC	32.15 ± 1.84
Light + 100 mM IAA	2.39 ± 0.12
Dark + 100 mM IAA	24.02 ± 0.85
Light + 100 mM IAA + 0.5 mM ACC	2.72 ± 0.05
Dark + 100 mM IAA + 0.5 mM ACC	32.44 ± 2.09

(Horton and Saville, unpublished)
Light at a photon flux density of 180 µmol m^{-2} s^{-1}. Values are the means of three replicates ± 1 standard error.

ethylene release rates are not always directly related (Roberts and Osborne, 1981), auxin has been ascribed a specific role in the enhancement of ACC-synthase activity (Yang, 1980). However, there are other possibilities for auxin action including the regulation of the release of a cofactor for ethylene synthesis (Roberts and Osborne, 1981) or a direct effect on the retention of ethylene in plants. The latter idea appears consistent with the often-shown insensitivity to ethylene of auxin-enriched tissues (Abeles, 1973). However, the lag period for enhanced ethylene release from pea (*Pisum sativum* var. Thomas Laxton) leaf tissue is over 1 h (Horton and Saville, unpublished), a time period which is more consistent with an effect on synthesis than on retention. Our experiments with pea leaves show, however (*Table 5.3*), that auxin-induced ethylene release in the light, from both endogenous sources and from added ACC, is severely inhibited when carbon dioxide levels are depleted to the CO_2-compensation point.

CARBON DIOXIDE AND ETHYLENE ACTION

Our final understanding of the effect of carbon dioxide and other compounds on ethylene metabolism must be in terms of the physiology of growth and development not merely on isolated biochemical pathways. Ethylene has been given a role in many developmental systems (Abeles, 1973) and we have earlier attempted to relate the effects of carbon dioxide on ethylene metabolism to leaf senescence (Horton *et al.*, 1982). However, senescence is controlled by a multitude of factors other than ethylene (Thomas and Stoddart, 1980) and it is perhaps preferable to concentrate our efforts on such processes as epinasty (Dhawan, Bassi and Spencer, 1981; Veroustraete *et al.*, 1982), leaf abscission (Sexton and Roberts, 1982) or the growth of aquatic plants (Cookson and Osborne, 1978; Horton and Samarakoon, 1981; Malone and Ridge, 1983), where ethylene has a more clearly defined role. For aquatic plant growth, like some seed germination systems, it has been suggested that ethylene and carbon dioxide act together as promoters (Lieberman, 1979). This idea is largely derived from studies using etiolated tissues. In *R. sceleratus*, where petiole growth is promoted by submergence or by ethylene treatment (Horton and Samarakoon, 1981)—both ethylene and ACC levels rise in tissues on submergence (Samarkoon and Horton, 1984)—carbon dioxide treatment

Table 5.4 EFFECT OF CARBON DIOXIDE AND ETHYLENE ON PETIOLE GROWTH OF *RANUNCULUS SCELERATUS* LEAVES IN THE DARK

Carbon dioxide (ml l^{-1})	Growth (mm)		
		Ethylene (μl l^{-1})	
	0[a]	0.01	1.0
Ambient	3.5 ± 0.2	9.5 ± 0.5	20.6 ± 0.6
0[b]	4.3 ± 0.3	6.7 ± 0.4	22.9 ± 0.7
10	4.3 ± 0.5	—	23.0 ± 0.8
50	4.0 ± 0.3	—	16.8 ± 1.2
100	3.3 ± 0.2	5.9 ± 0.3	15.2 ± 0.6

(Samarakoon and Horton, unpublished)
Initial petiole length 25 mm. Each value represents the increase in petiole length (± 1 standard error) of 16 replicates over a 24-h period.
[a]In the presence of mercuric perchlorate as an ethylene absorbent.
[b]In the presence of KOH as a carbon dioxide absorbent.

does not enhance ethylene-promoted growth of light-grown leaves held in the dark (*Table 5.4*). Indeed it can be seen that carbon dioxide inhibits the ethylene-promoted growth of *R. scleratus* petioles. Gas exchange in aquatics has been widely studied in attempts to understand carbon dioxide and oxygen fluxes under water and both of these gases will influence ethylene synthesis and action (Yang, 1980). Therefore in these, and in other plant systems, an understanding of growth and developmental physiology will require experiments in which we measure and manipulate all three gases simultaneously.

Acknowledgements

These studies were supported by grants from the Natural Sciences and Engineering Research Council, Canada. I thank my colleagues Ananda Samarakoon, Lorna Woodrow, Ingrid Boesel, Barry Saville and Bernard Grodzinski for all their work and encouragement.

References

ABELES, F.B. (1973). *Ethylene in Plant Biology*. Academic Press, New York
ADAMS, D.O. and YANG, S.F. (1979). *Proceedings of the National Academy of Sciences (USA)*, **76**, 170–174
BASSI, P.K. and SPENCER, M.S. (1982). *Plant Physiology*, **69**, 1222–1225
BASSI, P.K. and SPENCER, M.S. (1983). *Plant Physiology*, **73**, 758–760
BEYER, E.M. (1975). *Nature*, **255**, 144–147
BEYER, E.M. (1979). *Plant Physiology*, **63**, 169–173
COOKSON, C. and OSBORNE, D.J. (1978). *Planta*, **144**, 39–47
DE LAAT, A.M.M., BRANDENBURG, D.C.C. and VAN LOON, L.C. (1981). *Planta*, **153**, 193–200
DHAWAN, K.R., BASSI, P.K. and SPENCER, M.S. (1981). *Plant Physiology*, **68**, 831–834
GEPSTEIN, S. and THIMANN, K.V. (1980). *Planta*, **149**, 196–199
GRODZINSKI, B., BOESEL, I. and HORTON, R.F. (1982a). *Journal of Experimental Botany*, **33**, 344–354
GRODZINSKI, B., BOESEL, I. and HORTON, R.F. (1982b). *Journal of Experimental Botany*, **33**, 1185–1193
GRODZINSKI, B., BOESEL, I. and HORTON, R.F. (1983). *Plant Physiology*, **71**, 588–593
HALL, M.A., CAIRNS, A.J., EVANS, D.E., SMITH, A.R., SMITH, P.G., TAYLOR, J.E. and THOMAS, C.J.R. (1982). In *Plant Growth Substances 1982*, pp. 375–383. Ed. by P.F. Wareing. Academic Press, London
HORTON, R.F. and SAMARAKOON, A.B. (1981). *Aquatic Botany*, **13**, 97–104
HORTON, R.F., WOODROW, L., BOESEL, I. and GRODZINSKI, B. (1982). In *Growth Regulators in Plant Senescence*, pp. 83–101. Ed. by M.B. Jackson, G. Grout and I.A. McKenzie. Monograph 8, British Plant Growth Regulator Group, Wantage, UK
KAO, C.H. and YANG, S.F. (1982). *Planta*, **155**, 261–266
LIEBERMAN, M. (1979). *Annual Review of Plant Physiology*, **30**, 533–591
LÜRSSEN, K., NAUMANN, K. and SCHRÖDER, R. (1979). *Zeitschrift für Pflanzenphysiologie*, **129**, 285–294
MALONE, M. and RIDGE, I. (1983). *Planta*, **157**, 71–73

McRAE, D.G., COKER, J.A., LEGGE, R.L. and THOMPSON, J.E. (1983). *Plant Physiology*, **73**, 784–790
OSBORNE, D.J. (1976). In *Proceedings in Life Sciences—Plant Growth Regulation*, pp. 161–171. Ed. by P.E. Pilet. Springer-Verlag, Heidelberg
RASCHKE, K. (1975). *Annual Review of Plant Physiology*, **26**, 237–258
ROBERTS, J.A. and OSBORNE, D.J. (1981). *Journal of Experimental Botany*, **32**, 875–889
SAMARAKOON, A.B. and HORTON, R.F. (1984). *Annals of Botany*, **54**, 263–270
SEXTON, R. and ROBERTS, J.A. (1982). *Annual Review of Plant Physiology*, **33**, 133–162
SHARKEY, T.D. and RASCHKE, K. (1981). *Plant Physiology*, **68**, 33–40
THOMAS, H. and STODDART, J.L. (1980). *Annual Review of Plant Physiology*, **31**, 83–111
VEROUSTRAETE, F., FREDERICQ, H., VAN WIEMEERSCH, I. and DE GREEF, J. (1982). *Photochemistry and Photobiology*, **35**, 261–264
YANG, S.F. (1980). *HortScience*, **15**, 172–180

6

THE EFFECT OF TEMPERATURE ON ETHYLENE PRODUCTION BY PLANT TISSUES

ROGER J. FIELD
Plant Science Department, Lincoln College, Canterbury, New Zealand

Introduction

The aims of this review are to describe current knowledge on the effect of temperature on ethylene production and to assess the implications of changed ethylene production on dependent physiological processes. Recent reviews on the physiology of ethylene by Osborne (1978) and Lieberman (1979) have presented very little information on temperature-induced manipulations of ethylene production. This is largely owing to the general paucity of information and until recently the lack of understanding of ethylene biosynthesis. This review does not attempt a complete coverage of all areas of temperature-manipulated ethylene production but rather a more detailed examination of a limited number of model systems, with an emphasis on presenting relevant experimental results. Considering the significance of temperature as a major environmental variable and its known influence on many physiological processes (Lyons, Graham and Raison, 1979) it is surprising that it has received relatively scant attention by plant hormone physiologists. As a corollary it is of some concern that relatively few reports on ethylene physiology include information on the temperature at the time of taking measurements, when Burg and Thimann (1959) and others have long established that small variations in temperature may bring about substantial changes in ethylene production (*Table 6.1*). Both apple fruit tissue (*Malus domestica* Borkh.) and bean leaf tissue (*Phaseolus vulgaris* L.) showed increases or decreases of at least 30% for divergencies of 5 °C from the reference temperature of 20 °C. Even smaller temperature fluctuations lead to significant changes in ethylene production, and

Table 6.1 PERCENTAGE CHANGES IN ETHYLENE PRODUCTION AT TEMPERATURES ABOVE AND BELOW 20 °C IN (a) APPLE FRUIT SECTIONS (DATA DERIVED FROM BURG AND THIMANN (1959)) AND (b) BEAN LEAF DISCS (FIELD, UNPUBLISHED)

(a) °C	% Change in ethylene production	(b) °C	% Change in ethylene production Wound	Basal
15	−43	15	−43.5	−31.5
22	+30	17.5	−23.5	−16.1
27	+35	22.5	+27.3	+21.5
32	+44	25	+53.1	+32.9

48 The effect of temperature on ethylene production by plant tissues

these changes could play an important role in increasing experimental variability and masking physiological events.

Effect of low temperature

The response of plants to lower temperatures than are optimal for growth may be directly linked to hormonal and metabolic control processes that include ethylene production. It is possible to recognize three categories of low temperature effect on plants;
(1) a non-damaging reduction in temperature;

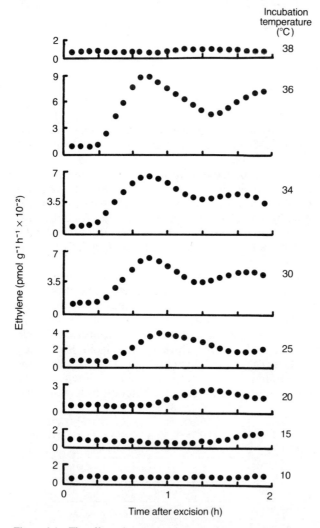

Figure 6.1 The effect of temperature on wound-induced ethylene production by subapical stem sections of etiolated pea seedlings. (After Saltveit and Dilley, 1978b, reproduced by permission of the publishers)

(2) a response to chilling in those species that are sensitive to temperatures below 10–12 °C but above 0 °C; and finally
(3) the effect of a short or long duration of exposure to a freezing temperature of 0 °C or below.

All three categories of low temperature treatment may affect ethylene production.

Saltveit and Dilley (1978b) investigated the temperature dependence of wound ethylene produced by excised segments of etiolated pea (*Pisum sativum* L.) stems. They noted that lowering incubation temperatures below the plant growing temperature of 24 °C markedly reduced ethylene production and delayed the detection of the wound ethylene peak produced after excision. Thus the maximum ethylene production rate occurred at 55 min after excision at 25 °C, but this had extended to 120 min at 15 °C and was not detected after a similar period at 10 °C (*Figure 6.1*). Over the temperature range of 10–25 °C there was approximately a tenfold increase in ethylene production from 40–400 pmol $g^{-1} h^{-1}$. These workers did not provide an explanation for the low temperature-induced changes in ethylene production although reference was made to parallel changes in respiration and possible perturbations of cellular membranes.

Perhaps the greatest interest in low temperature induced ethylene production has been in those plants that are chilling-sensitive at temperatures below 10–12 °C. Studies by several groups including Mattoo *et al*. (1977) working with tomato (*Lycopersicon esculentum* Mill.); Field (1981a) with dwarf bean and Wang and Adams (1982) working with cucumber (*Cucumis sativus* L.) have yielded similar results. Although technically and statistically difficult to resolve, Field (1981a) showed significant ethylene production at 2.5 °C in freshly cut and aged leaf discs of dwarf beans that had been grown at 25 °C (*Figure 6.2*). Similarly Mattoo *et al*. (1977) established ethylene production from tomato fruit plugs incubated at 2 °C. Thus, in the short term, incubation of chilling-sensitive tissues at subchilling temperature does not eliminate ethylene production. However in view of the fact that the incubation procedures were carried out at high humidity it is doubtful whether either system strictly resembled the conditions under which chilling damage is normally induced (Simon, 1974). Wright (1974) showed that chilling damage to primary leaves of dwarf bean, induced by exposure to 5 °C at 20% relative humidity, was associated with increased ethylene production and electrolyte leakage, indicating at least a partial change in membrane function. While chilling symptoms were not observed by Field (1981a) the pattern of ethylene production reflected on associated biochemical changes and disturbances in membrane function. Chilling sensitivity has been linked to partial perturbation of cell membranes and a phase-transition below a critical chilling-sensitive temperature (Lyons and Raison, 1970; Lyons, Graham and Raison, 1979). The changed membrane state induced below the critical temperature affects the activity of membrane-bound enzymes, increasing the activation energy of the rate-limiting step in the overall reaction. These changes can be identified by presenting the data in the form of an Arrhenius plot, where typically there is a discontinuity at the critical temperature leading to two different activation energies for the process. When applied to ethylene production data there is a marked discontinuity at 11.4 °C for dwarf bean (Field, 1981a) (*Figure 6.3*) and at 12 °C for tomato (Mattoo *et al*., 1977). While there are similarities in the transition temperatures for the two plant species the activation energies for tomato are much lower than for dwarf

The effect of temperature on ethylene production by plant tissues

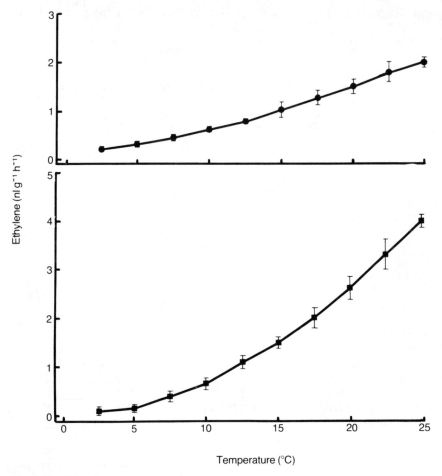

Figure 6.2 The effect of low temperature on basal (●) and peak wound (■) ethylene production by leaf discs of dwarf bean. Vertical bars indicate standard error of the mean. (From Field, 1981a, reproduced by permission of the publishers)

bean. The activation energy above the transition temperature for dwarf bean leaf discs is close to values calculated for apple fruit plugs (Apelbaum et al., 1981) and etiolated pea stem sections (Saltveit and Dilley, 1978b), over similar temperature ranges; being equal to 40–55 kJ mol^{-1}, or equivalent to a Q_{10} of 1.7–2.8 (*Table 6.2*).

It is perhaps not surprising that while there is some comparability between tissue systems in activation energies calculated for the higher temperature range there is a greater spread of values for the lower range (*Table 6.2*). The difficulties in determining the rather meagre ethylene production at temperatures around 5 °C, the frequent dependence on measurements of production at a limited number of temperatures and the lack of attention to the possible physical changes in ethylene diffusion from the tissue and its measurement all make for substantial experimental variation and error. Small deviations in the slopes of the Arrhenius plot can lead to significant changes in activation energies and may explain the extreme values calculated from the data of Saltveit and Dilley (1978b) (*Table 6.2*).

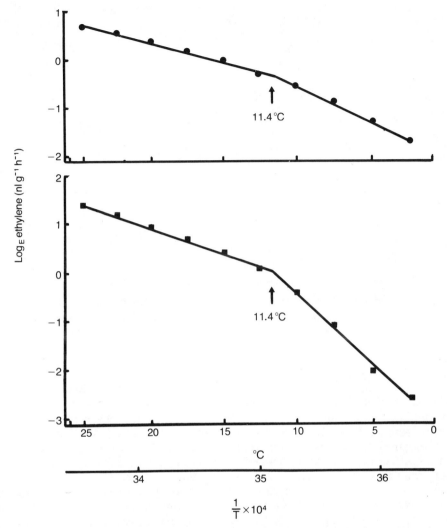

Figure 6.3 Arrhenius plots of basal (●) and wound (■) ethylene production by leaf discs of dwarf bean. The mean values for ethylene production at each temperature are superimposed on fitted regression lines. (From Field, 1981a, reproduced by permission of the publishers)

It has been suggested that discontinuous Arrhenius plots need not indicate a membrane phase change in chilling-sensitive species (Bagnall and Wolfe, 1978). Thus the presumption that discontinuous Arrhenius plots imply that the rate-limiting step in ethylene synthesis involves a membrane bound enzyme that is perturbed at a critical temperature in chilling-sensitive species has been questioned. The data in *Table 6.2* suggest that some normally chilling-insensitive plants, such as pea and apple, can demonstrate similar temperature breaks to chilling-sensitive species. It is possible that these non-acclimatized, chilling-insensitive plants could either show a lack of rapid biochemical adaptation, or simply have a major decline in ethylene synthesis below a certain temperature. However, experimental results

Table 6.2 SUMMARY OF LOW TEMPERATURE EFFECTS ON ETHYLENE PRODUCTION AND THE ACTIVATION ENERGY FOR THE PROCESS

Plant tissue	Temperature range (°C)		Activation energy (kJ mol^{-1}) (equivalent Q_{10} value)	
	Above transition point	Below transition point	Above transition point	Below transition point
Apple fruit sections[a]		10–25		74.0[e] (2.8)
Apple fruit plugs[b]	15–30	2.5–10	40.9 (1.7)	77.8 (2.7)
Bean leaf discs (basal)[c]	11.4–25	2.5–11.4	55.4 (2.1)	99.6 (3.8)
Bean leaf discs (wound)[c]	11.4–25	2.5–11.4	73.8 (2.8)	192.2 (8.3)
Pea stem sections[d]	15–36	10–15	50.6[e] (2.0)	(>18)[e]
Tomato fruit[b]	12–20	2.5–12	18.6 (1.2)	65.7 (2.7)

[a]Burg and Thimann (1959) (no transition point determined)
[b]Mattoo *et al.* (1977) (precise transition point not determined)
[c]Field (1981a)
[d]Saltveit and Dilley (1978b)
[e]Recalculated values

have shown the continued synthesis of ethylene at temperatures around 5 °C and some comparability with rates of reaction and activation energies for processes, such as respiration, that are known to involve membrane-bound enzymes (Burg and Thimann, 1959).

It is not surprising that ethylene production associated with freezing temperatures is poorly documented. For example, it is not possible to detect ethylene production by leaf discs of dwarf bean at 0 °C and it is reasonable to assume nil or negligible synthesis by other plant systems, although Hansen (1945) reported that certain apple varieties produced ethylene at 0 °C. Nichols (1966) observed ethylene production from cut flowers of carnation (*Dianthus caryophyllus* L.) at 1.6 °C, but no determinations were made at lower temperatures. Other workers have examined the effect of subfreezing temperatures on ethylene production, but only after returning tissues to ambient conditions (Young and Meredith, 1971).

Cell damage associated with freezing increases ethylene production from adjacent, undamaged cells (Elstner and Konze, 1976; Kimmerer and Kozlowski, 1981). For instance, point freezing of sugar beet (*Beta vulgaris* L.) leaf discs with a cold probe (−186 °C), increased ethylene production, provided that the percentage of undamaged cells did not decline below approximately 25% (*Figure 6.4*) (Elstner and Konze, 1976). In this case the changes in ethylene production with freezing-induced cell damage represent a wound response of the non-decompartmentalized, but physiologically perturbed cells, adjacent to the decompartmentalized, killed cells. In this experiment ethane production increased linearly with the percentage of total leaf disc area frozen, indicating a lack of dependence on intact compartmentalization in the leaf cells and a requirement for cellular disorder. This has been interpreted to indicate that ethylene and ethane are derived from different sources. As suggested by Elstner and Konze (1976) ethane production and the ethylene-to-ethane ratio, which shows a logarithmic dependence when plotted against the percentage of frozen leaf disc area (*Figure 6.4*), may be taken as an indicator of leaf tissue integrity. In another context ethane production and the ethylene-to-ethane ratio have been used as indicators of anaerobiosis (Curtis, 1969).

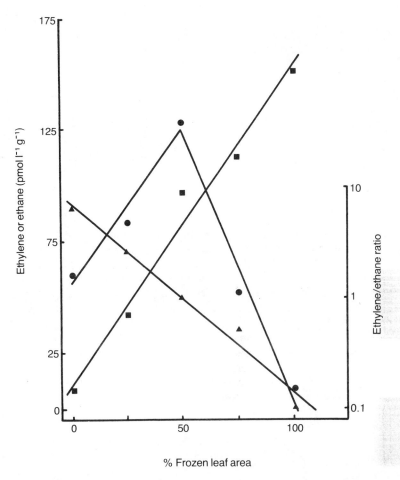

Figure 6.4 The effect of point freezing damage ($-186\,°C$) on ethylene (●) and ethane (■) production and the ethylene:ethane ratio (▲) by sugar beet leaf tissue. (From Elstner and Konze, 1976; reproduced by permission of the editor of *Nature* and the publishers Macmillan Journals Ltd)

Effect of high temperature

Ethylene production may be promoted by increases in temperature above those normally employed to grow experimental plants of temperate species (15–25 °C). The data compiled in *Table 6.3* reveal a temperature optimum of 30–35 °C for most species, with detectable production in intact tissue systems continuing to approximately 40 °C. The experimental verification of continuous ethylene production at temperatures above 30 °C is technically demanding. First, it is particularly difficult to maintain stable plant–water relations in intact plants and to prevent even small water potential deficits. Secondly in explant systems such as leaf discs where enclosure may maintain adequate water relations there is the likelihood of creating partial or total anaerobiosis because of rapid oxygen depletion by elevated

54 The effect of temperature on ethylene production by plant tissues

Table 6.3 SUMMARY OF HIGH TEMPERATURE EFFECTS ON ETHYLENE PRODUCTION

Optimum temperature (°C)	Maximum temperature at which production was measured (°C)	Plant tissue	Reference
32	40	Apple fruit sections	Burg and Thimann (1959)
35	42.5	Bean leaf discs	Field (1981b)
30	—	Apple fruit plugs	Mattoo et al. (1977)
20	—	Tomato fruit plugs	Mattoo et al. (1977)
30	>50	Pea shoot homogenate	Konze and Kende (1979)
36	38 (wound) >38 (basal)	Pea etiolated subapical stem sections	Saltveit and Dilley (1978b)
29	>35	Apple fruit discs	Apelbaum et al. (1981)
30	40	Apple fruit plugs	Yu, Adams and Yang (1980)
30	40	Mungbean hypocotyl	Yu, Adams and Yang (1980)

respiration. Frequently parallel measurements of ethylene production and respiration rate are not taken and there is a certain amount of guesswork associated with optimizing the duration of the enclosure period. This is particularly difficult at temperatures above 40 °C where ethylene production is reported to be minimal (Field, 1981b; Saltveit and Dilley, 1978b), but oxygen consumption by respiration is undoubtedly high. Although ethane production is a possible indicator of reduced oxygen tensions and lowered respiration it cannot be used conclusively because of observed inconsistencies (Saltveit and Dilley, 1978b). Also there is a lack of understanding of the mechanisms involved and the linkage between ethane and ethylene syntheses (Elstner and Konze, 1976).

When leaf discs, excised from dwarf bean plants grown at 25 °C, are incubated at temperatures from 25–47.5 °C the pattern of ethylene production is similar for

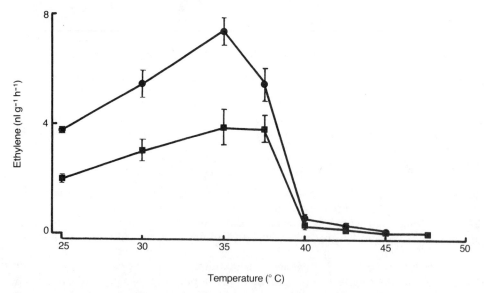

Figure 6.5 The effect of high temperature on basal (■) and peak wound (●) ethylene production by leaf discs of dwarf bean. Vertical bars indicate standard error of the mean. (From Field, 1981b, reproduced by permission of the publishers)

freshly cut discs (wound ethylene) and those aged for 16 h prior to measurement (basal ethylene) (*Figure 6.5*) (Field, 1981b). The systems are very labile and the dramatic decline in ethylene production above 37.5 °C is similar to the rapid reduction between 36 and 38 °C reported for etiolated pea stem sections (Saltveit and Dilley, 1978b). The imposition of a temperature regime higher than the growing temperature could be regarded as a wound stimulus with a consequent increase in ethylene synthesis (*Figure 6.5*), or simply as an increase in biochemical activity as a function of temperature. The corresponding increase in ethylene production from aged discs could more realistically be considered as a wound response to temperature. The close correspondence between the patterns of change of basal and wound ethylene production with increasing temperature suggests that the temperature effect is superimposed on the two basic levels of ethylene synthesis. The 10 °C rise in temperature between 25 and 35 °C approximately doubled the rate of ethylene production in both systems, giving activation energies for fresh discs of 51.9 kJ mol^{-1} and for aged discs of 48.6 kJ mol^{-1}, values which correspond to a Q_{10} of 1.9–2.0, and are similar to those quoted by Saltveit and Dilley (1978b). In no experiments covering the 25–37.5 °C range did ethylene production from aged discs reach the values for freshly-cut discs (*Figure 6.5*); suggesting that two separate mechanisms, but not necessarily distinctive biochemical pathways existed. The physical wounding and temperature effects may simply be additive in freshly-cut discs, which within the limits defined by experimental variation is possible for temperatures up to 37.5 °C (*Figure 6.5*). It is unclear whether the results are compatible with the suggestion of Hanson and Kende (1976) that substrate availability may be a key factor that differentially regulates basal and wound ethylene production.

Temperature has a pronounced effect on the rate and timing of wound-induced ethylene synthesis. Using subapical stem sections of etiolated peas, Saltveit and Dilley (1978a and b) showed that the lag period was in excess of 40 min at 20 °C but was reduced to a more or less constant period of 25 min at 30–36 °C, while the time to first maximum wound production was 90 min and 45–50 min respectively (*Figure 6.1*). The rate of maximum wound ethylene production increased fourfold over the 20–36 °C temperature range. At temperatures above 25 °C there were two pronounced peaks of wound ethylene production; which at 30 °C occurred at 45 and 105 min. It is suggested that the oscillations in wound-induced ethylene may involve a negative feedback control by endogenous ethylene. High levels of endogenous ethylene from the initial rise may deactivate part of the wound ethylene synthesizing system and thereby temporarily decrease rates of production; while the decline following the second peak of production may be due to depletion of a limiting substrate. However these proposed explanations were made without the benefit of knowledge of the behaviour of all the major intermediates of ethylene biosynthesis, as proposed by Adams and Yang (1979) and their involvement in wounding (Yu and Yang, 1980).

The rapid decline in ethylene production above the 36–38 °C temperature range (Field, 1981b; Saltveit and Dilley, 1978b) (*Table 6.3*, *Figures 6.1* and *6.5*), suggests a loss of integrity of the ethylene-synthesizing system. More specifically the high temperature may perturb membrane structure, leading to increases in the activation energy of membrane-bound enzymes and a reduced rate of ethylene synthesis (Field, 1981b). The precise cellular site of ethylene production is not known, but if located at a membrane-cell wall complex as suggested by Mattoo and Lieberman (1977), then electrolyte leakage through the plasmalemma may be a reasonable guide to the relationship between ethylene production and the functional state of the cell membrane. Field (1981b) showed that there was a marked increase in total electrolyte leakage from bean leaf discs at temperatures above 40 °C where there was complete cessation of ethylene production.

Changes in ethylene production following tissue transfer from low or high temperature

The low levels of ethylene production at 5 °C or below and above 40 °C, are frequently associated with non-permanent tissue damage and a resumption of higher levels of ethylene production when tissue is transferred to more normal plant growing temperatures in the 15–25 °C range. For instance, Field (1981a and b) found that exposure of bean leaf discs to 5 °C or temperatures in the 40–45 °C range for 1 h or more resulted in rapid ethylene production on transference of the tissue to 25 °C, usually at rates exceeding wound production at 25 °C (*Figures 6.6*

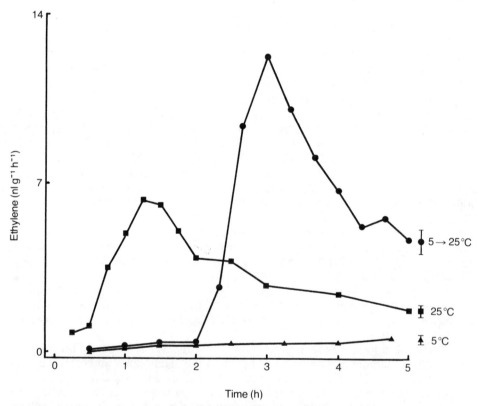

Figure 6.6 The production of wound ethylene at 5 °C (▲), 25 °C (■), and following transfer of leaf discs of dwarf bean from 5–25 °C after 2 h (●). Vertical bars indicate standard error of the mean. (From Field, 1981a, reproduced by permission of the publishers)

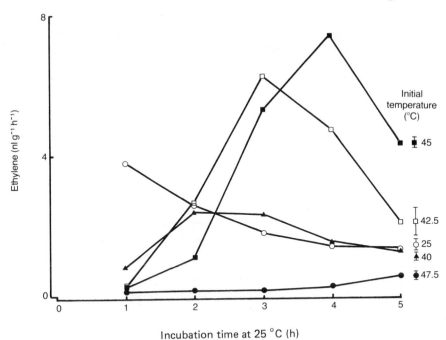

Figure 6.7 The effect of 1 h incubation at high temperature (25–47.5°C) on subsequent ethylene production at 25°C by leaf discs of dwarf bean. Vertical bars indicate standard error of the mean. (From Field, 1981b, reproduced by permission of the publishers)

and 6.7). The ethylene production at 25 °C following a 2 h exposure to 5 °C was virtually lag-free (*Figure 6.6*). The corresponding experiments involving pre-incubation at high temperature were not lag-free and resumption of detectable ethylene production was delayed for at least 1 h at the highest temperature treatments (42.5–45 °C) (*Figure 6.7*). However, the level of wound ethylene induced by high temperature was related to the initial pre-incubation temperature, with more ethylene produced following treatment at 45 °C than at 40 °C (*Figure 6.7*). Wang and Adams (1982) demonstrated similar results for cucumber fruit tissue that had been exposed to a chilling temperature of 2.5 °C for several days. Transference from 2.5–25 °C resulted in rapid ethylene production provided low temperature exposure did not exceed four days, whereas a similar experiment involving initial incubation at a nonchilling 13 °C resulted in no significant increase in ethylene production on transfer to 25 °C.

A very detailed characterization of high temperature effects on ethylene production in subapical sections of peas was provided by Saltveit and Dilley (1978b). They reported that a 1 min exposure to 40 °C was sufficient to prolong the lag period of wound ethylene production by 7 min, from 26–33 minutes at 30 °C. Exposure of the plant material to 40 °C for 5 min shifted the time of maximum wound ethylene production from 50–80 min, broadened the wound peak and reduced the maximum rate of ethylene production from 300 to 140 pmol g^{-1} h^{-1}.

Prior to the clearer understanding of the pathway of ethylene biosynthesis (Adams and Yang, 1979), it was more difficult to interpret the rapid reversibility of ethylene production following transfer of plant tissue from extreme temperatures to more normal growing temperatures. Some workers (Mattoo *et al.*, 1977; Saltveit and Dilley, 1978b; Field, 1981a and b) proposed that low or high temperature interfered with the activity of membrane-bound enzymes that were essential for normal ethylene biosynthesis, but that the systems were readily reversible. The scant information on the specific membrane fractions associated with ethylene biosynthesis did not assist interpretation (Saltveit and Dilley, 1978b). Field (1981b) showed that electrolyte leakage from cells of bean leaf discs was correlated with ethylene production and that high temperature induction of membrane leakiness resulted in no ethylene production, while membrane repair and the reversibility of electrolyte leakage was associated with a resumption of ethylene production. None of these workers could adequately explain the accumulation of a potential for 'overshoot' ethylene production when tissue was incubated at extreme temperatures of 5 and 40 °C. However, consideration was given to the production of an unidentified cytoplasmic factor (Field, 1981b), and the mechanism could have been similar to that proposed by Lieberman (1979), to explain the overshoot in ethylene production following the release of tissue from anaerobiosis. Although a complete explanation has yet to emerge, the recent characterization of ethylene biosynthesis has focused attention on the behaviour of pathway intermediates and particularly the role of ACC (1-aminocyclopropane-1-carboxylic acid). Further consideration of the role of ACC will be presented in the following section.

Hall *et al.* (1980) have suggested that plant tissues are capable of storing ethylene in specific cellular compartments and that the air space and cell concentrations of ethylene may not be related to the partition coefficient in water. Heating at 60 °C, but not freezing and thawing, caused the release of ^{14}C-labelled ethylene that had previously been incorporated into excised cotyledon tissue of *Phaseolus vulgaris* (Jerie, Shaari and Hall, 1979). It is clear that if increasing temperature releases stored and compartmentalized ethylene then increased ethylene production will not rely solely on *de novo* synthesis. The rapid decline in ethylene production above approximately 40 °C (*Table 6.3*), suggests either an effect on ethylene synthesis or an exhaustion of stored ethylene, or both. The ability of tissues to resume ethylene production after transfer from high temperature (>40 °C) to 25 °C (*Figure 6.7*), suggests that at least some synthesis is occurring.

The influence of temperature on the pathway of ethylene biosynthesis

The confirmation of S-adenosylmethionine (SAM) and ACC as intermediates in the biosynthesis of ethylene from methionine (Lurssen, Naumann and Schroder, 1979; Yang *et al.*, 1980) has offered the opportunity to determine which steps in the pathway are temperature-sensitive, or more specifically the possibility of identifying the rate-limiting step in ethylene production that follows temperature-induced perturbations. Several early workers have attempted to identify rate-limiting steps but it is the more recent work of Apelbaum *et al.* (1981), Field (1984) and Wang and Adams (1982) that is important to our current understanding.

Wang and Adams (1982) used cucumber fruit tissue and showed that the low ethylene production at a chilling temperature (2.5 °C) was associated with correspondingly low levels of ACC and ACC synthase, suggesting that the synthesis of ACC was the rate-limiting step in ethylene production. In bean leaf tissue Field (1984) has shown that enhanced ethylene production occurs at 5 °C, following exogenous application of ACC (*Figure 6.8*), and unpublished work shows that

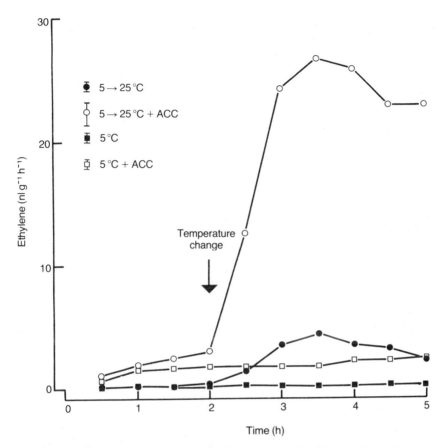

Figure 6.8 The effect of temperature transfer (5 → 25 °C) and ACC (0.1 mM) supplied at 0 h on ethylene production by dwarf bean leaf discs. Vertical bars indicate standard error of the mean. (From Field, 1984, reproduced by permission of the publishers)

ethylene production at low temperature is still ACC concentration-dependent. It appears that the conversion of ACC to ethylene is significantly inhibited by subchilling and low temperatures, but it is not the rate limiting step that determines the activation energy for ethylene production in chilling sensitive species (*Figure 6.3*). In the latter case the increased activation energy for ethylene production below a critical temperature has been interpreted as a perturbation of membrane-bound enzymes, although whether membrane integrity is linked to the ACC to ethylene step is unclear as this step may occur in tissue homogenates, following major disruption of membrane integrity (Konze and Kende, 1979). In further experiments with tissue homogenates, Mayak, Legge and Thompson (1981) isolated a microsomal membrane fraction from senescing carnation flowers and concluded that not all of the ACC to ethylene converting activity was membrane-bound. The microsomal membrane fraction had a temperature optimum of 35 °C for conversion of ACC to ethylene, which is similar to the 30 °C quoted by Konze and Kende (1979) for pea homogenates. However the point is far from straightforward as Apelbaum *et al.* (1981) considered the conversion of ACC to ethylene to be membrane-associated as it did not fully function following treatment of apple fruit discs with either the surface-active agent, Triton-X-100, or by imposing an osmotic shock. A further treatment involving a cold shock at 3°C for 10 min also reduced ethylene production from exogenous ACC but in view of the complexities of tissue response to short-term temperature treatments (Saltveit and Dilley, 1978b; Field, 1981a), the result can hardly be interpreted as providing evidence for a membrane-bound, enzymatic conversion of ACC to ethylene. Lürssen, Naumann and Schroder (1979) categorically state that the conversion of ACC to ethylene involves membrane-bound enzymes.

Perhaps the most interesting point to emerge from the work of Apelbaum *et al.* (1981) is that addition of 1 mM ACC to apple fruit tissue does not change the qualitative pattern of ethylene production over a 4–35°C temperature range (*Figure 6.9*); with ACC increasing ethylene production about 2.5 times over the complete temperature range. When plotted according to the Arrhenius equation both treatments showed similar changes in slope at 13 and 29°C, and the increases in ethylene production following ACC treatment were similar at all temperatures. The significance of this finding is unclear but it could be interpreted as indicating

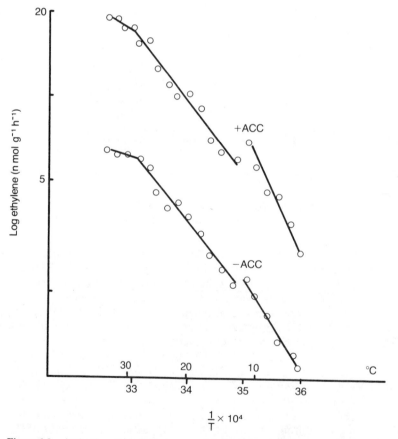

Figure 6.9 Arrhenius plots of ethylene production by apple fruit tissue, with and without added ACC (1 mM). (From Apelbaum *et al.*, 1981, reproduced by permission of the publishers)

that conversion of ACC to ethylene is not markedly affected by temperature and more particularly that ethylene production at the temperature extremes (4 and 35 °C) is no more affected than production at intermediate temperatures, providing there is a source of ACC. Unequivocal confirmation that ethylene production from exogenously supplied ACC utilizes the normal *in vivo* pathway and is subject to similar biochemical and physiological constraints is critical to the interpretation of the data of Apelbaum *et al.* (1981) and Field (1984) and for the establishment of the rate limiting step in ethylene production at extremes of temperature.

In the apple fruit system used by Yu, Adams and Yang (1980) there was a marked reduction in ethylene production between 30 and 35 °C; a higher temperature than that determined by Apelbaum *et al.* (1981). The decline in ethylene production was associated with a continual increase in ACC content of the tissue up to 40 °C; the highest temperature examined. The results were interpreted as showing that the conversion of ACC to ethylene is the primary site of high temperature inactivation. In a similar experiment with mungbean (*Vigna radiata* L.) hypocotyls the ACC-dependent and auxin-dependent ethylene production was impaired at 40 °C, indicating high temperature vulnerability of the reaction converting ACC to ethylene (Yu, Adams and Yang, 1980). In the mungbean hypocotyl system unlike the apple fruit system, there was a marked reduction in endogenous ACC levels between 35 and 40 °C, that paralleled the decline in ethylene production. In both tissue systems the higher rates of ethylene production and higher levels of ACC at 30 °C, as compared to 25 °C, may indicate that over an intermediate temperature range both ACC synthesis and degradation of ACC to ethylene were increased as temperature was raised but that the increase in ACC synthesis exceeded the degradation of ACC to ethylene, resulting in a net increase of ACC. However, in contrast, when the temperature was raised from 30 to 35 °C, ethylene production declined but the ACC content increased, suggesting that the conversion of ACC to ethylene is more sensitive to high temperature inactivation than is the synthesis of ACC. The ability of high temperatures (≥ 40 °C to prevent oxidation of ACC to ethylene has been investigated in homogenates of etiolated pea shoots by Konze and Kende (1979). In this system ethylene formation was only saturated at high ACC concentrations and the results were interpreted in two ways. Firstly that the enzyme converting ACC to ethylene needs a heat-stable cofactor that has a low affinity for its substrate, ACC, and secondly, that there may be a chemical reaction between ACC and the product of a reaction between an enzyme and a heat-stable co-substrate, in the following manner:

$$\text{heat-labile protein} + \text{heat-stable co-factor} + O_2 \rightarrow \text{enzyme reaction product} \xrightarrow{+ ACC} \text{ethylene}$$

More recent considerations of the *in vivo* enzymatic conversion of ACC to ethylene (Apelbaum *et al.*, 1981) have not provided evidence to support the suggestions of Konze and Kende (1979), and the biochemical basis for high temperature inhibition of this reaction has yet to be unequivocally established. Finally, Field (1981b) could not detect ethylene production above 42.5 °C in bean leaf tissue (*Figure 6.5*), while recent unpublished work by the same author shows that addition of exogenous ACC initiated ethylene production at 45 °C. Whether such ethylene is produced by the normal *in vivo* biosynthetic pathway, or by a non-enzymatic route has yet to be determined (Konze and Kende, 1979; Mayak, Legge and Thompson, 1981; Yu, Adams and Yang, 1980).

The mechanism of overshoot ethylene production following temperature transfer

The transfer of bean leaf tissue from a short (≤ 2 h) exposure to extreme temperatures (5 °C, 40 °C) results in a substantially increased wound response that is related to the duration of initial exposure and is virtually lag-free, particularly after low temperature exposures (*Figures 6.6* and *6.7*) (Field, 1981a and b). Some aspects of the mechanism for accumulation of overshoot ethylene production have recently been examined by Field (1984) and Wang and Adams (1982). The first point at issue was to determine if the potential for production involved accumulation of ethylene biosynthesis intermediates or specific blocks to their production. As the control of the synthesis and breakdown of ACC appear to be key steps in the control of ethylene production (Yang *et al.*, 1980), this is where most attention has been focused. The addition of ACC to bean leaf tissue held at 5 °C results in increased ethylene production, particularly upon transfer to 25 °C (*Figure 6.8*) (Field, 1984). Furthermore, when the ACC synthase inhibitor aminoethoxyvinylglycine (AVG) is introduced at the end of a 2 h incubation at 5 °C there is no

Table 6.4 EFFECT OF TEMPERATURE AND THE APPLICATION OF AVG (0.1 mM) ON THE PRODUCTION OF ETHYLENE

Initial temperature (°C) (0–120 min)	5	5	5	5	
Final temperature (°C) (120–345 min)	5	25	25	25	
Control (no AVG)	✓	✓			
AVG (at zero time)			✓		
AVG (at 110 min)				✓	
Time (min)		Ethylene (nl g^{-1} h^{-1})			s.e.
0–120	0.07	0.13	0.05	0.08	0.05
120–195	0.07	2.60	0.10	0.23	0.36
195–270	0.11	5.17	0.08	0.24	0.99
270–345	0.15	3.72	0.10	0.15	0.83

Table 6.5 EFFECT OF DURATION OF EXPOSURE TO 25 °C ON THE PRODUCTION OF ACC, FOLLOWING AN INITIAL INCUBATION FOR 2 h AT 5 °C

Period of exposure to 5 °C (h)	Subsequent period of exposure to 25 °C (h)	ACC (nmol g^{-1})
0	—	0.15
2	0	0.09
2	1	0.80
2	2	1.29
2	4	1.05
		s.e. 0.20

increase in ethylene production upon transfer to 25 °C, suggesting that accumulation of ACC is not responsible for the build up in potential for overshoot production (*Table 6.4*). The corollary is that ACC synthesis is rapidly induced after temperature transfer and this has been shown to occur in bean leaf tissue, where the changes in endogenous ACC levels (*Table 6.5*), follow the trend in overshoot ethylene production (*Figures 6.6* and *6.8*). A similar result was obtained by Wang and Adams (1982) using fruit tissue of cucumber but experiments were carried out over a much longer time period, with initial low temperature treatments at 2.5 °C being for four days.

It seems reasonable to conclude that accumulation of overshoot potential is associated with rapid ACC production upon transfer of tissue to a higher temperature and this could be achieved in at least two ways. Firstly, the regulation of ACC synthase activity is seen as the major pacemaker reaction in ethylene synthesis (Yang *et al.*, 1980) and is sensitive to low temperature (Field, 1984; Wang and Adams, 1982). At intermediate temperatures it is stimulated by IAA (Jones and Kende, 1979), but this promotive effect is largely eliminated at 5 °C (*Figure 6.10*) (Field, 1984). Addition of IAA during cold temperature exposure promotes overshoot ethylene production upon transfer to 25 °C (*Figure 6.10*), suggesting a

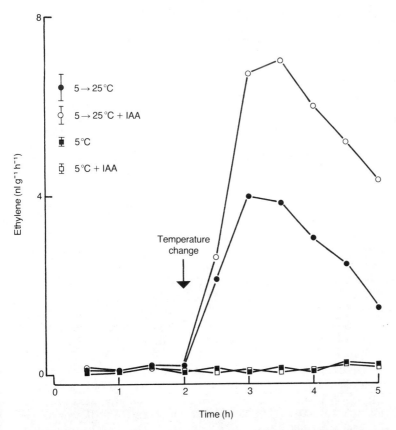

Figure 6.10 The effect of temperature transfer (5 → 25 °C) and IAA (0.1 mM) supplied at 0 h on ethylene production by dwarf bean leaf discs. Vertical bars indicate standard error of the mean. (From Field, 1984, reproduced by permission of the publishers)

major stimulation of ACC synthase. If ACC synthase activity is the limiting step then accumulation of overshoot potential could reside in a temperature-sensitive cofactor such as that described by Roberts and Osborne (1981) which, unlike other components of the enzyme system, is not required to accumulate but is readily activated upon transfer to higher temperatures. The second possible explanation for accumulation of an overshoot potential is experimentally unproven and relates to the interconversion of ACC with its conjugated derivative, 1-(malonylamino)-cyclopropane-1-carboxylic acid (MACC) (Hoffman, Fu and Yang, 1983). Hoffman, Liu and Yang (1983) have shown that water-stressed leaves of wheat (*Triticum aestivum* L.) accumulated MACC, while there was a corresponding reduction in ACC levels, suggesting conversion of ACC to MACC. Obviously this type of interconversion would not assist in interpreting low temperature-induced overshoot potential, but a clearer understanding of alternative pathways of MACC synthesis and accumulation and the role of other conjugates might lead to the suggestion that release of ACC from conjugated forms could be a primary mechanism in the rapid production of ethylene upon temperature transfer, with the conjugates providing a mechanism for chemically storing ethylene production potential.

Physiological manifestations of temperature effects

The observed effects of temperature on ethylene production are frequently substantial and it would not be surprising if they contributed to the initiation of a number of major physiological changes. The plant processes that are known to be triggered or controlled by ethylene, such as abscission, fruit ripening and some aspects of germination and seedling development are likely to be influenced by the increases in ethylene production associated with elevated temperature or rapid overshoot induction following low temperature exposure.

ABSCISSION

Premature abscission of fruits and leaves often follows a damaging exposure to freezing temperatures (Abeles, 1973). Several workers have suggested that the temperature-induced overshoot in ethylene production is responsible for prematurely initiating the normal abscission process described by Osborne (1973). A detailed study of freezing-induced abscission was carried out by Young and Meredith (1971) on ethylene production from leaves of seedling orange (*Citrus aurantium* Linn.). After a 4 h treatment at $-6.7\,°C$ leaves were transferred to ambient temperature and this resulted in no net ethylene production and in the inhibition of abscission, while a slightly higher temperature of $-6.1\,°C$ did not irreversibly stop ethylene production and enhanced abscission. However the duration and time-lag of the response in this experiment was considerable with elevated ethylene production occurring after 12 h at ambient temperature and rising steadily over several days. The avoidance of premature flower petal senescence and abscission has shown a definite linkage to temperature induced ethylene production in cut flowers of carnation (Nichols, 1966; Maxie *et al.*, 1973) and container-grown seed-geraniums (*Pelargonium* × *hortorum* Bailey) (Armitage *et al.*, 1980). Such studies are of some commercial significance in the development of cut-flower and nursery production.

There are no substantial studies on temperature-induced changes in the abscission process (Sexton and Roberts, 1982). A current study (Field, unpublished) using the model system of explants of the distal abscission zone from the primary leaves of dwarf bean (Jackson and Osborne, 1970), has shown that a 1–3 h exposure to 5 °C, at 24 or 48 h after excision reduces the time to 50% abscission by at least 24 h.

RIPENING

Early research by Hansen (1945) established that exposure of apple fruits to temperatures below 10 °C could result in elevated ethylene production on returning fruit to higher temperatures and that this had implications for storage and post-storage shelf-life and quality. A more recent study by Eaks (1980) demonstrated that a 4–12 week exposure of light green lemon fruit (*Citrus limon* L.), to 0 or 5 °C resulted in increased ethylene production and respiration on placement at 20 °C, while initial exposure to a non-chilling temperature of 12.8 °C had no marked effect. These results were associated with a study of biochemical markers, such as ethylene production, that could be used to indicate the extent of chilling injury during fruit storage.

Enhancement of fruit ripening by low temperature induction of ethylene was measured by Cooper, Rasmussen and Waldon (1969) in attached, mature-green grapefruit fruit (*Citrus paradisa* Macf.). A diurnal temperature regime of 20/5 °C was sufficient to elevate the ethylene concentration in the air space under the peel to 100 ppb after 14 days, compared to 4 ppb in fruit held at 25/20 °C. At the end of the 14 day incubation period the 20/5 °C treated fruit were ripening rapidly and had turned yellow, whereas fruit maintained at 25/20 °C remained green for two months. Similar results were obtained for harvested tangerine (*Citrus reticulata* Blanco) fruit (Cooper, Rasmussen and Waldon, 1969), and for pear fruit (*Pyrus communis* L.) that were initially stored at 5 °C before transfer to higher temperatures (Hansen, 1966; Sjakiotakis and Dilley, 1974). In some cases low temperature induction of ethylene production is complicated by associated chilling damage. Thus Cooper *et al.* (1969) showed that in a range of avocado (*Persea americana* Mill.) cultivars there was an association between high ethylene production by fruit at 5 °C and increased chilling-sensitivity of certain cultivars, while a chilling-tolerant cultivar showed low ethylene production at 5 °C, when compared to similar fruit held at 20 °C.

Ripening of post-harvest avocado fruits occurs most effectively at storage temperatures below 30 °C, where there is maximal ethylene production (Eaks, 1978). At temperatures between 30 and 40 °C ripening was either slow or abnormal and there was negligible ethylene production. Buescher (1979) correlated poor ripening and colour development of breaker stage tomato fruit with low level ethylene production following storage at 33 °C. Transfer of fruit from a 2–6 day treatment at 33 to 20 °C increased ethylene production and colour development while respiration and loss of fruit acid declined. Similar results were obtained by Ogura *et al.* (1976).

In the limited number of examples available there appears to be a close interaction between temperature, ethylene production and the ripening and storage characteristics of fruit. As far as can be judged the biochemical mechanisms are likely to be the same as those previously described for non-fruit systems. The implications of controlling ethylene production by fruit, particularly in understanding the extent of overshoot production following low temperature exposure during immediate pre-harvest or storage are critical aspects of post-harvest physiology and are deserving of further study.

ETHEPHON EFFICACY IN ABSCISSION AND RIPENING

The synthetic ethylene-releasing agent, ethephon, does not use the same biosynthetic pathway as endogenous ethylene production (Lurssen, 1982), but ethylene release from ethephon is sensitive to temperature and this has implications for its activity in ripening and abscission phenomena (Lougheed and Franklin, 1972; Wittenback and Bukovac, 1973). In describing some of the abscission-promoting characteristics of ethephon on leaves of sour cherry (*Prunus cerasus* L.) Olien and Bukovac (1978) showed that ethylene production increased almost 20-fold between 18 and 33 °C. The high level production at 33 °C more rapidly depleted the source of ethephon in the plant tissue and transfer to 18 °C resulted in an immediate reduction in production. Ethylene production from ethephon showed a similar temperature dependency to endogenous ethylene production although the activation energy for production increased to 134 kJ mol^{-1} from the endogenous production at 38 kJ mol^{-1} in sour cherry leaves (Olien and Bukovac, 1978).

Lougheed and Franklin (1972) applied ethephon to foliage or fruit of tomato and found that ethylene production from fruit was low ($<10\,\mu l\,kg^{-1}\,h^{-1}$) for the four days following treatment but significantly higher at 21 °C than at either 10 or 32 °C. Treated leaves showed similar production at 13 and 21 °C (approximately 150–250 $\mu l\,kg^{-1}\,h^{-1}$) for 76 h, whereas an initial burst of ethylene production after 4 h at 32 °C ($210\,\mu l\,kg^{-1}\,h^{-1}$) declined rapidly to $37\,\mu l\,kg^{-1}\,h^{-1}$ at 76 h. From these results, and others cited by Olien and Bukovac (1978), it is apparent that consistent release of ethylene from ethephon only occurs within a limited temperature range and that uneven and abnormal ripening may occur if ethephon is applied to foliage or fruit at temperatures outside a 15–30 °C range.

It is perhaps unfortunate that a study of synthetic and endogenous ethylene releasing compounds made by Lürssen (1982) did not include consideration of temperature as a major factor influencing their efficacy. Using the model systems described by Lürssen (1982) it may be possible to determine a difference in the temperature dependency of the mechanisms of ethylene production from the ACC-related chemicals such as N-formyl-1-aminocyclopropane-1-carboxylic acid (SDF 1664) and those based on the structure of ethephon. The presently available information does not describe these differences adequately, nor their practical significance.

GERMINATION AND SEEDLING DEVELOPMENT

Many seeds, including the classic lettuce (*Lactuca sativa* L.) seed experimental system, are sensitive to the effects of ethylene on germination (Abeles, 1973). The high temperature thermodormancy of lettuce is partially reversed by ethylene or cytokinin, although the thermodormancy reversal induced by ethylene occurs only in the presence of red light or gibberellin (Dunlap and Morgan, 1977; Keys et al., 1975). The role of ethylene is unclear but it is significant that transfer of lettuce (cv. Premier Great Lakes) seed from 36 to 22 °C rapidly induced germination (Dunlap and Morgan, 1977) and possibly ethylene evolution, although this was not measured. Earlier Burdett (1972) had shown that the inhibition of lettuce (c.v. Grand Rapids) seed germination by high temperature inhibition was due to restricted ethylene production. In this case thermodormancy was overcome by giving seeds a cold shock at 2 °C, which resulted in increased ethylene production and germination. The use of seed treatments that increase the germination of wholly or partially thermodormant seed is of some practical importance (Weaver, 1972). Ethylene production during seed vernalization has been observed (Suge, 1977), but its significance has not been determined.

Ethylene and other volatile organic components of the soil atmosphere (Smith and Dowdell, 1974) may stimulate the germination of buried seed (Holm, 1972; Taylorson, 1979). For example, soil ethylene concentrations may stimulate the germination of *Amaranthus retroflexus* L. seeds, although there is a marked temperature interaction (Schonbeck and Egley, 1981).

The well established ability of ethylene to inhibit elongation growth, but promote lateral expansion of the main or secondary plant axes (Abeles, 1973), has major implications for normal seedling development following germination. There are few instances where temperature-regulated ethylene production has been linked to abnormal seedling development. However shorter, more laterally expanded hypocotyls were found in certain cultivars of soybean (*Glycine max.* L.) that had been grown at 25 °C (Burris and Knittle, 1975; Samimy and LaMotte, 1976; Seyedin et al., 1982). The symptoms were similar to those of the so-called 'triple response' and were associated with a marked increase in ethylene production at 25 °C, that was not found in normally developing cultivars (Samimy and LaMotte, 1976). In a later report, Samimy (1978) showed that imbibition of seeds in an aqueous solution containing cobalt ions removed the anomalous growth effects, by inhibiting ethylene production.

Conclusions

This review of the effect of temperature on ethylene production highlights the lack of understanding of ethylene-temperature interactions, both at the biochemical and physiological level. The recent upsurge of interest in ethylene physiology and the associated role of ACC has rapidly enhanced our understanding of the mechanism by which temperature influences ethylene production. In a number of areas, particularly in relation to abscission, senescence and ripening phenomena, the results are of crucial practicability and argue for the continued development of both ethylene-releasing compounds and those that either inhibit ethylene biosynthesis or protect plant tissues from expressing ethylene-induced physiological changes.

An appreciation and understanding of temperature-influenced ethylene production provides a problem for all physiologists. An endeavour must be made to control, and indicate, the temperature at which experiments are carried out; a feature that is notably lacking in much of the current literature. In addition the applied physiologist must appreciate that pre-harvest and post-harvest temperature changes may substantially influence endogenous ethylene production and thus such processes as frost and heat stress-induced abscission and the duration of post-storage fruit quality.

References

ABELES, F.B. (1973). *Ethylene in Plant Biology*. Academic Press, New York
ADAMS, D.O. and YANG, S.F. (1979). *Proceedings National Academy of Science, USA*, **76**, 170–174
APELBAUM, A., BURGOON, A.C., ANDERSON, J.D., SOLOMOS, T. and LIEBERMAN, M. (1981). *Plant Physiology*, **67**, 80–84
ARMITAGE, A.M., HEINS, R., DEAN, S. and CARLSON, W. (1980). *Journal American Society of Horticultural Science*, **105**, 562–564
BAGNALL, D.J. and WOLFE, J.A. (1978). *Journal of Experimental Botany*, **29**, 1231–1242
BUESCHER, R.W. (1979). *Lebensmittel-Wissenschaft und Technologie*, **12**, 162–164
BURDETT, A.N. (1972). *Plant Physiology*, **50**, 201–204
BURG, S.P. and THIMANN, K.V. (1959). *Proceedings National Academy of Science, USA*, **45**, 335–344
BURRIS, J.S. and KNITTLE, K.H. (1975). *Crop Science*, **15**, 461–462
COOPER, W.L., RASMUSSEN, G.K. and WALDON, E.S. (1969). *Plant Physiology*, **44**, 1194–1196
CURTIS, R.W. (1969). *Plant Physiology*, **44**, 1368–1370
DUNLAP, J.R. and MORGAN, P.W. (1977). *Plant Physiology*, **60**, 222–224
EAKS, I.L. (1978). *Journal American Society of Horticultural Science*, **103**, 576–578
EAKS, I.L. (1980). *Journal American Society of Horticultural Science*, **105**, 865–869
ELSTNER, E.F. and KONZE, J.R. (1976). *Nature*, **263**, 351–352
FIELD, R.J. (1981a). *Annals of Botany*, **47**, 215–223
FIELD, R.J. (1981b). *Annals of Botany*, **48**, 33–39
FIELD, R.J. (1984). *Annals of Botany*, **54**, 61–67
HALL, M.A., ACASTER, M.A., BENGOCHEA, T., DODDS, J.H., EVANS, D.E., JONES, J.F., JERIE, P.H., MUTUMBA, G.C., NIEPEL, B. and SHAARI, A.R. (1980). In *Plant Growth Substances 1979*, pp. 199–207. Ed. by F. Skoog. Springer-Verlag, Berlin
HANSEN, E. (1945). *Plant Physiology*, **20**, 631–635
HANSEN, E. (1966). *Annual Review of Plant Physiology*, **17**, 459–480
HANSON, A.D. and KENDE, H. (1976). *Plant Physiology*, **57**, 538–541
HOFFMAN, N.E., FU, J.R. and YANG, S.F. (1983). *Plant Physiology*, **71**, 197–199
HOFFMAN, N.E., LIU, Y. and YANG, S.F. (1983). *Planta*, **157**, 518–523
HOLM, R.E. (1972). *Plant Physiology*, **50**, 293–297
JACKSON, M.B. and OSBORNE, D.J. (1970). *Nature*, **225**, 1019–1022
JERIE, P.H., SHAARI, A.R. and HALL, M.A. (1979). *Planta*, **144**, 503–507
JONES, J.F. and KENDE, H. (1979). *Planta*, **146**, 649–656
KEYS, R.D., SMITH, O.E., KUMAMOTO, J. and LYON, J.L. (1975). *Plant Physiology*, **56**, 826–829
KIMMERER, T.W. and KOZLOWSKI, T.T. (1982). *Plant Physiology*, **69**, 840–847
KONZE, J.R. and KENDE, H. (1979). *Planta*, **146**, 293–301
LIEBERMAN, M. (1979). *Annual Review of Plant Physiology*, **30**, 533–591
LOUGHEED, E.C. and FRANKLIN, E.W. (1972). *Canadian Journal of Plant Science*, **52**, 769–773
LÜRSSEN, K. (1982). In *Chemical Manipulation of Crop Growth and Development*, pp. 67–78. Ed. by J.S. McLaren. Butterworths, London

LURSSEN, K., NAUMANN, K. and SCHRODER, R. (1979). *Zeitschrift für Pflanzenphysiologie*, **92**, 285–294
LYONS, J.M., GRAHAM, D. and RAISON, J.K. (Editors) (1979). *Low Temperature Stress in Crop Plants*. Academic Press, London
LYONS, J.M. and RAISON, J.K. (1970). *Plant Physiology*, **45**, 386–389
MATTOO, A.K., BAKER, J.E., CHALUTZ, E. and LIEBERMAN, M. (1977). *Plant and Cell Physiology*, **18**, 715–719
MATTOO, A.K. and LIEBERMAN, M. (1977). *Plant Physiology*, **60**, 794–799
MAXIE, E.C., FARNHAM, D.S., MITCHELL, F.G., SOMMER, N.F., PARSONS, R.A., SNYDER, R.G. and RAE, H.L. (1973). *Journal American Society of Horticultural Science*, **98**, 568–572
MAYAK, S., LEGGE, R.L. and THOMPSON, J.E. (1981). *Planta*, **153**, 49–55
NICHOLS, R. (1966). *Journal of Horticultural Science*, **41**, 279–290
OGURA, N., HAYASHI, R., OGISHIMA, T., YUKO, A., NAKAGAWA, H. and TAKEHANA, H. (1976). *Nippon Nogeikagaku Kaishi*, **50**, 519–523
OLIEN, W.C. and BUKOVAC, M.J. (1978). *Journal American Society of Horticultural Science*, **103**, 199–202
OSBORNE, D.J. (1973). In *Shedding of Plant Parts*, pp. 125–144. Ed. by T.T. Kozlowski. Academic Press, New York
OSBORNE, D.J. (1978). In *Phytohormones and Related Compounds: A Comprehensive Treatise*, Volume 1, pp. 265–294. Ed. by D.S. Letham, P.B. Goodwin and T.J.V. Higgins. Elsevier/North Holland, Amsterdam
ROBERTS, J.A. and OSBORNE, D.J. (1981). *Journal of Experimental Botany*, **32**, 875–887
SALTVEIT, M.E. and DILLEY, D.R. (1978a). *Plant Physiology*, **61**, 447–450
SALTVEIT, M.E. and DILLEY, D.R. (1978b). *Plant Physiology*, **61**, 675–679
SAMIMY, C. (1978). *Plant Physiology*, **62**, 1005–1006
SAMIMY, C. and LaMOTTE, C.E. (1976). *Plant Physiology*, **58**, 786–789
SCHONBECK, M.W. and EGLEY, G.H. (1981). *Plant, Cell and Environment*, **4**, 237–242
SEXTON, R. and ROBERTS, J.A. (1982). *Annual Review of Plant Physiology*, **33**, 133–162
SEYEDIN, N., BURRIS, J.S., LaMOTTE, C.E. and ANDERSON, I.C. (1982). *Plant and Cell Physiology*, **23**, 427–431
SIMON, E.W. (1974). *New Phytologist*, **73**, 377–420
SJAKIOTAKIS, E.M. and DILLEY, D.R. (1974). *Hortscience*, **9**, 336–337
SMITH, K.A. and DOWDELL, R.J. (1974). *Journal of Soil Science*, **25**, 217–230
SUGE, H. (1977). *Plant and Cell Physiology*, **18**, 1167–1171
TAYLORSON, R.B. (1979). *Weed Science*, **27**, 7–10
WANG, C.Y. and ADAMS, D.O. (1982). *Plant Physiology*, **69**, 424–427
WEAVER, R.J. (1972). *Plant Growth Substances in Agriculture*. Freeman and Co., San Francisco
WITTENBACK, V.A. and BUKOVAC, M.J. (1973). *Journal American Society of Horticultural Science*, **98**, 348–351
WRIGHT, M. (1974). *Planta*, **120**, 63–69
YANG, S.F., ADAMS, D.O., LIZADA, C., YU, Y., BRADFORD, K.J., CAMERON, A.C. and HOFFMAN, N.E. (1980). In *Plant Growth Substances 1979*, pp. 219–229. Ed. by F. Skoog. Springer-Verlag, Berlin
YOUNG, R.E. and MEREDITH, F. (1971). *Plant Physiology*, **48**, 724–727
YU, Y-B., ADAMS, D.O. and YANG, S.F. (1980). *Plant Physiology*, **66**, 286–290
YU, Y-B. and YANG, S.F. (1980). *Plant Physiology*, **66**, 281–285

7

THE RELATIONSHIP BETWEEN POLLINATION, ETHYLENE PRODUCTION AND FLOWER SENESCENCE

A.D. STEAD
Department of Botany, Royal Holloway & Bedford Colleges, Egham, Surrey, UK

Introduction

The functional lifespan of the perianth differs greatly between species, varying from just a few hours in species such as *Hibiscus trionum*, *Portulaca oleracea* and *Ipomoea tricolor* to weeks or even perhaps months in other species such as orchids (Molisch, 1928). In the flowers of the longer-lived species pollination may cause the senescence of the perianth, whereas in the flowers of the short-lived species perianth senescence is not usually affected by pollination. For instance in *Ipomoea tricolor* detached corollas and even isolated rib segments of corollas will collapse simultaneously with those of similar-aged attached flowers (Kende and Hanson, 1976). Clearly in a case such as this pollination is not involved in the control of corolla senescence. Corolla senescence is however regulated by ethylene; exogenous ethylene will accelerate corolla senescence of fully open flowers and the natural senescence is associated with increased endogenous rates of ethylene production (Kende and Baumgartner, 1974).

Effect of pollination on flowers

In those species in which successful pollination restricts the functional lifespan of the flower it is usually by accelerating the normal sequence of flower senescence; by doing this further visits from pollination vectors will be prevented and therefore pollen wastage should be kept to a minimum. The expenditure of energy on

Table 7.1 THE EFFECT OF POLLINATION ON ETHYLENE PRODUCTION BY THE PISTIL OF *DIGITALIS*

	Hours after pollination			
	1	*2*	*3*	*4*
Unpollinated	0.48 ± 0.45	0.36 ± 0.18	0.24 ± 0.11	0.21 ± 0.19
Pollinated	2.40 ± 2.10	1.26 ± 0.62	1.59 ± 0.57	3.28 ± 1.17

Values are the means ± s.e. of hourly rates of production (nl pistil^{-1} h^{-1}) calculated from sequential measurements of the atmosphere surrounding four individually enclosed pistils. (For method *see Figure 7.3*)

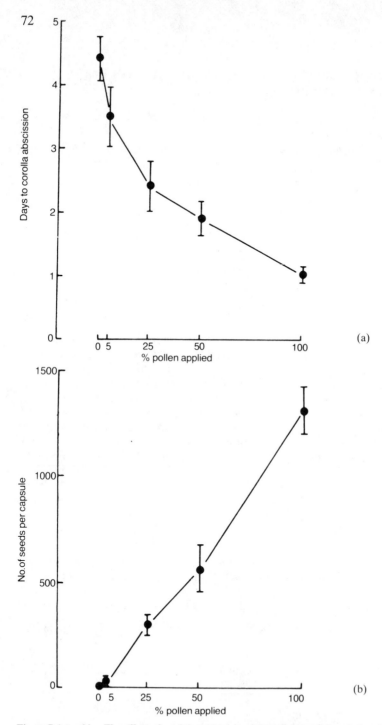

Figure 7.1a and b The effect of applying mixtures of pollen and powdered glass (w/w) to the stigma of intact *D. purpurea* flowers. (a) The effect upon flower longevity (days from pollination to corolla abscission). (b) The effect upon seed set. Standard errors represented by vertical bars

maintaining large elaborate floral structures should also be minimized. Pollination may therefore induce changes in the colour of the floral parts, e.g. *Lantana carnara* (Mathur and Mohan Ram, 1978),*Lupinus arizonicus* and *L. sparsiflorus* (Wainwright, 1978), induce corolla wilting, e.g. *Phalenopsis* sp. (Curtis, 1943), *Dianthus caryophyllus* (Nichols, 1971) or *Petunia hybrida* (Gilissen, 1976, 1977), or cause rapid corolla abscission, e.g. *Digitalis purpurea* (Stead and Moore, 1977). This last strategem is particularly common amongst the natural flora. In many of these studies ethylene has been implicated as having a major role in the control of perianth senescence, thus exogenous ethylene induces anthocyanin accumulation in *Cymbidium* orchids (Arditti, Hogan and Chadwick, 1973), petal wilting in carnations (Nichols, 1968) and *Vanda* orchids (Burg and Dijkman, 1967) and can induce corolla abscission in *Digitalis purpurea* (foxgloves) (Stead and Moore, 1983). In *Digitalis purpurea* pollination induces significant weakening of the corolla abscission zone within 8 h and complete separation approximately 24 h after pollination (Stead and Moore, 1979). Increased ethylene production can be detected approximately 5 h after pollination from isolated whole flowers (unpublished data). By removing the corolla, which produces very little ethylene either before or after

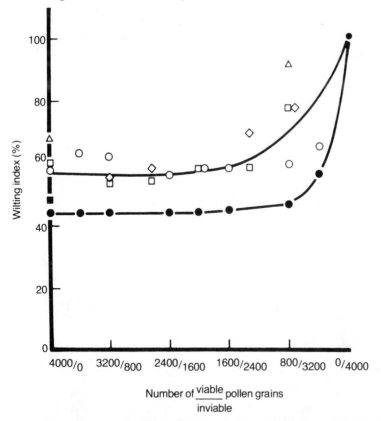

Figure 7.2 Wilting of flowers of *Petunia* (clone W166H) following pollination with viable self-pollen plus irradiated self or *Nicotiana* pollen (○, □, ◇, △), or pollinated with cross-pollen plus irradiated *Nicotiana* pollen (●). The wilting index following pollination is represented as the percentage of the wilting index of the unpollinated control plotted against the ratio of viable/killed pollen. (From Gilissen, 1977)

pollination, the pistil (flower minus corolla) can be enclosed in a very much smaller chamber. Using such isolated pistils it is possible to detect increased rates of ethylene production within 2 h, and in some cases within 1 h of pollination (*Table 7.1*). Thus increased ethylene production, as a result of pollination, precedes any noticeable reduction in the force required to detach the corolla.

The length of time between pollination and corolla abscission from intact flowers is related to the amount of pollen applied to the stigma (*Figure 7.1a*). Abscission is slowest in those flowers receiving no pollen and occurs more rapidly as the amount of pollen applied to the stigma increases. As would be expected the number of seeds per capsule is also related to the amount of pollen applied with application of pure pollen producing the most seeds and lesser amounts producing fewer seeds (*Figure 7.1b*).

In *Petunia hybrida* Gilissen (1977) concluded that pollination-induced corolla wilting was an 'all or nothing' response, for when 1600 or more viable pollen grains were applied to the stigma maximal stimulation of corolla wilting occurred. When fewer than 1600 viable pollen grains were applied either no effects were observed or the results were inconsistent (*Figure 7.2*). In this species, however, each capsule produces only 800 seeds on average, with a maximum of approximately 1200, it is therefore not surprising that addition of 1600 or more pollen grains results in a maximal response and it would perhaps be worthwhile to investigate in more detail the effect of applying fewer than 1600 grains.

Effect of pollination on ethylene production

In *Digitalis* the degree of stimulation of ethylene production is also related to the amount of pollen applied, the effect being most noticeable between 4 and 18 h of pollination (*Figure 7.3*). The effect of pollination upon the weakening of the corolla abscission zone is not however detectable until at least 8 h after pollination. The relationship between the ethylene production of *Digitalis* flowers and the force required to detach the corolla is shown in *Figure 7.4*. At any one time after pollination, those flowers producing most ethylene required least force to detach the corolla.

Although the rate of ethylene production and the time to corolla abscission is proportional to the amount of pollen applied the total amount of ethylene produced, per flower, prior to corolla abscission, is remarkably constant for each of the pollen dilutions used (*Table 7.2*). It therefore seems that pollination merely accelerates the production of ethylene and that in turn ethylene accelerates abscission zone weakening. The simplest explanation of this would be that an ethylene precursor, for instance 1-aminocyclopropane-1-carboxylic acid (ACC), which could be abundant in the style and ovary, is slowly converted to ethylene in unpollinated flowers, but that after pollination the rate of conversion is accelerated. The possibility of such a compound being ACC however seems unlikely, since all parts of *Digitalis* flowers convert exogenously applied ACC to ethylene very rapidly. This rapid conversion also occurs in the corolla which normally does not produce appreciable amounts of ethylene even after wounding. The style in particular is very responsive to exogenously applied ACC (*Figure 7.5*), a situation which has also been reported for carnation styles (Manning, Chapter 8). In this tissue the older styles are more sensitive than those from younger flowers.

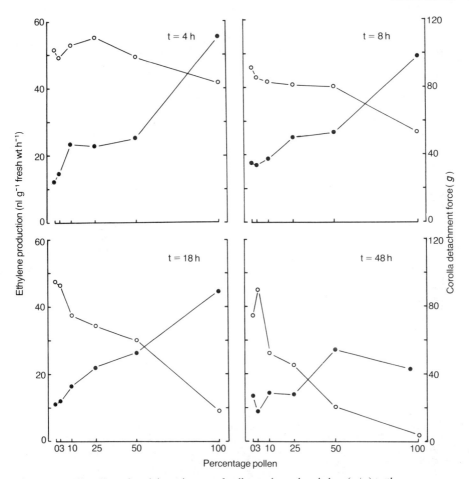

Figure 7.3 The effect of applying mixtures of pollen and powdered glass (w/w) to the stigma of isolated *D. purpurea* flowers. The effect upon ethylene production (●) and corolla detachability (○) either 4, 8, 18 or 48 h after pollination. Ethylene production from the pistil after corolla removal was measured by sampling the atmosphere after 1 h enclosure. Samples analysed on a Pye 104 GC fitted with an alumina-filled glass column operating at 100°C, with a nitrogen flow rate of 30 ml min^{-1}. Corolla detachability measured by attaching a force gauge (Halda, Sweden) to the corolla via a 'bulldog' clip and removing the corolla with a single straight pull. Each point is the mean of at least eight replicates

Analysis of ACC in the pistil has shown that there is very little difference in the content of ACC of unpollinated and pollinated flowers regardless of the time of pollination; the levels of ACC measured are approximately in the range 0.4–0.6 nmol per flower. Since each flower produces approximately 250 nl of ethylene prior to corolla abscission, which would require a minimum of 11 nmol ACC, the production of ACC must occur continuously. This observation is supported by the fact that the ethylene biosynthesis inhibitor aminoethoxyvinylglycine (AVG) retards corolla abscission zone weakening in pollinated flowers but is less effective at preventing the gradual weakening of the corolla abscission zone in unpollinated flowers (*Table 7.3*).

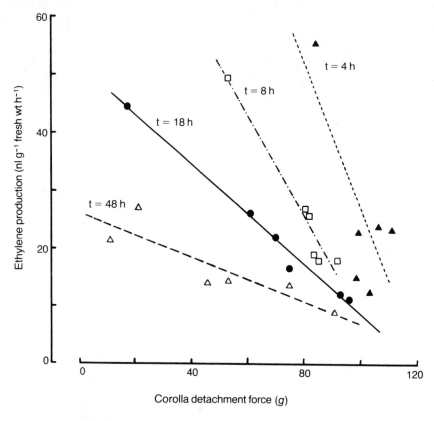

Figure 7.4 The relationship between ethylene production from the pistil of *D. purpurea* and the force required to remove the corolla of flowers pollinated for 4 h (▲), 8 h (□), 18 h (●) or 48 h (△). Each point is the mean of at least eight replicates for a particular dilution of pollen. Data derived from *Figure 7.3*

Table 7.2 ESTIMATES OF THE TOTAL AMOUNT OF ETHYLENE PRODUCED BY PARTIALLY POLLINATED *D. PURPUREA* PISTILS (nl g^{-1} fresh wt)

Time period (h after pollination)	0	3	Pollen (%) 10	25	50	100
0–5	60	75	115	110	110	275
5–8	54	52	57	75	81	150
8–18	110	120	165	220	260	450
18–48	370	270	450	420	810	(260)
48–96	250	430	(600)	(200)		
96–144	410	(200)				
Total	1254	1147	1387	1025	1261	1135

Data derived from *Figure 7.3*.
Notes
(1) Bracketed figures indicate that corolla abscission had occurred before the completion of the time period.
(2) Average pistil fresh weight approximately 250 mg.

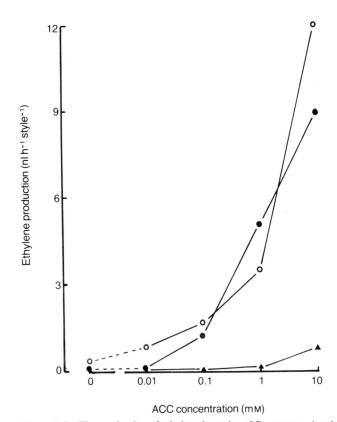

Figure 7.5 The production of ethylene by styles of *D. purpurea* incubated in a range of concentrations of ACC. The styles were from unpollinated flowers in which the corolla was about to abscise naturally (○); in which the corollas were just opening (●) and from young flower buds seven to nine days prior to flower opening (▲). In each case only the cut end of the style was in contact with the ACC. Each point is the mean of at least eight replicates and ethylene production measured as described in *Figure 7.3*

Table 7.3 (a) THE FORCE REQUIRED (g ± s.e.) TO REMOVE THE COROLLA OF *D. PURPUREA* FLOWERS THAT HAD BEEN REMOVED FROM THE FLOWERING SPIKE AND PLACED FOR 18 h WITH THE PEDICEL IMMERSED IN EITHER WATER OR 10 μM AVG. WHERE APPROPRIATE FLOWERS WERE CROSS-POLLINATED IMMEDIATELY AFTER REMOVAL FROM THE FLOWERING SPIKE[1]

Solution	Unpollinated	Pollinated
Water	61.97 ± 5.40	8.52 ± 3.31
10 μM AVG	86.10 ± 7.15	39.40 ± 7.90

(b) SIMILAR BUT USING 2.5 μM AVG AND GIVING THE FLOWERS A 48 h PULSE BEFORE POLLINATION. COROLLA DETACHABILITY AGAIN MEASURED 18 h AFTER POLLINATION[1]

Solution	Unpollinated	Pollinated
Water	76.75 ± 10.93	11.40 ± 3.83
2.5 μM AVG	89.13 ± 8.18	42.25 ± 10.05

[1]Force required to remove corollas of flowers immediately after removal from the flowering spike was in excess of 120 g.

Source of the pollination-induced ethylene production

The nature of the stimulus which triggers the increased rates of ethylene biosynthesis after pollination has been investigated in very few species. In *Vanda* orchids Burg and Dijkman (1967) suggested that pollen-held auxin could initiate increased ethylene production in the tissues closest to the stigma, and because ^{14}C-IAA did not move throughout the petals, this ethylene in turn induced further ethylene production autocatalytically. In *Digitalis* application of IAA directly to the abscission zone did induce partial abscission zone weakening (unpublished data) and placing isolated flowers with the pedicel in IAA solutions did induce increased ethylene production and abscission zone weakening (*Figure 7.6*). Application of up to 138 nmol IAA to the stigmatic lobes however, did not reduce the time to corolla abscission (Stead and Moore, 1979).

In *P. hybrida* Gilissen (1976, 1977) suggested that the style reacts to the mechanical damage caused by the growth of pollen tubes and that this induces corolla wilting. Wilting of *Petunia* corollas can be induced by ethylene and it has recently been shown that both mechanical damage to the stigma or style and pollination induce increased ACC and ethylene production (Nichols, personal

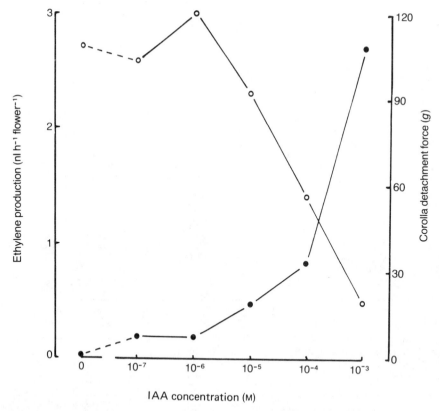

Figure 7.6 The effect of placing isolated flowers of *D. purpurea* with the pedicel immersed in IAA on the production of ethylene by the flowers (●) and upon corolla detachability (○) 18 h after placing in IAA. Each point is the mean of at least ten replicates. Method as described in *Figure 7.3*

Table 7.4 THE EFFECT OF REMOVING THE STIGMA, OR STIGMA PLUS HALF OF THE STYLE, ON THE FORCE REQUIRED TO DETACH THE COROLLA OF ISOLATED UNPOLLINATED *D. PURPURA* FLOWERS

Treatment	Force required (g ± s.e.)
Intact control	97.87 ± 4.67
Stigma removed	94.62 ± 7.55
Stigma plus half of style removed	91.20 ± 11.01

All measurements taken after 18 h at 20 ± 1 °C.

Table 7.5 THE LEVEL OF ACC EITHER RELEASED FROM POLLEN BY WASHING OR EXTRACTED FROM POLLEN BY TRICHLOROACETIC ACID (TCA)

Species	$nmol\ g^{-1}$	Reference
(1) ACC released from pollen after washing with a germination media		
Petunia hybrida	296.0	Whitehead *et al.* (1983)
Lathyrus odorata	50.0	Whitehead *et al.* (1983)
Dianthus caryophyllus	22.8	Whitehead *et al.* (1983)
Papaver nudicaule	1.8	Whitehead *et al.* (1983)
Anthirrhinum majus	1.7	Whitehead *et al.* (1983)
(2) ACC extracted from pollen with 0.2 M TCA		
Pinus contorta	2.9	
Dianthus caryophyllus	292.9	Hill *et al.* (unpublished)
Digitalis purpurea	2.9	
Petunia hybrida	428.0	Nichols (unpublished)
Nicotiana tabacum cv. *Samsun*	6512.5	
N. glutinosa	463.6	
N. rustica	997.4	
Solanum melongena	1109.0	Nichols (unpublished)
Lycopersicum esculentum cv. *sonatine*	388.0	Nichols (unpublished)
Lilium henryi	113.6	

In each case ACC assayed by the liberation of ethylene by alkaline NaOCl (Lizada and Yang, 1979; Bufler *et al.*, 1980).

communication). In *Digitalis* however, no reduction in the time to corolla abscission is seen when the style and stigma are physically removed (Stead and Moore, 1979). Furthermore, although the style does produce ethylene in response to wounding, no significant difference in the force required to detach the corolla was found 18 h after removing the stigma and/or style (*Table 7.4*). Therefore, neither auxin nor mechanical damage appear to trigger the rapid pollination-induced corolla abscission found in *Digitalis*, with the recent discovery of ACC in the pollen of *Petunia* (Hoekstra *et al.*, 1982) the possibility exists that pollen-held ACC could act as the trigger for the autocatalytic production of ethylene in *Digitalis* or even as the source of the pollination-induced ethylene. Analysis of the level of ACC in *Digitalis* pollen, however, shows it to contain one of the lowest concentrations of ACC so far reported (*Table 7.5*) although Whitehead, Fujino and Reid (1983) have recorded very low levels of ACC released from *Antirrhinum* pollen, another member of the Scrophulariceae, during a 1 min wash with germination media. Other species, particularly those of the Solanaceae, contain very much higher levels of ACC at the time of anther dehiscence. Comparison of the total extractable levels of ACC to the amounts washed off during a brief wash suggests, however, that not all of the pollen-held ACC readily diffuses away from the pollen. In *Petunia* for example, more than 50% may be lost from the pollen grains quite rapidly, but with carnations Whitehead, Fujino and Reid (1983) obtained less than 10% of the value we have obtained in our laboratory for the total extractable levels of ACC. It remains to be seen if more of the total extractable ACC diffuses away from the pollen after germination or during the growth of the pollen tube. In *Digitalis*, with its very low level of ACC in pollen, pollen-held ACC cannot be the source of the pollination-induced ethylene production, neither does it seem that pollen-held ACC is the sole trigger since applications of ACC equivalent to 5 mg pollen do not induce autocatalytic ethylene production, nor corolla abscission. Preliminary experiments with carnations also suggest that the levels of ACC in carnation pollen are not sufficient to induce any of the recognized pollination-induced responses in either the style or petals. It does seem likely, however, that germinating pollen produces ACC and thus the levels of pollen-produced ACC in the style may be much greater than the level of extractable ACC in dry pollen. The exceptionally high levels of ACC in the pollen of the Solanaceae, especially *Nicotiana*, may therefore be the result of the synthesis of ACC prior to pollen germination, whereas in other species it may occur after pollen germination.

Conclusion

Whatever mechanism mediates pollination-induced corolla abscission in *Digitalis* it is undoubtedly a very sensitive system, since it responds quantitatively to the amount of pollen applied. The increased ethylene production observed after pollination appears to be due to an acceleration of synthesis of the gas and in response to this there is a proportional reduction in the corolla abscission zone break strength. The system therefore integrates the response of the tissue with the quantity of ethylene produced and would support the contention that ethylene is not the trigger for abscission, but merely coordinates the phenomenon.

References

ARDITTI, J., HOGAN, N.M. and CHADWICK, A.V. (1973). *American Journal of Botany*, **60**, 883–888
BUFLER, G., MOR, Y., REID, M.S. and YANG, S.F. (1980). *Planta*, **150**, 439–442
BURG, S.P. and DIJKMAN, M.J. (1967). *Plant Physiology*, **42**, 1648–1650
CURTIS, J.T. (1943). *American Orchid Society Bulletin*, **11**, 258–260
GILISSEN, L.J.W. (1976). *Planta*, **131**, 201–202
GILISSEN, L.J.W. (1977). *Planta*, **133**, 275–280
HOEKSTRA, F.A., WEGES, R., VAN ROEKEL, G.C., DE LAAT, A.M.M. and BRUINSMA, J. (1982) 11th International Conference on Plant Growth Substances, p. 43. (Abstract only)
KENDE, H. and BAUMGARTNER, B. (1974). *Planta*, **116**, 279–289
KENDE, H. and HANSON, A.D. (1976). *Plant Physiology*, **57**, 523–527
LIZADA, M.C.C. and YANG, S.F. (1979). *Analytical Biochemistry*, **100**, 140–145
MATHUR, G. and MOHAN RAM, H.Y. (1978). *Annals of Botany*, **42**, 1473–1476
MOLISCH, H. (1928). *Die Lebensdaver de Pflanze*, pp. 226. Translated by E.H. Fulling (1938). New York Botanical Garden, New York
NICHOLS, R. (1968). *Journal of Horticultural Science*, **43**, 335–349

NICHOLS, R. (1971). *Journal of Horticultural Science*, **46**, 323–332
STEAD, A.D. and MOORE, K.G. (1977). *Annals of Botany*, **41**, 283–292
STEAD, A.D. and MOORE, K.G. (1979). *Planta*, **146**, 409–414
STEAD, A.D. and MOORE, K.G. (1983). *Planta*, **157**, 15–21
WAINWRIGHT, C.M. (1978). *Bulletin of the Torrey Botanical Club*, **105**, 24–38
WHITEHEAD, C.S., FUJINO, D.W. and REID, M.S. (1983). *Scientia Horticulturae*, **21**, 291–297

8
THE ETHYLENE FORMING ENZYME SYSTEM IN CARNATION FLOWERS

K. MANNING
Glasshouse Crops Research Institute, Littlehampton, West Sussex, UK

Introduction

The senescence of cut carnation flowers (*Dianthus caryophyllus* L. cv 'White Sim') ends with inrolling of petals and is characterized by a rise in respiration associated with a sharp increase in ethylene production (Nichols, 1966). At the onset of wilting, petals and styles account for practically all of the ethylene produced, although when expressed on a fresh weight basis ethylene emanation by the styles is by far the greater (Nichols, 1977).

Senescence of carnation flowers can be modified by various treatments. Pollination promotes ethylene production by all flower parts and causes accelerated petal wilting within two to three days (Nichols, 1977). Treatment of cut flowers with ethylene or propylene produces wilting symptoms after a few hours (Mayak, Vaadia and Dilley, 1977) whereas carbon dioxide delays wilting (Smith, Parker and Freeman, 1966). Pretreatment of flowers with silver thiosulphate prevents the climacteric rise in ethylene production (Veen, 1979) and delays senescence (Reid *et al.*, 1980). The surge in ethylene accompanying senescence is an autocatalytic phenomenon in which ethylene stimulates its own biosynthesis. Treatments that accelerate or delay natural senescence also enhance or inhibit ethylene biosynthesis, respectively.

The pathway of ethylene synthesis in several higher plants has been established as L-methionine → S-adenosylmethionine → 1-aminocyclopropane-1-carboxylic acid (ACC) → ethylene (Adams and Yang, 1979; Lürssen, Naumann and Schröder, 1979). The formation of ACC from S-adenosylmethionine by the enzyme ACC synthase and the conversion of ACC to ethylene by the ethylene forming enzyme (EFE) system have been shown to be important steps in the control of ethylene production (*see* review by Yang, 1980). In the carnation both ACC content and ethylene formation are low before the start of flower senescence. When the petals inroll, ACC content increases by 30-fold and ethylene production by 1000-fold suggesting that the rate of formation of ACC and its rate of conversion to ethylene may both determine the overall rate of ethylene production (Bufler *et al.*, 1980).

Application of ACC to many plant tissues that normally produce low amounts of ethylene can increase ethylene levels dramatically indicating that in these cases the rate-limiting step in the pathway is the conversion of S-adenosylmethionine to ACC

(Cameron et al., 1979). Freshly cut carnation flowers supplied with ACC through the stem senesce earlier than control flowers in water, the time to petal wilting being dependent on the concentration of ACC (Veen and Kwakkenbos, 1983). Petals detached from flowers four days after opening senesce at the same time as petals left on the parent flower; in comparison, petals removed from freshly cut flowers wilt irreversibly some time later (Mor and Reid, 1981). Isolated petals wilt prematurely in ACC solutions (Mor and Reid, 1981; Mor, Spiegelstein and Halevy, 1983; Sacalis, Wulster and Janes, 1983) and provide a useful experimental system for studying the regulation of ethylene production and flower senescence.

This chapter discusses the distribution of the EFE system within the flower and its induction by ACC and ethylene. A model system capable of forming ethylene from ACC involving peroxidase is described and its relevance to the *in vivo* EFE system is also discussed.

The ethylene forming enzyme in freshly cut and wilting carnations

The EFE in flower tissues was assayed by determining the maximum rate of ethylene production by the tissue at 25 °C in the presence of saturating levels of ACC. Preliminary experiments showed that maximum EFE activity was obtained when the ACC concentration in the external solution was at least 1 mM. This value is in general agreement with that reported for several other plant tissues. A linear rate of ethylene production was usually attained 2–4 h after the addition of ACC. However, in some experiments ethylene levels gradually increased after 6 h. This was due to induction of the EFE system rather than a slow diffusion of ACC into the tissue. The possibility of inducing ethylene production when assaying EFE was avoided by restricting incubation times to less than 6 h.

The ability of the various parts of the flower to produce ethylene was examined by incubating the tissue in distilled water at 25 °C in a closed tube. Ethylene production was monitored by sampling the gas in the tube after 1 h. Tissue was then transferred into 1 mM ACC and ethylene production monitored over the following 2–4 h. Results showing the mean ±s.e. from six flowers are given in *Table 8.1*.

Styles (including stigmas) from flowers cut at the stage when the outer petals were reflexed perpendicularly to the stem axis, i.e. fresh, produced significantly ($P<0.05$) more ethylene than any of the other flower parts on a fresh weight basis. When given ACC, the ethylene production by the stylar tissue increased more than tenfold. In contrast ethylene production by the petals, ovary and receptacle was only stimulated to a small extent in the presence of ACC. Styles from flowers sampled at the same stage of development varied considerably in size, weight and ethylene production. Seasonal variations in styles were also observed, those from

Table 8.1 ETHYLENE PRODUCTION FROM PARTS OF FRESH AND WILTING FLOWERS WITH AND WITHOUT EXOGENOUS ACC

Flower part	Ethylene production (nl h^{-1} g fresh wt^{-1})			
	Fresh		Wilting	
	− ACC	+ ACC	− ACC	+ ACC
Outer petal	0.499 ± 0.174	2.04 ± 0.49	25.7 ± 11.3	113 ± 46.0
Styles	9.54 ± 3.88	109 ± 47	72.6 ± 70.8	876 ± 110
Ovary	1.29 ± 0.89	2.24 ± 0.78	3.34 ± 1.51	3.88 ± 1.78
Receptacle	0.849 ± 0.729	2.67 ± 2.35	6.91 ± 4.19	24.1 ± 10.4

plants sampled between April and September were generally larger and more developed and had much higher EFE activity. This value was sometimes in excess of 3500 nl h^{-1} g fresh wt^{-1}, and is one of the highest values so far reported for plant tissues (Aharoni and Yang, 1983).

Petals from flowers aged in water for seven days at 20 °C as described by Nichols (1971) and just showing inrolling symptoms, produced 50-fold more ethylene than fresh petals and were further stimulated fivefold by ACC. Although at this stage styles had increased their ethylene production to a lesser degree than petals (sevenfold) they showed a greater response to ACC (12-fold). In comparison the ovary and receptacle from these wilting flowers produced relatively little ethylene and had correspondingly low EFE activities. At the peak of ethylene production EFE activity in petals had increased by 1000-fold over that in fresh tissue.

Styles therefore differed from the other floral tissues in having substantial EFE activity well before natural senescence began. Immature styles removed from flowers at very early stages of flower opening were able to convert ACC to ethylene at appreciable rates in comparison with all other flower parts (*Figure 8.1*).

Flower part	Ethylene production (nl h^{-1} (g fresh wt)$^{-1}$)			
	Stage 1	Stage 2	Stage 3	Stage 4
Petals, ovary, receptable	<1	<1	<1	<1
Styles	64 ± 13	55 ± 26	102 ± 36	200 ± 59

Figure 8.1 Ethylene production during flower development. Flowers were cut at stages shown in the photograph (styles displayed above each flower) and EFE activity determined in flower parts incubated with 1 mM ACC. Values are means ± s.e. of six samples

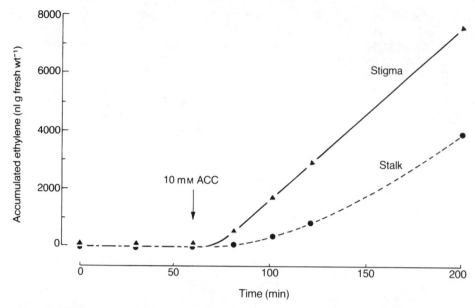

Figure 8.2 Ethylene production by parts of the style. Styles from freshly cut flowers were divided into stigmas and stalks and each incubated with water in a sealed tube at 25 °C. At 61 min a solution of ACC was injected (arrow) to give a final concentration of 10 mM. Consecutive gas samples were removed at intervals for ethylene determination. Points represent the mean of five samples

In the style, as in petals (Nichols, 1977; Wulster, Sacalis and Janes, 1982; Sacalis, Wulster and Janes, 1983), ethylene production was found to vary in different parts of the tissue. In the presence of exogenous ACC more than 70% of whole style ethylene was produced by the stalk, by virtue of its greater weight. However, when expressed per unit weight the stigma had 50% higher EFE activity. Ethylene production increased from both the stigma and stalk within 5 min of supplying ACC and reached a linear rate after 20 min and 1 h, respectively (*Figure 8.2*).

Nichols *et al.* (1983) reported that carnation styles are unusual in that they produce substantial quantities of ethylene yet contain less ACC than other parts of the flower. The high EFE activity of styles probably accounts for this observation since any ACC formed would be more rapidly converted to ethylene.

The greater responsiveness of the style, and particularly the stigma, to ACC indicates a special function for this tissue. It has recently been demonstrated (Whitehead, Fujino and Reid, 1983) that pollen contains high concentrations of ACC and that application of pollen to the stigma of carnation flowers causes an early increase in ethylene production by the gynoecia. It is possible that the source of this early increase in ethylene is derived from the ACC in the pollen. Thus the constitutive EFE activity in styles may enable the flower to respond to pollination.

Induction of the ethylene forming enzyme by ACC

Mor and Reid (1981) suggested that senescing carnation petals acquired a greater capacity to convert ACC to ethylene because the enzyme responsible for this step

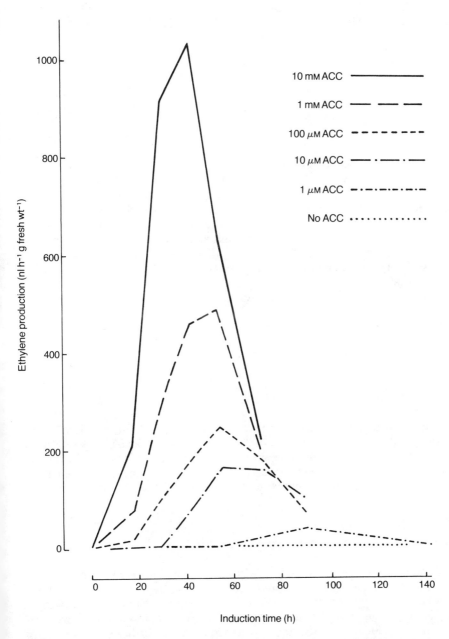

Figure 8.3 Induction of ethylene production by detached petals in ACC. Petals from freshly cut flowers were incubated in the dark at 20 °C in open tubes containing different concentrations of ACC. Ethylene production from each petal was followed at intervals by sealing the tube and withdrawing a gas sample after 30–60 min. Experimental points (omitted for clarity) occur at points of inflexion on the curves and represent the mean of six samples

was induced. It has also been shown that ACC promotes early senescence in isolated carnation petals (Mor, Spiegelstein and Halevy, 1983; Sacalis, Wulster and Janes, 1983). Auxins such as indole-3-acetic acid (IAA) and 2,4-dichlorophenoxyacetic acid (2,4-D) have been shown to cause premature wilting in whole flowers (Nichols, 1971; Sacalis and Nichols, 1980) and in detached petals (Wulster, Sacalis and Janes, 1982) and were suggested to act by enhancing ACC synthesis (Yu and Yang, 1979).

The effect of different concentrations of ACC on the induction of ethylene formation in petals detached from freshly cut flowers is shown in *Figure 8.3*. Control petals in water produced ethylene at a steady low rate during the period of the experiment (six days). Concentrations of ACC ranging from 1 μM to 10 mM promoted ethylene production. This increased to a maximum, depending on the level of ACC, and subsequently declined rapidly. Ethylene production, peaked after 42 h when petals were incubated in 10 mM ACC, and this peak occurred progressively later for petals in lower concentrations of ACC. This experiment was repeated but when petals reached their peak of ethylene formation 10 mM ACC was added in order to determine the maximum EFE activity. It was found that this was the same in all samples regardless of the initial ACC concentration. Ethylene production by detached petals was therefore limited by the availability of externally supplied ACC. Those petals in low concentrations of ACC produced much less ethylene than petals allowed to senesce on the flower indicating that more ACC is available to petals remaining on the flower.

Induction of EFE in petals was inhibited by 100 μM cycloheximide (CHI) indicating that protein synthesis at the translational level was essential. Chloramphenicol (100 μg ml^{-1}), added to prevent bacterial growth, did not suppress increases in EFE activity so this induction was unlikely to be mediated by plastids or mitochondria.

Extractable protein was determined in petals induced by ACC. Half-petals, obtained by dividing the petal longitudinally, were used to reduce sample variability. A preliminary experiment showed in a paired comparison that the variability in the amount of protein extracted from petal halves was less than one-sixth of that from whole petals. In petals induced with 1 mM ACC there was a highly significant ($P<0.005$) decline in extractable protein 6 h before maximum ethylene formation occurred. Protein levels continued to fall after this period. During senescence loss of EFE activity may be due to increased proteolysis together with tissue disorganization.

Induction of the ethylene forming enzyme by ethylene

There are numerous reports in the literature of ethylene stimulating its own production whilst triggering ripening and senescence (*see* review by Abeles, 1973). This 'autocatalysis' is generally believed to be regulated by the induction of ACC synthase leading to increased synthesis of ACC. However, few studies have been reported on the induction of EFE in the absence of ACC synthesis (Hoffman and Yang, 1982; Riov and Yang, 1982, Chalutz *et al.*, 1984) although several metabolic inhibitors are now available to block this step (Amrhein and Wenker, 1979).

The effect of ethylene on the induction of EFE was examined in carnation petals detached from freshly cut flowers (*Figure 8.4*). Control petals not exposed to ethylene produced less than 2 nl h^{-1} g fresh wt^{-1} of ethylene for the duration of the

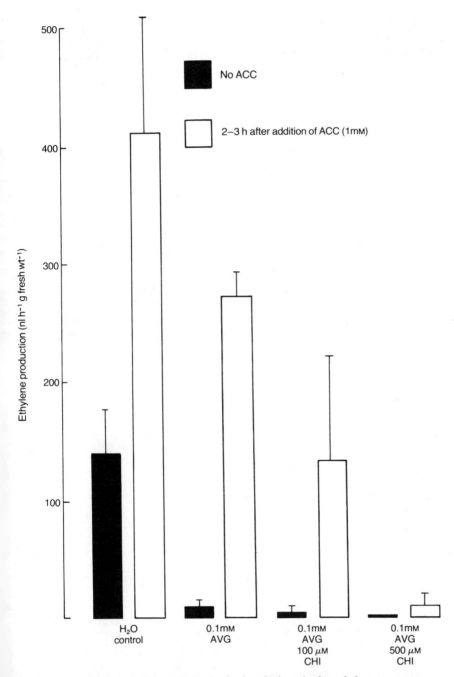

Figure 8.4 Induction of ethylene production by detached petals after ethylene treatment. Petals from freshly cut flowers were pretreated with solutions as shown. After 24 h the petals were transferred to a 5 litre container with 2 μl l^{-1} ethylene and kept in the dark at 20 °C. After a further 24 h the petals were removed and ethylene production determined in the presence and absence of 1 mM ACC. Values shown are means (bar indicates standard error) of six samples

experiment. Control petals in water treated with ethylene produced high amounts of ethylene which were enhanced threefold by 1 mM ACC. Petals pre-incubated with 0.1 mM aminoethoxyvinylglycine (AVG), an inhibitor of ACC synthesis, and subsequently treated with ethylene, produced ethylene at 7% of the control rate. However, ACC stimulated ethylene from these AVG treated petals 27-fold, approaching the rate for controls given ACC without AVG. Thus under conditions in which ACC formation was severely limited by AVG, EFE was strongly induced. Protein synthesis was necessary for induction of the EFE since ethylene formation was substantially reduced by 100 µM CHI and almost totally inhibited by 500 µM CHI.

In the carnation flower, and probably in most plant tissues responsive to ethylene, the induction of EFE activity appears to be regulated by ethylene directly as well as through the enhancement of ACC synthesis.

IAA-oxidase model system

Since the discovery that ethylene is formed from ACC there has been interest in the enzyme system regulating it (*see* Kende, Acaster and Guy, Chapter 3). Attempts to study the system *in vitro* have been hindered by the difficulty of extracting the enzyme in an active form. Some '*in vitro*' systems converting ACC to ethylene have been described, including microsomal membranes (Mayak, Legge and Thompson, 1981; McRae, Baker and Thompson, 1982), mitochondria (Vinkler and Apelbaum, 1983), a cell-free system from etiolated pea stems (Konze and Kende, 1979), IAA-oxidase (Vioque, Albi and Vioque, 1981; Shimokawa, 1983) and peroxidase (Rohwer and Mäder, 1981). Hoffman *et al.* (1982) have shown that EFE in a number of plant tissues can differentiate between the four stereoisomers of 1-amino-2-ethylcyclopropane-1-carboxylic acid (AEC). They have suggested that the validity of an *in vitro* system may be tested using these isomers. McKeon and Yang (1984) have used such a test to demonstrate that the pea epicotyl system was probably not related to the *in vivo* EFE.

The conversion of ACC to ethylene and AEC isomers to 1-butene was examined in carnation styles, in an IAA-oxidase system involving purified horseradish (HRP) and in the chemical oxidation system of Lizada and Yang (1979) using hypochlorite (*Table 8.2*). The composition of the IAA oxidase mixture was optimized to produce maximum ethylene from ACC. In contrast to other IAA-oxidase systems it did not contain pyridoxal phosphate, and p-coumaric acid (20 µM) replaced 2,4-dichlorophenol. Orthophosphate stimulated ethylene formation, thus a 100 mM phosphate buffer at pH 6.2 was used. In the presence of 1 mM ACC, 0.6 mM IAA,

Table 8.2 OLEFIN PRODUCTION FROM ACC, (±) ALLOCORONAMIC ACID AND (±) CORONAMIC ACID IN THREE SYSTEMS

Substrate	Olefin production (nmol)		
	Hypochlorite (100 nmol substrate 0°C, 5 min)	IAA-Oxidase (1 mM substrate 25°C, 8 min)	Styles (1 mM substrate 20 h)
ACC	93.15	0.489	18.12
(±) Allocoronamic acid	0.51	0.126	0.326
(±) Coronamic acid	2.40	0.469	0.0318

0.2 mM Mn^{2+} and 1.6 µg ml^{-1} HRP this system generated ethylene at 7.5 nmol h^{-1} mg peroxidase^{-1}.

In agreement with findings for other plant tissues (Hoffman *et al.*, 1982), (±)-allocoronamic acid, which contains the isomer preferentially converted to 1-butene, was more readily converted to 1-butene than was (±)-coronamic acid in carnation styles. The IAA-oxidase system used here resembled EFE in being heat denaturable, oxygen-dependent and inhibited by Co^{2+} and free-radical scavengers, such as n-propyl gallate (results not shown). It differed, however, from the *in vivo* system in one important respect, namely, that it preferentially oxidized (±)-coronamic acid. It is believed that free radicals generated by the enzymic oxidation of IAA reacted non-enzymically and without stereo discrimination with AEC. A similar mechanism was earlier proposed by Yang (1967) to account for ethylene formation from methional by peroxidase.

Attempts to prepare protoplasts and subcellular fractions from styles have resulted in low recoveries of ethylene-forming activity. In addition to being highly labile, the EFE system may also be inhibited by endogenous substances in the flower (Mayak, Legge and Thompson, 1981).

Microsomal fractions from carnation petals at all stages of senescence were reported to be capable of forming ethylene from ACC (Mayak, Legge and Thompson, 1981). This is in contrast to the situation in intact petals described here in which EFE activity in fresh petals was negligibly low compared with wilting petals. In pea microsomal membranes superoxide radicals were found to be involved in the conversion of ACC to ethylene (McRae, Baker and Thompson, 1982). If the carnation microsomal system of Mayak, Legge and Thompson (1981) produced ethylene by a similar mechanism, it may not be related to the *in vivo* system.

Conclusions

The autocatalytic rise in ethylene production during senescence of the carnation flower is associated with a dramatic increase in the ability of the petals to form ethylene from ACC. This large increase in the EFE depends on *de novo* protein synthesis and is induced by ethylene directly and through ACC synthesis. In contrast to petals, the EFE in styles is active in freshly cut flowers. This constitutive EFE activity of styles might have a role in pollination. In common with other plant tissues so far examined EFE in the carnation can discriminate in favour of the (±)-allocoronamic acid isomer mixture of AEC. Although the function of IAA-oxidase/peroxidase enzymes during ripening and senescence is unknown, on the basis of the model system it does not appear that they are involved directly in the conversion of ACC to ethylene.

References

ABELES, F.B. (1973). *Ethylene in Plant Biology*, pp. 161–178. Academic Press, New York

ADAMS, D.O. and YANG, S.F. (1979). *Proceedings of the National Academy of Sciences, USA*, **76**, 170–174

AHARONI, N. and YANG, S.F. (1983). *Plant Physiology*, **73**, 598–604

AMRHEIN, N. and WENKER, D. (1979). *Plant and Cell Physiology*, **20**, 1635–1642
BUFLER, G., MOR, Y., REID, M.S. and YANG, S.F. (1980). *Planta*, **150**, 439–442
CAMERON, A.C., FENTON, C.A.L., YU, Y., ADAMS, D.O. and YANG, S.F. (1979). *HortScience*, **14**, 178–180
CHALUTZ, E., MATTOO, A.K., SOLOMOS, T. and ANDERSON, J.D. (1984). *Plant Physiology*, **74**, 99–103
HOFFMAN, N.E. and YANG, S.F. (1982). *Plant Physiology*, **69**, 317–322
HOFFMAN, N.E., YANG, S.F., ICHIHARA, A. and SAKAMURA, S. (1982). *Plant Physiology*, **70**, 195–199
KONZE, J.R. and KENDE, H. (1979). *Planta*, **146**, 293–301
LIZADA, M.C.C. and YANG, S.F. (1979). *Analytical Biochemistry*, **100**, 140–145
LÜRSSEN, K., NAUMANN, K. and SCHRÖDER, R. (1979). *Zeitschrift für Pflanzenphysiologie*, **92**, 285–298
McKEON, T.A. and YANG, S.F. (1984). *Planta*, **160**, 84–87
McRAE, D.G., BAKER, J.E. and THOMPSON, J.E. (1982). *Plant and Cell Physiology*, **23**, 375–383
MAYAK, S., LEGGE, R.L. and THOMPSON, J.E. (1981). *Planta*, **153**, 49–55
MAYAK, S., VAADIA, Y. and DILLEY, D.R. (1977). *Plant Physiology*, **59**, 591–593
MOR, Y. and REID, M.S. (1981). *Acta Horticulturae*, **113**, 19–25
MOR, Y., SPIEGELSTEIN, H. and HALEVY, A.H. (1983). *Plant Physiology*, **71**, 541–546
NICHOLS, R. (1966). *Journal of Horticultural Science*, **41**, 279–290
NICHOLS, R. (1971). *Journal of Horticultural Science*, **46**, 323–332
NICHOLS, R. (1977). *Planta*, **135**, 155–159
NICHOLS, R., BUFLER, G., MOR, Y., FUJINO, D.W. and REID, M.S. (1983). *Journal of Plant Growth Regulation*, **2**, 1–8
REID, M.S., PAUL, J.L., FARHOOMAND, M.B., KOFRANEK, A.M. and STALBY, G.L. (1980). *Journal of the American Society for Horticultural Science*, **105**, 25–27
RIOV, J. and YANG, S.F. (1982). *Plant Physiology*, **70**, 136–141
ROHWER, F. and MÄDER, M. (1981). *Zeitschrift für Pflanzenphysiologie*, **104**, 363–372
SACALIS, J.N. and NICHOLS, R. (1980). *HortScience*, **15**, 499–500
SACALIS, J.N., WULSTER, G. and JANES, H. (1983). *Zeitschrift für Pflanzenphysiologie*, **112**, 7–14
SHIMOKAWA, K. (1983). *Phytochemistry*, **22**, 1903–1908
SMITH, W.H., PARKER, J.C. and FREEMAN, W.W. (1966). *Nature, London*, **211**, 99–100
VEEN, H. (1979). *Planta*, **145**, 467–470
VEEN, H. and KWAKKENBOS, A.A.M. (1983). *Scientia Horticulturae*, **18**, 277–286
VINKLER, C. and APELBAUM, A. (1983). *FEBS Letters*, **162**, 252–256
VIOQUE, A., ALBI, M.A. and VIOQUE, B. (1981). *Phytochemistry*, **20**, 1473–1475
WHITEHEAD, C.S., FUJINO, D.W. and REID, M.S. (1983). *Scientia Horticulturae*, **21**, 291–297
WULSTER, G., SACALIS, J. and JANES, H.W. (1982). *Plant Physiology*, **70**, 1039–1043
YANG, S.F. (1967). *Archives of Biochemistry and Biophysics*, **122**, 481–487
YANG, S.F. (1980). *HortScience*, **15**, 238–243
YU, Y.-B. and YANG, S.F. (1979). *Plant Physiology*, **64**, 1074–1077

ETHYLENE BIOSYNTHESIS IN *PENICILLIUM DIGITATUM* INFECTED CITRUS FRUIT*

ODED ACHILEA, EDO CHALUTZ, YORAM FUCHS and ILANA ROT
Department of Fruit and Vegetable Storage, Agricultural Research Organization,
The Volcani Center, Bet Dagan 50250, Israel

Introduction

Citrus fruit are usually classified as non-climacteric, based on the fact that the mature fruits do not exhibit a rise in either respiration or ethylene production throughout their normal pre- and post-harvest life (Biale, 1960; Sinclair, 1972). However, upon excision, citrus peel discs evince a peak in ethylene production. This wound-induced ethylene production is 1-aminocyclopropane-1-carboxylic acid (ACC)-dependent and originates from methionine (Hyodo, 1977) and is thus produced by the normal biosynthetic pathway of ethylene in higher plants (Adams and Yang, 1979; Yang *et al.*, Chapter 2).

Penicillium digitatum Sacc. is a common post-harvest pathogen of citrus. The fungus produces ethylene at high rates during growth on different media (Ilag and Curtis, 1968; Chalutz, Lieberman and Sisler, 1977). This ethylene production originates from glutamate and α-ketoglutarate (Chou and Yang, 1973), but under certain culturing conditions *P. digitatum* can also use methionine as a precursor for ethylene production (Chalutz, Lieberman and Sisler, 1977).

Although most isolates of *P. digitatum* produce ethylene at high rates, one has been found that produces the gas at a very low rate. This isolate is as pathogenic to citrus fruit as the normal, high-ethylene-producing one (Chalutz, 1979).

When citrus fruit are infected with *P. digitatum*, high rates of ethylene are produced by the diseased fruit (Schiffmann-Nadel, 1974). The objective of the work reported here was to study this process, and with the aid of different isolates of *P. digitatum* elucidate the contribution of the host and pathogen to the ethylene produced at different stages during the pathogenesis.

Plant and fungal material

Grapefruits (*Citrus paradisi* Macf. cv. 'Marsh Seedless') used for this work were harvested in mid-season and stored at 11 °C (85% relative humidity) until use. For inoculations, a single spore culture of *Penicillium digitatum* Sacc. was used. It was

---

*Contribution from the Agricultural Research Organization, The Volcani Center, Bet Dagan, Israel, No. 1038-E 1984 series

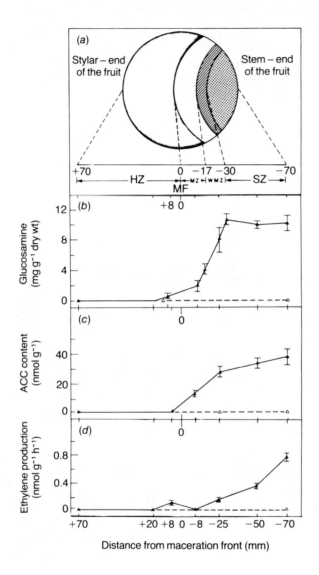

Figure 9.1 Morphological, biochemical and physiological changes in the peel of *Penicillium digitatum*-infected grapefruit. (a) Definitions of regions on a grapefruit, six days after inoculation and their graphic presentation as a function of their distance from the maceration front; (b) to (d) Changes in biochemical and physiological parameters in an infected fruit as related to the distance from the maceration front; (b) Glucosamine content; (c) ACC content; and (d) Ethylene production. All assays were carried out on the flavedo part of the peel, taken from healthy or *P. digitatum*-infected grapefruits, six days after inoculation of the fruit near the stem-end. Bars indicate s.d. △: healthy fruit; ▲: infected fruit. HZ: healthy zone; MZ: maceration zone; WMZ: white mycelium zone; SZ: green spores zone; and MF: maceration front

either the common, ethylene-producing isolate (ATCC No. 10030), or a non-ethylene-producing isolate. The cultures were propagated on potato dextrose agar slants.

Prior to inoculation, the fruit were washed with water and surface-sterilized with 70% ethanol. They were inoculated by introducing a 5 × 5 × 1 mm piece of inoculum into a V-shaped, 3 mm-deep cut, 20 mm away from the stem-end of the fruit. The wound was then covered with masking tape, and the fruit was stored at 23 °C in a ventilated dark chamber.

For the various tests, samples of peel were removed with a scalpel or a cork borer, and the albedo and flavedo parts were separated. The peel samples were taken from several defined regions of the inoculated fruit (*Figure 9.1a*). A minimum number of nine fruit was used as replicates for the determination of each parameter studied.

Biochemical and physiological assays

The quantification of fungal mycelium or spores in the fruit peel was based on the determination of glucosamine (Elson and Morgan, 1933), the monomer of chitin—a compound present in fungal cell wall material but not in higher plant tissues. The method of Netzer, Kritzman and Chet (1979) was used to relate glucosamine content to the amount of fungus present in the tissue.

Production of ethylene by the peel tissues was determined as follows: peel discs, 12 mm in diameter, were cut from the fruit with a cork-borer, and separated into the albedo and flavedo parts. Samples of five discs of the same tissue were then weighed, placed inside 21 ml glass vials and incubated at 24 °C in the dark. The vials were sealed with rubber serum caps, and the ethylene content of the gas phase was determined 60 min later, by gas chromatography. During this time period significant amounts of wound-induced ethylene were not detectable, since such production does not start until 6 h after excision (Chalutz, 1979; Achilea *et al.*, unpublished). The ability of the fruit peel tissue to convert exogenously applied ACC to ethylene was tested by directly applying a 5 mM solution of ACC to individual albedo discs cut from different zones of the infected fruit. The solution was immediately absorbed by the disc.

For tracer studies, ten discs were excised from each zone of the infected fruits. The flavedo part was weighed and placed inside 25 ml Erlenmeyer flasks to which was added 1.3 ml of water (control) or labelled precursor solution: $8.0\,\mu\text{Ci flask}^{-1}$ of ^{3}H-(G)-glutamic acid ($45\,\text{Ci mmol}^{-1}$) (Amersham), or $3.5\,\mu\text{Ci flask}^{-1}$ of ^{14}C-(3,4), methionine ($57\,\text{mCi mmol}^{-1}$) (Commisariat a l'Energie Atomique, France). The flasks were incubated in a shaking bath ($120\,\text{strokes min}^{-1}$) at 24 °C in the dark. Labelled ethylene was trapped according to the method described by Hyodo (1977).

ACC content of the peel was determined by the method of Lizada and Yang (1979). Peel extract was obtained by homogenizing 1 g of tissue with 10 ml of water at 4 °C for 1 min, using an Ultra-Turrax. The homogenate was then centrifuged at $27\,000 \times g$ for 10 min and the supernatant was used as the tissue extract in which ACC content was determined.

Development of infection

Six days after fungal inoculation, several well-defined zones could be observed on the surface, as follows (*Figure 9.1a*): Green spore zone (SZ) where mature fungal spores had developed; white mycelium zone (WMZ), distinguishable by the dense white mycelium which developed on the peel; maceration zone (MZ), distinguishable by the soft peel; a healthy, non-infected area of the peel (HZ), consisting of the rest of the fruit; and the maceration front (MF), a distinct line between the MZ and the HZ. The distance of any part of the fruit peel from the MF can be used as a reference reflecting the extent of fungal invasion. Data concerning the various physiological changes taking place in the infected fruit are expressed in relation to this value (*Figure 9.1a*), with MF referred to as the zero point. Sites in the HZ were referred to as $+X$, while sites in the infected zones were referred to as $-X$, where X represents the distance (in mm) of the site from the MF. All results were expressed on a peel tissue fresh-weight basis, except for those of glucosamine content, which were expressed on a dry-weight basis.

Unless otherwise indicated, the results were obtained with fruits inoculated with the common ethylene-producing isolate of *P. digitatum*.

FUNGAL PRESENCE IN THE PEEL

The distribution of fungal mycelium in the various regions of the fruit peel is shown in *Figure 9.1b*. The fungus was found not only in the zones showing infection symptoms, but also in the adjacent, apparently healthy tissue, up to $+15$ mm away from the MF. Fungal mycelium presence increased continuously in the MZ and the WMZ, and maintained a steady high level throughout the SZ. Similar values were obtained with the non-ethylene-producing isolate of *P. digitatum*.

Physiological changes associated with the infection

The effect of fungal infection on ACC content of the peel is shown in *Figure 9.1c*. The basal level of ACC in healthy fruit as well as in the distal parts of the HZ of an infected fruit was low (0.27 to $0.32\,\text{nmol}\,\text{g}^{-1}$), with ACC content of the peel increasing due to the infection process. The content reached $3\,\text{nmol}\,\text{g}^{-1}$ at the site of $+8$, and $29\,\text{nmol}\,\text{g}^{-1}$ at the site of -25, with the highest ACC values found at the centre of the SZ. Similar values were found when the fruit was infected by the non-ethylene producing isolate of *P. digitatum*.

The rate of ethylene production by healthy peel discs, shortly after excision from the fruit, was very low ($0.02\,\text{nmol}\,\text{g}^{-1}\,\text{h}^{-1}$, *Figure 9.1d*). In infected fruit a small but distinct increase in ethylene production was detected at the proximal HZ, and a pronounced increase started in the WMZ, reaching a peak of $0.8\,\text{nmol}\,\text{g}^{-1}\,\text{h}^{-1}$ at the centre of the SZ. The non-ethylene producing isolate of *P. digitatum* affected ethylene production similarly in all regions between $+70$ and -8, but no increase in ethylene production was found in the region between -8 and -70.

Isolated *P. digitatum* mature spores produced ethylene at a rate of $3.6\,\text{nmol}\,\text{g}^{-1}\,\text{h}^{-1}$, while spores from a non-ethylene producing isolate failed to produce detectable amounts of the gas.

When peel discs were incubated with uniformly labelled ^{3}H-glutamate (*Figure 9.2a*), labelled ethylene production during the first hour of incubation was low in the apparently healthy part of the fruit (sites $+70$ and $+8$) and high in the zones showing symptoms of the disease (sites -8 and -70). The same pattern, with higher values in the diseased region of the fruit, was manifested during the third

Table 9.1 THE CONTENT OF 1-AMINOCYCLOPROPANE-1-CARBOXYLIC ACID AND ITS CONVERSION TO ETHYLENE BY ALBEDO DISCS CUT FROM DIFFERENT SITES ON *PENICILLIUM DIGITATUM*-INFECTED GRAPEFRUIT

Assayed site	Control		ACC-treated
	ACC content (nmol g^{-1})	Ethylene production (nl g^{-1} h^{-1})	Ethylene production (% of control)
+70	0.5 ± 0.03	0.2 ± 0.06	994 ± 130
+ 8	8.8 ± 2.7	2.3 ± 0.8	142 ± 31
− 8	18.1 ± 3.0	0.2 ± 0.08	102 ± 17
−70	32.4 ± 4.2	22.0 ± 0.6	91 ± 17

Albedo discs were excised from the indicated sites (*see Figure 9.1a*) of *P. digitatum*-infected grapefruit. Five discs from each site were treated with water (control) or ACC (5 mM) and incubated at 24 °C for 6 h in the dark. ACC content and ethylene production were then determined. ACC content of the ACC-treated discs ranged from 141 to 180 nmol g^{-1}. Each value in the table is the mean of five replicates with its s.d.

hour of incubation. However, peel discs from the same sites incubated with ^{14}C-(3,4) methionine exhibited a different pattern of labelled ethylene production (*Figure 9.2b*). ^{14}C-ethylene production during the first hour of incubation was very low from all regions. During the third hour of incubation, however, all regions produced ^{14}C-ethylene at high rates, with discs cut from the healthy side of the MF producing ethylene at 2.5 to 3 times the rate of production by discs from other zones.

Ethylene production by ACC-treated albedo discs cut from the HZ (+70) increased at a rate nearly ten times that of the control (*Table 9.1*), whereas the increase was only 42% of the control in discs cut from the +8 zone. Discs cut from the SZ (−70) or from the MZ (−8) did not produce ethylene at a rate higher than the control discs, following ACC treatment.

Discussion

The findings presented here indicate that ethylene production in *P. digitatum*-infected grapefruit involves the contribution of both the host and the pathogen. The non-infected part of the fruit, i.e. where no glucosamine is detected (*Figure 9.1b*), maintains a steady low rate of ethylene production (*Figure 9.1d*). Ethylene production from plant origin increases at site +8, possibly as a result of the physiological stress that the invading fungal hyphae impose on the plant tissue. Barmore and Brown (1979) have shown that fungal infection of citrus fruit causes the cells to plasmolyse and the cell walls to swell in the apparently healthy zone. Physical and biochemical stresses are known to induce ACC synthesis (Yu and Yang, 1980; Hyodo and Nishino, 1981; Bradford and Yang, 1981). Such a phenomenon of increase in ACC synthesis and ethylene production in a region close to the invading pathogen was recently reported by De Laat and Van Loon (1983) in virus-infected tobacco leaves. In fungus-infected citrus fruit the peak of ethylene production at the +8 site (*Figure 9.1d*) could be detected also when the fruit was infected with the non-ethylene producing isolate of the fungus. Therefore, ethylene production at this zone is likely to originate from the fruit.

The results obtained with labelled ethylene precursors provide additional support for this hypothesis (*Figure 9.2a* and *9.2b*). The contribution of the fungus

98 Ethylene biosynthesis in Penicillium digatatum infected citrus fruit

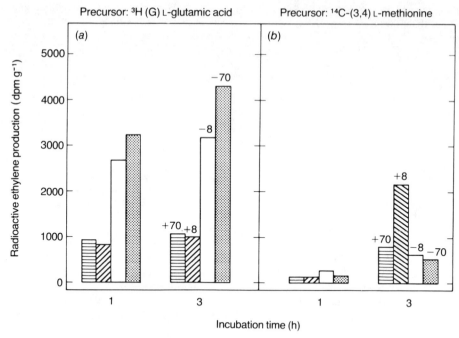

Figure 9.2 Radioactive ethylene production from peel discs excised from *Penicillium digitatum*-infected grapefruit. Ten flavedo discs, weighing 2.2 g, were excised from each of the four zones of a grapefruit, six days after infection (*Figure 9.1a*). The discs were incubated in 25 ml Erlenmeyer flasks to which 1.3 ml of the labelled precursor solution was added. Ethylene production was determined during the first and third hours of incubation. The four different zones are indicated on the right hand side of each pair of columns

to ethylene production at the +8 site was found to be relatively small, as shown by the low amounts of radioactive ethylene produced during the incubation of peel discs with labelled glutamate (*Figure 9.2a*). A clear increase in ethylene production from methionine was found at the +8 site during the third hour of incubation (*Figure 9.2b*), corresponding to the increase in total ethylene produced (*Figure 9.1d*). Fungal ethylene production increased in the infected regions (−8 and −70), as indicated by the large amounts of labelled ethylene produced at these sites from radioactive glutamate (*Figure 9.2a*) but not from methionine (*Figure 9.2b*). On the other hand, the methionine data suggest that fungal infection inhibits ethylene production of host origin at this zone (*Figure 9.2b*).

The decrease in ethylene production of plant origin in the infected regions, which occurs in spite of the increasing levels of ACC present at these sites (*Figure 9.1c*), suggests that infection by the pathogen may inhibit the conversion of ACC to ethylene. This effect would seem to be contrary to the induction of ACC by the fungus at the early stages of infection, as shown here (*Figure 9.1c*) and by others (De Laat and Van Loon, 1983). However, the results of this study show that the infected tissues do gradually lose their capacity to convert exogenous ACC to ethylene during the infection process (*Table 9.1*). Thus, the suppression of host ethylene production by the developing fungus (*Figure 9.2b*), while ACC is still being synthesized, and therefore accumulating (*Figure 9.1c*), appears to result from the impairment of the conversion step of ACC to ethylene. This phenomenon may arise as a result of a change in membrane integrity (Mattoo et al., 1982). A similar account of ACC accumulation and decreased ethylene production has been reported for post-climacteric avocado fruit (Hoffman and Yang, 1980). Thus, since the host ethylene-producing mechanism seems to be suppressed at the advanced stage of fungal infection (*Figure 9.1b* and *9.2a*), the elevated levels of ethylene produced at this time are most likely to originate primarily from the fungus. Further support for this hypothesis comes from the observation that the later stages of infection in citrus peel resulting from pathogenic attack by the non-ethylene producing isolate are not associated with elevated ethylene production.

Conclusions

Based on the findings reported here, the following scheme is suggested as to the origin of the ethylene produced during the infection of grapefruit by *P. digitatum*. At the initial phases of infection, the enhanced production of ethylene originates from the peel cells and is ACC mediated. This biosynthetic pathway is apparently impaired as the infection proceeds, since the tissue exhibits a decreasing ability to convert ACC to ethylene, and ACC accumulates. The increased production of ethylene at these stages is therefore likely to be of fungal origin, and does not use ACC as the precursor. This hypothesis is in agreement with earlier suggestions (Schiffmann-Nadel, 1974; Chalutz, 1979) indicating that the pathogen is responsible for the high rates of ethylene produced in this interaction.

Acknowledgement

This research was supported in part by a grant from the US–Israel Binational Agricultural Research and Development Fund (BARD).

References

ADAMS, D.O. and YANG, S.F. (1979). *Proceedings of the National Academy of Science USA*, **76**, 170–174
BARMORE, C.R. and BROWN, G.E. (1979). *Phytopathology*, **69**, 675–678
BIALE, J.B. (1960). *Handbuch der Pflanzen Physiologie*, **12**, 539–592. Springer Verlag, Berlin
BRADFORD, K.J. and YANG, S.F. (1981). *HortScience*, **16**, 25–30
CHALUTZ, E. (1979). *Physiological Plant Pathology*, **14**, 259–262
CHALUTZ, E., LIEBERMAN, M. and SISLER, H.D. (1977). *Plant Physiology*, **60**, 402–407
CHOU, T.W. and YANG, S.F. (1973). *Archives of Biochemistry and Biophysics*, **157**, 73–82
DE LAAT, A.M.M. and VAN LOON, L.C. (1983). *Physiological Plant Pathology*, **22**, 261–273
ELSON, L.A. and MORGAN, W.I. (1933). *Biochemical Journal*, **27**, 1824–1828
HOFFMAN, N.E. and YANG, S.F. (1980). *Journal of the American Society of Horticultural Science*, **105**, 492–495

HYODO, H. (1977). *Plant Physiology*, **59**, 111–113
HYODO, H. and NISHINO, T. (1981). *Plant Physiology*, **67**, 421–423
ILAG, L. and CURTIS, R.W. (1968). *Science*, **159**, 1357–1358
LIZADA, M.C.C. and YANG, S.F. (1979). *Analytical Biochemistry*, **100**, 140–145
MATTOO, A.K., ACHILEA, O., FUCHS, Y. and CHALUTZ, E. (1982). *Biochemical and Biophysical Research Communications*, **104**, 765–770
NETZER, D., KRITZMAN, G. and CHET, I. (1979). *Physiological Plant Pathology*, **14**, 47–55
SCHIFFMANN-NADEL, M. (1974). *Colloques Internationaux du Centre National de la Recherche Scientifique*, No. 238
SINCLAIR, W.B. (1972). The Grapefruit, its Composition, Physiology and Products. *University of California, Division of Agricultural Science, Riverside, California*
YU, Y.B. and YANG, S.F. (1980). *Plant Physiology*, **66**, 281–285

10
ETHYLENE BINDING

A.R. SMITH and M.A. HALL
Department of Botany and Microbiology, University College of Wales, Aberystwyth, Dyfed, UK

Introduction

The classical receptor concept envisages that the action of any plant growth substance is mediated by an interaction between the regulator and a site or sites, within a plant cell. This interaction initiates some biochemical event or events which ultimately results in a developmental response. This concept has originated from research on animals where receptors of the nature described above have been characterized for hormones such as insulin (Cuatrecasas, 1972a and b) and steroids (Towle *et al.*, 1976). In plants however, progress has been impeded by our lack of knowledge of the primary biochemical events initiated by plant growth substances. This has necessitated a different approach characterized by a search for 'binding sites' possessing the anticipated properties of a receptor in terms of affinity and specificity.

Work on binding sites in plants began using auxins and auxin analogues (Lembi *et al.*, 1971; Batt, Wilkins and Venis, 1976; Dohrmann, Hertel and Kowalik, 1978; Jacobs and Hertel, 1978; Vreugdenhil, Burgers and Libbenga, 1979). Since then studies have extended to all the other endogenous growth regulators (Berridge, Ralph and Letham, 1972; Gardner, Sussman and Kende, 1978; Polya and Davis, 1978; Stoddart, 1979a,b) and indeed to other naturally occurring substances such as fusicoccin (Dohrmann *et al.*, 1977; Aducci *et al.*, 1980), which although having profound effects on higher plants are not produced by them.

In the case of ethylene it seemed unlikely for technical reasons that a receptor would be readily identified. Many ethylene responses show short lag times; that is, they are switched on relatively soon after the growth regulator is applied and are switched off equally rapidly when the growth regulator is removed. If we apply classical kinetics to such a problem the implication is that the hormone-receptor interaction will have high rate constants of association and dissociation, i.e. the ligand binds rapidly to, and dissociates rapidly from, the receptor (other explanations such as rapid turnover of the receptor are also possible but examples are lacking). This is the case with the auxin binding sites so far characterized but the problem of assay is overcome by sedimenting the (membrane-bound) sites in the presence of ligand, then removing the unbound ligand in the resultant supernatant and assessing the bound ligand in the pellet. Clearly, in the case of an ethylene binding site of the type described above, the first of these procedures would be

102 Ethylene binding

difficult and the second impossible since the bound ethylene would diffuse away from the pellet when the unbound ligand was removed.

However, binding sites for ethylene have been isolated and described but *only* because, as we shall see, sites exist with low rate constants of dissociation.

Isolation and properties of ethylene binding sites

KINETICS

Ethylene binding sites were discovered independently and simultaneously in two laboratories (Sisler, 1979; Bengochea *et al.*, 1980a,b). In our own case the discovery arose unexpectedly from work designed to determine whether ethylene is translocated in higher plants (Jerie, Zeroni and Hall, 1978a). It soon became clear that while significant translocation did not occur, the behaviour of the gas—particularly in relation to its relative concentrations in the gaseous and aqueous (i.e. plant) phases—did not follow the expected pattern (Jerie *et al.*, 1978b). In particular, in two species investigated in detail, namely *Vicia faba* and *Phaseolus vulgaris*, the concentration of the growth regulator in the tissue was much higher than that expected from its solubility characteristics, i.e. the tissue appeared to 'accumulate' ethylene (Jerie and Hall, 1978; Jerie, Shaari and Hall, 1979). In *Vicia*

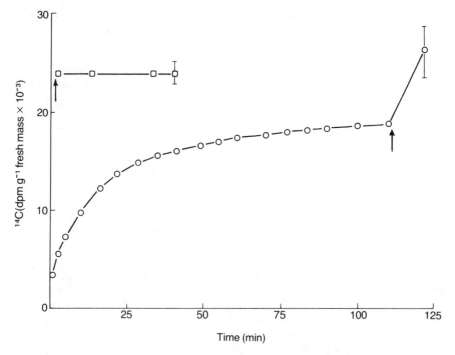

Figure 10.1 The elution of radioactivity from single $^{14}C_2H_4$ pretreated bean cotyledons into toluene at room temperature. Treatments involved heating the tissue in toluene — PPO to 60 °C for 2 h (arrows) at zero time (□) and after 110 min (○). The cotyledons for both treatments were taken at random from a single pool. Each point shows the mean of four replicates, recounted at the times indicated. Vertical bars are standard deviations (after Jerie, Shaari and Hall, 1979)

faba the effect was explained by the existence of an ethylene metabolizing system which produced a product, namely ethylene oxide, which was much more soluble in water than ethylene (*see* Beyer, Chapter 12). However, in *Phaseolus vulgaris* all the ^{14}C-ethylene incorporated into the tissue could be recovered, either by heat-killing the tissue, or by allowing the labelled growth regulator to emanate. The kinetics of release are shown in *Figure 10.1*. Further studies *in vivo* indicated that the system which accumulated ethylene had a high affinity for the gas and that physiologically active analogues of ethylene such as propylene and vinyl chloride could competitively inhibit accumulation of the natural growth regulator to an extent roughly proportional to their relative effectiveness in mimicking ethylene in the stimulation of developmental events.

Subsequently, work showed that tissue treated with ^{14}C-ethylene could be homogenized and subjected to separation by ultracentrifugation with little loss of radioactivity. Moreover, the radioactivity separated at discrete locations in the sucrose gradients used for the studies. If untreated tissue was separated in the same way and fractions from the gradients treated with ethylene, then binding was observed in the same fractions as in the experiments where tissue was prelabelled.

The cell-free system did not discriminate between ^{12}C- and ^{14}C-ethylene and Scatchard analyses (*Figure 10.2*) of the results of binding experiments indicated the presence of a single class of sites with a dissociation constant (K_D) of about 6×10^{-10} M (liquid phase = $0.13\,\mu l\,l^{-1}$ ethylene gas phase). The concentration of sites varied considerably during seed development in the range 5–100 pmol g^{-1} fresh mass.

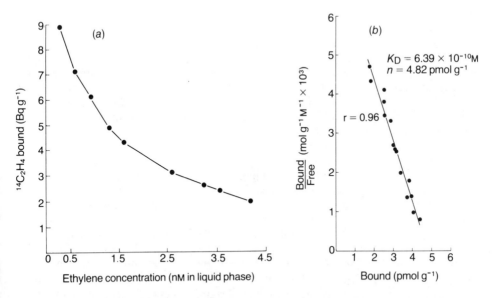

Figure 10.2 (a) The effect of unlabelled ethylene on binding of ^{14}C$_2$H$_4$. 16.7 Bq ^{14}C$_2$H$_4$ ml^{-1} air were present in all incubations 1 pmol ^{14}C$_2$H$_4$ = 4.4 Bq. Samples were incubated for 24 h at 25 °C. Results representative of three expriments. (b) Scatchard plot of data obtained by incubating binding preparation with a range of concentrations of ^{14}C$_2$H$_4$ from 4×10^{-10}M to 5.4×10^{-9}M (in the liquid phase). Incubations were carried out for 24 h at 25 °C. Results representative of ten experiments (after Bengochea *et al.*, 1980a)

It is worth noting here that if the interaction between ethylene and the binding site is of the form

$$R + E \rightleftharpoons RE \tag{1}$$

then

$$K_D = \frac{[R][E]}{[RE]} \tag{2}$$

where $[R]$, $[E]$ and $[RE]$ are the concentrations of binding site, ethylene and ethylene binding site complex respectively at equilibrium.

Clearly, if R is a functional receptor and the magnitude of the response evoked is proportional to the percentage occupancy of receptor sites then when $K_D = [E]$ half of the sites are occupied (i.e. $R = RE$) and the K_D should therefore correspond to the concentration of ethylene bringing about a half-maximal response. The figure obtained in the work described does in fact closely correspond to the concentration of ethylene bringing about such a half-maximal response in many developmental systems (see e.g. Abeles, 1973).

It should be noted however that the K_D obtained is only 'apparent' since the simple kinetics used depend on the assumption that $[R]$ is at least an order of magnitude smaller than $[E]$; where $[R] \geqslant [E]$ then the correct formulation is given by

$$K_{D_{APP}} = K_{D_{REAL}} + \frac{[R]}{2} \tag{3}$$

where $K_{D_{APP}}$ and $K_{D_{REAL}}$ are the apparent and actual dissociation constants. A plot of $[R]$ versus $K_{D_{APP}}$ should yield a straight line of intercept $K_{D_{REAL}}$. Using such an approach we were able to demonstrate that $K_{D_{REAL}}$ in the *Phaseolus vulgaris* system is 8.8×10^{-11} M (*Figure 10.3*). This phenomenon is potentially of more than academic interest since it provides a mechanism whereby the affinity of the site for ethylene could be changed by altering binding site concentration. Thus, in effect, the same binding site could operate in the control of varying processes with different dose responses to ethylene.

Measurement of the rate constants of association (k_1) and dissociation (k_{-1}) in membrane preparations yielded figures of 3.18×10^4 $M^{-1}s^{-1}$ and $1.84 \times 10^{-5}s^{-1}$ respectively, these low values confirming the impression gained from work with intact tissue and the separation studies. These findings also provided independent corroboration of the value for the K_D obtained by Scatchard plots since $k_{-1}/k_1 = K_D = 5.8 \times 10^{-10}$ M.

The approach taken by Sisler and his coworkers in investigating ethylene binding was markedly different to our own but has yielded very similar results overall. The initial work on *Nicotiana tabacum* (Sisler, 1979) was carried out by exposing intact leaves to ^{14}C-ethylene in a large volume and then transferring them to another vessel and adding ^{12}C-ethylene. Ethylene binding was assessed by measuring the amount of ^{14}C-ethylene displaced. A similar technique was used for studies on the effects of structural analogues of ethylene. Using this technique ethylene binding with a K_D of 0.27 µl l^{-1} (1.2×10^{-9} M liquid phase) was demonstrated. Equally, physiologically active analogues of ethylene competed with the growth regulator for binding sites.

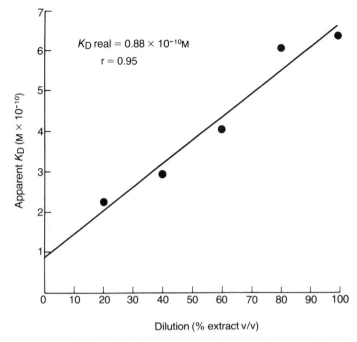

Figure 10.3 Effect of binding site concentration upon the apparent dissociation constant for ethylene. Each point is derived from a Scatchard analysis of ethylene binding at different dilutions of binding site (after Bengochea *et al.*, 1980a)

The rates of association and dissociation were not measured, but from consideration of the times taken to equilibrate it is clear that although the processes are faster in *Nicotiana tabacum* than in *Phaseolus vulgaris* the rate constants are also very low in the former. The work on tobacco has since been extended to *Phaseolus aureus* (mungbean) and a particulate fraction prepared from this tissue. This particulate fraction has sedimentation characteristics similar to those of the system in *Phaseolus vulgaris* (Sisler, 1980).

In both species of *Phaseolus*, binding is heat labile and partially destroyed by treatment with proteolytic enzymes. The pH optimum for binding in *Phaseolus aureus* is a broad peak around 6 whereas it is between 7.5 and 9.5 in *Phaseolus vulgaris*. Further comparison of the membrane-bound sites is difficult since most of Sisler's subsequent work has been carried out with solubilized preparations (*see below*).

EFFECTS OF STRUCTURAL ANALOGUES AND INHIBITORS

The ability of structural analogues of ethylene to compete with the growth regulator for binding site occupancy has been investigated in a number of species with intact tissues, membrane-bound preparations and solubilized fractions. Since in general no significant differences in specificity have been observed between different types of preparation we will treat them together although it should be borne in mind that not all the analogues have been tested in each type of preparation.

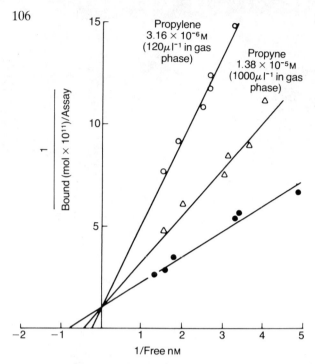

Figure 10.4 Lineweaver–Burk plots for ethylene binding to cell-free preparations from *Phaseolus vulgaris*. Samples were treated with $^{14}C_2H_4$ at a range of concentrations in the presence or absence of propylene or propyne at 25 °C overnight. Control (●); propylene (○); propyne (△). (After Bengochea *et al.*, 1980b)

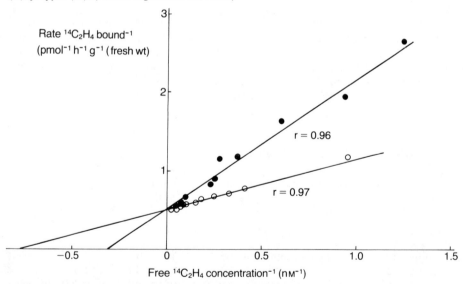

Figure 10.5 Lineweaver–Burk plots of ethylene binding to *Phaseolus vulgaris* cell-free preparations solubilized with 0.5% Triton X-100 showing competitive inhibition by 2,5-norbornadiene (2160 μl l^{-1} gas phase). Samples incubated with $^{14}C_2H_4$ at a range of concentrations (0.7–20.0 nM liquid phase) in the presence (●) or absence (○) of cycloalkene. (After Smith *et al.*, 1984)

On the whole, physiologically active structural analogues of ethylene show competitive inhibition (e.g. *Figures 10.4* and *10.5*) whereas physiologically inactive analogues do not. A summary of the data obtained so far is given in *Table 10.1*.

The reasons for such differences as do occur between species or within the same species are as yet unresolved. In *Phaseolus vulgaris* the affinity of intact tissue for ethylene is much lower than is the case with membrane preparations. However, the concentration of binding sites in intact tissue is much higher than in membrane preparations suspended in buffer (5×10^{-7} M and 10^{-9} M respectively) and this difference will of itself result in a marked disparity in apparent affinity (*see* equation 3). These differences are also reflected in the respective values for propylene and vinyl chloride.

Comparing different species the major area of disagreement lies between the values for cycloalkenes in *Phaseolus vulgaris* and *Phaseolus aureus* in that the K_i values are much higher in the former.

In both *Phaseolus vulgaris* and *Phaseolus aureus* CO_2 has no effect on binding to solubilized preparations but in membrane preparations from *Phaseolus vulgaris* CO_2 shows competitive inhibition. This last result must, however, be treated with some caution since at the high concentration of CO_2 used, substantial buffering is required. Moreover the concentration of CO_2 as HCO_3^- and CO_3^{2-} obtaining in a buffered membrane preparation will not be the same as that in the cell for a given concentration of CO_2 in the gas phase.

One interesting recent observation concerns the effects of different isomers of butene upon binding and pea stem extension. 1-Butene will mimic the action of ethylene, albeit at much higher concentrations (Burg and Burg, 1967), whereas *cis*- and *trans*-2-butene do not. However, Sisler and Yang (1983) demonstrated that in peas both 1-butene and *cis*-2-butene inhibited ethylene binding with similar K_is. Moreover, when applied to pea epicotyls 1-butene inhibited extension, *cis*-2-butene promoted and *trans*-2-butene was without effect. Sisler and Yang (1983) ascribe the difference between their results and those of Burg and Burg (1967) by suggesting that in the latter case, ethylene production by the tissue was inhibited by the inclusion of Co^{2+}. Thus, controls would have been growing at a maximal rate, unhindered by endogenous ethylene. If *cis*-2-butene acts, as suggested, by occupying the binding site but failing to bring about the necessary conformational change in the receptor then an effect would only be observed where endogenous ethylene occupying some of the sites could be displaced by the analogue.

Sisler and Yang (personal communication) also suggest a similar mode of action for 2,5-norbornadiene (which can also reverse ethylene effects) but the potential usefulness of this compound in reversing or preventing ethylene effects is much diminished by its carcinogenic properties.

A word of caution is appropriate here in relation to work with structural analogues. In general, workers in the field of ethylene physiology tend to use and compare concentrations in the gas phase. In the main, this practice is acceptable, although it is perhaps appropriate to remind oneself that the solubility of gases varies markedly with temperature and that at 15 °C one is not dealing with the same tissue ethylene concentration as at 25 °C—even if the gas phase concentration is identical. However, serious errors can arise if comparisons are made between the activities of ethylene and its analogues in the gas phase since these various substances have very different solubilities. Thus, whereas on a gas phase comparison carbon monoxide and acetylene have much the same activity relative to ethylene, in fact the latter is much less active since it is much more soluble than the

Table 10.1 COMPARISON OF DISSOCIATION AND INHIBITOR CONSTANTS FOR ETHYLENE AND ITS STRUCTURAL ANALOGUES IN VARIOUS ETHYLENE BINDING SITE SYSTEMS

Species	Type of preparation	Analogues	K_D or K_i (gas)[a] (M)	K_i (gas) relative	Concentration for half maximal activity[b] (relative)	Reference
Phaseolus vulgaris	Intact developing cotyledons	Ethylene	4×10^{-8}	1	1	Jerie, Shaari and Hall (1979)
		Propylene	5.4×10^{-7}	13.5	128	
		Vinyl chloride	4.5×10^{-6}	112	1400	
Nicotiana tabacum	Leaves	Ethylene	1.2×10^{-8}	1	1	Sisler (1979)
		Propylene	1.9×10^{-6}	158	167	
		Carbon monoxide	1.2×10^{-5}	1000	887	
Citrus sinensis	Leaves	Ethylene	6.8×10^{-9}			Sisler and Goren (personal communication)
Ligustrum japonicum	Leaves	Ethylene	1.4×10^{-8}			
Phaseolus vulgaris	Membrane preparations	Ethylene	8.15×10^{-9}	1	1	Bengochea et al. (1980b)
		Propylene	1.04×10^{-6}	100	128	
		Vinyl chloride	3.8×10^{-6}	466	1400	
		Carbon monoxide	8.7×10^{-6}	1068	2700	
		Acetylene	8.25×10^{-6}	1013	2800	
		Vinyl fluoride	9.3×10^{-6}	1139	4300	
		Propyne	2.16×10^{-5}	2651	8000	
		Vinyl methyl ether	1.11×10^{-3}	136196	100000	
		1-Butene	4.9×10^{-3}	601227	270000	
		Carbon dioxide	1.15×10^{-3}	141104	300000	

Species	Preparation	Compound	Concentration (μl l⁻¹)			Reference
Phaseolus vulgaris	Triton X-100 solubilized preparations	Ethylene	5.5×10^{-9}		1	Thomas, Smith and Hall (unpublished); Smith *et al.* (1984)
		Propylene	1.71×10^{-7}	31	128	
		Acetylene	1.29×10^{-5}	2345	2800	
		Cyclopentene	9.41×10^{-5}	10490	1100	
		2,5-Norbornadiene	9.79×10^{-5}	10920	170	
		1,3-Cyclohexadiene	1.55×10^{-4}	17250	4650	
		1,3-Cycloheptadiene	1.90×10^{-4}	21130	870	
		Carbon dioxide			300000	
		Norbornene			360	
Phaseolus aureus	Triton X-100 solubilized preparations	Ethylene	4.53×10^{-9}	1	1	Sisler (1982a)
		Propylene	1.8×10^{-7}	40	100	
		2,5-Norbornadiene	3.3×10^{-7}	73	170	
		Norbornene	∞	∞	360	
Phaseolus aureus	Triton X-100 solubilized preparations	1,3 cyclohexadiene	5.4×10^{-6}	1192	—	Beggs and Sisler (1983)
		Cyclopentene	8.2×10^{-6}	1810	1100	
		2-vinylpyridine	7.8×10^{-5}	17220	—	
		1,4 cyclohexadiene	1.5×10^{-4}	33112	4650	
		Cyclohexene	3.5×10^{-4}	77263	6060	
		Carbon monoxide	65, 100	—	2700	
		Carbon dioxide	0, 100000	—	300 000	

[a] Some absolute values are unavailable for the Triton X-100 solubilized preparations from *Phaseolus aureus*, instead the inhibition of binding obtained (%) and the concentration used (μl l⁻¹) respectively are given where appropriate.
[b] The values here are for those of Burg and Burg (1967) for pea growth except for *Nicotiana tabacum* (Sisler, 1979) and the figures on cycloalkenes (pea growth, Sisler and Yang, 1983, personal communication)

former. The figures in *Table 10.1* are given as gas phase values for comparative purposes only but in work with cell-free preparations or isolated proteins absolute affinities should always be quoted in terms of the liquid phase since this is the phase in immediate contact with the binding site.

The effects of a number of other inhibitors upon ethylene binding have been investigated in *Phaseolus vulgaris* and *Phaseolus aureus*. Both sulph-hydryl reactive agents and thiols such as *p*-chloromercuribenzoate and dithiothreitol inhibit binding indicating the involvement of a cysteine residue and a disulphide group at the binding site in both species. The inhibition of binding by thiourea and diethyldithiocarbamate are of interest because it has been suggested that $Cu^{(1)}$ may be a metal involved in ethylene binding (Beyer, 1976; Burg and Burg, 1967).

In this connection it has been demonstrated that $Cu^{(1)}$ ethylene coordination chemistry is consistent with the proposed role of copper in ethylene binding sites (Thompson, Harlow and Whitney, 1983). A series of $Cu^{(1)}$ alkene complexes was synthesized showing the binding characteristics necessary for biological systems. In these complexes forward donation or ϵ-binding between cuprous ion and the coordinate alkene is the dominant interaction. This may then be extrapolated to explain the observed decrease in biological activity relative to ethylene for alkenes with electron-withdrawing groups. These electron-withdrawing groups would alter the interactions between the metal and the ligand; the presence of such groups enhancing the π-back binding contribution from metal to alkene and weakening the ϵ-donation to the metal ion from the alkene. The effect of steric interactions was also implicated in explaining the decrease in biological activity with alkenes of increasing molecular size. These model systems may very well prove to be of great value in comparative studies with the ethylene binding complex in plant tissues.

One further interesting observation by Sisler (1982a) was that iodide inhibited binding very considerably. It is well known (Herrett *et al.*, 1962) that iodide can mimic the effects of ethylene on leaf abscission and the effect of iodide on binding may well provide an explanation of such observations.

In this connection a general mechanism whereby ethylene and π-acceptor compounds (such as iodide) could bind to a metal-containing receptor site has been suggested (Sisler, 1977). The proposal is that a structural change in the receptor protein could be brought about by a *trans* effect leading to the initiation of the primary response. Sisler has examined the properties of several π-acceptor compounds not hitherto considered as analogues of ethylene (e.g. phosphorus trifluoride, cyclohexyl isocyanide) and many of these do seem to be able to mimic the action of the growth regulator in some developmental systems (Sisler, 1977). So far, a wide range of such substances have not been tested in binding systems but such a screening seems worthwhile.

It would be premature to extrapolate the results with inhibitors too far, especially since some of the effects may be on that part of the molecule not directly involved in binding. On the other hand such work provides valuable pointers to future work on both the further characterization and the purification of the binding site.

SUBCELLULAR LOCALIZATION OF BINDING SITES

Thus far, binding site localization has only been attempted in *Phaseolus vulgaris*. Two approaches have been used (Evans *et al.*, 1982a,b) and both lead to the same conclusions. In the first study binding activity was fractionated by rate-zonal and

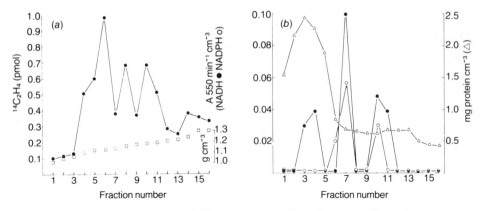

Figure 10.6 Distribution of ethylene binding, enzyme activities and protein on isopycnic centrifugation. (a) $^{14}C_2H_4$ bound (●); sucrose density (□)*; (b) antimycin-insensitive NADH-cytochrome c reductase (●); NADPH-cytochrome c reductase (○), total protein (△)*. *All values expressed per cm^3 gradient fraction. (After Evans et al., 1982a)

isopycnic centrifugation and the distribution of binding activity in such separations compared with that of putative marker enzymes for cellular organelles. An example of a separation on an isopycnic gradient is shown in *Figure 10.6*. Consideration of both types of separation technique and of the marker enzymes associated with the binding activity led to the conclusion that the latter was associated with elements of the endomembrane system and protein body membranes. In the case of the endoplasmic reticulum it was shown that binding was associated with the membranes rather than the ribosomal component since dissociation of the latter by treatment with EDTA and separation by centrifugation did not diminish binding to the membrane component.

In a parallel study, binding site localization was investigated by means of high resolution autoradiography. This was rendered feasible by two unique properties of the system. Firstly it was possible to saturate all the binding sites in pieces of tissue with ^{14}C-ethylene and then, because of the low rate constant of dissociation of

Table 10.2 ELECTRON MICROSCOPE AUTORADIOGRAPHY: SPECIFIC ACTIVITY VALUES

Organelle	Specific activity[a]
Endoplasmic reticulum	3.4
Protein body membrane	1.5
Protein body contents	0.6
Cytoplasm	0.9
Cell wall/plasma membrane	0.4
Mitochondria	0.7
Chloroplasts and nuclei	0.1
Bare plastic/other	0.5
Starch/lipid bodies	0.3

[a]Specific activity = frequency of silver grains over organelle/area of organelle
(From Evans et al., 1982b)

ethylene from the site, to remove all unbound ethylene rapidly. Secondly, it was possible to fix the bound ethylene in the tissue by treatment with osmium tetroxide which forms an insoluble complex with alkenes. Both light-microscope and electron-microscope autoradiography revealed a non-random distribution of silver grains in tissue sections and extensive statistical analysis of the data (Williams, 1969) showed that radioactivity was predominantly associated with endoplasmic reticulum and protein membranes (*Table 10.2*).

Purification and properties of solubilized binding site preparations

Extensive studies have now been carried out on solubilized ethylene binding site preparations from *Phaseolus aureus* and *Phaseolus vulgaris* (Sisler, 1984; Hall *et al.*, 1984; Thomas, Smith and Hall, 1984). In both cases the only technique giving high recovery and stability involves the use of detergents, in the absence of which the binding site precipitates, indicating that the binding protein is highly hydrophobic—a common property of membrane proteins. Of the detergents used initially, Triton X-100 proved most suitable and most of the work described below was carried out with this substance. However, octyl glucoside appears to be as effective and we anticipate that this will be used in future work, especially since it can be removed readily from samples whereas Triton X-100 cannot.

Although extraction with different detergents results in variable yields, the affinity of the site for ethylene is little affected (*Figure 10.7*). This is also true for analogues of ethylene other than CO_2.

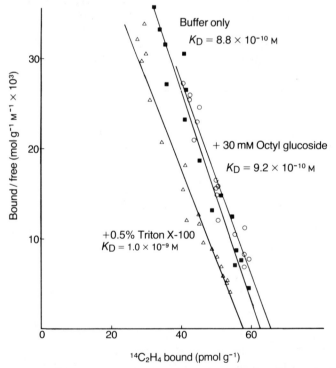

Figure 10.7 Scatchard plots of ethylene binding to cell-free preparations from *Phaseolus vulgaris* solubilized in buffer (■), 30 mM octyl glucoside (○) and 0.5% Triton X-100 (△)

Solubilization of binding site preparations from both *Phaseolus vulgaris* and *Phaseolus aureus* results in a marked downward shift in the pH optimum for binding to between pH 4 and pH 7. The sedimentation coefficient of the ethylene binding site (EBS) detergent complex derived from isokinetic sucrose gradients (Noll, 1967) was found to be 2.25s and a Stoke's radius of 6.1 nm was obtained from gel permeation chromatography on Sephacryl S-300. Calculation of molecular weight from the results obtained with each method leads to values differing by more than an order of magnitude. However, the frictional ratio (Siegel and Monty, 1966) was calculated to be 2.4 indicating that the EBS-detergent complex is asymmetric, accounting for the disparity in the results obtained by the two methods. If the results from both approaches are combined, but taking the asymmetry of the molecule into account, a molecular weight of between 52 000 and 60 000 daltons may be calculated (Thomas, Smith and Hall, 1985). Both the molecular weight determination and the apparent degree of asymmetry of the protein must be treated with some caution. In the former case because the observed molecular weight is for the EBS-detergent complex and hence is an overestimate as a result of the Triton molecules attached to the protein and the asymmetry may not be a reflection of the natural state of the protein in the membrane before solubilization.

Further support for the hypothesis that the presence of detergent induces changes in protein conformation come from the observation that the rate constants of association and dissociation are even lower for solubilized EBS than for membrane bound EBS.

Both Sisler and ourselves have adopted a number of approaches in attempting to purify the solubilized EBS. While most are conventional techniques of protein separation it has often been necessary to modify them substantially because of complications introduced by the presence of detergent in the preparations. Although neither group has established a definite protocol, present efforts revolve around combinations of four or five techniques namely, gel permeation chromatography, detergent partitioning, pH precipitation, isoelectric focusing and FPLC. Since the EBS protein has not yet been purified to homogeneity it is inappropriate here to describe the detailed results obtained with the various techniques, especially since they are fully described elsewhere (Sisler, 1984; Hall *et al.*, 1984; Thomas, Smith and Hall, 1984). Suffice to say that rigorously purified EBS proteins from both species of *Phaseolus* are likely to be obtained in the near future.

Conclusions

Ethylene binding sites in two separate systems have been investigated in great detail. The kinetic properties of these sites in terms of affinity and specificity are consistent with those one would expect of ethylene receptors. Nevertheless, the problem remains that as yet no connection has been established between these sites and the initiation of a biochemical response characteristic of an ethylene effect upon plant growth and development. Both Sisler's group and ourselves have demonstrated that there are changes in binding site concentrations during ontogeny (seed development, seed germination and seedling growth) but these changes have not even been related to tissue sensitivity. Indeed, as far as *Phaseolus vulgaris* and *Phaseolus aureus* seeds are concerned it is difficult to see how such a correlation can be obtained since there is no known effect of ethylene on these seeds during ontogeny.

There are however ethylene sensitive systems where at least some of the primary biochemical events are well characterized, for example ripening fruit (Grierson and Tucker, 1983) and leaf abscission (Sexton and Roberts, 1982). Both Sisler (1982b) and ourselves have evidence for ethylene binding in such systems. However, in both cases it is evident that the concentration of binding site in the tissue is very low and this poses a serious practical difficulty. The ^{14}C-ethylene used in our work is almost at the maximum specific activity possible (ca. 4.4 TBq mol^{-1}, 11.87 × 10^9 dmin^{-1} cm^{-3}). Assuming a stoichiometry of one, then 1 pmol of binding site saturated with ^{14}C ethylene will only yield about 266 dpm. In developing seeds where binding site concentration may be as high as 100 pmol g^{-1} fresh mass detection presents no problem but we have evidence that in abscission zones of *Phaseolus vulgaris* the concentration of binding sites is two to three orders of magnitude lower than in the seeds. Clearly, even with intact tissue, accurate measurement is difficult even if it were practical to prepare 10 g of abscission zone tissue for each of several treatments. The situation is exacerbated further if work with cell-free preparations is envisaged, where some loss of activity is inevitable.

It is self-evident that progress will only be made if much greater sensitivity can be obtained in assays. This is not an uncommon situation with animal hormones such as insulin. Here the problem is overcome by iodinating the insulin molecule with I^{125}. While a directly analogous approach is impossible with ethylene, the availability of purified EBS protein may well provide a means of overcoming the problem. It is our intention to prepare antibodies to the protein and to utilize isotope- or enzyme-linked antibodies in the localization and quantitation of binding sites where they are present in low concentrations.

This may indirectly open up other avenues of study. Thus, we have already noted that studies on binding sites for ethylene have only been rendered possible by their low rate constants of dissociation. However, as discussed earlier, it seems likely that there are present in some tissues sites with high rate constants of dissociation. Now, since ethylene controls a wide range of physiological processes it seems probable that there exist many kinds of receptor. While, because they initiate a wide range of biochemical events, these are likely to be very diverse in some respects it seems rather unlikely that the actual site at which the ligand becomes attached will show great variability. If, therefore, antibodies specific to the point of attachment can be obtained then it should be possible to use these to assess binding site concentration—even if such sites could not otherwise be quantified because they have high rate constants of dissociation.

One further area where the ability to detect like proteins may prove valuable relates to parallel studies in our laboratory and that of Beyer. It is inappropriate here to discuss these studies exhaustively since they are dealt with elsewhere in this volume (Beyer, Chapter 12) but suffice to say that ethylene metabolism has been implicated in the mode of action of ethylene and some of the enzymes involved are soon likely to be available in a relatively pure state. Since these enzymes are derived from genera closely related to *Phaseolus*, e.g. (*Vicia*, *Pisum*) and may, therefore, show similarities, it seems timely to compare their structures and immunological cross-matching is one avenue well worth exploring.

Little more can be said in conclusion except that the field is advancing rapidly and that unless unforeseen circumstances intervene it is not unreasonable to predict that a number of the approaches noted above are likely to come to fruition in the near future.

References

ABELES, F.B. (1973). *Ethylene in Plant Biology*. Academic Press, London
ADUCCI, P., CROSETTI, G., FEDERICO, R. and BALLIO, A. (1980). *Planta*, **148**, 208–210
BATT, S., WILKINS, M.B. and VENIS, M.A. (1976). *Planta*, **130**, 7–13
BEGGS, M.J. and SISLER, E.C. (1983). *Plant Physiology* (Supplement), **72**, 40 no. 226
BENGOCHEA, T., DODDS, J.H., EVANS, D.E., JERIE, P.H., NIEPEL, B., SHAARI, A.R. and HALL, M.A. (1980a). *Planta*, **148**, 397–406
BENGOCHEA, T., ACASTER, M.A., DODDS, J.H., EVANS, D.E., JERIE, P.H. and HALL, M.A. (1980b). *Planta*, **148**, 407–411
BERRIDGE, M.V., RALPH, R.K. and LETHAM, D.S. (1972). In *Plant Growth Substances*, p. 248. Ed. by D.J. Carr. Springer Verlag, Berlin
BEYER, E.M. (1976). *Plant Physiology*, **58**, 268–271
BURG, S.P. and BURG, E.A. (1967). *Plant Physiology*, **42**, 144–152
CUATRECASAS, P. (1972a). *Proceedings of the National Academy of Sciences, USA*, **69**, 1277–1281
CUATRECASAS, P. (1972b). *Journal of Biological Chemistry*, **247**, 1980–1991
DOHRMANN, U., HERTEL, R., PESCI, P., COCUCCI, S., MARRÈ, E., RANDATTO, G. and BALLIO, A. (1977). *Plant Science Letters*, **9**, 291–299
DOHRMANN, U., HERTEL, R. and KOWALIK, H. (1978). *Planta*, **140**, 97–106
EVANS, D.E., BENGOCHEA, T., CAIRNS, A.J., DODDS, J.H. and HALL, M.A. (1982a). *Plant, Cell and Environment*, **5**, 101–107
EVANS, D.E., DODDS, J.H., LLOYD, P.C., AP GWYNN, I. and HALL, M.A. (1982b). *Planta*, **154**, 48–52
GARDNER, G., SUSSMAN, M.R. and KENDE, H. (1978). *Planta*, **143**, 67–73
GRIERSON, D. and TUCKER, G. (1983). *Planta*, **157**, 174–179
HALL, M.A., CAIRNS, A.J., EVANS, D.E., SMITH, A.R., SMITH, P.G., TAYLOR, J.E. and THOMAS, C.J.R. (1982). In *Plant Growth Substances*, pp. 375–383. Ed. by P.F. Wareing. Academic Press, London
HALL, M.A., SMITH, A.R., THOMAS, C.J.R. and HOWARTH, C.J. (1984). In *Ethylene–Biochemical, Physiological and Applied Aspects*, pp. 55–63. Ed. by Y. Fuchs. Martinus Nijhoff; Dr W. Junk, The Hague
HERRETT, R.A., HATFIELD, H.H., CROSBY, D.G. and VLITOS, A.J. (1962). *Plant Physiology*, **37**, 358–363
JACOBS, M. and HERTEL, R. (1978). *Planta*, **142**, 1–10
JERIE, P.H. and HALL, M.A. (1978). *Proceedings of the Royal Society, London. Series B*, **200**, 87–94
JERIE, P.H., ZERONI, M. and HALL, M.A. (1978a). *Pesticide Science*, **9**, 162–168
JERIE, P.H., SHAARI, A.R., ZERONI, M. and HALL, M.A. (1978b). *New Phytologist*, **81**, 499–504
JERIE, P.H., SHAARI, A.R. and HALL, M.A. (1979). *Planta*, **144**, 503–507
LEMBI, C.A., MORRÉ, D.J., THOMSON, K. and HERTEL, R. (1971). *Planta*, **99**, 37–45
NOLL, H. (1967). *Nature*, **215**, 360–363
POLYA, G.M. and DAVIS, A.W. (1978). *Planta*, **139**, 139–147
SEXTON, R. and ROBERTS, J.A. (1982). *Annual Review of Plant Physiology*, **33**, 133–62
SIEGEL, L.M. and MONTY, K.J. (1966). *Biochimica et Biophysica Acta*, **112**, 346–362
SISLER, E.C. (1977). *Tobacco Science*, **21**, 43–45
SISLER, E.C. (1979). *Plant Physiology*, **64**, 538–543
SISLER, E.C. (1980). *Plant Physiology*, **66**, 404–406

SISLER, E.C. (1982a). *Journal of Plant Growth Regulation*, **1**, 211–218
SISLER, E.C. (1982b). *Journal of Plant Growth Regulation*, **1**, 219–226
SISLER, E.C. and YANG, S.F. (1983). *Plant Physiology* (supplement), **72**, 40 no. 225
SISLER, E.C. (1984). In *Ethylene–Biochemical, Physiological and Applied Aspects*, pp. 45–54. Ed. by Y. Fuchs. Martinus Nijhoff/Dr W. Junk, The Hague
SMITH, A.R., HOWARTH, C.J., SANDERS, I.O. and HALL, M.A. (1984). In *Ethylene–Biochemical, Physiological and Applied Aspects*, pp. 99–100. Ed. by Y. Fuchs. Martinus Nijhoff/Dr W. Junk, The Hague
STODDART, J.L. (1979a). *Planta*, **146**, 353–361
STODDART, J.L. (1979b). *Planta*, **146**, 363–368
THOMAS, C.J.R., SMITH, A.R. and HALL, M.A. (1984). *Planta* (in press)
THOMAS, C.J.R., SMITH, A.R. and HALL, M.A. (1985). *Planta* (in press)
THOMPSON, J.S., HARLOW, R.L. and WHITNEY, J.F. (1983). *Journal of the American Chemical Society*, **105**, 3522–3527
TOWLE, H.C., TSAI, M.H., HIROSE, M., TSAI, S.Y., SCHWARTZ, R.J., PARKER, M.G. and O'MALLEY, B.W. (1976). In *The Molecular Biology of Hormone Action*, p. 107. Ed. by J. Papaconstantinou. Academic Press, New York
VREUGDENHIL, D., BURGERS, A. and LIBBENGA, K.R. (1979). *Plant Science Letters*, **16**, 115–121
WILLIAMS, M.A. (1969). In *Advances in Optical and Electron Microscopy*, pp. 219–272. Ed. by R. Barer and V.E. Cosslett. Academic Press, London

11
ETHYLENE BINDING IN *PHASEOLUS VULGARIS* L. COTYLEDONS

C.J. HOWARTH, A.R. SMITH and M.A. HALL
Department of Botany and Microbiology, University College of Wales, Aberystwyth, Dyfed, UK

Introduction

Previous work has established the existence of a binding site with a high affinity and specificity for ethylene in developing cotyledons of *Phaseolus vulgaris* (Bengochea *et al.*, 1980a,b). This binding occurs without modification of the ethylene (Jerie and Hall, 1978). The behaviour of the binding site suggests that it is proteinaceous in nature (Bengochea *et al.*, 1980a) and binding activity appears to be localized on the membranes of the endoplasmic reticulum and protein bodies (Evans *et al.*, 1982a,b).

It has proved possible to solubilize binding site activity using detergents (Thomas, Smith and Hall, 1984) and the work described here reports the further characterization and purification of this binding site.

Isolation of a cell free ethylene binding system

The characteristics of the membrane-bound ethylene binding system (EBS) from *Phaseolus vulgaris* L. cv Canadian Wonder are well established (Bengochea *et al.*, 1980a). It is possible to pellet the EBS from immature cotyledons of *P. vulgaris* by differential ultracentrifugation without loss of activity and subsequently to solubilize the EBS using the non-ionic detergents Triton X-100 and Triton X-114 (Thomas, Smith and Hall, 1984). Solubilization can also be achieved using 30 mM octyl glucoside in 0.02 M Tricine, 8% (w/v) sucrose pH 7. Detergent extraction does not affect greatly the affinity of the EBS for ethylene as shown in *Figure 11.1*, and the characteristics of the EBS in the presence and absence of Triton detergents are similar (Thomas, Smith and Hall, 1984). The K_D of the EBS for ethylene is of the order 10^{-9}–10^{-10} M. Detergent solubilization of an integral membrane protein such as the EBS permits the use of purification and characterization procedures possible with soluble proteins. However, it must be remembered that the characteristics obtained are that of a protein-detergent complex. Due to its extreme hydrophobicity, the binding protein precipitates in the absence of detergent.

Ethylene binding in Phaseolus vulgaris L. cotyledons

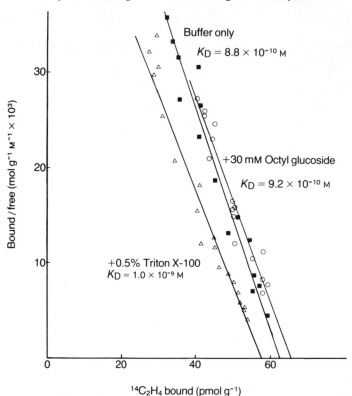

Figure 11.1 Scatchard plots of ethylene binding to cell-free preparations from *Phaseolus vulgaris* cotyledons solubilized in buffer (■), 0.5% Triton X-100 (△) or 30 mM octyl glucoside (○). Samples incubated with $^{14}C_2H_4$ at a range of concentrations (0.6–13.2 nM liquid phase) for 18 h

Effects of cycloalkenes on the ethylene binding site

Sisler and Pian (1973) reported that cycloalkenes were capable of counteracting an ethylene-induced rise in respiration in tobacco leaves and later found that the effect appeared to depend on competition for the EBS present in *Phaseolus aureus* (Beggs and Sisler, 1983). The potential of the use of such inhibitors in purifying the EBS by affinity chromatography has been suggested and in order to investigate this possibility the kinetic parameters of the effects of cycloalkenes on ethylene binding in *P. vulgaris* cotyledons have been investigated.

Cyclopentene, 2,5-norbornadiene, 1,3-cyclohexadiene and 1,3-cycloheptadiene all competitively inhibited binding of $^{14}C_2H_4$ in *P. vulgaris* cell-free preparations. A representative example of a Lineweaver–Burk plot showing inhibition of ethylene binding by 2,5-norbornadiene is presented in *Figure 11.2*. The concentrations of cycloalkenes used were in the range of 2000–3000 µl l^{-1} (gaseous phase). However, neither norbornene nor the saturated hydrocarbon, norbornane, at concentrations up to 10^3 µl l^{-1} in the gaseous phase had any effect on $^{14}C_2H_4$ binding.

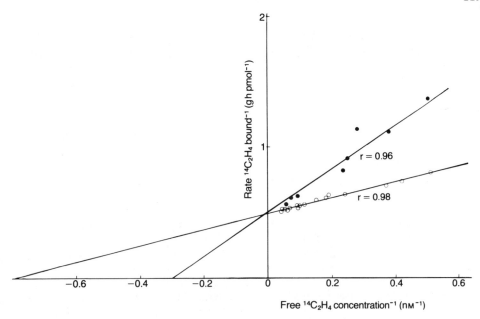

Figure 11.2 Lineweaver–Burk plots of ethylene binding to *Phaseolus vulgaris* cell-free preparations solubilized with 0.5% Triton X-100 showing competitive inhibition by 2,5-norbornadiene (2160 µl l^{-1} gas phase) at a range of concentrations of $^{14}C_2H_4$ (0.7–20.0 nM liquid phase) in the presence (●) or absence (○) of vapourized cycloalkene

Table 11.1 COMPARISON OF INHIBITOR CONSTANTS FOR CYCLOALKENES IN THE *PHASEOLUS VULGARIS* COTYLEDON CELL-FREE ETHYLENE BINDING SYSTEM

		Partition coefficient	K_i (gas) (µl l^{-1})	K_i (gas) relative	K_i (liquid) (µl l^{-1})	K_i (liquid) relative
Ethylene	$CH_2 = CH_2$	0.11	0.198	1	0.022	1
Cyclopentene		0.30	2077	10490	630	28606
2,5-Norbornadiene		0.98	2162	10920	2127	91670
1,3-Cyclohexadiene		1.43	3415	17250	3453	156975
1,3-Cycloheptadiene		1.70	4184	21130	7113	323310

Samples incubated with $^{14}C_2H_4$ at a concentration range of 0.6×10^{-9}–13.0×10^{-9}M (liquid phase) in the presence or absence of vapourized cycloalkene. K_i values calculated from Scatchard plots. The K_i (liquid) was calculated from the K_i (gas) using the partition coefficient of the inhibitor in the assay medium
Norbornane and norbornene over a concentration range of 4000–10000 µl l^{-1} (gas phase) did not inhibit ethylene binding

Table 11.1 shows the inhibitor constant (K_i) for each of the cycloalkenes tested in the *P. vulgaris* cotyledon cell-free binding system prior to detergent extraction. Each result is representative of at least three experiments with each analogue. The apparent dissociation constants in the absence and presence of the inhibitor (K_D and K_P respectively) were calculated from Scatchard plots of the data. The K_D obtained for ethylene binding in the *P. vulgaris* system was similar to those previously reported. *Table 11.1* also expresses the K_i values in the liquid phase. These were calculated using the respective partition coefficients and represent the concentrations in the immediate environment of the binding site. When the cell-free preparation was solubilized with Triton X-100 the K_i obtained for 2,5-norbornadiene was 1405 μl l^{-1} as opposed to 2162 μl l^{-1} in the absence of Triton X-100, indicating that the cycloalkene is more soluble in solutions containing Triton X-100. Even at a concentration of 8000 μl l^{-1}, 2,5-norbornadiene does not completely inhibit ethylene binding. At these concentrations of inhibitor the presence of any impurities must be considered as they themselves might be eliciting a response. The 2,5-norbornadiene used contains 2,6-ditert-butyl-4-methylphenol present as a stabilizer at a concentration of 0.05% and this substance was tested alone but had no effect on ethylene binding.

The K_i gives a measure of the affinity of the inhibitor for the binding site and thus indicates its effectiveness relative to ethylene. The closer the K_i is to the K_D for ethylene, the more effective will the competitor be in reducing binding of the true ligand to the binding site (Dodds and Hall, 1980). The K_is obtained for each of the cycloalkenes tested in the *P. vulgaris* system are all very high (*see Table 11.1*) and when they are compared to the K_D for ethylene binding their relative effectiveness as inhibitors can be seen to be very low. It would appear that none of the cycloalkenes tested would be of value in affinity columns for the purification of the ethylene binding site or in providing further information into the mechanism of ethylene action. If the K_i values of the appropriate analogues in the liquid phase are considered then their relative effectiveness as inhibitors is even less. These results differ to those previously reported (Beggs and Sisler, 1983) which, considering the gas phase values only, show much lower K_i values using a mungbean seed extract.

Purification of solubilized ethylene binding system by fast protein liquid chromatography

A number of approaches are possible to the problem of purifying the solubilized EBS whilst maintaining its ethylene binding activity. One of the most successful methods employed so far has been the use of Fast Protein Liquid Chromatography (FPLC) using a Pharmacia Mono Q HR 5/5 anion exchange column. The column was equilibrated with 0.02 M Tris pH 8.9, 0.5% Triton X-100 and elution was carried out using a 0–100% linear gradient of 0.35 M NaCl at a flow rate of 2 cm^3 min^{-1} and a pressure of 3 MPa. When required, desalting of fractions to remove NaCl was carried out using a Pharmacia PD 10 column prepacked with Sephadex G-25. The results from a typical separation of a Triton X-100 solubilized prelabelled 96 000 × *g* pellet are shown in *Figure 11.3*. This technique yields substantial purification (at least 11-fold) and has the added advantage of rapidity, high recovery and high resolution. With prelabelled samples, the distribution of radioactivity in the fractions eluted from the column can be monitored and two peaks are usually obtained. The first peak (fraction 1) elutes in the void volume

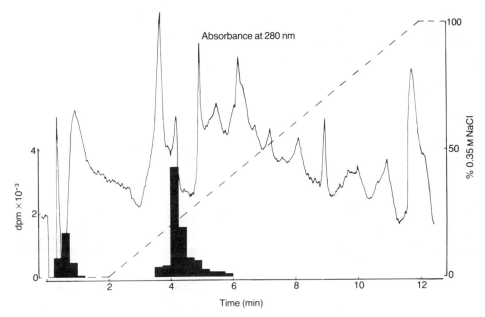

Figure 11.3 FPLC of Triton X-100 solubilized cell-free preparation on a Mono Q column. 0.5 cm^3 prelabelled sample (incubated for 18 h with 4.1×10^4 dpm cm^{-3} $^{14}C_2H_4$ liquid phase) was applied to the column and 0.5 cm^3 fractions collected. Protein (——); salt concentration (– – –); binding activity (■).

whereas the second (fraction 5) is eluted at a concentration of approximately 25% 0.35 M NaCl. If the protein content of the sample is increased, the proportion of ^{14}C in the first peak increases. This fraction can be rechromatographed on the Mono Q column where again two peaks of ^{14}C are obtained. Similarly, fraction 5, after desalting, can be rechromatographed on the Mono Q column. Although some of the ^{14}C elutes early, fraction 5 of this rechromatogram represents further purification of the EBS. Protein levels were monitored continuously at 280 nm but Triton X-100 interferes with ultraviolet spectroscopy and, in addition, the protein present in fractions collected was determined using a modified Lowry method (Markwell et

Table 11.2 SUMMARY OF PURIFICATION OF THE ETHYLENE BINDING SITE IN *P. VULGARIS* COTYLEDONS

	Specific activity (dpm mg^{-1} protein)	Relative to filtered homogenate	Relative to 96 000 g × pellet
Filtered homogenate	398	1	—
96 000 × g pellet	2694	7	1
FPLC Mono Q of 96 000 g × pellet: Fraction 5	29 824	75	11
Fraction 5 desalted and rechromatographed on FPLC Mono Q: Fraction 5	33 253	84	12
Triton X-114 partitioning of 96 000 g × pellet: detergent phase	11 188	28	4
FPLC Mono Q of detergent phase: Fraction 5	57 941	146	22

al., 1981). From these results it is possible to determine the specific activities of the applied sample and of the fractions eluted. *Table 11.2* indicates the purification achieved by these procedures.

The solubilized EBS has a very low rate constant of dissociation for ethylene (Thomas, Smith and Hall, 1984) and this property has made feasible the use of prelabelled samples. It is important also to be able to postlabel fractions to ensure that ethylene binding activity is retained. Optimum separation on the Mono Q column is obtained using Tris buffer at pH 8.9 but these are not optimum conditions for ethylene binding so buffer exchange into Tricine buffer at pH 6 is necessary to ensure that post-labelling of fractions will occur after chromatography on Mono Q. A similar distribution of $^{14}C_2H_4$ binding activity is found as in prelabelled samples.

Purification of ethylene binding system by detergent partitioning

Integral membrane proteins such as the EBS are extremely hydrophobic and this property can be exploited in purification procedures. One such method is detergent

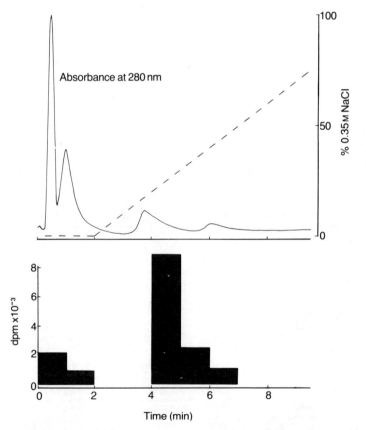

Figure 11.4 FPLC of Triton X-114 partitioned EBS. Sample prelabelled by incubating with 3.6×10^4 dpm cm^{-3} $^{14}C_2H_4$ (liquid phase) for 18 h before partitioning. Detergent phase diluted 1 in 4 in start buffer, 0.5 cm^3 applied to Mono Q column and 2 cm^3 fractions collected. Protein (——); salt concentration (– – –); binding activity (■)

partitioning using the non-ionic detergent Triton X-114 (Bordier, 1981). At 30 °C an aqueous solution of Triton X-114 separates into an aqueous phase and a detergent phase; hydrophilic proteins partition exclusively into the aqueous phase and hydrophobic, integral membrane proteins into the detergent phase. Thus, when a 96 000 × g pellet containing the EBS was solubilized with Triton X-114 in 0.02 M Tricine 50 mM NaCl pH7 and centrifuged at 10 000 × g for 30 min at 30°C, over 90% of the ethylene binding activity was recovered in the detergent phase. Most of the protein in the sample remained in the aqueous phase. This represents a fourfold increase in specific activity over an untreated sample (*Table 11.2*). It was possible to both prelabel and postlabel the detergent phase indicating that heating the EBS to 30 °C for 30 minutes had no effect either on binding activity or on the dissociation of ethylene already bound. A prelabelled detergent phase sample was subsequently chromatographed on the FPLC Mono Q column (*Figure 11.4*); over 60% of the ^{14}C present in the sample was recovered in fraction 5 but only 6% of the applied protein. This represents a 22-fold purification over the 96 000 × g pellet (*Table 11.2*) with 55% of the ethylene bound initially recovered in this fraction. This two-step procedure yields the greatest purification of the EBS so far achieved and work is now in progress to examine the purity of this fraction by electrophoresis.

References

BEGGS, M.J. and SISLER, E.C. (1983). *Plant Physiology*, **72**, S226
BENGOCHEA, T., DODDS, J.H., EVANS, D.E., JERIE, P.J., NIEPEL, B., SHAARI, A.R. and HALL, M.A. (1980a). *Planta*, **148**, 397–406
BENGOCHEA, T., ACASTER, M.A., DODDS, J.H., EVANS, D.E., JERIE, P.H. and HALL, M.A. (1980b). *Planta*, **148**, 407–411
BORDIER, C. (1981). *Journal of Biological Chemistry*, **256**, 1604–1607
DODDS, J.H. and HALL, M.A. (1980). *Science Progress*, **66**, 513–535
EVANS, D.E., BENGOCHEA, T., CAIRNS, A.J., DODDS, J.H. and HALL, M.A. (1982a). *Plant, Cell and Environment*, **5**, 101–107
EVANS, D.E., BENGOCHEA, T., CAIRNS, A.J., DODDS, J.H. and HALL, M.A. (1982b). *Planta*, **154**, 48–52
JERIE, P.H. and HALL, M.A. (1978). *Proceedings of the Royal Society London. Series B*, **200**, 87–94
MARKWELL, M.A.K., HAAS, S.M., TOLBERT, N.E. and BIEBER, L.L. (1981). *Methods in Enzymology*, **72**, 296–303
SISLER, E.C. and PIAN, A. (1973). *Tobacco Science*, **17**, 68–72
THOMAS, C.J.R., SMITH, A.R. and HALL, M.A. (1984). *Planta*, **160**, 474–479

12

ETHYLENE METABOLISM

E.M. BEYER, JR.
E. I. du Pont de Nemours & Co., Inc., Agricultural Chemicals Department, Building 402, Experimental Station, Wilmington, Delaware, USA

Introduction

The early scepticism concerning the metabolism of ethylene in plants was overturned in 1975 with the discovery that peas (*Pisum sativum*) oxidize ethylene to simple oxidation products such as ethylene oxide, ethylene glycol and CO_2 (Beyer, 1975a). Today this metabolism is recognized as an important part of the biochemistry of ethylene. The following working model (*Figure 12.1*), involving a monooxygenase, has been suggested based on $^{18}O_2$ studies and ethylene metabolites identified in peas and *Vicia faba* cotyledons (Blomstrom and Beyer, 1980; Dodds and Hall, 1982).

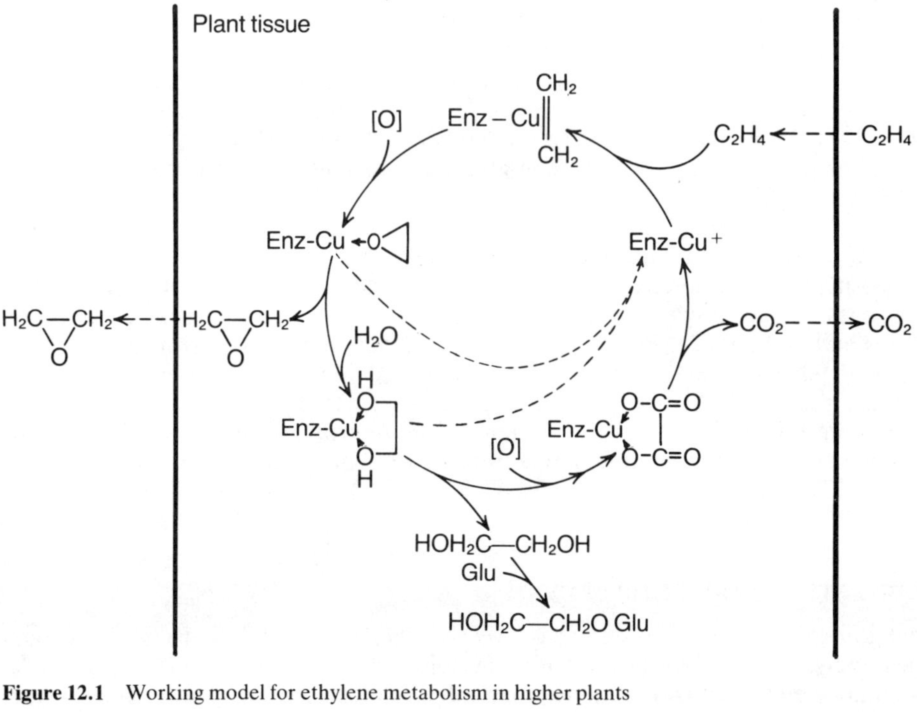

Figure 12.1 Working model for ethylene metabolism in higher plants

Some tissues such as peas produce both ethylene oxide and CO_2 (Beyer, 1981). Others produce predominantly CO_2 (Beyer, 1980) while some, like *Vicia faba* cotyledons, produce ethylene oxide but little or no CO_2 (Beyer, 1980). In addition to these gaseous products, free ethylene glycol and its glucose conjugate are also formed plus oxalate and a variety of other metabolites, many of which have not yet been identified (Giaquinta and Beyer, 1977; Blomstrom and Beyer, 1980; Dodds and Hall, 1982). Based principally on studies in non-biological systems (Jerie and Hall, 1978; Thompson, Harlow and Whitney, 1983), copper (Cu^+) is thought to be involved in ethylene coordination and oxidation (Beyer, 1980, 1981; Beyer and Blomstrom, 1980).

To date, over a dozen tissues have been reported to oxidize ethylene. The type and distribution of products formed depend on several factors including the tissue, growth stage and exposure conditions. The apparent widespread occurrence of this hormone metabolic system in plants coupled with its apparent high degree of regulation suggests that it may have a physiological role. This chapter highlights some of the more important features of this system and examines its possible physiological function in plants.

Key features of ethylene metabolism

PRODUCTS OF ETHYLENE METABOLISM

Except for ethylene oxide and CO_2, positive identification of the products of ethylene metabolism has been limited to studies with dark grown pea seedlings (Giaquinta and Beyer, 1977; Blomstrom and Beyer, 1980) and *Vicia faba* cotyledons (Dodds and Hall, 1982). In peas 60–70% of the ^{14}C-tissue metabolites derived from $^{14}C_2H_4$ were found to be water-soluble after a 24 h exposure period with lesser amounts in the protein (10–15%), lipid (1%) and insoluble (1–2%) fractions. Ion exchange chromatography of the water-soluble ^{14}C-metabolites into basic, neutral and acidic fractions revealed a 50:40:10 distribution pattern, respectively. So far, only two of these components have been positively identified. They are the neutral metabolites, ethylene glycol and its glucose conjugate (Blomstrom and Beyer, 1980).

In the *Vicia faba* cotyledonary system (Jerie and Hall, 1978), which metabolizes ethylene much more rapidly than do peas and most other tissues, ethylene oxide was positively identified as the major gaseous metabolite. Significantly, and unlike peas, little or no ethylene was metabolized to CO_2 in this system (Beyer, 1980; Dodds and Hall, 1982). Refluxing the *Vicia* cotyledonary tissue with 95% ethanol removed 90% of the ^{14}C-tissue metabolites and ion-exchange chromatography revealed that 65–75% was present in the basic fraction, 15–25% in the neutral fraction and 5–10% in the acidic fraction. The neutral fraction contained mainly ethylene glycol while ethanolamine represented more than 60% of the basic fraction. The acidic fraction contained significant amounts of oxalate and glycollate.

SENSITIVITY TO VARIOUS TREATMENTS

Heat killing, homogenization, anaerobiosis, chelation and inhibitors of various types (CS_2, COS, 2,4-dinitrophenol, cycloheximide, sodium cyanide, N-ethylmaleimide) either severely inhibit or completely block metabolism *in vivo* (Beyer

Table 12.1 EFFECT OF TEMPERATURE ON ETHYLENE METABOLISM[a]

Metabolism rate (DPM 0.1 g dry wt^{-1} 20 h^{-1})	Exposure temperature 18 °C	28 °C	Increase
Gaseous metabolites (recovered in NaOH trap)[b]	4160	14280	3.4
Tissue metabolites (recovered in tissue after drying)	4010	5035	1.3

[a]Excised 10 mm pea tips from four-day-old etiolated seedlings were equilibrated for 1 h at the two temperatures and then exposed to 4.0 µl l^{-1} of $^{14}C_2H_4$ (110 mCi mmol^{-1}) for 24 h.
[b]In addition to $^{14}CO_2$, ethylene oxide is also trapped in NaOH since upon coming into contact with alkali it is rapidly hydrolysed to ethylene glycol. Dissolved $^{14}C_2H_4$ is readily removed from the alkaline solution with an air stream while $^{14}CO_2$ can be released with acid. Neither treatment, however, removes ^{14}C-ethylene glycol

and Blomstrom, 1980). Temperature markedly affects ethylene metabolism. Over the temperature range of 18–28 °C the ^{14}C-gaseous metabolites trapped in alkali were found to increase by 3.4-fold (*Table 12.1* and see Beyer and Blomstrom, 1980). This was accompanied by a much smaller increase in ^{14}C-tissue metabolites. Interestingly, simple detachment of the cotyledons from the root-shoot axis of pea seedlings dramatically reduces ethylene metabolism without altering natural ethylene production or respiration (Beyer, 1975b). Collectively, these results demonstrate that ethylene metabolism is a natural metabolic function of healthy tissue.

CHANGES DURING GROWTH AND DEVELOPMENT

In all cases examined so far, ethylene metabolism has been found to vary markedly during growth and development. Recent examination of changes in ethylene metabolism during the first five days of seedling growth in alfalfa (*Medicago sativa*), wheat (*Triticum aestivum*), soybeans (*Glycine max*), cucumber (*Cucumis sativus*), Black Valentine beans (*Phaseolus vulgaris*), Canadian Wonder beans (*Phaseolus vulgaris*) and cotton (*Gossypium hirsutum*) has shown that ethylene metabolism increases during the first few days following imbibition and then eventually declines. This is seen in *Figure 12.2* for alfalfa seedlings exposed to 5 µl l^{-1} of $^{14}C_2H_4$ (110 mCi mmol^{-1}). These seedlings have recently been found to have the highest rates of ethylene metabolism so far observed. In comparative tests, peak rates on a dry weight basis were even higher than those observed in *Vicia* cotyledons (data not shown). In alfalfa and *Vicia*, metabolism is so rapid that radioactive ethylene is not even needed for detecting it. Metabolism can be detected by monitoring the disappearance of non-labelled ethylene using standard GC techniques.

In addition to the seedlings mentioned above, fluctuations in ethylene metabolism have also been observed in peas (Beyer, 1975a,b), ripening tomato fruit (Beyer and Blomstrom, 1980), cotton (Beyer, 1979a) and bean abscission zones (unpublished data, Beyer) and senescing carnation (Beyer, 1977) and morning glory flowers (Beyer, 1978a). Peaks in ethylene metabolism are often associated with similar peaks in natural ethylene production but rarely do the two exactly coincide. For example, in pea seedlings (Beyer, 1975a) natural ethylene production reached a peak on day 2 following imbibition whereas ethylene metabolism reached a peak on day 3. In carnation flowers (Beyer, 1977) the surge in natural ethylene production occurred on day 7 after placing buds in the tight bud stage in water

128 *Ethylene metabolism*

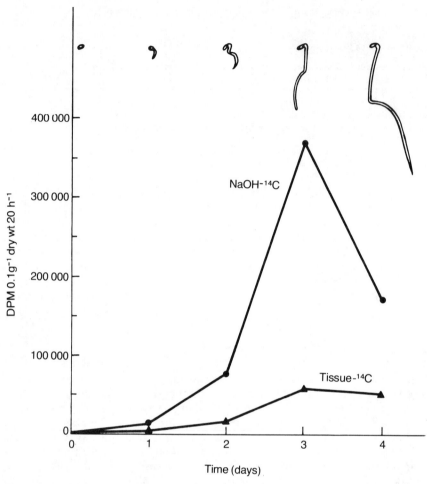

Figure 12.2 Ethylene metabolism in alfalfa seedlings

whereas tissue incorporation of ethylene in the reproductive and receptacle tissues peaked on day 5 and oxidation of ethylene to carbon dioxide peaked on day 10.

STRUCTURAL ANALOGUES

Alkanes (e.g. methane, ethane, propane) are not metabolized and they do not compete with ethylene for the ethylene metabolism site (Dodds, Heslop-Harrison and Hall, 1980). In contrast, alkenes, alkynes, vinyl halides and ethers, carbon monoxide and carbon dioxide inhibit ethylene metabolism to varying degrees. Some have been found to be metabolized by the same system and to display classical competitive inhibitor kinetics (Beyer, 1978b; Dodds, Heslop-Harrison and Hall, 1980; Dodds and Hall, 1982). For example, like ethylene, propylene is metabolized by pea seedlings to propylene glycol (Beyer, 1978b). The rate of metabolism, however, is much faster and the distribution of products is quite different. A relatively good correlation has been found between the effectiveness of these analogues to compete with ethylene in metabolism and their ability to mimic ethylene action (Dodds, Heslop-Harrison and Hall, 1980).

KINETIC PARAMETERS

A comparison of the K_ms for ethylene metabolism in *Vicia faba* and peas indicates the values differ by four orders of magnitude thus explaining the much higher rate of ethylene metabolism in *Vicia* (Hall *et al.*, 1982; Evans *et al.*, 1984). Surprisingly, the V_{max}s for ethylene in both tissues are very similar. Even more surprising is that the affinity of the *Vicia* and pea systems for propylene are very similar. This much higher affinity of the pea system for propylene as compared to ethylene accounts for its more rapid metabolism in this tissue.

EFFECT OF ETHYLENE OXIDE AND GLYCOL ON METABOLISM

When ethylene oxide and ethylene glycol are added exogenously to excised pea tips even at the very high concentrations of 500 µl l^{-1} and 10^{-3} M, respectively, the rate of $^{14}C_2H_4$ metabolism is unaltered (*Table 12.2*). Moreover, when similar tissue was

Table 12.2 EFFECT OF ETHYLENE OXIDE AND ETHYLENE GLYCOL ON ETHYLENE METABOLISM[a]

Treatment	^{14}C-Gaseous metabolites (recovered in NaOH trap)	^{14}C-Metabolites (recovered in tissue after drying)
	(DPM 0.1 g dry wt^{-1} 20 h^{-1})	
$^{14}C_2H_4$ alone	5630	2812
$^{14}C_2H_4$ + ethylene oxide (500 µl l^{-1})	5592	2788
$^{14}C_2H_4$ + ethylene glycol (10^{-3} M)	5602	2875

[a]Excised 10 mm pea tips from five-day-old etiolated seedlings were exposed to 2.0 µl l^{-1} of $^{14}C_2H_4$ (110 mCi mmol^{-1}) with and without ethylene oxide and ethylene glycol for 20 h. Distilled water was allowed to equilibrate with 500 µl l^{-1} of ethylene oxide overnight and 3 ml were added to sealed, 62 ml incubation flasks (Beyer, 1975b) for ethylene oxide treatments. All gas concentrations were determined at the end of the experiment by GC analysis

exposed to ^{14}C-ethylene oxide, insignificant amounts of ^{14}C-ethylene glycol and $^{14}CO_2$ were formed as compared to when an equivalent amount of $^{14}C_2H_4$ was applied. In view of the general reactivity of ethylene oxide, this was somewhat surprising. These results strongly suggest that ethylene glycol and $^{14}CO_2$ are only formed in significant amounts in a tightly coupled system which starts with ethylene (*see Figure 12.1*). Once released, ethylene oxide and ethylene glycol do not readily re-enter the oxidation sequence.

In contrast, application of ^{14}C-ethylene glycol to pea tissue results in the rapid formation of its glucose conjugate. Thus, the glycol conjugation reaction does not appear to be tightly coupled to the overall ethylene oxidation process.

IN VITRO ACTIVITY

An *in vitro* system capable of oxidizing ethylene to ethylene oxide has been isolated from *Vicia faba* cotyledons (Dodds *et al.*, 1979; Dodds and Hall, 1982). Like the *in vivo* system, ethylene is not converted to CO_2. Active preparations are particulate with more than half the activity sedimenting after 45 min between 3000 and 10 000 × g. Only recently have *in vitro* rates been obtained which approximate those observed *in vivo* (M.A. Hall, personal communication). Various protein reactive reagents (dithiothreitol, dithioerythritol and *p*-chloromercuribenzoate) even at μM concentrations significantly inhibit oxidizing activity (Dodds, Heslop-Harrison and Hall, 1980). This system has a requirement for molecular oxygen, a high affinity for ethylene (K_m of 4.2×10^{-10} M or $0.1\,\mu l\,l^{-1}$ in the gas phase) and is destroyed by heating to 100°C for 5 min. It was suggested that an oxygenase is involved and this is supported by the recent findings that ^{18}O labelled ethylene oxide is formed when *Vicia faba* cotyledons are exposed to ethylene in the presence of $^{18}O_2$ (Beyer, Morgan and Yang, 1984; Beyer, unpublished data).

Physiological role

Assigning a physiological role to ethylene metabolism is a major challenge facing current and future research in this area. It is counterproductive to suggest at this time that ethylene metabolism serves no useful purpose. Not only does this seem highly unlikely but even if true other possibilities must first be eliminated. Based on current data, the following three roles for ethylene metabolism in plants need to be carefully examined. First, ethylene metabolism may serve to reduce endogenous ethylene levels; secondly, the oxidation of ethylene at the hormone binding site may be a requirement for ethylene action; and thirdly, the product(s) of ethylene oxidation may serve to alter tissue sensitivity to ethylene. The evidence for and against these three possibilities is briefly reviewed.

ETHYLENE REMOVAL

There is one recent report which provides some evidence in favour of this idea (Dodds, Smith and Hall, 1982, 1983). This study examined the relationship between resistance to waterlogging and ethylene metabolism in 21 different cultivars of *Vicia faba*. Some correlation was found to exist between the ability of *Vicia faba* cultivars to withstand waterlogging and to metabolize ethylene in the cotyledonary tissue. It was suggested that resistance to waterlogging may be related to the plant's ability to effectively metabolize (remove) the increased amounts of ethylene produced in response to waterlogging. Unfortunately, measurements of ethylene oxidation to ethylene oxide were made only in the cotyledons. Thus, further work is needed to determine if the greater capacity of resistant cultivars to metabolize ethylene also extends to roots and shoots where ethylene removal would be especially critical.

There are several reasons for believing that ethylene metabolism does not generally represent a system for regulating endogenous ethylene levels. First, the total amount of ethylene metabolized appears too small to constitute an effective removal system (Beyer, 1980, 1981; Beyer and Blomstrom, 1980). However, in

Table 12.3 RATES OF ETHYLENE PRODUCTION AND ETHYLENE METABOLISM[a]

Tissue	Production rate	Metabolism rate	Amount metabolized (%)
	(nl g fresh wt^{-1} h^{-1})		
Pea seedlings	0.6	0.06	10
Morning glory flowers	8.0	0.02	0.3
Cotton abscission zones	0.4	0.02	5.0
Vicia faba seedlings	0.8	0.7	88
Alfalfa seedlings	2.0	1.7	85
Soybean seedlings	0.5	0.01	2
Cotton seedlings	0.5	0.01	2
Wheat seedlings	0.3	0.01	3
Black Valentine beans	1.8	0.02	1
Canadian Wonder beans	1.0	0.02	2

[a]Metabolism rates represent total metabolism but at elevated $^{14}C_2H_4$ concentrations ($\geqslant 5.0\,\mu l\,l^{-1}$). Since endogenous ethylene levels would normally be much less than $5\,\mu l\,l^{-1}$ in these tissues these metabolism rates overestimate this component. Thus, the percent of the natural ethylene normally metabolized is probably significantly less than that shown in the last column

those tissues where the rates of metabolism are exceptionally high, such as in alfalfa seedlings (*Figure 12.2*) and *Vicia faba* cotyledons (Jerie and Hall, 1978), this needs to be examined carefully. Typically, the amount of natural ethylene metabolized is estimated to be 10% or less (*Table 12.3*). The only exceptions so far observed are *Vicia* and alfalfa seedlings. In some tissues, such as morning glory flowers, it is only about 0.3% or less even during periods of maximum production (*see also* Beyer, 1978a). Experiments with CS_2 support this idea (*Table 12.4*). For example, CS_2 at $10\,\mu l\,l^{-1}$ markedly inhibited ethylene metabolism without affecting ethylene production. If metabolism were removing a significant amount of the ethylene being produced, such treatments should result in an apparent increase in the ethylene production rate. That this does not occur indicates that ethylene metabolism does not normally remove a significant amount of natural ethylene in peas and probably most other tissues.

Secondly, ethylene production rates and periods of peak production do not always correlate well with ethylene metabolism. Such a relationship would be expected if the purpose of ethylene metabolism were to reduce ethylene tissue levels.

Thirdly, ethylene readily diffuses out of most tissues. Thus, regulation of ethylene biosynthesis provides an effective means of controlling ethylene tissue levels making an inactivation system appear unnecessary.

Table 12.4 EFFECT OF CARBON DISULPHIDE (CS_2) ON ETHYLENE PRODUCTION AND METABOLISM IN PEAS[a]

Rate (nl g fresh wt^{-1} h^{-1})	Treatment		% Inhibition
	None	CS_2 ($10\,\mu l\,l^{-1}$)	
Ethylene production	0.48	0.44	8
Ethylene metabolism	0.11	0.02	82

[a]$^{14}C_2H_4$ ($110\,mCi\,mmol^{-1}$) was applied at $15\,\mu l\,l^{-1}$ accounting for the higher rate than shown in *Table 12.3*. Intact dark grown pea seedlings were two days old at the time of treatment

METABOLISM—ACTION HYPOTHESIS

It has been suggested that ethylene metabolism may be required for ethylene to carry out its biological function (Beyer, 1975b and 1979b). The principal arguments against this idea are the following. First, ethylene action and metabolism do not have similar dose response curves. Most ethylene responses saturate at ethylene concentrations between 1 and $10\,\mu l\,l^{-1}$. In marked contrast, the rate of ethylene metabolism increases fairly linearly up to $100\,\mu l\,l^{-1}$ (Beyer, 1975b). This means that, if ethylene action and metabolism were related, then some step beyond the initial ethylene oxidation step(s) would have to become rate limiting above 1–$10\,\mu l\,l^{-1}$. Alternatively, if the ratio of ^{14}C-tissue metabolites to ^{14}C-gaseous products produced were critical, this would explain the apparent anomaly (Beyer, 1979b). This ratio rapidly changes and even reverses over the concentration range of 0.1–$1.0\,\mu l\,l^{-1}$ (Beyer, 1975b).

Secondly, in peas propylene is metabolized much more rapidly than ethylene yet it is about 100 times less effective than ethylene in inducing characteristic ethylene responses (*Table 12.5* and Beyer, 1978b). Several major differences do exist

Table 12.5 ETHYLENE VERSUS PROPYLENE METABOLISM IN PEAS

^{14}C-Olefin[a] ($1\,\mu l\,l^{-1}$)	Total metabolism (nmol g dry wt^{-1} 24 h^{-1})
Ethylene	0.094
Propylene	3.04

[a]Initial amount of olefin added was 6 nmol g^{-1} dry wt

however between ethylene and propylene metabolism. Most notably, the products of propylene metabolism are mainly three carbon compounds (propylene oxide, propylene glycol and its conjugate) and the distribution of products is very different from that observed with ethylene.

Thirdly, while the affinity of the ethylene metabolism system in *Vicia* corresponds well with concentrations required to induce a half-maximal response in the pea-stem growth assay ($K_D = 4.2 \times 10^{-10}$ M or $0.1\,\mu l\,l^{-1}$ in gas phase), the affinity of the system in peas is several orders of magnitude lower (Hall *et al.*, 1982; Hall, 1983). This discrepancy is difficult to reconcile unless ethylene metabolism were to alter ethylene tissue sensitivity (*see below*) or receptor affinity. It is puzzling that the affinity for ethylene metabolism in peas should differ so markedly from that in *Vicia*, whereas the affinity for propylene is similar in both tissues. Equally puzzling is the lack of the ethylene to CO_2 forming system in *Vicia*.

Countering the evidence against the ethylene metabolism–action hypothesis is a substantial amount of data which suggests that such a relationship might exist. Firstly, and perhaps most noteworthy, is the quantitative relationship that has been demonstrated to exist between the effects of the powerful ethylene antagonist, Ag^+, on ethylene action and metabolism (*Figure 12.3* and Beyer, 1979b). In peas Ag^+ reduced ethylene-induced growth inhibition and metabolism (i.e. in terms of the ^{14}C-metabolites recovered in the tissue) in a remarkably parallel fashion. The inhibitory effect of Ag^+ on ethylene metabolism was recently confirmed in the same tissue (Evans *et al.*, 1984). Secondly, the other antagonist of ethylene action, CO_2, also modifies ethylene metabolism (Beyer, 1979b, Evans *et al.*, 1984).

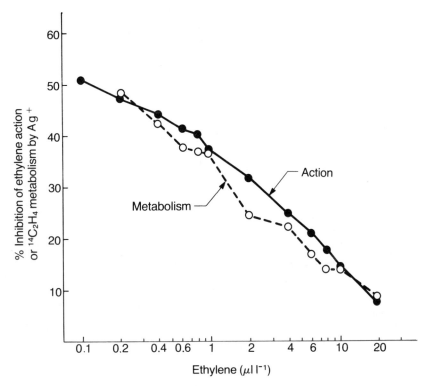

Figure 12.3 Effect of Ag^+ on ethylene metabolism and action in peas. Four-day-old etiolated seedlings were sprayed with either 100 mg l^{-1} of $AgNO_3$ in distilled water or with distilled water only. After two days of growth in ethylene-free air or various concentrations of ethylene epicotyl growth was determined and the percent reduction in ethylene induced growth inhibition caused by Ag^+ calculated. Ethylene metabolism was determined by excising 10 mm pea epicotyl tips from four-day-old seedlings of the same planting treated with water or 100 mg l^{-1} $AgNO_3$ and exposing them in sealed flasks to various concentrations of $^{14}C_2H_4$

Increasing the ambient CO_2 concentration from 4 to 10% has been found to inhibit the conversion of ethylene to CO_2 (*Figure 12.4*). Interestingly, the ^{14}C-metabolites recovered from the tissue were unaffected up to 5% CO_2. Above this level there was a marked stimulation.

These studies, plus those with Ag^+, indicate that certain aspects of ethylene metabolism can be inhibited by two of the most potent, specific known inhibitors of ethylene action. The significance of the ability of Ag^+ and CO_2 to inhibit different parts of the ethylene metabolic system is not known but could explain the greater potency and non-competitive nature of the Ag^+ effect. The ability of CO_2 to either increase or decrease certain aspects of ethylene metabolism is particularly interesting since CO_2 is known to inhibit ethylene action in some situations and to mimic or enhance it in others (Abeles, 1973).

Thirdly, a number of correlations have been found to exist between changes in tissue sensitivity to ethylene and changes in ethylene metabolism. Morning glory flower buds which do not respond to ethylene also have been found to lack the ability to metabolize ethylene (Beyer, 1978a). However, just as the buds become

134 Ethylene metabolism

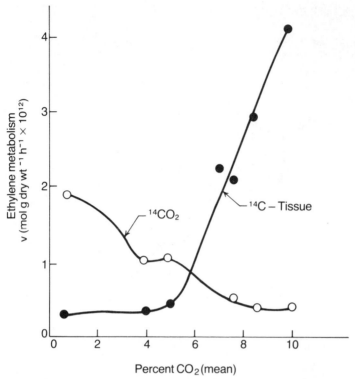

Figure 12.4 Effect of CO_2 on ethylene metabolism in etiolated pea (from Evans et al., 1984)

responsive, this ability appears and then increases rapidly. Similar parallel relationships have been noted in abscission zone tissue during cotton leaf abscission (Beyer, 1979a). A constant rate of ethylene metabolism was observed in the separation zone tissue prior to abscission induction. Induction of abscission by deblading resulted in over a sixfold increase in metabolism and this increase preceded by one day the first signs of abscission. Hormone treatments that delayed or stimulated abscission had a parallel effect on ethylene metabolism.

ALTERED TISSUE SENSITIVITY

Since ethylene oxide is central to the overall ethylene oxidation process, its effects on ethylene metabolism (*see above*) and ethylene action are of considerable interest. In contrast to early work (Abeles, 1973) which suggested that ethylene oxide may be an ethylene antagonist, recent studies (Beyer, 1980; Beyer, unpublished data) clearly demonstrate that the opposite is true, i.e. ethylene oxide increases tissue sensitivity to ethylene.

Ethylene oxide will not mimic ethylene when applied alone. At high concentrations ($>250\,\mu l\,l^{-1}$) it typically arrests growth and development. Visual toxic symptoms are generally lacking unless very high concentrations are applied ($>1000\,\mu l\,l^{-1}$). Undoubtedly, the earlier reports suggesting that ethylene oxide blocks ethylene action in flowers and fruits were due to a general suppression of cellular metabolism resulting from the high ethylene oxide concentrations employed (500–7500 µl l^{-1}). These concentrations tend to 'freeze' plant metabolic functions making normal responses to ethylene impossible.

Three different systems have been used to demonstrate the synergistic effect of ethylene oxide, namely the 'triple response' in peas, leaf abscission in cotton and growth stimulation in rice. Importantly, ethylene oxide does not stimulate ethylene production in these tissues. In dark grown pea seedlings all of the characteristic effects of ethylene were enhanced when ethylene oxide (28–220 µl l^{-1}) was combined with ethylene (0.25 µl l^{-1}) in a continuous gas-flow system (*Figure 12.5*).

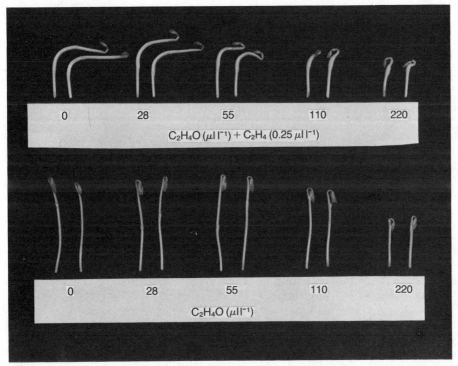

Figure 12.5 Effect of ethylene oxide alone and in combination with ethylene on etiolated pea seedling growth

Seedlings were three days old at the time of exposure and were treated continuously for three days. Concentrations of ethylene oxide of 55 µl l^{-1} enhanced the 'triple response'. Note that when 55 µl l^{-1} of ethylene oxide was added alone there was no growth inhibition. Ethylene oxide reduced growth above 55 µl l^{-1} but without added ethylene did not promote stem swelling. Similarly, in cotton abscission (*Table 12.6*), ethylene oxide (200 µl l^{-1}) in combination with ethylene (6.0 µl l^{-1}) significantly accelerated abscission. With rice seedling (*Table 12.6*) ethylene oxide (3 µl l^{-1}) was found to significantly increase the effectiveness of ethylene (7.0 µl l^{-1}) in stimulating rice coleoptile growth.

These and other similar studies raise the question of whether or not the products of ethylene metabolism such as ethylene oxide might modulate tissue sensitivity to ethylene. Clearly, the levels of ethylene oxide used in most studies exceed those

Table 12.6 SYNERGISTIC EFFECTS OF ETHYLENE OXIDE WITH ETHYLENE IN COTTON ABSCISSION AND STIMULATION OF RICE COLEOPTILE GROWTH[a]

Cotton abscission	Percent abscission (day 5)
Air	0
C_2H_4O (200 µl l^{-1})	0
C_2H_4 (6.0 µl l^{-1})	25
C_2H_4 (6.0 µl l^{-1}) + C_2H_4O (200 µl l^{-1})	76

Rice coleoptile growth	Percent increase in coleoptile growth over air controls
C_2H_4O (3 µl l^{-1})	5
C_2H_4 (7 µl l^{-1})	98
C_2H_4O (3 µl l^{-1}) + C_2H_4 (7 µl l^{-1})	145

[a]Cotton plants (Stoneville 213) were four weeks of age at the time of treatment. Rice seedlings (NATO) were three days old at the time of treatment. One ml of water was added to all 62 ml treatment flasks. In the case of ethylene oxide treatments, the water was taken from a container equilibrated with 3 µl l^{-1} C_2H_4O overnight

considered to be physiological based on rates of ethylene metabolism in most tissues. However, this possibility cannot be ruled out since ethylene oxide or other metabolites formed *in vivo* may be much more effective than when applied exogenously. It is known for example that in peas ethylene oxide is not readily converted to ethylene glycol or CO_2 when applied exogenously, yet *in vivo* both are readily formed provided the precursor is ethylene.

BURGS' ACTION HYPOTHESIS

In addition to the various relationships mentioned above, it should be noted that the scheme proposed by Burg and Burg in 1965 for ethylene action (Burg and Burg, 1965) has certain elements remarkably similar to the ethylene metabolism–action hypothesis. Based on their mechanism of action studies, they proposed that ethylene action involves the binding of ethylene and oxygen to the metal of a metalloenzyme. Ethylene was viewed strictly as a dissociable activator molecule. If this scheme were modified slightly so that ethylene reacted with the oxygen to form ethylene oxide, thereby initiating the ethylene oxidation sequence, the basic concepts that have emerged from work on ethylene metabolism and ethylene action would be reconciled into one scheme. Interestingly, the resulting enzyme proposed by the Burgs' would be the mono-oxygenase involved in ethylene metabolism.

Conclusion

Ethylene metabolism in plants has several possible physiological functions. There are data for and against each possibility. It would be premature at this time to discard any of them. Ethylene metabolism should be viewed as a relatively new doorway through which future investigations might shed light on the regulatory nature of ethylene. This is also true for ethylene binding (Hall, 1983). It is the view of the author that at some point these two lines of investigation will merge and provide new insight into the mechanism by which ethylene so profoundly affects plant growth and development.

References

ABELES, F.B. (1973). *Ethylene in Plant Biology*, pp. 302. Academic Press, New York
BEYER, E.M., Jr. (1975a). *Nature*, **255**, 144–147
BEYER, E.M., Jr. (1975b). *Plant Physiology*, **56**, 273–278
BEYER, E.M., Jr. (1977). *Plant Physiology*, **60**, 203–206
BEYER, E.M., Jr. (1978a). *Plant Physiology*, **61**, 896–899
BEYER, E.M., Jr. (1978b). *Plant Physiology*, **61**, 893–895
BEYER, E.M., Jr. (1979a). *Plant Physiology*, **64**, 971–974
BEYER, E.M., Jr. (1979b). *Plant Physiology*, **63**, 169–173
BEYER, E.M., Jr. (1980). In *Aspects and Prospects of Plant Growth Regulators*, DPGRG/BPGRG. Monograph 6, pp. 27–38
BEYER, E.M., Jr. (1981). In *Recent Advances in the Biochemistry of Fruits and Vegetables*, pp. 107–121
BEYER, E.M., Jr. and BLOMSTROM, D.C. (1980). *Proceedings of the Tenth International Conference on Plant Growth Substances (1979)*, pp. 208–218. Ed. by F. Skoog. Springer-Verlag, Berlin
BEYER, E.M., Jr., MORGAN, P.W. and YANG, S.F. (1984). In *Advanced Plant Physiology*, pp. 111–126. Ed. by M.B. Wilkins. Pitman Books, London
BLOMSTROM, D.C. and BEYER, E.M., Jr. (1980). *Nature*, **283**, 66–68
BURG, S.P. and BURG, E.A. (1965). *Science*, **148**, 1190–1196
BUXTON, G.V., GREEN, J.C. and SELLERS, R.M. (1976). *Journal of the Chemical Society, Dalton Transactions*, 2160–2165
DODDS, J.H. and HALL, M.A. (1982). *International Reviews of Cytology*, **76**, 299–325
DODDS, J.H., HESLOP-HARRISON, J.S. and HALL, M.A. (1980). *Plant Science Letters*, **19**, 175–180
DODDS, J.H., MUSA, S.K., JERIE, P.H. and HALL, M.A. (1979). *Plant Science Letters*, **17**, 109–114
DODDS, J.H., SMITH, A.R. and HALL, M.A. (1982/83). *Plant Growth Regulation*, **1**, 203–207
EVANS, D.E., SMITH, A.R., TAYLOR, J.E. and HALL, M.A. (1984). *Plant Growth Regulation*, **2**, 187–196
GIAQUINTA, R.T. and BEYER, E.M., Jr. (1977). *Plant and Cell Physiology*, **18**, 141–148
HALL, M.A. (1983). In *Receptors in Plants and Cellular Slime Moulds*. Ed. by C.M. Chadwick and D.R. Garrod. Marcel Dekker, New York (in press)
HALL, M.A., EVANS, D.E., SMITH, A.R., TAYLOR, J.E. and AL-MUTAWA, M.M.A. (1982). In *Growth Regulators in Plant Senescence*, British Plant Growth Regulator Group, Monograph 8, pp. 103–111
JERIE, P.H. and HALL, M.A. (1978). *Proceedings of the Royal Society, London*, B, **200**, 87–94
THOMPSON, J.S., HARLOW, R.L. and WHITNEY, J.F. (1983). *Journal of the American Chemical Society*, **105**, 3522–3527

13

ETHYLENE METABOLISM IN *PISUM SATIVUM* L. AND *VICIA FABA* L

A.R. SMITH, D.E. EVANS, P.G. SMITH and M.A. HALL
Department of Botany and Microbiology, University College of Wales, Aberystwyth, Dyfed, UK

Introduction

Until relatively recently the consensus of opinion was that higher plants did not metabolize ethylene. However, much evidence has accumulated over the past ten years that plant tissues do have the ability to metabolize ethylene and this work has been comprehensively reviewed by Beyer (Chapter 12).

Dark grown seedlings of pea (*Pisum sativum* L. cv. Alaska) metabolize ethylene to tissue incorporated material and NaOH-soluble volatile compounds (Beyer, 1975). The products of tissue incorporation are predominantly ethylene glycol and its glucose conjugate, while the NaOH-soluble product of oxidation has been identified as CO_2 (Blomstrom and Beyer, 1980). Rates of metabolism at physiological concentrations of ethylene are apparently very low in this tissue, hence studies to date have been performed *in vivo*. Another ethylene metabolizing system has been demonstrated in developing cotyledons of *Vicia faba* (Jerie and Hall, 1978; Dodds *et al.*, 1979) but this system differs in two major ways from that in *Pisum*. Firstly, ethylene is metabolized very rapidly in *Vicia* at physiological concentrations, and secondly only one primary product is formed, namely ethylene oxide. However, in *Pisum* seedlings exposed to ^{14}C-ethylene, only part of the radioactivity trapped in the NaOH is released upon acidification the remainder proving to be ethylene glycol. Since ethylene oxide is readily converted to ethylene glycol by strong base, it seems that in *Pisum* as in *Vicia* ethylene oxide may be the primary product of ethylene metabolism.

Data on the kinetics of ethylene metabolism in *Vicia* have previously been reported. These have shown that the relative effectiveness of structural analogues in inhibiting ethylene metabolism closely parallels their relative effectiveness in mimicking developmental responses to ethylene (Dodds, Heslop-Harrison and Hall, 1980). Due to the high level of ethylene oxidizing activity in *Vicia* cotyledons, investigation of the enzyme system in cell-free preparations has been permitted. The work presented here describes *in vitro* studies on the enzyme activity in *Vicia*, and *in vivo* studies on the kinetics of ethylene metabolism in *Pisum*.

Ethylene metabolism in *Pisum* seedlings

Seeds of *Pisum sativum* L. cv. Alaska were sterilized and germinated using the method of Evans *et al.* (1984). Incubations and assays for ethylene metabolism were carried out as described by Jerie, Shaari and Hall (1979) and Evans *et al.* (1984). Gases were purified by preparative gas chromatography as described by Bengochea *et al.* (1980).

The results in *Figure 13.1* illustrate the incorporation of ^{14}C-ethylene into *Pisum* seedlings in the presence and absence of propylene. From these Lineweaver–Burk plots, it can be seen that tissue incorporation of ^{14}C-ethylene showed Michaelis–Menten kinetics and propylene acted as a competitive inhibitor of such incorporation. The K_m for ethylene is 1.6×10^{-6} M and the K_i for propylene is 3.7×10^{-7} M.

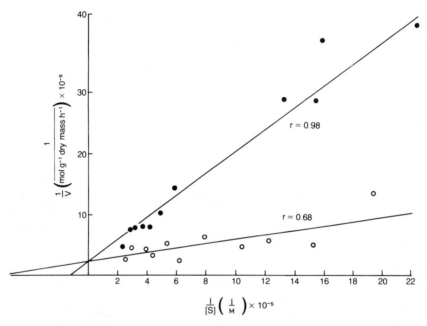

Figure 13.1 Lineweaver–Burk plots for incorporation of ^{14}C-ethylene into tissue by two day old etiolated *Pisum* seedlings. Seedlings were treated with a range of ethylene concentrations in the dark in the presence (●) or absence (○) of 3.3×10^{-6}M propylene (gas phase). r = correlation coefficient

In contrast, oxidation to $^{14}CO_2$, although also inhibited by propylene, appears to be unsaturable and the process does not appear to be O_2 requiring. The mechanism whereby $^{14}CO_2$ is produced in this system is currently under investigation.

Carbon dioxide and Ag^+ have been shown to antagonize ethylene action in a variety of systems (e.g. Beyer and Sundin, 1978) and in pea seedlings inhibition of ethylene oxidation to CO_2 by 7% CO_2 and inhibition of tissue incorporation by Ag^+ have been demonstrated (Beyer, 1979). Our results show (*Figure 13.2*) that the presence of 7% CO_2 caused not only an inhibition of ethylene oxidation to CO_2, but also resulted in an increased affinity of the tissue incorporation system for ethylene. On the other hand, Ag^+ when applied as a 0.6mM $AgNO_3$ solution to

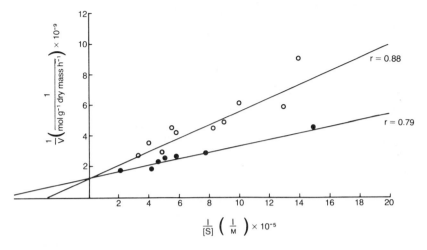

Figure 13.2 Lineweaver–Burk plots for incorporation of ^{14}C-ethylene into tissue by two day old etiolated *Pisum* seedlings in the presence (●) or absence (○) of 7% CO_2 (gas phase). r = correlation coefficients

Pisum seedlings inhibited tissue incorporation non-competitively (data not presented). It is of interest that while Ag^+ inhibits tissue incorporation and seems to promote oxidation of ethylene to CO_2, the affinity of the enzyme system for ethylene does not seem to be affected.

The fact that the presence of 7% CO_2 increased the affinity of tissue incorporation for ethylene is also of interest although the physiological significance of this is unknown. Nevertheless, it is clear from the effect of differing concentrations of CO_2 on tissue incorporation and oxidation to CO_2, that CO_2 itself is able to modulate ethylene metabolism in this system. It has also been shown that CO_2 has an effect on ethylene biosynthesis (Kao and Yang, 1982).

Ethylene metabolism in extracts from *Vicia* cotyledons

Vicia faba L. cv. Aquadulce plants were grown and cotyledons obtained as described previously (Dodds *et al.*, 1979). Cotyledons were homogenized in two volumes (w/v) of cold 10% (w/v) sucrose, 50 mM Tris-HCl, pH 8.5, 1 mM $MgCl_2$ extraction medium. The brei was filtered through nylon cloth and then centrifuged at 500 g for 15 min. Where possible manipulations were made in a cold room at 4 °C. Supernatant samples were desalted on a column of Sephadex G-25 equilibrated with extraction medium. Elutions with the same medium gave appropriate volumes of excluded fraction and low molecular weight fraction. The membrane pellet was prepared by centrifuging the 500 g supernatant at 40 000 g for 15 min and decanting the resultant supernatant. Incubations and assays of ethylene metabolism were as in *Pisum* seedlings.

Earlier studies have demonstrated that extracts of *Vicia* cotyledons retained the capacity to oxidize ethylene but that the V_{max} *in vitro* was only approximately 1%

of the rate attained *in vivo*. In these cell-free preparations, application of cofactors such as NADH, NADPH and several cations, were found to have no significant effect on the rate of metabolism (Dodds *et al.*, 1979). However, during initial purification studies, it was discovered that upon desalting crude enzyme extracts on Sephadex G-25 columns, enzyme activity was consistently lost, and that if the retarded low molecular weight fraction was added back to the void volume fraction activity could be restored. This result indicated the presence of a necessary cofactor for enzyme activity which was within the exclusion limit of Sephadex G-25. The effect of this low molecular weight fraction on the enzyme activity of the membrane pellet was then determined. In the presence of the low molecular weight fraction the ethylene metabolizing activity of the membrane pellet was increased by up to tenfold.

Experiments were performed to determine something of the nature of this cofactor by the addition of cofactors which are used by comparable enzymes, for example NADPH. From the amount of ^{14}C-ethylene converted it was found that the low molecular weight fraction stimulated enzyme activity about sevenfold while NADPH caused a 20-fold increase (*Table 13.1*). When the ability of NADH and NADPH to stimulate enzyme activity in desalted extracts was compared, NADPH alone was about twice as effective as NADH (*Table 13.2*). At the concentrations tested, there was no evidence of synergism when both cofactors were applied.

Table 13.1 COMPARISON BETWEEN ENDOGENOUS COFACTOR AND NADPH IN STIMULATING ENZYME ACTIVITY IN A 40000 g MEMBRANE PELLET FROM *VICIA* COTYLEDONS

Sample	dpm sample^{-1} h^{-1}	pmol ^{14}C$_2$H$_4$ converted (mg^{-1} h^{-1})
Pellet in extraction medium (EM)	415	0.5
Pellet in low MW fraction from Sephadex G-25	3848	3.9
Pellet in EM + 1 mM NADPH	9707	10.7

Table 13.2 PYRIDINE NUCLEOTIDE REQUIREMENT FOR ENZYME ACTIVITY IN DESALTED EXTRACTS OF *VICIA* COTYLEDONS

Fraction	dpm sample^{-1} h^{-1}	pmol ^{14}C$_2$H$_4$ converted (mg^{-1} h^{-1})
500 g supernatant	3775	3.9
Excluded fraction (EF) in standard medium	46	0.1
EF + NADH (0.55 mM)	3465	9.9
EF + NADPH (0.55 mM)	6037	17.2
EF + NADH + NADPH	5318	15.1

In order to determine the K_m for NADPH, a rate curve for the desalted extract was prepared using 41 µl l^{-1} ethylene and a range of NADPH concentrations (*Figure 13.3*). From the data obtained an apparent K_m of 3.97×10^{-5} M was calculated. Using saturating concentrations of NADPH (500 µM) and a range of ethylene concentrations, Lineweaver–Burk plots for ethylene oxidation by desalted extracts were prepared (*Figure 13.4*). The K_m for ethylene was calculated to be 2×10^{-8} M. This compares with the value of 4.2×10^{-10} M for the *in vivo* system (*Table 13.3*). Although NADPH seems to satisfy the cofactor requirement, the identity of the endogenous cofactor remains unknown.

In addition to NADPH, there seems to be a requirement for O_2 as enzyme activity was reduced by over 60% when extracts were incubated in vials which had

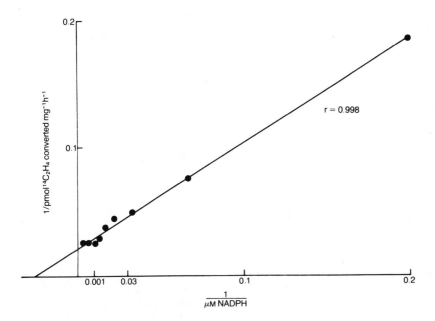

Figure 13.3 Lineweaver–Burk plot of the effect of a range of NADPH concentrations on the rate of ethylene oxidation by desalted extracts from *Vicia* cotyledons using 41 μl l^{-1} ethylene

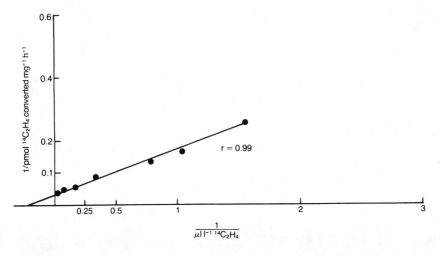

Figure 13.4 Lineweaver–Burk plot of the effect of 0.5 mM NADPH concentrations on the rate of ethylene oxidation by desalted extracts from *Vicia* cotyledons treated with a range of ethylene concentrations

Table 13.3 COMPARISON OF THE *IN VIVO* KINETIC PROPERTIES OF THE ETHYLENE METABOLIZING SYSTEMS IN *VICIA* AND *PISUM*

	Vicia	Pisum	
		NaOH soluble volatile compounds	Tissue incorporated compounds
K_m for ethylene (M)	4.2×10^{-10}	0.9×10^{-6}	1.6×10^{-6}
V_{max} mol g^{-1} d.m. h^{-1}	6.4×10^{-10}	2.4×10^{-10}	4.5×10^{-10}
K_i propylene (M)	5.0×10^{-6}	7.0×10^{-6}	3.7×10^{-7}

been flushed with N_2. Recently, Beyer (Chapter 12) using heavy oxygen has demonstrated incorporation of ^{18}O into ethylene oxide in *in vivo* experiments with *Vicia* cotyledons. The evidence to date then suggests that the enzyme is a mono-oxygenase and this is substantiated by the fact that the apparent K_m for NADPH is within the range of K_m values, albeit in the upper range, reported for other plant mono-oxygenases. We propose that the reaction may be summarized as follows:

$$C_2H_4 + O_2 + NADPH + H^+ \rightarrow C_2H_4O + NADP^+ + H_2O$$

The enzyme in *Vicia* cotyledons appears to be predominantly microsomal co-sedimenting with heat-labile NADPH-cytochrome-c-reductase. Progress in the precise localization of the enzyme has been hampered by the fact that sucrose and other polyhydroxy compounds commonly used in the fractionation of subcellular components drastically inhibit enzyme activity.

Relationship between ethylene metabolism and binding

Ethylene metabolism may be directly related to the mode of action of ethylene and, in principle, there is no reason why a metabolizing system should not act as a functional receptor. Nevertheless, there are certain difficulties in proposing ethylene metabolizing systems as candidates for ethylene receptors. It has been demonstrated that *Pisum* seedlings can metabolize propylene to 1,2-propanediol and its glucose conjugate (Blomstrom and Beyer, 1980) and it appears from chromatographic evidence that *Vicia* cotyledons can metabolize propylene to propylene oxide. However, when propylene was applied to *Pisum* seedlings, incorporation into tissue products was greater than that observed at equivalent concentrations of ethylene. Hence we have compared the *in vivo* kinetic properties of the ethylene metabolizing systems in both *Vicia* cotyledons and *Pisum* seedlings. It is apparent from the K_m values given in *Table 13.3* that tissue incorporation in *Vicia* saturates at a concentration of ethylene three orders of magnitude lower than in *Pisum*. It is significant that the V_{max} values in each of these three instances are very similar. Therefore, we have some explanation as to why the rates of metabolism in *Pisum* and *Vicia* are so different at physiological ethylene concentrations.

Conclusion

The data presented in *Figure 13.1* confirm the results of Beyer (1978) that propylene is a competitive inhibitor of ethylene metabolism. But, it appears that the affinity in *Pisum* for propylene is higher than that for ethylene but very similar to that for propylene in *Vicia* (*Table 13.3*). Thus the situation arises that in two different plant tissues the affinities of metabolizing systems for an unnatural analogue are more or less the same, whereas those for the natural substrate differ by three to four orders of magnitude.

At present there is no obvious positive explanation for these results and it must be emphasized that it is the kinetics of two *in vivo* systems which are being compared. The experimental approach taken up until now is based on the assumption that two independent systems are operating in *Pisum* and this may not be a true reflection of the real situation if a mechanism such as that suggested by Beyer (Chapter 12) is in operation.

References

BENGOCHEA, T., ACASTER, M.A., DODDS, J.H., EVANS, D.E., JERIE, P.H. and HALL, M.A. (1980). *Planta*, **148**, 407–411
BEYER, E.M. (1975). *Nature*, **255**, 144–147
BEYER, E.M. (1978). *Plant Physiology*, **61**, 893–895
BEYER, E.M. (1979). *Plant Physiology*, **63**, 169–173
BEYER, E.M. and SUNDIN, O. (1978). *Plant Physiology*, **61**, 896–899
BLOMSTROM, D.C. and BEYER, E.M. (1980). *Nature*, **283**, 66–68
DODDS, J.H., MUSA, S.K., JERIE, P.H. and HALL, M.A. (1979). *Plant Science Letters*, **17**, 109–114
DODDS, J.H., HESLOP-HARRISON, J.S. and HALL, M.A. (1980). *Plant Science Letters*, **19**, 175–180
EVANS, D.E., SMITH, A.R., TAYLOR, J.E. and HALL, M.A. (1984). *Plant Growth Regulation*, **2**, 187–196
JERIE, P.H. and HALL, M.A. (1978). *Proceedings Royal Society London B.*, **200**, 87–94
JERIE, P.H., SHAARI, A.R. and HALL, M.A. (1979). *Planta*, **144**, 503–507
KAO, C.H. and YANG, S.F. (1982). *Planta*, **155**, 261–266

14

REGULATION OF THE EXPRESSION OF TOMATO FRUIT RIPENING GENES: THE INVOLVEMENT OF ETHYLENE

D. GRIERSON, A. SLATER, M. MAUNDERS, P. CROOKES
Department of Physiology and Environmental Science, University of Nottingham
Faculty of Agricultural Science, Sutton Bonington
G.A. TUCKER
Department of Applied Biochemistry and Food Science, University of Nottingham
Faculty of Agricultural Science, Sutton Bonington
W. SCHUCH and K. EDWARDS
ICI Corporate Bioscience and Colloid Laboratory, Runcorn, Cheshire

Introduction

During ripening most fleshy fruit undergo changes in composition that affect various physical and chemical attributes such as colour, texture, flavour and aroma (*Table 14.1*). These changes have evolved naturally as a means of making fruit attractive to potential consumers and they play an important role in seed dispersal. The control of ripening is of great commercial significance since it has important implications for fruit harvesting, transport, storage life and quality for the human consumer. It is also of considerable scientific interest. Many diverse changes in physiological and biochemical activity are required for normal ripening. These events affect several different subcellular compartments and yet they are highly coordinated. This raises fundamental questions about the nature and regulation of ripening changes and communication both within and between cells.

There are many different types of fleshy fruit, derived from anatomically distinct parts of the reproductive apparatus. They follow a variety of patterns of growth and development, involving both cell division and expansion (Coombe, 1976). The processes causing changes in physicochemical attributes at maturity are also quite varied. For example, colour development in tomato is due to carotenoid synthesis

Table 14.1 PHYSIOLOGICAL AND BIOCHEMICAL CHANGES THAT OCCUR DURING THE RIPENING OF TOMATO FRUIT

Degradation of starch
Loss of chlorophyll
Accumulation of lycopene as the chloroplasts are transformed into chromoplasts
Increase in soluble pectin resulting from wall softening and degradation
Production of flavour and aroma compounds
Increase in citric and glutamic acids
Breakdown of the toxic alkaloid α-tomatine
Increase in respiration
Increase in ethylene synthesis
Increase in membrane permeability
Disappearance of some enzymes and synthesis of new ones
Synthesis of new mRNAs

throughout the fruit, whereas in some apples it is due to the production of anthocyanins in cells near the surface. Furthermore, softening in bananas is due partly to degradation of large amounts of stored starch whereas in tomatoes it is caused mainly by solubilization of cell wall pectin. There are also differences in respiratory behaviour and ethylene production during ripening (Rhodes, 1980). In climacteric fruit, such as apples, pears, bananas and tomatoes, respiratory output rises to a 'climacteric peak', when ripening changes are occurring most rapidly, and subsequently declines. In contrast, there is no change in respiration during ripening of non-climacteric fruit such as oranges, lemons and strawberries. In addition to the rise in respiration there is a dramatic increase in ethylene production during ripening of climacteric fruit. This has been shown to be due to enhanced synthesis of the gas, brought about by the 'activation' of the pathway involving 1-aminocyclopropane-1-carboxylic acid (ACC), discovered by Adams and Yang (1979). Ethylene production is autocatalytic in climacteric fruit: that is, the presence of the gas causes the fruit cells to synthesize more ethylene, provided the cells are sufficiently mature and competent to respond. It is possible that ethylene is involved in the initiation of ripening but an alternative view is that it simply accelerates the process. Application of ethylene to unripe climacteric fruit certainly stimulates ripening (McGlasson, Wade and Adato, 1978) and it even affects some ripening parameters in non-climacteric fruits such as degreening of citrus. Since supplying ethylene to many plant tissues, including non-climacteric fruit, induces a rise in respiration, it is possible that the only difference between climacteric and non-climacteric fruit is that the latter lack autocatalytic ethylene production (Rhodes, 1980).

Although the details of the ripening process may vary in different types of fruit there are nevertheless many common features. Ripening generally occurs after cell division and expansion is completed, it involves concomitant changes in a variety of biochemical pathways, and the physiological and biochemical behaviour of the cells is synchronized. Ripening undoubtedly involves changes in enzyme activity and leads ultimately to cell deterioration and senescence (Sacher, 1973; Grierson, 1984). There are two main theories of ripening control that attempt to explain these metabolic changes. On the one hand it has been argued that ripening involves a change in the 'organizational resistance' of the cell. This, it is proposed, is caused by an alteration in membrane permeability which leads to leakage of ions and metabolites and the release or activation of hydrolytic enzymes. An alternative explanation, that ripening involves an initial sequence of biosynthetic changes that depends upon the expression of specific genes, has recently been gaining support. The existence of specific mutations that affect various aspects of tomato ripening (*Table 14.2*) provides clear evidence in favour of the gene expression hypothesis. In

Table 14.2 CHARACTERISTICS OF SOME TOMATO RIPENING MUTANTS

Name	Chromosome	Phenotype of fruit homozygous for the mutation
Ripening inhibitor (*rin*)	5	Fruit turn yellow and only soften slightly. There is no rise in respiration or ethylene synthesis. Little or no polygalacturonase enzyme is synthesized.
Never-ripe (*Nr*)	7	Fruit turn orange and soften slowly. Synthesis of lycopene and polygalacturonase is much reduced
Greenflesh (*gf*)	8	Ripe fruit appear a red-brown in colour because the fruit retain some chlorophyll as they ripen. Polygalacturonase and lycopene synthesis are normal.

this article we review the results of experiments concerning the identification and function of ripening genes in tomatoes and consider the role of ethylene in regulating their expression.

RNA synthesis during fruit development and ripening

Labelling studies *in vivo* have shown that RNA is synthesized at high rates during the growth and development of tomatoes and then declines when fruit reach the mature-green stage (Rattanapanone, Grierson and Stein, 1977; Grierson, 1983). At the onset of ripening, when ethylene synthesis begins (Sawamura, Knegt and

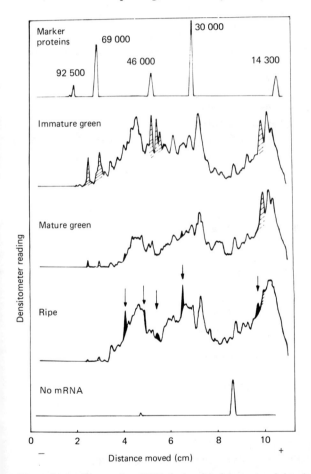

Figure 14.1 Changes in mRNA during development and ripening of tomato fruit. RNA was purified from fruit at different stages of development and translated *in vitro* in a rabbit reticulocyte lysate in the presence of ^{35}S-methionine. The radioactive proteins synthesized in response to tomato mRNA were fractionated by polyacrylamide gel electrophoresis and detected by X-ray film, which was scanned with a densitometer. Some mRNA translation products that disappear during fruit maturation are shown hatched. Some other mRNA translation products that accumulate during ripening are indicated by arrows. (From Grierson *et al.*, 1984)

Bruinsma, 1978), there is a marked increase in the rate of RNA synthesis and turnover. It has been demonstrated that the newly-synthesized molecules include high-molecular-weight rRNA, soluble RNA and polyA-containing mRNA (Rattanapanone, Grierson and Stein, 1977; Rattanapanone, Speirs and Grierson, 1978). Studies on the nature of the tomato fruit mRNA required its purification and the analysis of radioactive proteins synthesized by translation *in vitro*. These experiments provided the first evidence for the occurrence of a small number of changes in mRNA related to fruit ripening (Rattanapanone, Speirs and Grierson, 1978; Grierson, Tucker and Robertson, 1981a,b). The results of later experiments with avocado supported this conclusion (Christoffersen, Warme and Laties, 1982).

In the original experiments only two mRNA changes were detected during tomato ripening (Rattanapanone, Speirs and Grierson, 1978). In more recent work employing more sensitive procedures (Grierson, 1984) it has been possible to show that the changes are more extensive than thought previously (*Figure 14.1*). On the basis of these results we have classified tomato fruit mRNAs into three groups.

(1) Many of the mRNAs present in mature-green fruit remain throughout the ripening period.
(2) At least six abundant mRNAs that are present in immature-green tomatoes decline in quantity during maturation and ripening.
(3) Several other mRNAs either appear or increase greatly in quantity during ripening.

Analysis of the results of many experiments similar to those shown in *Figure 14.1* (Grierson *et al.*, 1984) indicates that there are between four and eight increases in fruit mRNA during ripening that can be detected by translation *in vitro* (*Table 14.3*). Experiments involving the analysis of mRNA changes by cDNA cloning

Table 14.3 MOLECULAR WEIGHTS OF RIPENING-RELATED mRNA TRANSLATION PRODUCTS

Major translation products	Minor translation products
190 000	80 000
55 000	57 000
48 000	44 000
35 000	20 000

procedures (*see below*), which are even more sensitive, suggest that the number may be substantially greater than this.

It is clear that there are significant alterations in the quantities of particular mRNAs during fruit maturation and ripening. This could arise by enhanced transcription or reduced rates of degradation of specific mRNAs and further experiments on RNA synthesis *in vitro* will be necessary before the possibility of transcriptional control can be tested unequivocally.

The function of the ripening-related mRNAs is of considerable interest. At least some of them are present in polyribosomes of ripening fruit and are presumably translated *in vivo* (Speirs *et al.*, 1984). We have shown that one mRNA that is translated *in vitro* to give a 48 000 molecular weight protein code for the cell-wall-softening enzyme polygalacturonase (*see below*). Another ripening-related mRNA, that codes for a protein with a molecular weight of 55 000, is present in small quantities in green fruit and accumulates substantially during ripening (*Figure 14.1*). We speculate that this protein functions in green tissue but the demand for it increases substantially during ripening. Preliminary evidence (Smith and Grierson, unpublished) suggests that the production of this protein may be related to ethylene synthesis. There is no clear evidence as to the function of the mRNAs that decline in quantity during fruit development (*Figure 14.1*). They presumably code for proteins not required during ripening and it is possible that at least some of them are involved in the production of chloroplast proteins. There is certainly a decline in several plastid proteins during ripening (Bathgate, Purton and Grierson, unpublished) and the mRNAs coding for these presumably become redundant as the chloroplasts are converted to chromoplasts.

Synthesis and function of polygalacturonase

It is important to establish that the ripening-related mRNAs that we have detected actually function in directing the synthesis of proteins required for ripening before the gene expression hypothesis can be accepted. Furthermore, the relationship between mRNA changes and specific physiological or biochemical events must be unequivocally established before any significant progress can be made in studying the regulation of ripening. The cell wall softening process, which depends largely on the pectin-degrading enzyme polygalacturonase, represents one facet of ripening that is beginning to be understood at the molecular level (Grierson, Tucker and Robertson, 1981a,b; Grierson, 1983, 1984).

Polygalacturonase (PG) is not present in green tomato tissue but enzyme activity appears and accumulates in large quantities at the onset of ripening (*see* a review of early work by Hobson, 1979). Purification of PG activity from ripening tomato fruit results in the separation of two isoenzyme forms (PG1 and PG2) with different molecular weights, heat stabilities and densities in caesium chloride (Tucker, Robertson and Grierson, 1981). Enzyme activity is virtually absent from yellow *rin* fruit and is reduced in orange *Nr* tomatoes (*Table 14.2*). Although PG1 and PG2 have different properties they both react with an antibody raised against pure PG2, suggesting that the polypeptides of the isoenzymes are related. Comparison of the fragments resulting from partial proteolysis of the purified isoenzymes under controlled conditions shows, indeed, that they are very similar (Tucker, Robertson and Grierson, 1980). Furthermore, it is possible to convert *in vitro* PG2 into a form with several of the properties of PG1 (Tucker, Robertson and Grierson, 1981). PG2 has been separated by non-denaturing gel electrophoresis into two fractions, PG2a and PG2b (Ali and Brady, 1982; Crookes and Grierson, 1983) with similar molecular weights and antigenic properties. The conclusion from these experiments is that all the multiple forms of tomato PG are structurally related. All three isoenzyme forms are glycosylated. The basic polypeptide present in each isoenzyme has a molecular weight of 46 000 measured by sodium dodecyl sulphate-polyacrylamide gel electrophoresis. This polypeptide is either modified after synthesis to produce the different isoenzymes, or there are similar but not identical genes coding for each isoenzyme.

The purified PG isoenzymes have been used to study their role in degradation of tomato fruit cell walls. PG2 has been shown to liberate galaturonic acid *in vitro* from cell walls isolated from green tomato tissue. In related experiments it was demonstrated that the only major wall-degrading activity absent from yellow *rin*

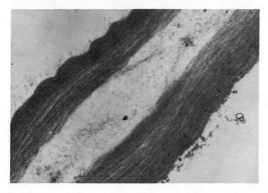

Figure 14.2 Effect of purified tomato polygalacturonase on the ultrastructure of fruit cell walls. The upper figure shows the appearance of green fruit cell walls in the electron microscope. The lower figure shows the effect on wall ultrastructure of incubating segments of green tomato tissue in purified polygalacturonase 2 overnight prior to preparation for electron microscopy

fruit is PG (Themmen, Tucker and Grierson, 1982). Both PG1 and PG2 are capable of solubilizing the pectin fraction of the cell wall when applied *in situ* to segments of green tomato tissue and the ultrastructural changes brought about are similar to those occurring during normal ripening (*Figure 14.2*; Crookes and Grierson, 1983). The fact that added PG attacks the cell walls in green tissue suggests that it is the availability of PG which is the main controlling factor in determining softening. However, cellulase activity may be significant in tomato (Hobson, 1968) and there may be inhibitors which reduce the effectiveness of PG under some circumstances (Brady, personal communication).

Four separate pieces of evidence indicate that PG is synthesized *de novo* during ripening.

(1) There is little or no PG enzyme activity, or any protein that reacts with antibodies raised against PG, present in green tomato tissue. Furthermore, the increase in enzyme activity during ripening is closely correlated with an increase in the number of protein molecules that react with PG antibody, measured by radio-immunoassay (Tucker, Robertson and Grierson, 1980; Tucker and Grierson, 1982; Brady *et al.*, 1982).

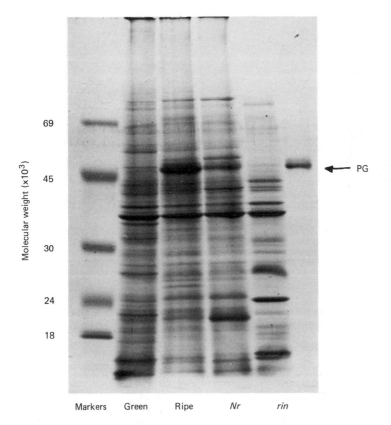

Figure 14.3 Synthesis of polygalacturonase during ripening. Cell-wall-associated proteins from normal and mutant tomatoes were fractionated by polyacrylamide gel electrophoresis in the presence of sodium dodecyl sulphate and stained with Coomassie blue. Marker proteins are shown on the left and a sample of purified polygalacturonase 2 on the right. Note the absence of polygalacturonase in green tomatoes, the massive accumulation during normal ripening, the reduction in *Nr* fruit and the almost complete absence of the protein in *rin* fruit

(2) There is an increase in the density of PG when ripening tomato fruit is incubated in deuterium oxide (Grierson, 1983).

(3) Electrophoresis of partially-purified tomato protein extracts shows that little or none of the 46 000 molecular weight PG polypeptide is present in green tissue but it increases dramatically during ripening (*Figure 14.3*; Tucker and Grierson, 1982).

(4) When discs of ripening tomato tissue are incubated in ^{35}S-methionine many radioactive proteins become labelled. One of these, with a molecular weight of 46 000, is specifically precipitated by PG antibody (Grierson *et al.*, 1984).

The mRNA coding for PG is not present in green tomato tissue before ethylene synthesis begins and only accumulates at the onset of ripening (Grierson *et al.*, 1984). The evidence for this is as follows. One of the ripening-related mRNAs codes for a protein with a molecular weight of approximately 48 000 (*Figure 14.1*,

154   Ethylene and fruit ripening genes

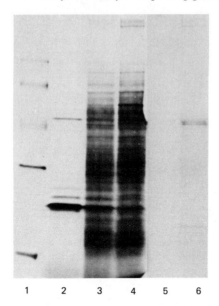

1   2   3   4   5   6

Figure 14.4   Immunoprecipitation of the polygalacturonase mRNA translation product *in vitro*. Messenger-RNA samples from mature-green and ripe tomatoes were translated *in vitro* in a rabbit reticulocyte lysate in the presence of ^{35}S-methionine. Aliquots of the mRNA translation products were challenged with polygalacturonase antibody. Radioactive proteins were fractionated by polyacrylamide gel electrophoresis and detected by X-ray film. Track 1, marker proteins with molecular weights (top to bottom) of 92 500; 69 000; 46 000; 30 000; 14 300. Track 2, reticulocyte lysate without added mRNA. Track 3, translation products of mRNA from green tomatoes. Track 4, translation products of mRNA from ripe tomatoes. Track 5, immunoprecipitate obtained by treating an aliquot of proteins from track 3 with polygalacturonase antibody. Track 6, immunoprecipitate obtained by treating an aliquot of proteins from track 4 with polygalacturonase antibody. (From Grierson *et al.*, 1984)

Table 14.3) which is specifically precipitated by an antibody raised against pure PG. It is thought to be a precursor form of PG translated *in vitro* and the amount of PG mRNA can be estimated by immunoprecipitation of the translation product. No translatable mRNA for PG is detectable in mature-green tomatoes but it does accumulate during ripening (*Figure 14.4*).

The explanation for cell wall softening during ripening is, therefore, that it is due, at least in part, to a small group of PG isoenzymes. These are not present in green fruit but are synthesized when the PG mRNA accumulates during ripening. It is possible that this involves transcriptional control but there is no good evidence for or against this proposal. Ethylene is implicated in the process but as discussed below, the *mechanism* of ethylene action is not clear.

Relationship between ethylene synthesis, the respiratory climacteric and the accumulation of polygalacturonase and its mRNA during ripening

Investigations of the ripening behaviour of different climacteric fruit show that in some ethylene synthesis occurs before the respiratory rise, in others the increase in production of ethylene and carbon dioxide are coincident and in yet others there is an increase in carbon dioxide output before any change in ethylene production can be detected (Biale and Young, 1981). It is not clear whether these differences reflect genuine variation in the order of the two events in different fruit or alternatively whether they are due to problems concerned with the measurement of gas production and diffusion in bulky organs. What is clear is that the application of ethylene hastens the ripening of many climacteric fruit (Biale and Young, 1981), whereas carbon dioxide does not.

In tomatoes a rise in the internal ethylene concentration has been shown to occur before the respiratory rise (Sawamura, Knegt and Bruinsma, 1978) and the application of ethylene to mature green tomatoes does stimulate ripening (*Figure 14.5*; Grierson and Tucker, 1983). This latter observation conflicts with the earlier

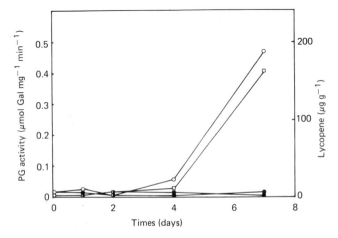

Figure 14.5 Effect of ethylene on the synthesis of lycopene and polygalacturonase by detached mature-green tomatoes. Green fruit were picked prior to the natural onset of ethylene synthesis and placed in air with 10 μl l^{-1} ethylene (○, □) or air plus mercuric perchlorate to absorb ethylene (●, ■). Samples of fruit were removed at intervals and polygalacturonase (○, ●) and lycopene (□, ■) measured. (From Grierson and Tucker, 1983)

conclusions of McGlasson, Dostal and Tigchelaar (1975) that immature green tomatoes are particularly resistant to ethylene. The differences between the two studies cannot be explained on the basis of threshold levels of ethylene required for ripening but are probably related to the developmental stage of the tomatoes used. Our results indicate that fruit respond to ethylene if treated at the mature-green stage just prior to the natural onset of ripening (Grierson and Tucker, 1983), whereas it seems that fruit at an earlier stage do not respond so readily (McGlasson, Dostal and Tigchelaar, 1975). We suggest that tomato fruit cells are only *competent* to respond to ethylene after a certain stage of development has been reached. This could be due to a change in sensitivity to ethylene, brought about by an alteration in the number of ethylene receptors or in the activity of coupling mechanisms leading to a response, or alternatively it could be explained by the decline in an inhibitor or antagonist of ethylene action. Evidence for the existence of an inhibitor of tomato ripening which is derived from the parent plant has been presented (Sawamura, Knegt and Bruinsma, 1978).

156 Ethylene and fruit ripening genes

It has been suggested that PG and not ethylene is the initiator of ripening in tomatoes (Tigchelaar, McGlasson and Buescher, 1978). According to this proposal the initial production of PG leads to cell wall degradation and the release of wall-bound enzymes which enter the cell and catalyse subsequent ripening changes. This theory has been tested by studying the relationship between the synthesis of ethylene and PG during fruit maturation and ripening (*Figure 14.6*). The results confirm that the natural production of ethylene precedes PG synthesis by at least 20 h (Grierson and Tucker, 1983) and there is general agreement that PG is not the initiator of ripening (Brady *et al.*, 1982).

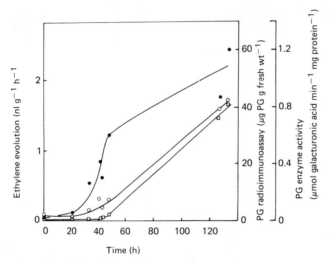

Figure 14.6 The relationship between the onset of enhanced ethylene production and the synthesis of polygalacturonase during tomato ripening. Mature-green fruit were harvested prior to the initiation of ripening and their ethylene production rates monitored individually over a period of several days until they began to ripen. Individual fruit at known times after the onset of enhanced ethylene production (●) were sampled for polygalacturonase activity, measured by enzyme assay (□) and radioimmunoassay (○). (From Grierson and Tucker, 1983)

Further experiments have shown that the internal ethylene concentration in tomatoes rises before PG synthesis can be detected and that it is correlated with increases in ACC synthase, which is one of the two controlling enzymes in the ethylene biosynthesis pathway (Su *et al.*, 1984).

When detached mature-green fruit are treated with ethylene in air they initiate the ripening programme and synthesize polygalacturonase and lycopene earlier than similar fruit held in ethylene-free air (Grierson and Tucker, 1983). Perhaps not surprisingly, ethylene also stimulates the appearance of mRNAs found to accumulate during ripening, including the PG mRNA (*Figure 14.7*). The effect of ethylene is not rapid, however, and the accumulation of ripening-related mRNAs can only be detected by translation *in vitro* of mRNA samples extracted 24–48 h after the start of treatment (*Figure 14.7*). Continued incubation in ethylene leads to the production of greater amounts of translatable mRNA than in normal ripening fruit. The mRNAs remain in substantial quantities for at least 14 days after ripening and appear to be undegraded even in soft red tissue. Although the accumulation of

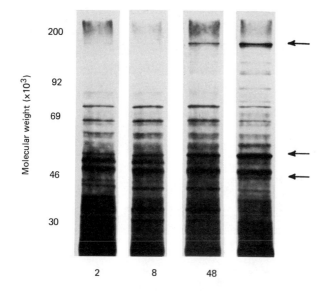

Figure 14.7 Effect of incubating mature-green tomatoes in ethylene on the production of ripening-related mRNAs. Mature-green fruit were harvested prior to the onset of enhanced ethylene production and incubated in air plus 10 μl l^{-1} ethylene. At intervals RNA was purified and translated *in vitro* in a rabbit reticulocyte lysate in the presence of ^{35}S-methionine. The radioactive proteins were fractionated by polyacrylamide gel electrophoresis and detected by X-ray film. The positions of marker proteins are shown on the left. The arrows show changes in three ripening-related mRNAs, corresponding to the proteins with molecular weights of 190 000, 55 000 and 48 000 (the polygalacturonase mRNA product) in *Table 14.3*

ripening-related mRNAs does seem to occur relatively slowly it is not correct to say that there is no rapid effect of ethylene on mRNA. We have noted the production of an mRNA coding for a 38 000 molecular weight protein within 8 h of treatment (Slater and Grierson, unpublished). However, this mRNA declines in quantity before the increase in the other ethylene induced mRNAs and we are not certain whether or not it is related to ripening initiation.

Cloning the ripening genes

In order to identify and characterize the genes responsible for the ripening associated changes in tomato, Poly A$^+$ RNA was isolated from ripe fruit. This mRNA was assessed for its message potential by *in vitro* translation (*see above*), and a sample was then reverse transcribed *in vitro* to yield cDNA copies of each mRNA species present. This single stranded cDNA was converted to double stranded DNA, size fractionated by passage down a column to remove small (incompletely transcribed) molecules, and then cloned into the Pst I site of the bacterial plasmid pAT 153. This cloned DNA was used to transform *E. coli* strain C600; transformants were identified by their tetracycline-resistant/ampicillin-sensitive phenotype. The many bacterial colonies produced in this way were

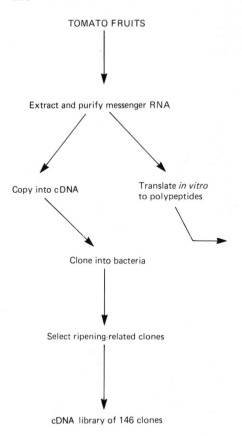

Figure 14.8 Scheme for the production and screening of cDNA copies of ripening-related mRNAs

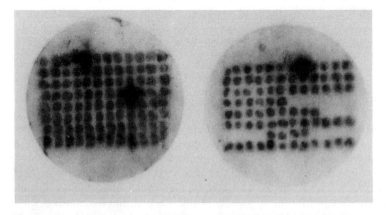

Figure 14.9 Selection of related cDNA clones and grouping into families by cross-hybridization. A radioactive cDNA copy of a ripening-related mRNA is hybridized to bacterial colonies containing copies of each of the 146 ripening-related clones. Bacteria containing cloned cDNAs similar to the radioactive probe are detected by X-ray film

individually 'spotted' onto nitrocellulose membranes, grown overnight, and then lysed and the DNA bound to the membrane by the method of Grunstein and Hogness (1975). The membrane-bound DNA was then hybridized to ^{32}P-labelled-nick-translated cDNA (which had not been cloned into pAT153) prepared from either ripe or mature green fruit. Those colonies which hybridized to the 'ripe' probe, but not to the 'green' probe (146 in all) were designated 'ripening-associated clones' (*Figure 14.8*).

This library of clones was then subdivided into families of related clones, again by the Grunstein and Hogness colony hybridization method, using individual cloned cDNA inserts as the probes (*Figure 14.9*). In this way nine families have been identified, containing two to 18 members. Several more clones are unique (eight have been identified so far) and appear to derive from low abundance mRNAs. The degree of homology between cross-hybridizing clones in the same family is being studied by restriction endonuclease mapping, to determine whether they derive from identical mRNAs or whether they are coded for by related but different genes.

The identity of each family of clones is being determined by hybrid-release translation: one of the cloned cDNAs is immobilized and hybridized to its complementary mRNA from a total RNA extract of ripe tomato fruit, thereby selecting out a single RNA. The isolated mRNA is then translated *in vitro* to yield the polypeptide product, hence enabling a cloned gene to be allotted to a specific polypeptide. Preliminary experiments on several clones have allowed a single polypeptide to be assigned to each with different products being coded for by each gene family.

Discussion

Ethylene plays an important role in the development of climacteric fruit. The production of the gas by the fruit cells is the first detectable sign of ripening and supplying ethylene to mature-green fruit hastens ripening. However, we have very little idea of the underlying mechanisms of ethylene action. The ripening response of the fruit to either endogenous or exogenous ethylene involves mRNA accumulation and enzyme synthesis. It is tempting to consider ripening simply as a result of a redirection in gene expression in response to ethylene. It is clear, however, that this is not the complete story. For instance it is difficult to see how ethylene *initiates* the increase in ACC synthase activity, which is probably involved in regulating ethylene synthesis (Su *et al.*, 1984), unless the operation of some other agent is involved. Also, tomato fruit are resistant to ethylene before the mature-green stage thus it seems that they are not *competent* to respond rapidly until a certain stage in development is reached. This suggests there are biochemical changes occurring during development that affect the sensitivity to, or the ability to respond to, ethylene. Thus ethylene may be involved in activating or accelerating a particular set of predetermined biochemical processes but this represents only one aspect of a more profound developmental programme.

Although the exact role of ethylene in controlling any changes in gene expression is unclear, it is obvious that these changes are important for fruit ripening. The results discussed in this article establish that a programme of mRNA accumulation and enzyme synthesis is intimately involved in the regulation of tomato ripening. The work of Tucker *et al.* (Chapter 15) has shown that the synthesis of new mRNAs is also important in avocado ripening. In tomato fruit some mRNAs decline in quantity as the fruit matures whereas others, perhaps coding for so-called 'housekeeping' proteins persist throughout ripening. In addition, new mRNAs appear or increase in amount at the onset of ripening. The capacity for protein synthesis persists throughout ripening and one of the mRNA species which appears has been shown to code for PG, a major softening enzyme that is synthesized *de novo* in the fruit at this time. At least some of the other ripening-related mRNAs have been shown to be present in polyribosomes (Speirs *et al.*, 1984) and presumably also function in directing the synthesis of proteins required for the various aspects of ripening (*Table 14.1*). About eight possible ripening-related mRNAs can be detected by translation *in vitro* (*Table 14.3*) but the cDNA cloning approach, which is more sensitive, indicates that there may be twice this number of ripening genes being expressed. It is clear that the ripening process in climacteric fruit results from changes in gene expression. It is likely that this statement could also apply to non-climacteric fruit but very little information is available concerning gene expression in these. A study such as the one reported here but using a non-climacteric fruit would be very useful.

One important task for the future is to identify and study the regulation of ripening genes. It is anticipated that the cDNA clones will be very useful for this purpose. They are being used to select individual mRNAs and hence establish which proteins they code for. They can also act as probes for assaying rates of RNA synthesis *in vitro* in order to test for transcriptional control. Finally, the ripening-specific cDNA clones are being used to identify the original genomic DNA sequence and thus allow the organization of these genes and the sequences of potential controlling regions to be determined.

Acknowledgements

This work was supported by grants from the Agriculture and Food Research Council, The Science and Engineering Research Council and by a Nuffield Foundation Science Research Fellowship to Don Grierson.

References

ADAMS, D.O. and YANG, S.F. (1979). *Proceedings of the National Academy of Science, USA*, **76**, 170–174

ALI, Z.M. and BRADY, C.J. (1982). *Australian Journal of Plant Physiology*, **9**, 171–178

BIALE, J.B. and YOUNG, R.E. (1981). In *Recent Advances in the Biochemistry of Fruit and Vegetables*, pp. 1–39. Ed. by J. Friend and M.J.C. Rhodes. Academic Press, London

BRADY, C.J., MACALPINE, G., MCGLASSON, W.B. and UEDA, Y. (1982). *Australian Journal of Plant Physiology*, **9**, 171–178

CHRISTOFFERSEN, R.E., WARM, E. and LATIES, G.G. (1982). *Planta*, **155**, 52–57

COOMBE, B.G. (1976). *Annual Review of Plant Physiology*, **27**, 207–228

CROOKES, P.R. and GRIERSON, D. (1983). *Plant Physiology*, **72**, 1088–1093

GRIERSON, D. (1983). In *Post-harvest physiology and crop preservation*, pp. 45–60. Ed. by M. Lieberman. Plenum Press, New York

GRIERSON, D. (1984). In *Cell aging and cell death*. Ed. by I. Davies and D.C. Sigee (in press). Cambridge University Press, Cambridge
GRIERSON, D. and TUCKER, G.A. (1983). *Planta*, **157**, 174–179
GRIERSON, D., TUCKER, G.A. and ROBERTSON, N.G. (1981a). In *Recent Advances in the Biochemistry of Fruit and Vegetables*, pp. 147–160. Ed. by J. Friend and M.J.C. Rhodes. Academic Press, London
GRIERSON, D., TUCKER, G.A. and ROBERTSON, N.G. (1981b). In *Quality in stored and processed vegetables and fruit*, pp. 179–191. Ed. by P.W. Goodenough and R.K. Atkin. Academic Press, London
GRIERSON, D., SLATER, A., SPEIRS, J. and TUCKER, G.A. (1984). *Planta*, (in press)
GRUNSTEIN, M. and HOGNESS, D.S. (1975). *Proceedings of the National Academy of Sciences, USA*, **72**, 3961–3965
HOBSON, G.E. (1968). *Journal of Food Science*, **33**, 588–592
HOBSON, G.E. (1979). *Current Advances in Plant Sciences*, **11**, 1–11
MCGLASSON, W.B., DOSTAL, H.C. and TIGCHELAAR, E.C. (1975). *Plant Physiology*, **55**, 218–222
MCGLASSON, W.B., WADE, N.L. and ADATO, I. (1978). In *Phytohormones and related compounds—a comprehensive treatise*, Vol. II, pp. 447–493. Ed. by D.S. Letham, P.B. Goodwin and T.J.V. Higgins. Elsevier/North-Holland, Amsterdam
RATTANAPANONE, N., GRIERSON, D. and STEIN, M. (1977). *Phytochemistry*, **16**, 629–633
RATTANAPANONE, N., SPEIRS, J. and GRIERSON, D. (1978). *Phytochemistry*, **17**, 1485–1486
RHODES, M.J.C. (1980). In *Senescence in Plants*, pp. 157–205. Ed. by K.V. Thimann. CRC Press, Boto Raton, Fl.
SACHER, J.A. (1973). *Annual Review of Plant Physiology*, **24**, 197–224
SAWAMURA, M., KNEGT, E. and BRUINSMA, J. (1978). *Plant and Cell Physiology*, **19**, 1061–1069
SPEIRS, J., BRADY, C.J., GRIERSON, D. and LEE, E. (1984). *Australian Journal of Plant Physiology*, **11**, 225–233
SU, L-Y, MCKEON, T., GRIERSON, D., CANTWELL, M. and YANG, S.F. (1984). *HortScience*, **19**, 576–578
THEMMEN, A.P.N., TUCKER, G.A. and GRIERSON, D. (1982). *Plant Physiology*, **69**, 122–124
TIGCHALAAR, E.C., MCGLASSON, W.B. and BUESCHER, R.W. (1978). *HortScience*, **13**, 508–513
TUCKER, G.A. and GRIERSON, D. (1982). *Planta*, **155**, 64–67
TUCKER, G.A., ROBERTSON, N.G. and GRIERSON, D. (1980). *European Journal of Biochemistry*, **112**, 119–124
TUCKER, G.A., ROBERTSON, N.G. and GRIERSON, D. (1981). *European Journal of Biochemistry*, **115**, 87–90

15
INDUCTION OF CELLULASE BY ETHYLENE IN AVOCADO FRUIT

MARK L. TUCKER*, ROLF E. CHRISTOFFERSEN†, LISA WOLL and
GEORGE G. LATIES
*Department of Biology and Molecular Biology Institute, University of California,
Los Angeles, USA*

Involvement of nucleic acid and protein turnover in fruit ripening

Several fruit show increases in protein synthesis during the early stages of ripening (Sacher, 1973). In particular, Richmond and Biale (1966) reported that, in avocado, amino acid incorporation into protein increased very early in the respiratory climacteric and diminished to less than preclimacteric levels at the respiratory peak. Tucker and Laties (1984) observed a change in polysome prevalence with time similar to that described by Richmond and Biale for amino acid incorporation into protein.

Although an increase in protein synthesis in several climacteric fruits has been demonstrated, its role in the ripening process has been questioned (Sacher, 1973). Brady and O'Connell (1976) exposed 6 mm thick cross-sections of banana fruit to $5\,\mu l\,l^{-1}$ ethylene for 6 h and observed an increase in protein synthesis without concomitant ripening; however, when the sections were exposed to ethylene for 12 h enhanced protein synthesis was accompanied by ripening. Upon analysing *in vivo* labelled protein samples on polyacrylamide gels they found that the increase in protein synthesis after 24 h of ethylene treatment was not limited to a few ripening specific proteins, but reflected a generic increase in protein synthesis. They suggested, however, that continued exposure to ethylene might initiate ripening-specific proteins.

In avocado a threefold increase in polysomal poly(A) + RNA during the early climacteric constitutes primarily a generic increase in constitutive mRNA, based on an *in vitro* translation assay (Tucker and Laties, 1984). However, specific changes in gene expression do occur during the course of the climacteric (Christoffersen, Warm and Laties, 1982; Tucker and Laties, 1984). An example of one of these is the appearance during avocado fruit ripening of a polysomal message which upon translation yields a 53 000 dalton polypeptide that is immunoprecipitable by antibody raised against avocado cellulase (Tucker and Laties, 1984).

---

*Present address: Department of Molecular Plant Biology, University of California, Berkeley, CA 94720, USA
†Present address: Mann Laboratory, University of California, Davis, CA 95616, USA

Changes in cellulase gene expression during avocado ripening

ACCUMULATION OF ANTIGENICALLY ACTIVE PROTEIN

The appearance of cellulase activity in fruit extracts during ripening correlates closely with softening of the fruit (Pesis, Fuchs and Zauberman, 1978; Awad and Young, 1979). To determine if this increase in enzyme activity is accompanied by an increase in cellulase protein Christoffersen, Tucker and Laties (1984) prepared extracts of total protein from ripening avocado fruit at three successive phases. Each extract was electrophoresed on an SDS-polyacrylamide gel, and then immunoblotted with cellulase antiserum (*Figure 15.1*). The results show that the extract of preclimacteric (unripe) fruit had a very low level of detectable antigen at 53 000 dalton, the molecular weight of purified cellulase. As ripening proceeded the amount of immunoreactive antigen increased.

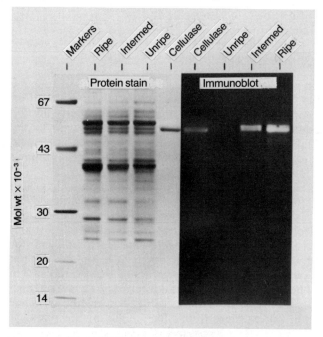

Figure 15.1 Immunoblot electrophoresis of avocado extracts separated by molecular weight and challenged with avocado cellulase antiserum. Avocado protein extracts and purified cellulase were electrophoretically separated on an SDS-polyacrylamide gel, then either stained with Coomassie blue or immunoblotted on to nitrocellulose and assayed by a double antibody using fluorescein conjugated goat anti-rabbit IgG as the second antibody. (Christoffersen, Tucker and Laties, 1984)

Total protein was extracted from fruit which had been treated with $10\,\mu l\,l^{-1}$ ethylene for 40 h to induce the climacteric. This protein was analysed by two dimensional gel electrophoresis (separated by isoelectric point and then by molecular weight) and immunoblotted with cellulase antiserum. The results shown in *Figure 15.2* indicate that in climacteric avocado there are at least three antigenically active polypeptides of approximately 53 000 having isoelectric points between 5.6 and 6.2. The apparent double band at approximately 53 000 in the

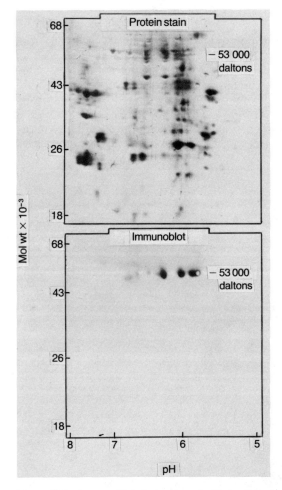

Figure 15.2 Immunoblot electrophoresis of avocado extracts separated by two-dimensional electrophoresis and subsequently challenged with avocado cellulase antiserum. Avocado protein extracted from ripe fruit was electrophoretically separated in two dimensions and either stained with silver or immunoblotted on to nitrocellulose and assayed for antibody binding to antigen by use of ^{125}I-labelled protein A

immunoblot shown in *Figure 15.1* may be due to the fact that the three or more polypeptides of approximately 53 000 in ripe fruit are only partially resolved by separation on the basis of molecular weight.

Coomassie staining of protein extracts after electrophoresis in one dimension do not show a large increase of the polypeptide at 53 000 during ripening (*Figure 15.1*). The cellulase protein therefore probably underlies a more abundant polypeptide that is present in the preclimacteric as well as early and late climacteric extracts. This is borne out by the meagre amount of protein detectable after staining a two-dimensional gel with silver at the locus where cellulase would be observed. The location of this position was determined by an immunoblot of a replicate gel (*Figure 15.2*).

166 Induction of cellulase by ethylene in avocado fruit

ACCUMULATION OF AN IMMUNOPRECIPITABLE *IN VITRO* TRANSLATION PRODUCT

The *in vitro* translation products of polysomal poly(A)+RNA isolated from avocado fruit treated with ethylene for 0, 10, 24 and 48 h respectively, showed an increase in a 53 000 dalton polypeptide (*Figure 15.3*). This protein had the same molecular weight as purified avocado cellulase (*Figure 15.1*). Furthermore, antiserum against avocado cellulase selectively immunoprecipitated a 53 000 translation product from ethylene treated fruit (*Figure 15.3*). The relative concentration of the 53 000 translation product can be expressed as the ratio of the absorbance of the 53 000 band on the fluorograph in *Figure 15.3* to the sum of the absorbances of all

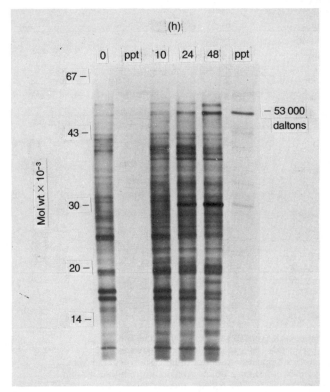

Figure 15.3 Fluorograph of electrophoretically separated *in vitro* translation products of avocado fruit polysomal poly(A)+ RNA. The number above each lane refers to hours of ethylene treatment. Antiserum made against avocado fruit cellulase was used to precipitate antigen in 0 and 48 h samples of *in vitro* translation products. (Ivarie and Jones, 1979)

the translation products. These values can then be used to quantify increases in the 53 000 polypeptide with time. These relative absorbance values are plotted in *Figure 15.4* in relation to the change in carbon dioxide evolution and polysome prevalence. The increase in the relative concentration of the 53 000 *in vitro* translation product (*Figure 15.4*) correlates with the increase in antigenically active protein (*Figure 15.3*) and the accumulation of cellulase activity (Awad and Lewis, 1980).

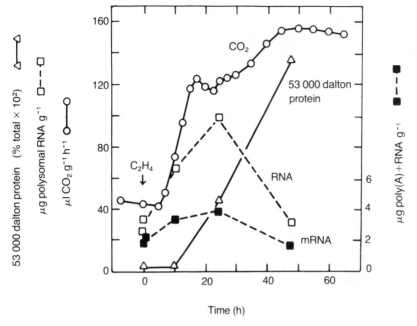

Figure 15.4 Effect of 10 μl l⁻¹ ethylene treatment of avocado fruit on respiration, polysome prevalence, poly(A)+ RNA extracted from polysomes, and the relative accumulation of a 53 000 dalton polypeptide in the *in vitro* translations of poly(A)+ RNA. Carbon dioxide curve is an average of three separate samples. All fruit were picked and kept at 25 °C in a continuous flow of water-saturated air for 24 hours prior to treatment with ethylene. Individual fruits were homogenized and RNA extracted after 0, 10, 24 and 48 hours of ethylene treatment. The two 0 h samples shown for RNA quantification were extracted 24 and 72 hours respectively after detachment from the tree

DETECTION OF mRNA CODING FOR CELLULASE BY USE OF A cDNA PROBE

Christoffersen, Tucker and Laties (1984) synthesized cDNA from poly(A)+RNA extracted from ripe avocado and subsequently cloned this cDNA library in pBR322. Ripening specific cDNA clones were selected by differential colony hybridization wherein replica nitrocellulose filters containing the bacterial colonies were probed with either unripe or ripe single strand cDNA radiolabelled with ^{32}P. One clone (pAV5), which gave a strong differential signal between the unripe and ripe fruit, was subsequently found to have homology to mRNA coding for an *in vitro* translation product of 53 000 which could be immunoprecipitated by cellulase antiserum (Christoffersen, Tucker and Laties, 1984).

The denatured cDNA probe, pAV5, was fixed to nitrocellulose and hybridized to poly(A)+RNA from 40 h ethylene treated fruit. The hybrid-selected mRNA was released and subsequent *in vitro* translation yielded a 53 000 polypeptide product (*Figure 15.5*). The specificity of this probe for mRNA coding for cellulase was ascertained by mixing the *in vitro* translation product of pAV5 hybrid-selected message (*Figure 15.5*, release) with translation products of total polysomal poly(A)+RNA from unripe fruit (*Figure 15.5*, 0 h), and subsequently challenging the mixture with antiserum against avocado cellulase. The translation product of

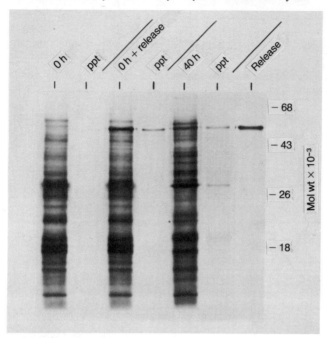

Figure 15.5 Characterization of pAV5 by immunoprecipitation of hybrid released *in vitro* translation products. Time refers to hours of ethylene treatment. Tracks depict fluorograph of *in vitro* translation products labelled with ^{35}S methionine. (0 h) mRNA from unripe fruit; (0 h + release) unripe fruit mRNA plus hybrid-released mRNA from ripe fruit; (40 h) mRNA from ripe fruit; (release) hybrid-released mRNA from ripe fruit

the hybrid-selected message was selectively immunoprecipitated (*Figure 15.5*, 0 h + release ppt). By contrast, translation products of total polysomal poly(A)+RNA from unripe (0 h) fruit alone showed no such immunoprecipitable band at 53 000 (*Figure 15.5*, 0 h ppt).

The relative concentrations of total cellular mRNA homologous to the probe in unripe and ripe avocado fruit were quantified by dot blot hybridizations (*Figure 15.6*). The difference in initial slopes of the titration curves in *Figure 15.6* is at least 50-fold. This apparent 50-fold induction of cellulase mRNA is considered a minimum value.

In vitro translation products of polysomal poly(A)+RNA from 40 h ethylene treated fruit were analysed by two dimensional electrophoresis and visualized by fluorography (*Figure 15.7*, total). As noted in the figure, there are at least two polypeptides at 53 000 having isoelectric points between 5.6 and 5.9. After 48 h of ethylene treatment, four or more polypeptides of approximately 53 000 are evident, with isoelectric points between 5.4 and 6.0 (Tucker and Laties, 1984). By contrast, the fluorograph of the *in vitro* translation products of the hybrid-selected mRNA from the same RNA population shows only a single polypeptide of 53 000 (*Figure 15.7*, release). The minor polypeptides of lower molecular weight in *Figure 15.7* (release) are thought to be degradation products of the 53 000 polypeptide, caused by a four-month storage period at $-20\,°C$.

Tucker and Laties (1984) have proposed that a family of cellulase genes are switched on during avocado ripening and give rise to a group of *in vitro* translation

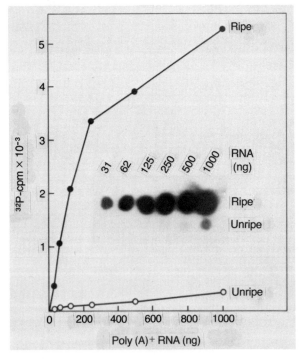

Figure 15.6 Dot hybridization of pAV5 with poly(A)+ RNA from unripe and ripe avocado fruit. Increasing amounts of poly(A)+ RNA extracted from either unripe or ripe fruit were applied to nitrocellulose filters as dots. Filters were hybridized with nick translated ^{32}P DNA from pAV5. Filters were autoradiographed and spots cut out for scintillation counting. The inset shows the pAV5 probe hybridized to the following amounts of RNA: 31, 62, 125, 250, 500, 1000 ng of poly(A)+ RNA from unripe and ripe avocado fruit. The figure shows the corresponding counts measured by scintillation counting for unripe and ripe RNA

products of 53 000 daltons with isoelectric points between 5.4 and 6.0. This proposal is supported by the immunodetection of at least three antigenically active polypeptides of approximately 53 000 in two-dimensional gels (*Figure 15.2*). The occurrence of several 53 000 polypeptides is deemed real, since it would seem unlikely for an artefact to occur in the two-dimensional electrophoresis of total protein extracts (*Figure 15.2*) and total *in vitro* translation products (*Figure 15.7*), and not occur in similar electrophoresis of translation products of hybrid-selected message (*Figure 15.7*, release). Thus it is proposed that the cDNA probe, pAV5, has homology to only one member of this cellulase gene family.

The pAV5 clone contains an insert of 640 base pairs. Electrophoresis of poly(A)+RNA from ripe fruit, followed by subsequent hybridization to labelled pAV5, indicates that the probe has homology with a single message of approximately 2000 bases (Christoffersen, Tucker and Laties, 1984). Accordingly, the cloned DNA insert contains no more than one-third of the full cellulase message sequence. The 2000 base mRNA homologous to pAV5 is long enough, if completely translated, to yield a polypeptide of 75 000. This calculation implies that approximately one-third of the message is not translated into protein. The previous proposal regarding a cellulase gene family is plausible in that the probe, which is

170 Induction of cellulase by ethylene in avocado fruit

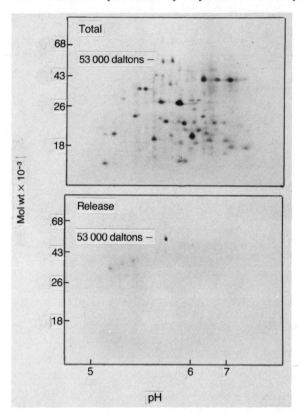

Figure 15.7 Fluorograph of a two-dimensional electrophoretic analysis of the *in vitro* translation products of poly(A)+ RNA from 40 h ethylene treated fruit (total, and pAV5 hybrid-selected message from the same 40 h poly(A)+ RNA extract (release)

one-third the length of the homologous mRNA, may be homologous to the untranslated non-conserved region of the message. Experiments are under way to synthesize a longer cDNA probe for cellulase mRNA to determine the presence or absence of a family of cellulase genes, some or all of which are expressed in avocado fruit ripening.

Concluding remarks

The build-up of cellulase enzyme activity in ripening avocado fruit is thought to be due to the accumulation of cellulase mRNA during the climacteric, which in turn controls the amount of cellulase protein. In this view the increase of cellulase activity during ripening is a manifestation of differential gene expression.

The means by which the level of cellulase mRNA is regulated remains to be established. Nichols and Laties (1984) have demonstrated by nuclear run-off transcription that several ethylene-induced mRNAs of carrot roots are regulated at the level of transcription. It seems probable therefore that the synthesis of cellulase mRNA which takes place during ethylene-promoted ripening of avocado fruit also occurs at the transcriptional level. However, the prospect of other levels of control must as yet remain open.

Acknowledgements

We wish to thank Laura DeFrancesco for invaluable technical assistance and discussion. We thank Lowell Lewis and Mary Durbin for the generous gift of avocado fruit cellulase antiserum.

References

AWAD, M. and LEWIS, L.N. (1980). *Journal of Food Science*, **45**, 1625–1628
AWAD, M. and YOUNG, R.E. (1979). *Plant Physiology*, **64**, 306–308
BRADY, C.J. and O'CONNELL, P.B.H. (1976). *Australian Journal of Plant Physiology*, **3**, 301–310
CHRISTOFFERSEN, R.E., TUCKER, M.L. and LATIES, G.G. (1984). *Plant Molecular Biology*, **3**, 384–392
CHRISTOFFERSEN, R.E., WARM, E. and LATIES, G.G. (1982). *Planta*, **155**, 52–57
IVARIE, R.D. and JONES, P.P. (1979). *Analytical Biochemistry*, **97**, 24–35
NICHOLS, S.E. and LATIES, G.G. (1984). *Plant Molecular Biology*, **3**, 393–402
PESIS, E., FUCHS, G. and ZAUBERMAN, G. (1978). *Plant Physiology*, **61**, 416–419
RICHMOND, A. and BIALE, J.B. (1966). *Biochemical and Biophysical Acta*, **138**, 625–627
SACHER, J.A. (1973). *Annual Review of Plant Physiology*, **24**, 197–224
TUCKER, M.L. and LATIES, G.G. (1984). *Plant Physiology*, **74**, 307–315

16

ETHYLENE AND ABSCISSION

R. SEXTON
Department of Biological Science, Stirling University, Stirling, UK
L.N. LEWIS
Molecular Plant Biology, University of California, Berkeley, California, USA
A.J. TREWAVAS and P. KELLY
Department of Botany, Edinburgh University, Edinburgh, UK

Introduction

There is a wealth of evidence from many different abscission systems that ethylene is a potent accelerator of this process. In this chapter we will attempt to review what is known about the physiology and biochemistry of this effect and address the problem of whether ethylene has any role in natural abscission. To facilitate this discussion it is first necessary to become familiar with the general features of abscission. The following account attempts merely to outline the more important aspects of the process and readers are referred to two recent reviews for further details and references (Addicott, 1982; Sexton and Roberts, 1982). Only papers that have appeared since these reviews will be quoted.

The general features of abscission

The term abscission is used to describe the process whereby various structures are shed from the parent plant. The process is characterized by cell wall breakdown at the point of weakening and involves the active participation of the cells either side of the fracture line.

If a wide range of higher plants are surveyed it is possible to find examples where virtually any aerial portion of the plant is abscised. In the majority of cases however the process is limited to the shedding of leaves, bud scales, immature flowers, fruits, petals and other floral structures. The point at which fracture will occur is probably genetically determined and takes place at predictable positions known as 'abscission zones'.

Examination of abscission zones reveals that they are anatomically similar to adjacent regions of the stem, petiole or pedicel through which they pass. The major difference is that the cells are generally smaller and the stele which usually branches in the zone, also lacks lignified fibres. Cell separation characteristically does not occur throughout the entire abscission zone but is limited to a discrete 1–3 cell wide 'separation layer' at the distal end. Prior to the induction of abscission, separation layer cells cannot usually be distinguished, however there is recent evidence suggesting that they are cryptically differentiated (Osborne, 1982).

174 *Ethylene and abscission*

The process of cell separation is generally induced experimentally by excising the abscission zone along with small amounts of the adjacent tissue. The 'explant' is then enclosed in a humid atmosphere to which ethylene (1–50 μl l^{-1}) is often added to accelerate and synchronize the process. The time course of abscission can be followed by determining the breakstrength or force necessary to rupture the abscission zone as it weakens. Typically there is a lag period directly after excision and this is followed by a progressive localized weakening of the separation layer (*Figure 16.1*). The length of the time course is very variable. Abscission zones from leaves and fruit, which are used in the majority of experiments, may take as long as 30–150 h, whilst flower buds and more exceptionally petals can complete the process in 2–4 h (Sexton, Struthers and Lewis, 1983).

During the lag phase the cells of the separation layer become distinguishable from those in the rest of the abscission zone since they accumulate cytoplasm and organelles. The rates of respiration and RNA and protein synthesis increase during this period. Inhibitor studies have shown that RNA and proteins made during this phase are essential if abscission is to occur.

The period of breakstrength decline is characterized by the breakdown of the cell walls in the separation layer. Wall hydrolysis is reported to start at one or two loci

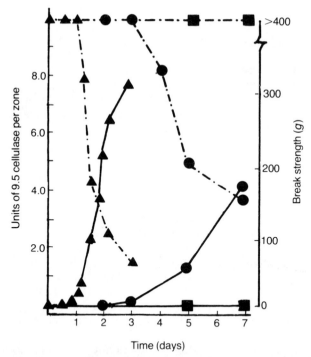

Figure 16.1 The effect of ethylene (50 μl l^{-1}) and hypobaric pressures (125 mmHg) on the abscission of bean leaf explants. The breakstrength ·—·—· of the lower primary leaf abscission zone is plotted after various incubation times in cabinets with ethylene (▲) or air (●) flowing through them. Note that the lag before weakening commences is shortened by ethylene and the rate of breakstrength decline is increased. The levels of 9.5 cellulase ——— assayed by immunoprecipitation (Durbin, Sexton and Lewis, 1981) are similarly increased by ethylene treatment. Hypobaric pressures (■) stop both breakstrength decline and cellulase production. (From Kelly *et al.*, 1984)

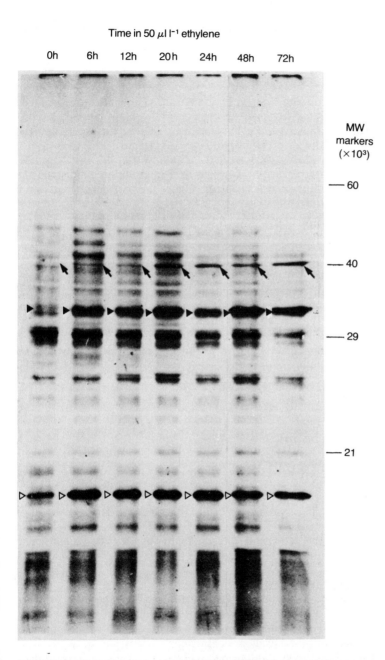

Figure 16.2 *In vitro* translation products of the mRNA extracted from bean abscission zones explants after various time periods in ethylene (50 µl l^{-1}). A product with a molecular weight of 42 000 appears within the first 6 h after excision (arrow) and two other products with molecular weights of 32 000 and 17 000 (▲△) increase substantially. (From Kelly *et al.*, 1984)

in the separation layer and spread laterally until all the cells are involved. Since the separation layer bisects the whole organ (i.e. stem, petiole, pedicel, etc.) it contains many different cell classes, however all the living cell types including sieve elements, degrade their walls.

Little is known of the chemistry of the wall changes though Ca^{++} is lost and pectins are solubilized. EM cytochemistry shows that the middle lamella becomes swollen and the staining properties of adjacent areas of the primary wall change to varying extents dependent on the cell type. Final fracture is facilitated by the development of mechanical stresses across the separation layer. Many mechanisms have evolved to produce these forces but in bean leaves for example, the contraction of the pulvinus distal to the fracture plane coupled with expansion of the petiole proximal to it produces stresses at the junction that aid final rupture. The freshly exposed scar surfaces are usually covered in rounded cells since the fracture line runs along the middle of the cell wall. Where the mechanical forces are large some cell rupture may also occur.

The breakdown of the wall is assumed to be enzymatically mediated since protein synthesis inhibitors effectively stop abscission. Two wall hydrolases have been shown in many abscission systems to appear in the separation layer in a way consistent with some role in cell separation. They are β1:4 glucan 4 glucan hydrolase E.C.3.2.1.4. (cellulase) and polygalacturonase E.C.3.2.1.15. Recently there has been a single paper suggesting a hemicellulase participation (Hashinaga *et al.*, 1981), although further study of uronic acid oxidase (Huberman and Goren, 1982) indicates this enzyme is probably not directly involved in abscission. The increase in cellulase in bean abscission zones is thought to be due to '*de novo*' synthesis of the enzyme. Kelly *et al.* (1984) have recently shown that new mRNA species are formed in bean abscission zones 6 h after excising the explant and exposing it to ethylene.

Ethylene and abscission: a historical review

The discovery that ethylene accelerates abscission can be traced back to the observations of Girardin (1864) and Kny (1871) both of whom reported that leaking illuminating gas caused defoliation of surrounding vegetation. Later Fitting (1911) who was investigating petal abscission found that the minute concentrations of the gas in laboratory atmospheres would cause the rapid loss of petals from several species. After the active component of coal gas was identified as ethylene both Harvey (1913) and Doubt (1917) showed that ethylene at very low concentrations (1–8 µl l^{-1}) would cause leaf and cotyledon abscission, a finding later extended to flowers, fruits and petals (Crocker, 1948).

The next advance came when Gane (1934) showed that fruits could synthesize ethylene and later many respiring plant tissues were found to have this capacity (Crocker, 1948). Ethylene produced by apples was used for the commercial defoliation of roses (Milbrath, Hansen and Hartman, 1940). In spite of these observations the notion that ethylene might be involved in natural abscission does not seem to have been given serious consideration and the effect was viewed as a curious artefact.

Much of the abscission research between 1930 and 1950 concentrated on the newly discovered inhibition of the shedding of both leaves and fruits by auxin. It had been established that removal of the leaf blade would hasten the abscission of the remaining petiole and that the inhibitory role of the lamina could be replaced either by IAA or a synthetic substitute. In general it was assumed that IAA production by the lamina or immature fruit normally inhibited abscission but as the leaf or fruit matured the supply dwindled and cell separation ensued (reviewed by Addicott, 1970, 1982).

Interest in the role of ethylene in abscission was rekindled as a consequence of the quest to find defoliants to facilitate harvesting. Gawadi and Avery (1950) proposed that one such substance ethylene chlorohydrin acted by accelerating the ageing of immature leaves. They also suggested that ethylene, which was known to be produced by senescing tissues, had a similar role in natural abscission. Their 'ethylene auxin balance' theory proposed that the induction of abscission was determined by the relative balance in the production of inhibitory auxin and accelerating ethylene in the subtending organ. Hall (1952) tested this theory by showing that if the ethylene concentration in young leaves was raised they would abscise in spite of the fact that they had high levels of endogenous auxin. He also showed that the abscission of petiole bases was dependent on relative concentrations of auxin and ethylene applied to them.

It seems surprising that so little was done to follow up these suggestions during the next decade. However, the development of the gas chromatographic measurement of ethylene and the refinement of Livingston's (1950) explant technique for the production of reproducible abscising material, set the scene for a resurgence of interest which started in the mid-1960s and continues to the present day.

The characteristics of the response

The acceleration of abscission by ethylene or ethylene generating sprays must be one of the most consistent and widely demonstrated growth substance responses. *Table 16.1* illustrates the wide range of plant genera and structures which show the effect. The threshold concentration necessary to speed abscission is generally $100\,\text{nl}\,\text{l}^{-1}$ while $10\,\mu\text{l}\,\text{l}^{-1}$ usually saturates the response (*Table 16.2*). The degree to which ethylene stimulates abscission varies, there are many examples where treated tissues have completely abscised before there is any evidence of abscission in air held controls (*Figure 16.1*). In some cases the effect of ethylene is less pronounced, for instance *Impatiens* leaf explants reach 50% abscission at 18 h in air and 13 h in $25\,\mu\text{l}\,\text{l}^{-1}$ ethylene. However it is important to bear in mind that ethylene additions will only be effective if the natural levels of the gas are subsaturating. Abeles (1973) only lists four examples where ethylene does not apparently accelerate abscission and since this review Osborne (1983) reported no effect on the shedding of grass seeds. Similarly Chan, Corley and Seth (1972) could find no acceleration of fruit abscission in oil palm. Once weakening of the abscission zone has started the continual presence of ethylene is necessary if the acceleration is to be maintained (dela Fuente and Leopold, 1969; Abeles and Leather, 1971; Stead and Moore, 1983).

The characteristics of the abscission response are similar to other ethylene effects such as those on pea stem growth, fruit ripening and flower senescence. Different unsaturated gases can substitute for ethylene with the same degrees of efficacy as they do in other systems (*Table 16.3*; Abeles and Gahagan, 1968). CO_2 is an inhibitor of ethylene induced abscission (Abeles and Holm, 1966; Jordan, Morgan

Table 16.1 EXAMPLES OF ABSCISSION SYSTEMS WHERE ETHYLENE OR ETHYLENE GENERATING SPRAYS HAVE BEEN SHOWN TO ACCELERATE ABSCISSION

	Senior author	Year	Reference
Flowers			
Begonia	Hanisch ten Cate Ch.	(1973)	*Acta Bot. Neerl.*, **22**, 675.
Tobacco	Henry, E.W.	(1974)	*Plant Physiol.*, **54**, 192.
Apple	Edgerton, L.J.	(1971)	*Hort. Science*, **6**, 26.
Ecballium	Wong, C.H.	(1978)	*Planta*, **139**, 103.
Pistachio	Crane, J.C.	(1982)	*J. Am. Soc. Hort. Sci.*, **106**, 14.
Zygocactus	Cameron, A.C.	(1981)	*Hort. Science*, **16**, 761.
Lilium	Durieux, A.J.	(1982)	*Sci. Hort.*, **18**, 267.
Leptospermum	Zieslin, N.	(1983)	*Physiol. Plant.*, **58**, 144.
Tomato	Roberts, J.A.	(1984)	*Planta*, **160**, 159
Petals			
Geranium	Sexton, R.	(1983)	*Protoplasma*, **116**, 179.
Foxglove	Stead, A.D.	(1983)	*Planta*, **157**, 15.
Ovary			
Orange	Goldschmidt, E.R.	(1971)	*Am. J. Bot.*, **58**, 14.
Style			
Lemon	Sipes, D.L.	(1982)	*Physiol. Plant.*, **56**, 6.
Young fruit			
Avocado	Davenport, T.L.	(1982)	*J. Exp. Bot.*, **33**, 815.
Apple	Edgerton, L.J.	(1971)	*Hort. Sci.*, **6**, 26.
Orange	Huberman, M.	(1979)	*Physiol. Plant*, **45**, 189.
Pecan	Wood, B.W.	(1983)	*Hort. Sci.*, **18**, 53.
Prune	Shaybany	(1977)	*J. Am. Soc. Hort. Sci.*, **102**, 501.
Mature fruit			
Orange	Goren, R.	(1976)	*Physiol. Plant*, **37**, 123.
Tomato	Biain de Elizalde M.	(1980)	*Phyton*, **38**, 71.
Peppers	Batal, K.M.	(1982)	*Hort. Sci.*, **17**, 944.
Grape	Lane, R.P.	(1979)	*Hort. Sci.*, **14**, 727.
Olive	Martin, G.C.	(1981)	*J. Am. Soc. Hort. Sci.*, **106**, 325.
Lemon	El-Zeftawi, B.W.	(1980)	*J. Hort. Sci.*, **55**, 207.
Cherry	Bukovac, M.J.	(1972)	*Hort. Sci.*, **6**, 387.
Ecballium	Jackson, M.B.	(1972)	*Canad. J. Bot.*, **50**, 1465.
Coffea	Browning, G.	(1970)	*J. Hort. Sci.*, **45**, 223.
Leaves			
Pecan	Martin, G.C.	(1980)	*J. Am. Soc. Hort. Sci.*, **105**, 34.
Bean	Rubinstein, B.	(1964)	*Q. Rev. Biol.*, **39**, 359.
Olive	Lavee, S.	(1981)	*J. Am. Soc. Hort. Sci.*, **106**, 14.
Orange	Ratner, A.	(1969)	*Plant Physiol.*, **64**, 1717.
Mung bean	Curtis, R.W.	(1981)	*Plant Physiol.*, **68**, 1249.
Cotton	Hall, W.C.	(1952)	*Bot. Gaz.*, **113**, 310.
Sambucus	Osborne, D.J.	(1976)	*Planta*, **130**, 203.
Coleus	Abeles, F.B.	(1967)	*Physiol. Plant.*, **20**, 442.
Rubber	Hashim, I.	(1980)	*Ann. Bot.*, **45**, 681.
Mulberry	Hashinaga, F.	(1981)	*Nip. Nogeik. Kaishi*, **55**, 1217.
Poplar	Riov, J.	(1979)	*Plant Cell Environ.*, **2**, 83.
Eucalyptus	Riov, J.	(1979)	*Plant Cell Environ.*, **2**, 83.
Stipules			
Philodendron	Marousky, F.J.	(1979)	*J. Am. Soc. Hort. Sci.*, **104**, 876.
Bud scales			
Aesculus	Sexton, R.	(1983)	Unpublished
Plantlets			
Lemnacea	Negbi, M.	(1972)	*Isr. J. Bot.*, **21**, 108.
Stem			
Oak	Chaney, W.R.	(1972)	*Canad. J. For. Res.*, **2**, 492.
Bean	Webster, B.D.	(1972)	*Bot. Gaz.*, **133**, 292.
Mistletoe	Livingston, W.H.	(1983)	*Plant Disease*, **67**, 909.
Axillary shoots			
Bean	Sexton, R.	(1983)	Unpublished

Table 16.2 THE MINIMUM AND SATURATING CONCENTRATIONS OF ETHYLENE ($\mu l\ l^{-1}$) NECESSARY TO ACCELERATE ABSCISSION

Organ	Min.	Sat.	Senior author	Date	Reference
Begonia flowers	0.1	—	Ten Cate Ch.H.	1975	*Physiol. Plant*, **33**, 280
Foxglove corolla	0.01	10	Stead, A.D.	1983	*Plant*, **157**, 15
Geranium petals	0.1	10	Armitage, A.M.	1980	*J. Am. Soc. Hort. Sci.*, **105**, 562
Avocado fruit	1.0	100	Davenport, T.L.	1982	*J. Exp. Bot.*, **35**, 815
Bean leaves	0.1	10	Abeles, F.B.	1968	*Plant Physiol.*, **43**, 1255
Cotton leaves	1.0	90	Beyer, E.M.	1971	*Plant Physiol.*, **48**, 208
Sambucus leaves	0.1	10	Osborne, D.J.	1976	*Planta*, **132**, 197
Plantlets spirodela	0.1	10	Negbi, M.	1972	*Isr, J. Bot.*, **21**, 108
Cotton fruit	0.1	0.5	Lipe, J.A.	1973	*Plant Physiol.*, **51**, 949
Leptospermum flowers	1.0	10	Zieslin, N.	1983	*Physiol. Plant.*, **58**, 114

Table 16.3 THE RELATIVE EFFECTIVENESS OF ETHYLENE AND OTHER UNSATURATED GASES ON BEAN LEAF ABSCISSION AND PEA STEM EXTENSION BASED ON THE CONCENTRATION NECESSARY TO OBTAIN HALF MAXIMAL STIMULATION (FROM ABELES AND GAHAGAN, 1968)

Compound	Abscission	Stem growth
Ethylene	1	1
Propene	60	100
Carbon monoxide	1250	2700
Acetylene	1250	2800
Vinyl fluoride	2500	4300
1-Butene	100 000+	270 000
1-3 Butene	100 000+	5 000 000

and Davenport, 1972; Marousky and Harbaugh, 1979), it probably acts competitively (Abeles and Gahagan, 1968). One unusual feature of abscission is that oxygen tension seems to have little effect on ethylene sensitivity (Abeles and Gahagan, 1968) though Marynick and Addicott (1976) did find oxygen levels above 20% stimulated cotton abscission. Silver ions are as effective at inhibiting abscission as they are with other ethylene responsive systems (Beyer, 1976; Sagee, Goren and Rieu, 1980; Cameron and Reid, 1981; Curtis, 1981; Zieslin and Gottesman, 1983). Methionine was known to stimulate abscission before its role as an ethylene precursor was established (Yager and Muir, 1958; Valdovinos and Muir, 1965; Abeles, 1967; Chatterjee and Chatterjee, 1971), and ACC has recently been shown to accelerate stylar abscission (Sipes and Einset, 1982). Surprisingly Lavee and Martin (1981) could get no effect of ACC on olive leaves though the authors suggest that this was because it was too rapidly converted to ethylene to give a response.

Ethylene sensitivity

Although ethylene accelerates abscission in most systems, responsiveness or sensitivity can be very variable (Abeles, 1973). *Lilium* flower buds provide a good illustration of the dramatic nature of these changes (Durieux, Kamerbeek and Van Meeteren, 1983). Buds that have been developing less than four weeks or more than six weeks will not abscise in ethylene while in the intervening period all buds respond. Differing sensitivities can also be demonstrated by placing whole plants in ethylene and showing that abscission of different organs is far from synchronous. Study of the same organ in different varieties similarly reveals considerable variations in response (Craker and Abeles, 1969b; Hanisch ten Cate and Bruinsma, 1973a). The physiological state of the tissue is also important. As a general rule the abscission zones of leaves, fruit and flowers become more sensitive as they get older (Abeles, 1973; Wong and Osborne, 1978; Biain de Elizalde, 1980; Stead and Moore, 1983). However there are a number of exceptions and partially expanded cotton leaves, for instance, are more sensitive than mature ones (Morgan and Durham, 1973).

Other factors besides physiological age can alter sensitivity, water stress makes cotton leaves more responsive (Jordan, Morgan and Davenport, 1972), while sprays of calcium salts usually decrease sensitivity (Poovaiah and Leopold, 1973; Iwahori and Oohata, 1980; Martin, Campbell and Carlson, 1980; Martin, Lavee and Sibbert, 1981). The presence and condition of the subtending organ also influences the abscission zone. If a leaf is delaminated it will usually abscise more rapidly in low levels of ethylene.

The changes in sensitivity of explant systems have also been the subject of much research. Firstly the age and physiological state of the plant at the time of excision can influence sensitivity (Jackson, Hartley and Osborne, 1973). Secondly once the explant itself has been cut its response to ethylene is thought to change dramatically. The simplest demonstration of this effect is to expose explants to short periods in ethylene (usually 12 h) at various times after excision and then evaluate their effectiveness at some fixed end point. Most authors using this type of approach have found that explants do not respond to ethylene added directly after excision but later additions accelerate abscission (Abeles and Rubinstein, 1964; Abeles, 1967; 1969; de la Fuente and Leopold, 1968; Jackson and Osborne, 1970; Hänisch ten Cate, and Bruinsma, 1973b; Webster and Leopold, 1972). These data have led to the idea that there is an ethylene insensitive stage just after explant preparation and many authors have left or 'aged' their explants after cutting to allow ethylene sensitivity to develop. The status of this insensitivity is in some doubt however as a result of Abeles, Craker and Leather's reinvestigations (1971). They showed that ethylene administered during the insensitive period produced no measurable response but it enhanced the rate of abscission if a second later application of ethylene was made. Their work showed that ethylene added in the 'insensitive' period brought forward the time known as stage 2 during which further additions of ethylene speeded wall degradation (Abeles, Craker and Leather, 1971). Not all explants show an insensitive period (Pollard and Biggs, 1970; Stead and Moore, 1983; Ismail, 1970) and as Burg (1968) pointed out this apparent insensitivity to ethylene could simply be due to the fact that the system is saturated already with wound ethylene which is produced as a result of excision.

The auxin status of the abscission zone is considered to be a major factor regulating its sensitivity to ethylene (Abeles, 1973; Riov and Goren, 1979). This conclusion has been reached primarily on the basis of extensions of Hall's (1952) original observation that if shoots were treated with auxin before being placed in ethylene, defoliation was dramatically reduced (Addicott, 1965; Abeles, Holm and Gahagan, 1967; dela Fuente and Leopold, 1968; Abeles, 1968; Craker and Abeles, 1969a; Hallaway and Osborne, 1969; Morgan, 1969, 1976; Craker, Chadwick and Leather, 1970; Jackson and Osborne, 1972; Osborne, 1973, 1982; Hänisch ten Cate and Bruinsma, 1973b; Riov, 1974; Osborne and Sargent, 1976b; Greenberg, Goren and Riov, 1975; Einset, Lyon and Johnson, 1981). Most authors seem to hold the view that the auxin status of the abscission zone is largely determined by supply from the subtending organ and thus factors which influence this supply may alter the sensitivity of the zone. In leaves the physiological state of the lamina is obviously important and if it senesces or is damaged so that the auxin supply is reduced then the ethylene sensitivity of the zone would be expected to increase. The importance of the auxin supply was illustrated by Morgan and Durham (1972) who showed that auxin transport inhibitors applied to the petiole considerably enhanced the ethylene sensitivity of the abscission zone. The removal of the source of auxin is also used to explain the increased sensitivity of explants after lamina excision. Ethylene is a potent inhibitor of auxin transport and thus might well elevate the sensitivity of the zone to itself by reducing the flow of auxin from the leaf blade (Beyer and Morgan, 1971; Beyer, 1973). This effect was illustrated by Beyer (1975) in a simple but very elegant experiment. He developed a system whereby the leaf blade and abscission zone could be exposed to ethylene independently. If the zone itself or leaf blade alone were treated, then no abscission occurred, however if both were treated together the leaf was shed. The importance of ethylene in reducing auxin supply from the lamina was deduced from further experiments either augmenting the auxin supply from a lamina held in ethylene, or by the use of auxin transport inhibitors on laminas held in air. Ethylene is thought to reduce the levels of auxin predominantly through a reduction of auxin transport (Beyer and Morgan, 1971; Beyer, 1973; Riov and Goren, 1979) although enhanced auxin deactivation (Gaspar *et al.*, 1978) and reduction of synthesis (Ernest and Valdovinos, 1971) may play some part.

There have been few detailed studies of the dose response interaction between IAA and ethylene. It seems that the two are antagonistic to the extent that ethylene in the range 0.1–$100\,\mu l\,l^{-1}$ increasingly reduces the inhibitory effect of IAA application and auxin in the range 10^{-5}–10^{-2} M increasingly inhibits the accelerating effect of ethylene (Addicott, 1965, 1970; Osborne and Sargent, 1976b). IAA can be very inhibitory at low ethylene concentrations increasing the time until 50% abscission several-fold; however most experiments have been terminated too quickly to be certain whether auxin totally blocks ethylene accelerated abscission.

It has been known for many years that there is only a limited period after explant preparation during which auxin additions are strongly inhibitory (Rubinstein and Leopold, 1963; Chatterjee and Leopold, 1963; Abeles and Rubinstein, 1964). This period is known as stage 1 which is followed by stage 2 in which auxin additions either have little effect or may actually speed abscission if the explants are held in air (*see below*). The duration of stage 1 gets shorter as the leaf from which the explant is cut gets older (Chatterjee and Leopold, 1965). It has been suggested that the duration of stage 1 reflects the time taken for the auxin levels initially present in the explant to fall to a non-inhibiting level (Jackson and Osborne, 1972). Adding auxin simply keeps the level up. Stage 1 still occurs if explants are kept in ethylene (*Figure 16.3*; Craker and Abeles, 1969a; Craker, Chadwick and Leather, 1970; Greenberg, Green and Riov, 1975; Hänisch ten Cate and Bruinsma, 1973b) though in bean its duration is decreased (Abeles *et al.*, 1971). In our experiments the transition from stage 1 to stage 2 (i.e. IAA sensitivity to insensitivity) is not as abrupt as often portrayed (*Figure 16.3*). In bean explants held in ethylene additions of 10^{-3} M IAA cause similar inhibitions if administered up to 15 h after explant preparation, but additions after 27 h have only a slight retarding influence.

182 Ethylene and abscission

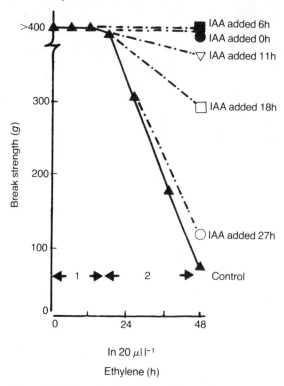

Figure 16.3 The effects of adding IAA on the responsiveness of bean explants kept in ethylene (20 μl l^{-1}). If IAA ·—· (10^{-3}M) is added to explants as soon as they are excised (●) then the explants do not weaken by 48 h even if held in ethylene (20 μl l^{-1}). If IAA additions are made after 6 h in ethylene (■) then the treatment is equally effective, however by 11 h (▽) and 18 h (□) some of the explant population escape from the inhibitory influence of IAA and limited weakening occurs. Additions at 27 h (○) produce weakening rates very similar to that of the controls ——. Stage I is defined as the period when IAA additions delay abscission (approximately 12–18 h) whilst stage 2 is the period when auxin has no effect (Sexton, Durbin and Lewis, unpublished)

Between 15 and 27 h we obtain mixed populations of explants the majority of which show intermediate states of inhibition (*Figure 16.3*). Jackson and Osborne's (1972) data similarly suggest an intermediate period in which IAA is still effective but only if added before ethylene.

Cytokinins are also able to inhibit abscission by extending the duration of stage 1 (Gorter, 1964; Chatterjee and Leopold, 1964) though the effect is less researched than the auxin response. Abeles, Holm and Gahagan (1967) produced evidence that cytokinins also reduced ethylene sensitivity of the tissue, however it should be noted that some authors have found cytokinins to produce no inhibition (Hänisch ten Cate and Bruinsma, 1973b).

It is clear that if we are to understand abscission we must be able to explain what causes these changes in responsiveness both to ethylene and to other abscission regulating substances. The first factor that has been identified concerns the differentiation of the separation layer. Wong and Osborne (1978) showed that *Ecballium* flower buds will not abscise until endopolyploid cells have differentiated in the abscission zone and when they have developed only the 8C cells respond to ethylene. Unfortunately not all separation layers are composed of such 8C cells but there is increasing evidence of the existence of a differentiated class of 'target cell' (Politio and Stallman, 1981; Osborne, 1982).

While the differentiation of the separation layer may explain the switching on or switching off of responsiveness to abscission regulating substances it is unlikely to account for relative changes in sensitivity. One possible explanation of the latter is that sensitivity depends on the interaction with receptors in the tissue and it is levels or affinities of the receptors that determine the extent of the response (Trewavas, 1982). In the case of the loss of IAA sensitivity during abscission preliminary evidence points in this direction (Jaffe and Goren, 1979). However it is difficult to envisage an increase in ethylene sensitivity being due to the synthesis or activation of a receptor alone. While an abscission zone is insensitive to ethylene to the extent that it does not weaken, it may still be responsive in other ways, perhaps in showing an epinastic response or by exhibiting a reduction in auxin transport in the presence of ethylene. So if the answer involves receptors, more than one type of ethylene receptor must be envisaged. A further explanation has been provided by Beyer (1979) who suggests that ethylene metabolism and particularly its oxidation may determine the extent of ethylene action. His results show that when cotton is debladed, a treatment which sensitizes the tissue to ethylene, ethylene oxidation in the abscission zone increases. Naphthalene acetic acid applied to the zone markedly inhibits both abscission and ethylene oxidation. Thus it is conceivable that it is the appearance of the ethylene oxidation pathway that determines the responsiveness of the zone.

An alternative possibility is that factors which change abscission zone sensitivity do so by an indirect effect on some aspect of the cell's metabolism that is important during abscission. Whatever the explanation, it is apparent that changes in sensitivity are just as important in regulating when abscission occurs as are changes in the concentrations of abscission regulating substances, such as ethylene.

Ethylene and natural abscission

The question arises as to whether ethylene has any role in natural abscission. At this stage we will not attempt to distinguish whether it is inducing the process or merely accelerating it (*see below*). The most common approach to the problem is to try and establish a parallel between increasing ethylene levels and the onset of abscission. A sample of the large number of positive correlations are listed in *Table 16.4*. Some are temporal correlations, where an increase in ethylene production precedes abscission and others establish significant differences between the behaviour of abscising and non-abscising populations. Not all authors have reported simple parallels (*Table 16.4*) some finding no change in ethylene levels during abscission (Hänisch ten Cate *et al.*, 1975b; Witztum and Keren, 1978). One problem with these types of data is that there is a very wide range of opinion as to what constitutes an acceptable correlation. For instance some authors consider a gap of several days between increased ethylene production and abscission acceptable, while others seem to require the two events to be almost synchronous. Attempts to overcome this difficulty with the use of statistical tests (Marynick, 1977) have done little to clarify the situation.

The abrupt changes in ethylene sensitivity complicate interpretation. They are used in the literature to explain why some bursts of ethylene production do not

Table 16.4 A: STUDIES WHERE A POSITIVE CORRELATION BETWEEN ETHYLENE LEVELS AND ABSCISSION IS CLAIMED

Tissue	Inductive treatment	Reference
Apple, Cherry, Flowers	Pollination	Blanpied (1972)
Apple, Raspberry, Fruit	Natural	Blanpied (1972)
Cotton cotyledons	Natural	Beyer and Morgan (1971)
Cotton fruit	Dim light	Guinn (1982)
Tumble weeds	Natural	Zeroni *et al.* (1978)
Bean leaves	Explant excision	Jackson and Osborne (1970)
Parthenocissus leaves	Natural	Jackson and Osborne (1970)
Lily buds	Dark period	Van Meeteren and de Proft (1982)
Digitalis corolla	Pollination	Stead and Moore (1983)
Citrus leaves	Bacterial infection	Gotto *et al.* (1980)
Peanut leaves	Fungal infection	Ketring and Melouk (1982)
Orange leaves	Explant excision	Ben Yehoshua and Aloni (1974)
Coleus leaves	Mercury vapour	Goren and Siegel (1976)
Young cotton fruit	Water deficit	Guinn (1976)
Citrus leaves	Cold	Young and Meredith (1971)
Avocado fruit	Nucellar senescence	Davenport and Manners (1982)
Ecballium fruits	Natural	Jackson *et al.* (1972)
Bean leaves	Explant excision	Jackson *et al.* (1973)
Peach fruit	Natural	Jerie (1976)
Cotton leaves	Fungal infection	Wiese and De Vay (1970)
Okra fruit	Natural	Lipe and Morgan (1973)
Cotton fruit	Dark	Vaughan and Bate (1977)

B: SIMILAR STUDIES SHOWING INCONSISTENCIES OR LACK OF CLEAR CORRELATION

Tissue	Inductive treatment	Reference
Geranium petals	Natural	Armitage *et al.* (1980)
Cotton leaves	Water stress	McMichael *et al.* (1972)
Digitalis flowers	Natural	Stead and Moore (1977)
Colonies *Spirodela*	UV light	Witzum and Keren (1978)
Pineapple orange fruit	Cold injury	Young (1972)
Begonia flowers	Excision	Hänisch ten Cate *et al.* (1975b)
Cherry fruit	Natural	Blanpied (1972)
Apple fruit	Natural	Blanpied (1972)
Cotton explants	Excision	Marynick (1977)

cause abscission and similarly to explain why unchanging ethylene levels still do not necessarily preclude the gas from having a role in the process. Thus it is difficult to draw any concrete conclusions from these types of data.

In spite of the problems of having to deal with both changing ethylene levels and changing sensitivity there are nonetheless some striking correlations between increased ethylene and abscission. For instance Ketring and Melouk (1982) reported that only peanut varieties which produce ethylene following infection by *Cercospora* become defoliated. Stead and Moore (1983) showed that corolla abscission in foxglove which starts 6 h after pollination, is preceded by an increased rate of ethylene production, which reaches 11 times that of the unpollinated controls. Dark induced abscission in lily buds is accompanied by an eightfold increase in ethylene production. This increase does not occur in light held plants nor in buds at developmental stages where abscission does not occur (van Meeteren and de Proft, 1982). There are several other very convincing parallels but it is important not to lose sight of the fact that they are merely correlations and do not provide proof of involvement.

To firmly implicate ethylene in the acceleration of abscission it is necessary to show that the endogenous ethylene level increases above the threshold concentrations of $0.1-1\,\mu l\,l^{-1}$ (*Table 16.2*) that are normally effective when added exogenously. Unfortunately it is not so easy, for as Guinn (1976) pointed out, 'Because the sensitivity of organs to ethylene varies with age, physiological condition and the levels of other hormones it is difficult to establish a threshold at which ethylene will definitely cause abscission'. A second complication is that the abscission zone produces ethylene at rates faster than adjacent tissues (Marynick, 1977; Aharoni, Lieberman and Sisler, 1979) and therefore probably contains higher concentrations of ethylene. The only way to determine the levels in the zone is to excise it and then the difficulties of getting meaningful values from tiny pieces of wounded material become formidable.

Some authors have measured internal levels directly but a more common approach is to measure rates of ethylene production and try and establish a rate which would be expected to generate internal ethylene levels that stimulate abscission. Burg (1968) suggested that rates in the range $3-5\,\mu l\,kg^{-1}$ fresh $wt^{-1}\,h^{-1}$ would produce saturating concentrations of ethylene in abscission tissues, and most of the studies in *Table 16.4* have produced values of this order. Using these methods either singly or in combination, Beyer and Morgan (1971), Morgan *et al.* (1972), Lipe and Morgan (1973), Swanson *et al.* (1975), Hänisch ten Cate *et al.* (1975b), Ben-Yehoshua and Aloni (1974), and Morgan and Durham (1980) have all shown internal levels that would stimulate abscission if applied exogenously. A clever alternative was used by Jackson, Hartley and Osborne (1973). They measured the rate of ethylene production in bean just prior to abscission and showed that if CEPA (2-chloroethyl phosphonic acid) was applied to petioles to generate similar rates of ethylene production, these would accelerate abscission.

Another approach is to try and remove endogenous ethylene from naturally abscising systems and show some inhibition. Early experimenters used permanganate and mercuric perchlorate to adsorb external ethylene and by increasing the diffusion gradient reduce internal levels (Jackson and Osborne, 1970; Young and Meredith, 1971). In both these cases abscission was delayed. A more effective approach is to use 'hypobaric' or low pressures to reduce internal ethylene levels. Hypobaric treatments are thought to cause (i) an immediate drop in ethylene partial pressure as gases vent from the tissue, (ii) an increase in diffusive loss from the tissue as the gas phase becomes less dense, (iii) a reduced retention due to low pressure effects on ethylene solubility, and (iv) a reduction in ethylene biosynthesis (and perhaps metabolism) due to limited O_2 availability (Nilsen and Hodges, 1983). A number of authors have shown that hypobaric conditions slow or completely inhibit abscission (*Figure 16.2*) (Jordan, Morgan and Davenport, 1972; Cooper and Horanic, 1973; Lipe and Morgan, 1973; Rasmussen, 1974; Morgan and Durham, 1980; Sipes and Einset, 1982). Of course it is important to show that these effects are specifically due to the removal of ethylene and this can be achieved by adding ethylene back into the hypobaric chambers. Cooper and Horanic (1973) did this successfully but in our hands the rate of abscission in hypobaric + ethylene treatments was slower and more erratic than that produced by the same ethylene concentrations at normal atmospheric pressure (Kelly *et al.*, 1984). Burg and Burg (1966) obtained a similar result with fruit ripening and provided experimental proof that another effect of low pressures is to decrease ethylene sensitivity.

Internal ethylene levels can also be reduced by the use of the ethylene synthesis inhibitor amino-ethoxyvinylglycine (AVG). This will slow natural abscission in

Citrus (Sagee, Goren and Riov, 1980; Einset, Lyon and Johnson, 1981; Sipes and Einset, 1982), apple (Bangerth, 1978; Williams, 1980; Greene, 1980), tomato (Roberts, Schindler and Tucker, 1984) and avocado (Davenport and Manners, 1982).

Ethylene responses are characterized by being inhibited by Ag^+ and CO_2 and both have been shown to delay abscission. Silver salts inhibit abscission of leaf explants (Sagee, Goren and Riov, 1980), infected leaves (Ketring and Melouk, 1982) and young flowers (Cameron and Reid, 1981, 1983; van Meeteren and de Proft, 1982), though mature flowers of *Leptospermum* do not respond (Zieslin and Gottesman, 1983). CO_2 has been used to inhibit leaf explant abscission in several species (Abeles and Gahagan, 1968) and peach fruit (Jerie, 1976).

Thus there is a considerable body of evidence that ethylene is present at levels which accelerate abscission.

Ethylene mediation of other growth regulator responses

Abeles and Rubinstein (1964) first proposed that the abscission accelerating effects of other growth regulators might be mediated by their ability to stimulate ethylene synthesis. These authors confirmed Morgan and Hall's (1964) observation that auxin stimulates ethylene production and put forward the view that it was the production of ethylene that explained why auxin could accelerate abscission under certain conditions. For instance Rubinstein and Leopold (1963) had shown that auxin additions made soon after explant excision inhibited abscission (stage 1) while delayed auxin additions (made in what became known as stage 2) actually accelerated the processes. Abeles and Rubinstein (1964) provided experimental evidence to support the hypothesis that in stage 2 the tissue had lost its ability to be inhibited by auxin and was thus stimulated to abscise by auxin induced ethylene production. They demonstrated that the ability of different phenoxyacetic acid compounds to accelerate abscission closely paralleled their ability to stimulate ethylene production (Rubinstein and Abeles, 1965; Abeles, 1967). Similarly the acceleration of abscission caused by auxin additions made on the stem (proximal) side of the zone (Addicott, 1970) can also be explained in terms of the stage 1, stage 2 effect. Briefly, auxins applied distal to the zone moved rapidly towards it by polar transport and thus arrive in stage 1 while auxin applied proximally moved slowly by diffusion opposed by polar transport and thus do not reach the zone till stage 2. If steps are taken to get the proximally applied auxin to the zone in stage 1 by reducing the diffusion path or increasing the concentration, abscission is blocked (Abeles, 1967). While these explanations seem to satisfy most workers they have been challenged by others (Addicott, 1982).

The idea that ethylene was involved in other abscission stimulating treatments followed. Certain amino acids, potassium iodide, endoxohexahydrophthalic acid, morphactin and gibberellic acid were all shown to stimulate ethylene formation and abscission (Abeles, 1967; Curtis, 1968; Roberts in Sexton and Roberts, 1982). ABA is also known to stimulate abscission and ethylene production (Cooper *et al.*, 1968; Cracker and Abeles, 1969a; Jackson and Osborne, 1972; Rasmussen, 1974; Sagee, Goren and Riov, 1980). However, the hypothesis that it too has its effect via ethylene mediation is more contentious (Addicott, 1982). The notion is supported by studies which show no additional effect of ABA on an ethylene saturated system (Jackson and Osborne, 1972) and that ABA has no effect if added in the presence of AVG (Sagee, Goren and Riov, 1980). On the other hand several investigators report that ABA can stimulate abscission without affecting ethylene production (Marynick, 1977; Dörffling *et al.*, 1978). Further support for the direct action of ABA comes from experiments showing it is still effective under hypobaric conditions (Cooper and Horanic, 1973; Rasmussen, 1974), in the presence of CO_2, and in saturating levels of ethylene (Craker and Abeles, 1969a). The uncharacterized abscission accelerator 'senescence factor' is also thought to be effective by increasing ethylene levels (Osborne, Jackson and Milborrow, 1972).

The mechanism of ethylene action

It is quite clear that ethylene can accelerate abscission in a number of ways. Firstly it may act indirectly by altering the concentrations of other regulators that can modify the rate of abscission. Many of these indirect effects have been discussed above but a summary list with references to provide keys to the literature are included in *Table 16.5*. In this section we will review the circumstantial evidence that ethylene acts by directly modifying the processes that lead to wall breakdown.

Table 16.5

A   *Indirect effects of ethylene that could accelerate abscission*[1]
  (1) Inhibition of the auxin transport mechanism (Beyer, 1973; Riov and Goren, 1979).
  (2) Enhanced auxin destruction (Riov and Goren, 1979).
  (3) Increased auxin conjugation (Riov and Goren, 1979).
  (4) Decreased auxin synthesis (Ernest and Valdovinos, 1971).
  (5) Reduced Stage I, i.e. auxin sensitivity (Abeles *et al.*, 1971a).
  (6) Increased ABA levels (Shaybany *et al.*, 1977).
  (7) Increased ethylene production (de la Fuente and Leopold, 1968).
  (8) Increased ethylene sensitivity (Abeles *et al.*, 1971).
  (9) Enhanced senescence precipitating a decrease in 1AA and cytokinin and increase in senescence factor, ABA and ethylene.
  (10) Enhanced ripening of distal organ thus increasing ABA, senescence factor and ethylene.
B   *Direct effects thought to be involved in ethylene accelerated abscission*
  (1) Enhancing mRNA transcription and or translatability (Kelly *et al.*, 1984).
  (2) Enhancing synthesis or activation of enzymes where abscission is actinomycin D insensitive (Henry *et al.*, 1974).
  (3) Increasing enzyme secretion (Abeles and Leather, 1971).
  (4) Inducing cell expansion and thus producing tensions across the separation layer (Osborne, 1982).
  (5) Induction of tyloses.

[1]The references represent recent keys to the literature and do not necessarily represent the author's views.

The first evidence that supports the idea of direct involvement, is that abscission responses to ethylene are often so rapid that it is difficult to envisage a role for changes in intermediaries. Flower buds and petals can be completely abscised 2–4 h after adding ethylene (Henry, Valdovinos and Jensen, 1974; Sexton, Struthers and Lewis, 1983; Lieberman, Valdovinos and Jensen, 1982) and presumably the changes precipitating fracture must start considerably earlier. In one early paper Fitting (1911) claimed petal abscission could take place after 20 min exposure to illuminating gas. Even in the slower fruit and leaf abscission zones the rapidity of ethylene responses can be demonstrated. In general if ethylene is added to these systems it shortens the lag before breakstrength decline commences and increases the rate of weakening (*Figure 16.1*; Pollard and Biggs, 1970; Lewis and Varner,

1970; Henry, Valdovinos and Jensen, 1974; Reid et al., 1974; Lieberman, Valdovinos and Jensen, 1982). If ethylene is added to explants which are beginning to weaken in air the rate of breakstrength decline is increased 1–2 h later (de la Fuente and Leopold, 1969; Abeles and Leather, 1971). Similarly the continued presence of ethylene is necessary to maintain this rate of weakening (Greenberg, Goren and Riov, 1975; Stead and Moore, 1983) and if ethylene is withdrawn the rate slows within 30 min (de la Fuente and Leopold, 1969; Abeles and Leather, 1971).

The processes leading to breakstrength decline are dependent on protein synthesis, since inhibitors such as cycloheximide have been shown to prevent weakening in a range of abscission systems (list: Sexton and Roberts, 1982). Results using RNA synthesis inhibitors have been more variable but in general the slower fruit and leaf zones are inhibited (Abeles and Holm, 1966, 1967; Holm and Abeles, 1967; Ratner, Goren and Mouselise, 1969) while some of the faster floral systems are not (Henry, Valdovinos and Jensen, 1974; Hanisch ten Cate et al., 1975a).

To investigate if ethylene had any effect on protein synthesis Abeles and Holm (1966, 1967; Holm and Abeles, 1967) measured the rate of C^{14} leucine and P^{32} incorporation into protein and RNA. They found that the rate of incorporation into RNA and protein was increased 1 and 2 h respectively after adding ethylene. This enhanced synthesis occurred mainly in the separation layer and both ribosomal and mRNA fractions showed increased incorporation (Abeles, 1968). Based on the differential effects of 5-fluorouracil and actinomycin D the authors concluded that it was the synthesis of mRNA that was essential (Abeles, 1968). Since this ethylene enhanced protein synthesis was sensitive to actinomycin D it was proposed that ethylene regulated the synthesis of specific mRNA and other RNA fractions that in turn were necessary for abscission (Abeles, 1968; Abeles et al., 1971). This theory has gained wide acceptance in the literature however it is very important not to lose sight of the fact that ethylene can also stimulate abscission in other systems that are insensitive to transcriptional inhibitors (Henry, Valdovinos and Jensen, 1974; Riov, 1974; Hänisch ten Cate et al., 1975a or b). Thus ethylene may also control abscission at some translational or post-translational step.

Using a bean explant system similar to that employed by Abeles' group Kelly et al. (1984) have recently shown by in vitro translation that new mRNA species do appear in the abscission zone within 6 h of adding ethylene (*Figure 16.2*). These mRNA changes occur more slowly in air held controls but do not take place in explants held in hypobaric conditions. While these changes in the mRNA population are not found in the stem adjacent to the abscission zone they appear in the distal petiolar tissues. In this respect the choice of bean abscission zones may prove unfortunate since while enzymes such as cellulase which are thought to mediate abscission are found principally in the zone, they also are produced to a lesser extent in the petiole (Sexton et al., 1981). In addition adventitious abscission zones are formed very readily in these distal tissues (Poovaiah and Rasmussen, 1974; Morre, 1968; Osborne, 1982).

Kelly et al. (1984) have also demonstrated that two of the new mRNA species which appear during abscission, disappear again if the explants are transferred from ethylene to hypobaric conditions (*Figure 16.4*). Thus in this abscission system there is good evidence for ethylene mediated control of gene expression, though these genes are not necessarily directly concerned with abscission.

Studies of the cell biology of abscission have indicated that weakening of the

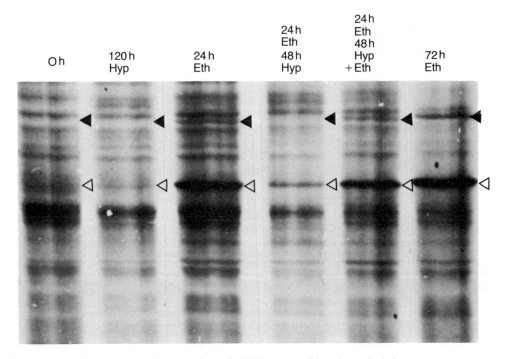

Figure 16.4 The *in vitro* translation products of mRNA extracted from bean abscission zones after various treatments. Lane 1, material collected directly after excision: Lane 2, zones exposed to hypobaric pressures for 120 h: Lane 3, zones held in ethylene (50 μl l^{-1}) for 24 h: Lane 4, zones held in ethylene for 24 h then transferred to hypobaric conditions for 48 h: Lane 5, explants treated as in Lane 4 except that ethylene was added to generate a partial pressure of 5×10^{-5} atm. Lane 6: zones kept for 72 h in ethylene at NTP. Note that the product with a molecular weight of 42 000 daltons (arrows) disappears if placed in hypobaric conditions without added ethylene. Another band with a MW of 32 000 daltons also decreases substantially with similar treatment (From Kelly *et al.*, 1984)

separation layer is dependent on (a) wall degradation by hydrolytic enzymes and (b) the development of stresses across the fracture line that facilitate rupture (Sexton and Roberts, 1982). In addition alterations of the physical environment within the wall (i.e. changes in pH, chelators, etc.) and reduction in the rates of wall synthesis may play some role (Abeles *et al.*, 1971).

Ethylene has been directly implicated in the increase in wall hydrolase activity. There are some 20 papers describing how it increases the rate of cellulase synthesis and shortens the lag before it appears (*Figure 16.2*) (e.g. Abeles, 1969; Ratner, Goren and Monselise, 1969; Wright and Osborne, 1974; Huberman and Goren, 1979; Lewis and Koehler, 1979). In those abscission zones where an increase in polygalacturonase accompanies abscission, ethylene also increases its rate of production (Riov, 1974; Greenberg, Goren and Riov, 1975; Huberman and Goren, 1979). Experiments reducing natural levels of ethylene with AVG or hypobaric pressures, have decreased the production of these hydrolases (Rasmussen, 1974; Sagee, Goren and Riov, 1980; Kelly *et al.*, 1984). Since both translational and in some cases transcriptional inhibitors block the effect of ethylene on wall hydrolase levels (Abeles, 1969; Ratner, Goren and Mouselise, 1969; Riov, 1974; Greenberg, Goren and Riov, 1975; Hanisch ten Cate *et al.*, 1975a), some authors have speculated that the gas is directly involved in the induction of these enzymes (Abeles *et al.*, 1971; Sagee, Goren and Riov, 1980).

Cellulase and polygalacturonase are not the only enzymes in the abscission zone that respond to ethylene. Chitinase, β1.3 glucanase (Abeles *et al.*, 1971) and uronic acid oxidase (Huberman and Goren, 1982) all increase and the peroxidase isozyme profiles also change (Poovaiah and Rasmussen, 1973; Wittenbach and Bukovac, 1975; Henry, Valdovinos and Jensen, 1974; Gaspar *et al.*, 1978). However there is no direct evidence that these changes are directly involved in abscission.

Ethylene has been shown to stimulate the secretion of wall hydrolases (Abeles and Leather, 1971). It also induced increases in rough endoplasmic reticulum (Iwahori and Van Steveninck, 1976; Lieberman, Valdovinos and Jensen, 1983) and changes in the buoyant density of the plasma membrane (Koehler and Lewis, 1979) which may be implicated in this response. It should be noted however that there is no gross collapse in cell permeability (Abeles and Leather, 1971).

In certain abscission zones ethylene may also cause enlargement of cells in and around the fracture line (Wright and Osborne, 1974; Osborne, 1982; Chapter 17). The expansion of these cells may prove important in generating the stresses which will lead to the rupture of tissues, such as xylem vessels and the cuticle, which are not degraded during abscission.

In summary, it seems that in all the range of abscission related processes investigated to date '*Ethylene always acts in a way likely to accelerate abscission*'. In processes as diverse as IAA transport, cell expansion, ethylene biosynthesis, enzyme secretion, senescence, ABA synthesis, its effect is always the same. Since all these elements which influence the rate of abscission involve a wide spectrum of different metabolic and physiological processes and since they take place in a range of different tissues it is clear that besides being an accelerator '*Ethylene acts to coordinate and synchronize the abscission programme*'.

Does ethylene induce abscission?

The literature reviewed above reveals an almost incontrovertible body of evidence suggesting that ethylene may often reach concentrations where it accelerates and coordinates abscission. Some authors go further and suggest it is an essential inducer of the process.

Data that have encouraged this hypothesis can usually be challenged on two grounds. Firstly it is difficult to exclude the possibility that ethylene does not have its effect via some intermediary growth regulator change. Secondly abscission is generally occurring in control experimental systems at a finite rate. Thus it is very difficult to distinguish if ethylene is inducing or merely accelerating an ongoing process. Similarly reduction of ethylene levels is seldom so effective that it totally stops abscission. Terms such as induction, repression and blocking which imply an on–off response usually come from short-term experiments. Here the slow abscission of the controls may well not have been apparent by the time the experiment was terminated, particularly if terminal fracture is used as an index of the rate of progress.

To be certain that ethylene is the inducer of abscission it is necessary to demonstrate an absolute dependence on the regulator and also to understand the

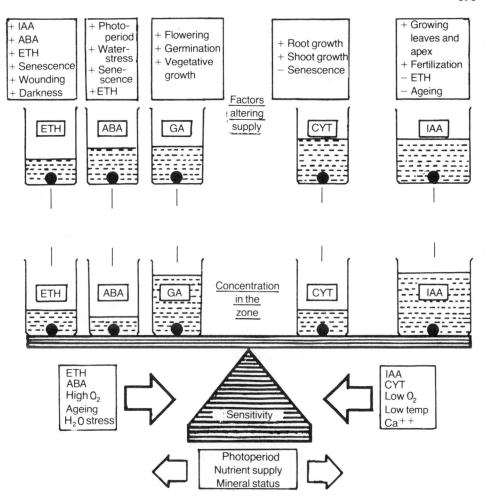

Figure 16.5 A diagrammatic representation of the possible multifactorial control of abscission. Abscission is envisaged as being determined by the relative concentrations of accelerators and inhibitors and the sensitivity of the zone to them. In this diagram the accelerators are shown on the left hand side (LHS) of a beam balance and they are counterbalanced by inhibitors on the right hand side (RHS). If the LHS of the balance goes down, abscission takes place. Ethylene and IAA have been positioned at the ends of the beam since changes in their concentration (weight) are relatively more effective than fluctuations in the regulators positioned nearer the fulcrum. The concentration of each regulator in the zone is determined by the rates of synthesis, transport and loss and this is depicted as a supply dripping into and out of the containers on the beam. Some of the possible factors that increase (+) and decrease (−) the concentrations of each regulator in the zone are shown above the supply tanks. The balance of the beam can also be affected by changing the sensitivity. This is depicted by being able to move the position of the fulcrum along the beam to the right, making the accelerators more effective and to the left decreasing their efficacy. Some of the factors that influence sensitivity are depicted together with arrows indicating how sensitivity is affected. This model oversimplifies the changes in sensitivity since it assumes that responsiveness to all the accelerators will change together; they may well be able to change independently of one another. ETH = ethylene, ABA = abscisic acid, IAA = auxins, CYT = cytokinins, GA = gibberellic acid

molecular events involved in its action. To make such a claim on the basis of present data is clearly premature.

Single or multifactor control?

We are conscious that in writing a review concentrating on the role of a single regulator in abscission we may be fuelling the common assumption that abscission is controlled by a single master hormone. These hypotheses usually envisage that changes in the concentration of either IAA, ABA or ethylene induce wall hydrolase production that in turn causes abscission. Such simple working hypotheses not only seriously underestimate the complexity of the abscission process but they direct attention away from the equally viable alternative, that abscission is under multifactorial control. In *Figure 16.5* we depict diagrammatically a model where abscission is not induced by changes in a single hormone, nor by only one set of hormonal conditions but is determined by a complex interplay of both growth regulator concentrations and tissue sensitivities. We believe that this scheme is in line with much growth regulator work suggesting that abscission can actually be induced or inhibited by a wide range of different treatments.

References

ABELES, F.B. (1967). *Physiologia Plantarum*, **20**, 442–454
ABELES, F.B. (1968). *Plant Physiology*, **43**, 1577–1586
ABELES, F.B. (1969). *Plant Physiology*, **44**, 447–452
ABELES, F.B. (1973). *Ethylene in Plant Biology*. Academic Press, New York, London
ABELES, F.B., CRAKER, L.E. and LEATHER, G.R. (1971). *Plant Physiology*, **47**, 7–9
ABELES, F.B. and GAHAGAN, H.E. (1968). *Plant Physiology*, **43**, 1255–1258
ABELES, F.B. and HOLM, R.E. (1966). *Plant Physiology*, **41**, 1337–142
ABELES, F.B. and HOLM, R.E. (1967). *Annals New York Academy of Sciences*, **144**, 367–373
ABELES, F.B., HOLM, R.E. and GAHAGAN, H.E. (1967). *Plant Physiology*, **42**, 1351–1356
ABELES, F.B. and LEATHER, G.R. (1971). *Planta*, **97**, 87–91
ABELES, F.B., LEATHER, G.R., FORRENCE, L.E. and CRAKER, L.E. (1971). *HortScience*, **6**, 371–376
ABELES, F.B. and RUBINSTEIN, B. (1964). *Plant Physiology*, **39**, 963–969
ADDICOTT, F.T. (1965). *Encyclopaedia of Plant Physiology*, **15/2**, 1094–1126
ADDICOTT, F.T. (1970). *Biological Reviews*, **45**, 485–524
ADDICOTT, F.T. (1982). *Abscission*. California University Press, Berkeley, California
AHARONI, N., LIEBERMAN, M. and SISLER, H.D. (1979). *Plant Physiology*, **64**, 796–800
ARMITAGE, A.M., HEINS, R., DEAN, S. and CARLSON, W. (1980). *Journal of the American Society of Horticultural Science*, **105**, 562–564
BANGERTH, F. (1978). *Journal of the American Society of Horticultural Science*, **103**, 401–404
BEN-YEHOSHUA, S. and ALONI, B. (1974). *The Botanical Gazette*, **135**, 41–44

BEYER, E.M. (1973). *Plant Physiology*, **52**, 1–5
BEYER, E.M. (1975). *Plant Physiology*, **55**, 322–327
BEYER, E.M. (1976). *Plant Physiology*, **58**, 268–271
BEYER, E.M. (1979). *Plant Physiology*, **64**, 971–974
BEYER, E.M. and MORGAN, P.W. (1971). *Plant Physiology*, **48**, 208–212
BIAIN DE ELIZALDE, M.M. (1980). *Phyton*, **38**, 71–79
BLANPIED, G.P. (1972). *Plant Physiology*, **49**, 627–630
BURG, S.P. and BURG, E.A. (1966). *Science*, **153**, 314–315
BURG, S.P. (1968). *Plant Physiology*, **43**, 1503–1511
CAMERON, A.C. and REID, M.S. (1981). *HortScience*, **16**, 761–762
CAMERON, A.C. and REID, M.S. (1983). *Scientia Horticulturae*, **19**, 373–378
CHAN, K.W., CORLEY, R.H.V. and SETH, A.K. (1972). *Annals of Applied Biology*, **71**, 243–249
CHATTERJEE, S. and CHATTERJEE, S.K. (1971). *Indian Journal of Experimental Biology*, **9**, 485–488
CHATTERJEE, S. and LEOPOLD, A.C. (1963). *Plant Physiology*, **38**, 268–273
CHATTERJEE, S.K. and LEOPOLD, A.C. (1964). *Plant Physiology*, **39**, 334–337
CHATTERJEE, S.K. and LEOPOLD, A.C. (1965). *Plant Physiology*, **40**, 96–101
COOPER, W.C. and HORANIC, G. (1973). *Plant Physiology*, **51**, 1002–1004
COOPER, W.C., RASMUSSEN, G.K., ROGERS, B.J., REECE, P.C. and HENRY, W.H. (1968). *Plant Physiology*, **43**, 1560–1576
CRAKER, L.E. and ABELES, F.B. (1969a). *Plant Physiology*, **44**, 1144–1149
CRAKER, L.E. and ABELES, F.B. (1969b). *Plant Physiology*, **44**, 1139–1143
CRAKER, L.E., CHADWICK, A.V. and LEATHER, G.R. (1970). *Plant Physiology*, **46**, 790–793
CROCKER, W. (1948). *Growth of Plants*. Reinhold Publishing Co., New York
CURTIS, R.W. (1968). *Plant Physiology*, **43**, 76–80
CURTIS, R.W. (1981). *Plant and Cell Physiology*, **22**, 789–796
DAVENPORT, T.L. and MANNERS, M.M. (1982). *Journal of Experimental Botany*, **33**, 815–825
DE LA FUENTE, R.K. and LEOPOLD, A.C. (1968). *Plant Physiology*, **43**, 1496–1502
DE LA FUENTE, R.K. and LEOPOLD, A.C. (1969). *Plant Physiology*, **44**, 251–254
DÖRFFLING, K., BOTTGER, M., MARTIN, V., SCHMIDT, D. and BOROWSKI, D. (1978). *Physiologia Plantarum*, **43**, 292–296
DOUBT, S. (1917). *The Botanical Gazette*, **63**, 209–224
DURBIN, M.L., SEXTON, R. and LEWIS, L.N. (1981). *Plant Cell and Environment*, **4**, 67–73
DURIEUX, A.J.B., KAMERBEEK, G.A. and VAN MEETEREN, U. (1983). *Scientia Horticulturae*, **18**, 287–297
EINSET, J.W., LYON, J.L. and JOHNSON, P. (1981). *Journal of the American Society of Horticultural Science*, **106**, 531–533
ERNEST, L.C. and VALDOVINOS, J.G. (1971). *Plant Physiology*, **48**, 402–406
FITTING, H. (1911). *Jahrbuch Wissenschaftliche Botanik*, **49**, 187–263
GANE, R. (1934). *Nature*, **134**, 1008–1009
GASPAR, T., GOREN, R., HUBERMAN, M. and DUBUCQ, M. (1978). *Plant Cell and Environment*, **1**, 225–230
GAWADI, A.G. and AVERY, G.S. (1950). *American Journal of Botany*, **37**, 172–180
GIRARDIN, J.P.L. (1864). *Jahresber. Agricult-Chem. Versuchssta, Berlin*, **7**, 199–200
GOREN, R. and HUBERMAN, M. (1976). *Physiologia Plantarum*, **37**, 123–130
GOREN, R. and SIEGEL, S.M. (1976). *Plant Physiology*, **57**, 628–631

GORTER, C.J. (1964). *Physiologia Plantarum*, **17**, 331-345
GOTTO, M., YAGUCHI, Y. and HYODO, H. (1980). *Physiological Plant Pathology*, **16**, 343-350
GREENBERG, J., GOREN, R. and RIOV, J. (1975). *Physiologia Plantarum*, **34**, 1-7
GREENE, D.W. (1980). *Journal of the American Society of Horticultural Science*, **105**, 717-720
GUINN, G. (1976). *Plant Physiology*, **57**, 403-405
GUINN, G. (1982). *Plant Physiology*, **69**, 349-352
HALL, W.C. (1952). *The Botanical Gazette*, **113**, 310-322
HALLAWAY, M. and OSBORNE, D.J. (1969). *Science*, **163**, 1067-1068
HÄNISCH TEN CATE, Ch.H. and BRUINSMA, J. (1973a). *Acta Botanica Neerlandica*, **22**, 666-674
HÄNISCH TEN CATE, Ch.H. and BRUINSMA, J. (1973b). *Acta Botanica Neerlandica*, **22**, 675-680
HÄNISCH TEN CATE, Ch.H., VAN NETTER, J., DORTLAND, J.F. and BRUINSMA, J. (1975a). *Physiologia Plantarum*, **33**, 276-279
HÄNISCH TEN CATE, Ch.H., BERGHOEFF, J., VAN DER HOORN, A.M.H. and BRUINSMA, J. (1975b). *Physiologia Plantarum*, **33**, 280-284
HARVEY, E.M. (1913). *The Botanical Gazette*, **56**, 439-442
HASHINAGA, F., IWAHORI, S., NISHI, Y. and ITOO, S. (1981). *Nipon Nogeikagaku Kaishi*, **55**, 1217-1223
HENRY, E.W., VALDOVINOS, J.G. and JENSEN, T.E. (1974). *Plant Physiology*, **54**, 192-196
HOLM, R.E. and ABELES, F.B. (1967). *Plant Physiology*, **42**, 1094-1102
HUBERMAN, M. and GOREN, R. (1979). *Physiologia Plantarum*, **45**, 189-196
HUBERMAN, M. and GOREN, R. (1982). *Physiologia Plantarum*, **56**, 168-176
ISMAIL, M.A. (1970). *Journal of the American Society of Horticultural Science*, **95**, 319-322
IWAHORI, S. and OOHATA, J.T. (1980). *Scientia Horticulturae*, **12**, 265-271
IWAHORI, S. and VAN STEVENINCK, R.F.M. (1976). *Scientia Horticulturae*, **4**, 235-246
JACKSON, M.B., HARTLEY, C.B. and OSBORNE, D.J. (1973). *New Phytologist*, **72**, 1251-1260
JACKSON, M.B., MORROW, I.B. and OSBORNE, D.J. (1972). *Canadian Journal of Botany*, **50**, 1465-1471
JACKSON, M.B. and OSBORNE, D.J. (1970). *Nature*, **225**, 1019-1022
JACKSON, M.B. and OSBORNE, D.J. (1972). *Journal of Experimental Botany*, **23**, 849-862
JAFFE, M.J. and GOREN, R. (1979). *The Botanical Gazette*, **140**, 378-383
JERIE, P.H. (1976). *Australian Journal of Plant Physiology*, **3**, 747-755
JORDAN, W.R., MORGAN, P.W. and DAVENPORT, T.L. (1972). *Plant Physiology*, **50**, 756-758
KELLY, P., TREWAVAS, A.J., SEXTON, R., DURBIN, M.L. and LEWIS, L.N. (1984). In preparation
KETRING, D.L. and MELOUK, H.A. (1982). *Plant Physiology*, **69**, 789-792
KNY, L. (1871). *Botanische Zeitung*, **29**, 852-854
KOEHLER, D.E. and LEWIS, L.N. (1979). *Plant Physiology*, **63**, 677-679
LAVEE, S. and MARTIN, G.C. (1981). *Plant Physiology*, **67**, 1204-1207
LEWIS, L.N. and KOEHLER, D.E. (1979). *Planta*, **146**, 1-5
LEWIS, L.N. and VARNER, J.E. (1970). *Plant Physiology*, **46**, 194-199
LIEBERMAN, S.J., VALDOVINOS, J.G. and JENSEN, T.E. (1982). *The Botanical Gazette*, **143**, 32-40

LIEBERMAN, S.J., VALDOVINOS, J.G. and JENSEN, T.E. (1983). *Plant Physiology*, **72**, 583–585
LIPE, J.A. and MORGAN, P.W. (1973). *Plant Physiology*, **51**, 949–951
LIVINGSTON, G.A. (1950). *Plant Physiology*, **25**, 711–721
MAROUSKY, F.B. and HARBAUGH, B.K. (1979). *Journal of the American Society of Horticultural Science*, **104**, 876–880
MARTIN, G.C., CAMPBELL, R.C. and CARLSON, R.M. (1980). *Journal of the American Society of Horticultural Science*, **105**, 34–37
MARTIN, G.C., LAVEE, S. and SIBBERT, G.S. (1981). *Journal of the American Society of Horticultural Science*, **106**, 325–330
MARYNICK, M.C. (1977). *Plant Physiology*, **59**, 484–489
MARYNICK, M.C. and ADDICOTT, F.T. (1976). *Nature*, **264**, 668–669
McMICHAEL, B.L., JORDAN, W.R. and POWELL, R.D. (1972). *Plant Physiology*, **49**, 658–660
MILBRATH, J.A., HANSEN, E. and HARTMAN, H. (1940). *Science*, **91**, 100
MORGAN, P.W. (1969). *Plant Physiology*, **44**, 337–341
MORGAN, P.W. (1976). *Planta*, **129**, 275–276
MORGAN, P.W. and DURHAM, J.I. (1972). *Plant Physiology*, **50**, 313–318
MORGAN, P.W. and DURHAM, J.I. (1973). *Plant Physiology*, **52**, 667–670
MORGAN, P.W. and DURHAM, J.I. (1980). *Plant Physiology*, **66**, 88–92
MORGAN, P.W. and HALL, W.C. (1964). *Nature*, **201**, 99
MORGAN, P.W., KETRING, D.L., BEYER, E.M. and LIPE, J.A. (1972). In *Plant Growth Substances 1970*, pp. 502–509. Ed. by D.J. Carr. Springer Verlag, Berlin
MORRE, D.J. (1968). *Plant Physiology*, **43**, 1545–1559
NILSEN, K.N. and HODGES, C.F. (1983). *Plant Physiology*, **71**, 96–101
OSBORNE, D.J. (1973). In *Shedding of Plant Parts*, pp. 125–148. Ed. by T.T. Kozlowski. Academic Press, London
OSBORNE, D.J. (1982). In *Plant Growth Substances*, pp. 279–290. Ed. by P.F. Wareing. Academic Press, London
OSBORNE, D.J. (1983). *News Bulletin of the British Plant Growth Regulator Group*, **6**, 8–11
OSBORNE, D.J., JACKSON, M.B. and MILBORROW, B.V. (1972). *Nature New Biology*, **240**, 98–101
OSBORNE, D.J. and SARGENT, J.A. (1976a). *Planta*, **130**, 203–210
OSBORNE, D.J. and SARGENT, J.A. (1976b). *Planta*, **132**, 197–204
POLITO, V.S. and STALLMAN, V. (1981). *Scientia Horticulturae*, **15**, 341–347
POLLARD, J.E. and BIGGS, R.H. (1970). *Journal of the American Society of Horticultural Science*, **95**, 667–673
POOVAIAH, B.W. and LEOPOLD, A.C. (1973). *Plant Physiology*, **51**, 848–851
POOVAIAH, B.W. and RASMUSSEN, H.P. (1973). *Plant Physiology*, **52**, 263–267
POOVAIAH, B.W. and RASMUSSEN, H.P. (1974). *American Journal of Botany*, **61**, 68–73
RASMUSSEN, G.K. (1974). *Journal of the American Society of Horticultural Science*, **99**, 229–231
RATNER, A., GOREN, R. and MONSELISE, S.P. (1969). *Plant Physiology*, **44**, 1717–1723
REID, P.D., STRONG, H.G., LEW, F. and LEWIS, L.N. (1974). *Plant Physiology*, **53**, 732–737
RIOV, J. (1974). *Plant Physiology*, **53**, 312–316
RIOV, J. and GOREN, R. (1979). *Plant, Cell and Environment*, **2**, 83–89
ROBERTS, J.A., SCHINDLER, B. and TUCKER, G.A. (1984). *Planta*, **160**, 159–163
RUBINSTEIN, B. and ABELES, F.B. (1965). *The Botanical Gazette*, **126**, 255–259

RUBINSTEIN, B. and LEOPOLD, A.C. (1963). *Plant Physiology*, **38**, 262–267
SAGEE, O., GOREN, R. and RIOV, J. (1980). *Plant Physiology*, **66**, 750–753
SEXTON, R., DURBIN, M.L., LEWIS, L.N. and THOMSON, W.W. (1981). *Protoplasma*, **109**, 335–347
SEXTON, R. and ROBERTS, J.A. (1982). *Annual Review of Plant Physiology*, **33**, 133–162
SEXTON, R., STRUTHERS, W.A. and LEWIS, L.N. (1983). *Protoplasma*, **116**, 179–186
SHAYBANY, B., WEINBAUM, S.A. and MARTIN, G.C. (1977). *Journal of the American Society of Horticultural Science*, **102**, 501–503
SIPES, D.L. and EINSET, J.W. (1982). *Physiologia Plantarum*, **56**, 6–10
STEAD, A.D. and MOORE, K.G. (1977). *Annals of Botany*, **41**, 283–292
STEAD, A.D. and MOORE, K.G. (1983). *Planta*, **157**, 15–21
SWANSON, B.T. Jr., WILKINS, H.F., WEISER, C.F. and KLEIN, I. (1975). *Plant Physiology*, **55**, 370–376
TREWAVAS, A.J. (1982). *Physiologia Plantarum*, **55**, 60–72
VALDOVINOS, J.G. and MUIR, R.M. (1965). *Plant Physiology*, **40**, 335–340
VAN MEETEREN, U. and DE PROFT, M. (1982). *Physiologia Plantarum*, **56**, 236–240
VAUGHAN, A.K.F. and BATE, G.C. (1977). *Rhodesian Journal of Agricultural Research*, **15**, 51–63
WEBSTER, B.D. and LEOPOLD, A.C. (1972). *The Botanical Gazette*, **133**, 292–298
WIESE, M.V. and DE VAY, J.E. (1970). *Plant Physiology*, **45**, 304–309
WILLIAMS, M.W. (1980). *Hortscience*, **15**, 77–78
WITTENBACH, V.A. and BUKOVAC, M.J. (1975). *Journal of the American Society of Horticultural Science*, **100**, 387–391
WITZTUM, A. and KEREN, O. (1978). *New Phytologist*, **80**, 107–110
WONG, C.H. and OSBORNE, D.J. (1978). *Planta*, **139**, 103–111
WRIGHT, M. and OSBORNE, D.J. (1974). *Planta*, **120**, 163–170
YAGER, R.E. and MUIR, R.M. (1958). *Proceedings of the Society of Experimental Biology and Medicine*, **99**, 321–323
YOUNG, R. (1972). *American Society of Horticultural Science*, **97**, 133–135
YOUNG, R. and MEREDITH, F. (1971). *Plant Physiology*, **48**, 724–727
ZIESLIN, N. and GOTTESMAN, V. (1983). *Physiologia Plantarum*, **58**, 114–118
ZERONI, M., HOLLANDER, E. and ARZEE, T. (1978). *The Botanical Gazette*, **139**, 299–305

17

TARGET CELLS FOR ETHYLENE ACTION

DAPHNE J. OSBORNE
Developmental Botany, Weed Research Organization, Oxford
MICHAEL T. McMANUS
Department of Biochemistry, University of Oxford
and JILL WEBB
Electron Microscopy, Weed Research Organization, Oxford

Introduction

It is very evident that when plants are exposed to ethylene their subsequent growth and development is modified and patterns of differentiation are changed. Because, as far as we know, all plants produce ethylene, one may concede that every cell of a plant is probably subject to regulation by ethylene at some stage in its life-cycle. But equally every cell must be influenced also by other hormonal inputs, by environmental signals such as light and temperature and, in certain circumstances, by nutrients.

It is also very evident that differentiation and development do not result solely from quantitative changes in the expression of the genome. As cells pass from one stage of maturation to another, phase specific molecules, usually proteins, are synthesized which are diagnostic molecular markers of the achievement of a particular phase of developmental determination.

An example is perhaps in order here. During embryogenesis in many dicotyledonous plants a highly specialized developmental programme unfolds in cells of those parts of the embryo that are destined to function as the storage organs of the seed, e.g. the cotyledons. Within these cells, protein synthesis becomes directed to producing massive amounts of embryo-specific proteins, globulins, lectins and specific protease inhibitors. The environmental or hormonal signal(s) that lead to this switch are as yet unknown but it is clear that they are precisely developmentally timed to occur at a particular stage after fertilization. Recognition of the signal by cotyledon cells results in the expression of genes which are not expressed in any other tissues of the plant at any period of growth. Messenger RNAs isolated from polysomes of soybean cotyledons contain unique sets of superabundant embryo-specific mRNAs not found in other tissues of the plant (Goldberg *et al.*, 1981a,b). It is inferred that the messages are transcribed from specific DNA sequences coding for seed proteins. The genes that code for cotyledon specific proteins must however be present in the genetic information of all cell-lines but are 'turned on' for a short period only during a particular stage of cotyledon development and remain 'turned off' during the rest of the soybean life-cycle.

Whatever signal evokes this gene expression requires a specific molecular complement within the cotyledon cells to ensure that they (but no other cells in the plant) are competent to recognize and respond to that signal. Cotyledon cells are, by this definition, specific targets for that signal. The discussion that now follows concerns physiological, biochemical and preliminary ultrastructural evidence which suggests that specific types of target cells for ethylene are differentiated in green plants.

Target cells for ethylene

In aerial shoots, modifications in the phenotypic expression and pattern of development by ethylene reflect recognizably distinct types of growth behaviour in immature cells. Because of the precise and specific nature of the expansion growth induced by ethylene in different cell types, each can be considered as an example of a particular ethylene target. In other words, each cell type possesses a particular competence for recognizing the ethylene signal and responding in a predetermined way. In a general classification, three major target cell types to ethylene have been described (Osborne, 1979, 1982).

(1) The first ones, designated Type 1, are by far the most numerous in plants and by far the most well studied. They are typical of the elongating subapical tissue of most aerial dicotyledonous shoots. As one example, when intact plants of *Pisum sativum* (either green or etiolated) are exposed to ethylene, elongation growth of the shoots is reduced but their lateral expansion growth is enhanced. Although the final volume of the enlarging cells is not significantly altered by ethylene, the orientation of deposition of new cell wall cellulose microfibrils and the composition of cell wall protein and polysaccharide are changed (Veen, 1970; Terry, Rubinstein and Jones, 1981) and the volume of the cell is accommodated in a more isodiametric shape (Osborne, 1976). By studying excised segments of immature shoot tissues it has been shown that ethylene has no growth stimulatory effect alone but will direct a reorientation of expansion growth that is induced either by auxin (Burg and Burg, 1966) or by gibberellin (Stewart, Lieberman and Kunishi, 1974). Pratt and Goeschl (1969) and more recently Eisinger (1983) have reviewed this type of cell growth in some detail. Although such cells can be recognized by the reorientated enlargement that is induced in response to ethylene, we have as yet no cytological or molecular markers to link the cells with this particular ethylene response. Other than the position that the cells occupy and the knowledge of the way that they will normally react to ethylene there are no known molecular determinants on which to base a *prediction* of this type of cell growth in ethylene.

(2) In marked contrast, another class of target cell, Type 3, shows enhanced expansion growth with ethylene and with auxin. Most of the cells in the stems and petioles of plants of semi-aquatic environments are of this type. Limited numbers also occur in land plants that exhibit pronounced epinasty (*Lycopersicon* and *Helianthus*; Palmer, 1976) and are differentiated in the adaxial tissues of the base of the petioles. In intact plants these cells exhibit enhanced elongation when the plant is exposed to ethylene and the asymmetric growth of the petiole results in the typical epinastic downward curvature. When the aerial parts of semi-aquatic plants are submerged the endogenously produced ethylene, which then accumulates within the tissues, is the signal for the promotion of cell elongation that brings the submerged shoot back to the water surface (Musgrave, Jackson and Ling, 1972; Cookson and Osborne,

1978). Ethylene does not initiate the growth response on its own for when excised segments are aged for several hours to deplete them of other endogenous hormones, the addition of ethylene fails to cause elongation. However when some auxin is supplied, the ethylene enhanced growth response by segments is restored (Walters and Osborne, 1979; Malone and Ridge, 1983). In some semi-aquatic plants, e.g. *Callitriche platycarpa*, the ethylene induced 'super-growth' can occur in conjunction with either gibberellin (Musgrave, Jackson and Ling, 1972) or auxin. Some cell growth appears to be an essential requirement for an ethylene-induced increase in cell elongation to operate. Although these cells display such a specific response to ethylene, so far we know of no biochemical markers that permit these kinds of cells to be identified as ethylene targets *before* an ethylene response is evoked.

(3) The main thrust of this discussion is now directed to another class of ethylene responsive target cell which is positionally differentiated at abscission zones in dicotyledons. These cells, designated Type 2, are produced in limited numbers only where ephemeral organs (e.g. leaves, floral parts, fruit) are shed from the main body of the plant. They may constitute a plate of one or at the most two rows of cells across the junction of the organ with the parent plant as in the distal pulvinus–petiole abscission zone of the primary leaf of *Phaseolus vulgaris*, or they may constitute a thin layer of tissue some 15 or more cells wide as in the rachis abscission zones of the compound leaf of *Sambucus nigra*.

In the immature condition the cells of *P. vulgaris* leaf abscission zones are characterized by enlargement and a loss of mutual adhesion in response to ethylene and by a repression of these enlargement and separation events by auxin (Wright and Osborne, 1974). Further, they are distinguished by the ethylene induction and auxin repression of specific enzymes, including a basic isozyme of β-1:4-glucanhydrolase. This isozyme of cellulase (pI 9.5) is associated with middle lamella dissolution and precisely localized loss of cell to cell adhesion that leads to the eventual shedding of the leaf (Horton and Osborne, 1967; Lewis and Koehler, 1979; Koehler *et al.*, 1981). The production and secretion of the novel abscission zone-specific cellulase isozyme results from an ethylene induced *de novo* synthesis of the protein (Lewis and Varner, 1970). The synthesis of novel mRNAs is also induced by ethylene (and repressed by auxin) in leaf abscission zones of *P. vulgaris* and the induction of these mRNAs is linked with the production of new proteins prior to cell separation (Sexton *et al.*, Chapter 16).

Clearly, abscission zone cells, like those of the developing cotyledon, are good candidates for the possession of specific molecular differentiation determinants marking their competence to recognize and respond to a specific signal. For abscission zone cells in dicotyledons, whether they be of fruit or leaves, all the present evidence points to ethylene being that signal, and as far as we know the pI 9.5 isozyme of β-1:4-glucanhydrolase is expressed only in those cells that exhibit the ethylene induced (and auxin repressed) enlargement and separation.

Specific markers for target cells

CYTOLOGICAL MARKERS

The first evidence for zone cell markers was sought at the cytological level in the abscission region of the female flower buds of a member of the *Cucurbitaceae*. In this family endoreduplication of nuclear DNA occurs as a normal event as the tissue ages. In unopened buds of *Ecballium elaterium*, the region between the base of the ovary and its adjoining pedicel will separate in $10\,\mu l\,l^{-1}$ ethylene but only after the bud has reached a certain critical stage of development (6 mm long). At this size, but not before, ethylene will induce enlargement in distinct pockets of cells in the presumptive separation region. The cells within the pockets start to separate and produce a discontinuous line of weakness like the perforations in a tear-off strip of paper (Wong and Osborne, 1978). As a result, the bud is shed.

The cells that are caused to enlarge by ethylene possess larger nuclei than those that do not enlarge. More importantly, they can be identified as potential ethylene responsive cells by an endoreduplicated (8C) nuclear DNA content when compared with their neighbour non-enlarging cells which contain 4C DNA. The evidence from these studies indicated that only those cells with 8C nuclei were target cells for ethylene and that abscission did not occur until cells in the position of the presumptive zone had attained the 8C phase of developmental determination. There was no evidence that amplification of the DNA was of itself instrumental in the abscission response, rather, the 8C state was a marker of the competence of a cell to respond to the ethylene signal.

BIOCHEMICAL MARKERS

Studies with the multilayer rachis abscission zone of *Sambucus nigra*, have shown that there is no specific morphological, cytological or ultrastructural feature by which to identify the zone cells *before* they are evoked by ethylene, other than their small size. Once they are induced to separate from each other by ethylene however, certain ultrastructural differences become apparent. These include enlargement of dictyosome stacks with the production of numerous secretory vesicles and considerable proliferation of rough endoplasmic reticulum. These responses to ethylene, suggestive of a high protein synthetic and secretory activity, are not seen in the adjacent non-separating cells. Nor are they seen in zone cells in which separation is repressed by auxin (Osborne and Sargent, 1976).

Secretory activity, cell enlargement and loss of cell adhesion appear as related events but the reasons for their precise localization is not known. Secreted proteins (e.g. β-1:4-glucanhydrolase) readily diffuse in the apoplast to cells remote from those in which they are synthesized. Nonetheless, only those cells that constitute the zone region are caused to enlarge and separate. This suggests that zone cells are also likely to differ from their neighbours in the structure and chemical organization of their cell walls and middle lamellae.

In order therefore to biochemically verify target status for abscission zone cells, it is necessary to demonstrate that they possess unique molecular (protein) determinants. Also, by analogy with many animal systems (e.g. the induction of vitellogenin by oestrogen in mammalian liver; Hayward, Mitchell and Shapiro, 1980) an abscission target cell once 'turned on' by the inducer (ethylene) should be subject to being 'turned off' by withdrawal of that inducer whilst retaining the competence to be 'turned on' again when the inducing agent is reintroduced. Evidence towards fulfilling these criteria for establishing a target status of abscission zone cells is now described.

PROTEIN DETERMINANTS SPECIFIC TO ABSCISSION ZONE CELLS

Although P. vulgaris is an easy plant to grow uniformly and in large numbers, there is difficulty in seeking specific protein determinants in leaf abscission zone cells when the 1–2 rows that constitute the zone comprise less than 20% of an 0.5 mm slice of tissue cut from the dome-shaped pulvinus–petiole junction. There is a more useful system in the rachis abscission zones of S. nigra leaves where just distal to insertion of the leaflets some 15–30 layers of zone cells are present and slices can be excised in which unevoked abscission zone cells (OZ) form the major proportion of the tissue removed. Non-abscinding, non-target cells (MR) can be obtained from

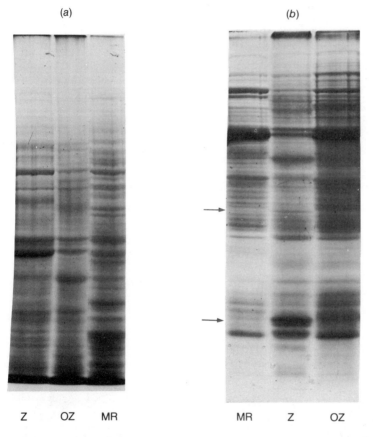

Figure 17.1 Electrophoretic fractionation in (a) 10% and (b) 15% polyacrylamide 2% SDS gels of 100–150 μg protein from S. nigra leaf evoked zone (Z) unevoked zone (OZ) and mid-rachis (MR) tissue. Tissues extracted in 50 mM Tris HCl, pH 7.6, 5 mM KCl, 10 mM EDTA, 0.15 M NaCl, 10 mM mercaptoethanol, 0.1 M phenylmethylsulphonylfluoride, 2% polyvinylpyrrolidone, filtered and centrifuged 12000 × g for 30 min. Supernatant proteins precipitated in 80% $(NH_4)_2SO_4$ were recovered, reconstituted in 10 mM Tris HCl pH 8.0 with 1.0 mM EDTA and 10 mM mercaptoethanol, dialysed and freeze dried. Samples reconstituted in 10 mM Tris HCl, 2% SDS, 1% mercaptoethanol, 1 mM EDTA were denatured at 95 °C, dialysed and centrifuged; 25–40 μl aliquots were subjected to electrophoresis at 30 mA constant current for (a) 4.5 h and (b) 3.5 h at 4 °C. Polypeptides visualized with Coomassie blue

202 Target cells for ethylene action

rachis tissue midway between the insertion of each pair of leaflets. Ethylene evoked zone cells (Z) can be collected as clusters of free but intact cells after an abscission zone has separated. When dialysed extracts of Z, OZ and MR are fractionated by electrophoresis in 10% polyacrylamide SDS gels, unique polypeptide bands (M_r c.110 000 and 95 000 dalton) can be resolved in Z and OZ (*Figure 17.1a*) which are not present in equivalent protein loadings of MR. In 15% gels Z specific polypeptides are discerned at MR c.43 000 and 28 000 dalton (*Figure 17.1b*).

IMMUNOLOGICAL MARKERS

Immunological assays can be more sensitive than biochemical fractionation for probing differences in protein complements. Antisera from rats immunized with

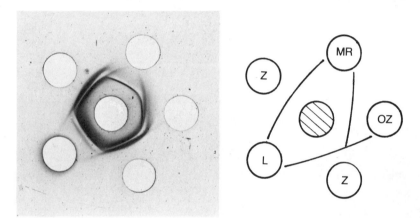

Figure 17.2 Ouchterlony double immunodiffusion. Antigenic determinants in *S. nigra* protein extracts (*see* legend to *Figure 17.1*) challenged against Z rat antiserum (Evoked zone (Z), unevoked zone (OZ), mid-rachis (MR) and leaf (L)). Antiserum well hatched. Precipitin bands visualized with amido black

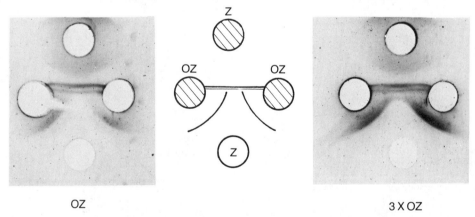

Figure 17.3 Intragel absorption. Antigenic determinants of *S. nigra* evoked zone (Z) protein extracts (*see* legend to *Figure 17.1*) challenged against Z rat antiserum across an OZ and 3 × OZ antiserum barrier. Precipitin bands visualized with amido black

protein preparations of Z and OZ have been used to recognize antigenic determinants in extracts of OZ and MR and leaf. Although Ouchterlony double diffusion visualizes many determinants common to all these tissues, it also demonstrates that Z and OZ possess common antigenic determinants not present in MR. As might be expected, Z also contains determinants not present in either OZ or MR (*Figure 17.2*).

Competition diffusion (intragel absorption) in which Z antigen must pass an OZ antiserum barrier before meeting the Z antiserum has confirmed the presence of unique Z precipitin bands which are not attenuated when the concentration of OZ antiserum is increased threefold (*Figure 17.3*).

Using immunoelectrophoretic techniques, fractionated Z, OZ and MR antigen preparations challenged with Z rat antiserum IgG show distinctive immunorecognition arcs for each tissue (*Figure 17.4a,b*). When electrophoretically fractionated Z and MR antigen preparations are challenged with Z antiserum IgG which is first competed with MR antigen, precipitin recognition arcs are eliminated from the MR lanes but recognition of unique Z determinants is confirmed (*Figure 17.4c*). That OZ possesses determinants that are different from MR is demonstrated by distinct precipitin recognition arcs in OZ compared with MR lanes when both antigens are challenged against rat OZ antiserum IgG (*Figure 17.4d*).

MAINTENANCE OF TARGET CELL COMPETENCE

Zone (Z) cells of *S. nigra* would seem therefore to be ideal tissue in which to verify whether β-1:4-glucanhydrolase activity that is 'turned on' by ethylene can be 'turned off' by withdrawal of ethylene (or the addition of auxin) and then 'turned on' again by ethylene. A requirement to test this is the maintenance of isolated Z cells in culture for periods long enough to perform the experiment. This we have not yet achieved. However, an alternative tissue is the distal abscission zones of leaves of *P. vulgaris*. Here, the 1–2 rows of zone cells separate only at their distal ends from the pulvinar cells above, leaving most still attached as a surface 'dome' over the petiole cells below them. The petiolar tissue functions as a nurse tissue for the dome of zone cells which can then be maintained viable under sterile conditions for many days. Because the zone cells cannot be collected separately from their non-target neighbours, biochemical experiments always reflect a combination of the two types of tissues. Thin slices of dome tissue (enriched with respect to the zone cells) can however be compared with petiole tissue alone, even though zone cells alone cannot be analysed.

It is clear that ethylene evokes different responses in the zone-enriched dome from those in the petiole. Firstly, exposure of petiole tissue to ethylene does not result in the induction of β-1:4-glucanhydrolase whereas high activity is induced in the dome cells (Horton and Osborne, 1967, and unpublished data). Secondly, after both tissues have been exposed to ethylene $4 \mu l \, l^{-1}$ for two days, the spectrum of polypeptides as separated in 12.5% polyacrylamide SDS denaturing gels is not the same (*Figure 17.5*). This indicates a difference in protein complement between the two cell types.

Critical to the concept of abscission zone cells as a specific target cell class is the maintenance of their determination and competence when they are cycled through a period of ethylene induction ('turning on') followed by a period of stimulus removal ('turning off') and a re-induction ('turning on' again) by ethylene. This

204 Target cells for ethylene action

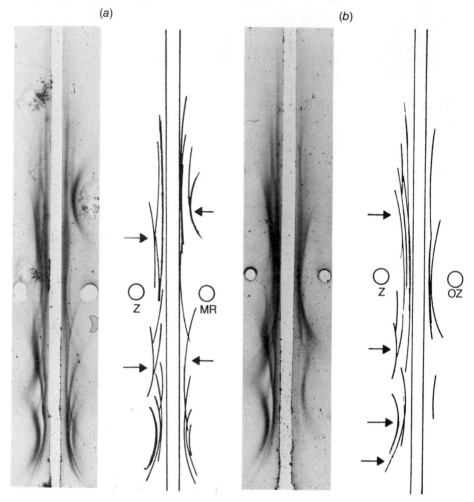

Figure 17.4 Antigenic determinants of *S. nigra* evoked zone (Z) unevoked zone (OZ) and mid-rachis (MR) protein extracts (*see Figure 17.1*) fractionated by electrophoresis in 1% agarose at 8 V/cm for 90 min and (a,b) challenged against Z rat antiserum IgG. (c) Z and MR challenged against Z antiserum IgG pre-competed with MR antigen. (d) OZ and MR challenged against OZ antiserum IgG. Precipitin arcs visualized with amido black

sequence has been demonstrated biochemically and ultrastructurally in *P. vulgaris* in the following way.

(a) Enzyme changes in P. vulgaris

The zone cells are 'turned on' by enclosing 1 cm explants containing the distal leaf abscission zone for three days in ethylene at $4\,\mu l\,l^{-1}$. The pulvinus is shed, leaving the evoked zone cells as a layer covering the dome of the petiole stump. The activity of β-1:4-glucanhydrolase extracts of the top 0.5 mm of the dome, was determined by viscometry using a carboxymethylcellulose substrate (Wright and

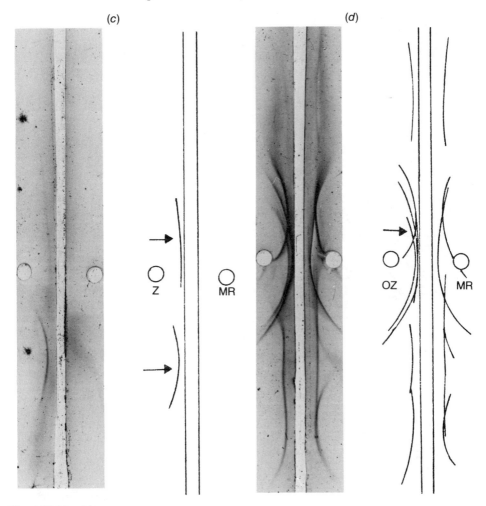

Figure 17.4 (cont.)

Osborne, 1974). The abscission domes are then 'turned off' either by treating with 1 μl drops of IAA (10^{-5} M or 10^{-3} M) every 8 h in the presence of 4 μl l^{-1} ethylene (controls treated with H$_2$O)(*Figure 17.6a*) or by similar drop treatments and enclosure with mercuric perchlorate (*Figure 17.6b*). β-1:4-Glucanhydrolase activity continues to rise in domes that are continually exposed to the ethylene stimulus alone, but withdrawal of the stimulus, or the addition of auxin arrests (and with time reduces) the level of active enzyme present in the dome tissues, e.g. the tissue response is progressively 'turned off' (*Figure 17.6a* and *b*). Such results are in general accord with the requirement for the continual presence of an hormonal stimulus to maintain a response in the cells of a target tissue.

That dome cells retain their determinant character when 'turned off' so that they can then be 'turned on' again even in the presence of auxin is shown by their response when given another exposure to ethylene. With 500 μl l^{-1}, increased β-1:4-glucanhydrolase activity is again induced (*Figure 17.6c*).

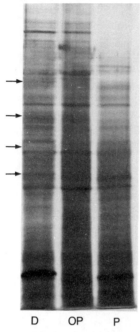

D OP P

Figure 17.5 Electrophoretic fractionation in 12.5% polyacrylamide 2% SDS gel of protein from *P. vulgaris* fresh petiole (OP), dome tissue of explants abscinded in ethylene 4 μl l^{-1} for 48 h (D) and tissue of petiole explants (P) similarly exposed to ethylene for 48 h. *See Figure 17.1* legend for methods. Visualized with silver stain

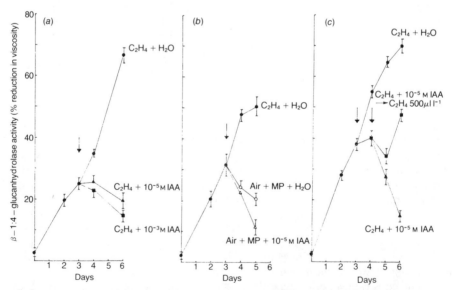

Figure 17.6 Activity of β-1:4-glucanhydrolase in extracts of dome explant tissue induced to separate ('turned on') by ethylene 4 μl l^{-1}, then 'turned off' by adding at day 3 (a) IAA 10^{-3} or 10^{-5} M in the presence of 4 μl l^{-1} ethylene or (b) transferring the explants to air and adding H$_2$O or IAA 10^{-5} M in the presence of mercuric perchlorate (MP). Dome tissue 'turned on' again by (c) transferring treated explants to 500 μl l^{-1} ethylene at day 4

(b) Ultrastructural changes in P. vulgaris

Whereas the sheet of abscission zone cells between the non-separating neighbour cells of pulvinus and petiole are particularly amenable for following positional differentiation, the tissue is not ideal (as stated earlier) for detailed biochemical studies because of the impossibility of isolating the zone cells from non-zone cells. It is our objective to follow the differentiation of the ethylene responsive target cells of the zone by immunological methods using antibodies linked to fluorescent dyes for their detection by light microscopy. This work is in progress. An alternative method for following changes in zone cells compared with their neighbours is to monitor differences in the ultrastructural organization that take place during a sequential 'turning on', 'turning off' and 'turning on' by ethylene and auxin. In this way one may gather further evidence for evaluating the concept that zone cells are distinct from their neighbour cells and are of a special determinant cell class with a particular competence to recognize and respond to an ethylene signal.

Since ethylene induces the production and secretion of specific enzyme proteins, it is reasonable to seek ultrastructural changes that could be linked to these events. We have therefore followed alterations in the shape and structure of organelles in zone cells of explant domes that were induced to separate in response to added ethylene. The cells were examined during a sequence of 'turning off' the β-1:4-glucanhydrolase activity (either by adding IAA or removal of ethylene) followed by a re-induction ('turning on') of enzyme activity in response to a second exposure to ethylene.

Small segments of each dome tissue were removed from the explants and processed for electron microscopy. Sections were cut from mesas of the blocks to include zone cells and, where possible, neighbouring cells below them and examined by transmission electron microscopy. Electron micrographs were taken at $\times$ 7000–8000 for general views of cell walls and cytoplasm, and the width and number of membranes in dictyosome stacks were selected as indicators of changes in the biochemical activity of the tissue. At magnification $\times$ 39 900, individual dictyosome profiles were photographed from representative cells from each treatment. The aim was to obtain examples of at least 20 dictyosomes from a minimum of five cells. Where the dictyosome numbers per cell were low as many as 20 cells were searched. With the exception of one sample in which only 13 dictyosomes were visible, a minimum of 18 were photographed per treatment.

Domes that were producing high levels of β-1:4-glucanhydrolase as judged by the viscometric assay (*Figure 17.6*) exhibited zone cell dictyosomes with wide stacks, wide intermembrane distances and dilated lacunae. This configuration also showed the presence of many vesicles at the edges of the stack and in the surrounding cytoplasm(*Figure 17.7*). The dictyosomes photographed were those which could be clearly distinguished as a stack. As a quantitative measure of their condition we have expressed the ratio of stack width over the number of membranes visible and calculated the mean and the standard error of the mean for all the stacks photographed in each treatment. The ratios were arranged in descending and ascending order of the values so obtained and compared with the values for extracted β-1:4-glucanhydrolase activity of the dome tissues as assayed by viscometry (*Figure 17.6*). In *Table 17.1a* and *b* the results have been grouped into 'on' or 'off' treatments.

Even with the relatively small numbers of dictyosomes that have been measured and the relatively small number of cells that it has been possible to scan, it is

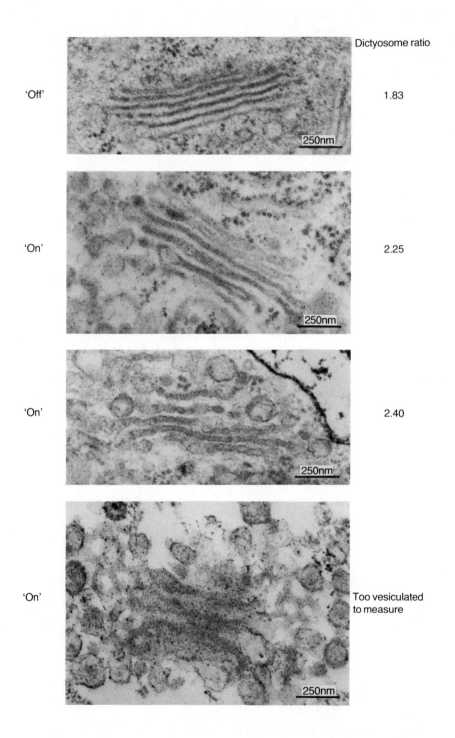

Table 17.1 VALUES FOR DICTYOSOME STACK WIDTH OVER MEMBRANE NUMBER IN ZONE CELLS OF DOMES OF *P. VULGARIS* EXPLANTS

IAA (M)	C_2H_4 (μl l^{-1})	Day	Stack width membrane number
(a) Values for 'turned on' treatments			
10^{-5}	500	5	2.97 ± 0.30
—	500	6	2.54 ± 0.20
—	4	3	2.50 ± 0.19
10^{-5}	500	6	2.32 ± 0.12
10^{-5}	500	6	2.27 ± 0.09
—	4	6	2.25 ± 0.08
—	4	5	2.24 ± 0.06
(b) Values for 'turned off' treatments			
10^{-5}	MP	6	1.59 ± 0.05
10^{-5}	4	6	1.80 ± 0.04
10^{-5}	4	4	1.85 ± 0.05
—	MP	5	1.88 ± 0.15
10^{-3}	4	6	1.99 ± 0.06
10^{-3}	MP	5	2.07 ± 0.06
10^{-3}	4	5	2.09 ± 0.11

Material fixed and analysed is part of the experiments to assay β-1:4-glucanhydrolase activity presented in *Figure 17.6*

evident that zone cells with high dictyosome ratios correspond to those in the 'turned on' condition. Cells in the 'off' condition are those with low dictyosome ratios. Of the 19 treatments from three separate experiments assessed for β-1:4-glucanhydrolase activity and for dictyosome ratios, all of the 'off' treatments have lower ratios than any of the 'on' treatments. Because dictyosome ratios could not be determined with accuracy in the most advanced stages of discharge where membranes were extremely vesiculated and their numbers in a stack could not be positively identified, the values for the most active 'on' treatments must tend to be underestimates.

Direct comparison of dictyosome ratios within a single experiment indicates that zone cells showing active secretion have a high ratio (2.50 ± 0.19) on the first day of separation (day 3) and revert to the turned 'off' low ratio category if ethylene is withdrawn (1.88 ± 0.15) and IAA is added (2.07 ± 0.06, *Figure 17.8*). IAA also 'turns off' β-1:4-glucanhydrolase activity when the dome explants are maintained in 4 μl l^{-1} ethylene a concentration that will, alone, maintain the induction of the enzyme (*see Figure 17.6a*). This 'turning off' of enzyme production is reflected in the reversion of the high dictyosome ratio from 2.50 ± 0.19 on day 3 to 1.99 ± 0.06 at day 6 (*Figure 17.9*). Zone cells of explants maintained in ethylene throughout retain an 'on' high dictyosome ratio (2.24 ± 0.06 and 2.25 ± 0.08 in *Figures 17.8* and *17.9* respectively), but we have observed that in these treatments the ratios regress from their maximum value with time. This may indicate an approach of senescence in the zone cells.

Figure 17.7 Range of dictyosome profiles from 'turned off' to 'turned on' condition. Thin segments of explant domes were fixed for 2–3 h in half strength Karnovsky's fixative in 0.1 M cacodylate buffer pH 7.2, washed three times over 2 h in buffer and post-fixed overnight in 2% osmium tetroxide in the 0.1 M cacodylate buffer. Tissue dehydrated through a graded alcohol series was transferred to propylene oxide, infiltrated with and polymerized in Spurr's resin. Sections (60–100 nm), were stained with uranyl acetate and lead citrate and examined in a transmission electron microscope at 60 kV

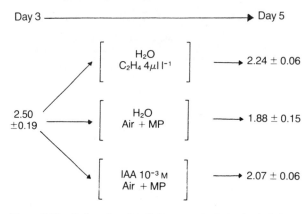

Figure 17.8 Ratio values for dictyosome stacks in abscinded zone cells of domes 'turned on' by ethylene 4 µl l^{-1} (day 3) and 'turned off' by removal of C_2H_4 and the addition of IAA

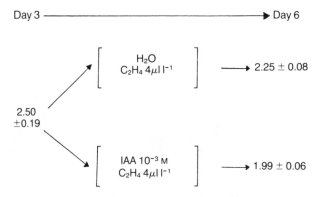

Figure 17.9 Ratio values for dictyosome stacks in abscinded zone cells of domes 'turned on' by ethylene 4 µl l^{-1} (day 3) and 'turned off' by addition of IAA in the presence of ethylene

Evidence that zone cells which are 'turned off' by adding IAA remain competent to be 'turned on' again, is indicated by dictyosome values in *Figure 17.10*. After 'turning off' for one day, the dictyosome ratio falls from 2.50 ± 0.19 to 1.85 ± 0.05 and remains low until day 6 (1.80 ± 0.04) even when levels of ethylene are present which would normally sustain β-1:4-glucanhydrolase production (4 µl l^{-1}). However, IAA treated cells do not lose their ethylene target competence; exposure to a high concentration of ethylene (500 µl l^{-1}) on day 4 again induces by day 6 the dilated form of the dictyosomes with a high ratio value (2.27 ± 0.09). Such treatments also show the 'turned on' production of β-1:4-glucanhydrolase (*Figure 17.6c*).

It is important to emphasize that during experiments with *P. vulgaris* the ultrastructural changes we describe were measured only in the 1–2 rows of cells that constitute the abscission zone; they were not observed in neighbour cells of the petiole below the zone. The IAA treatments ('turned off') induce some divisions in

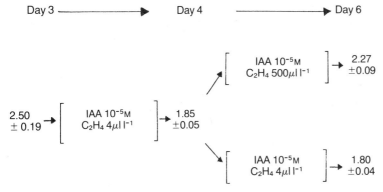

Figure 17.10 Ratio values for dictyosome stacks in abscinded zone cells of domes 'turned on' (day 3), 'turned off' by IAA in the presence of ethylene (day 4) and 'turned on' again by a high level of ethylene (500 μl l^{-1})

the cells below the zone surface layers. Although visibly divided cells were excluded from the dictyosome survey, early stages of division would not have been identified.

Summary

Three general types of ethylene target cells have been described based upon their recognizably distinct growth responses to ethylene and other hormones when immature.

Evidence has been presented that the cells positionally differentiated at abscission zones of *S. nigra* contain cell specific proteins and unique antigenic determinants that distinguish them from cells in neighbour leaf and rachis tissue. In response to ethylene, zone cells express new cell specific proteins and new antigenic determinants.

During abscission in *P. vulgaris* unique proteins can be demonstrated in response to ethylene in extracts of tissue that are enriched with zone cells. This tissue is induced to synthesize and secrete a novel β-1:4-glucanhydrolase. Enzyme induction by ethylene can be repressed by adding auxin but 'turned on' again by a higher level of ethylene so that the determination and competence of zone cells to respond to the ethylene signal is stable and retained.

Regulation of secretory activity in zone tissue as visualized by the ultrastructural conformation of dictyosome stacks occurs under ethylene 'turned on' and auxin 'turned off' regimes that also induce and repress β-1:4-glucanhydrolase activity. Using this technique the response is seen to be localized to specific cells of the abscission zone.

These results add further support to the target cell concept of abscission and argue in favour of the positional differentiation of cells (which we have called Type 2) which express specific molecular and antigenic determinants and a particular competence for a specific hormonal (ethylene) recognition and response.

Clearly, much further work is necessary to

(1) fully document the unique characteristics and antigenic components that constitute the differentiated abscission zone target cell, and

(2) to explain why only these cells express the specific gene complement that permits them to separate from their neighbours on receiving the appropriate (ethylene) signal.

Acknowledgements

We wish to thank Mrs Sheila Dunford for measuring the dictyosomes in electron micrographs and analysing the data for presentation and Mrs Rachael Daubney for expert assistance in the preparation of the manuscript.

References

BURG, S.P. and BURG, E.A. (1966). *Proceedings of the National Academy of Sciences, USA*, **55**, 262–269
COOKSON, C. and OSBORNE, D.J. (1978). *Planta*, **144**, 39–47
EISINGER, W. (1983). *Annual Review of Plant Physiology*, **34**, 225–240
GOLDBERG, R.B., HOSCHEK, G., TAM, S.H., DITTA, G.S. and BREIDENBACH, R.W. (1981a). *Developmental Biology*, **83**, 201–217
GOLDBERG, R.B., HOSCHEK, G., DITTA, G.S. and BREIDENBACH, R.W. (1981b). *Developmental Biology*, **83**, 218–231
HAYWARD, M.A., MITCHELL, T.A. and SHAPIRO, R.J. (1980). *Journal of Biological Chemistry*, **255**, 11308–11312
HORTON, R.F. and OSBORNE, D.J. (1967). *Nature*, **214**, 1086–1088
KOEHLER, D.E., LEWIS, L.N., SHANNON, L.M. and DURBIN, M.L. (1981). *Phytochemistry*, **20**, 409–412
LEWIS, L.N. and KOEHLER, D.E. (1979). *Planta*, **146**, 1–5
LEWIS, L.N. and VARNER, J.E. (1970). *Plant Physiology*, **46**, 194–199
MALONE, M. and RIDGE, I. (1983). *Planta*, **157**, 71–73
MUSGRAVE, A., JACKSON, M.B. and LING, E. (1972). *Nature (New Biology)*, **238**, 93–96
OSBORNE, D.J. (1976). In *Perspectives in Experimental Biology 2*, pp. 89–102. Ed. N. Sunderland. Pergamon Press, Oxford, New York
OSBORNE, D.J. (1979). *Scientific Horticulture*, **30**, 31–43
OSBORNE, D.J. (1982). In *Plant Growth Substances 1982*, pp. 279–290. Ed. P.F. Wareing. Academic Press, London, New York
OSBORNE, D.J. and SARGENT, J. (1976). *Planta*, **130**, 203–210
PALMER, J.H. (1976). *Plant Physiology*, **58**, 513–515
PRATT, H.K. and GOESCHL, J.D. (1969). *Annual Review of Plant Physiology*, **20**, 541–584
STEWART, R.N., LIEBERMAN, M. and KUNISHI, A.T. (1974). *Plant Physiology*, **54**, 1–5
TERRY, M.E., RUBINSTEIN, B. and JONES, R.L. (1981). *Plant Physiology*, **68**, 538–542
VEEN, B.W. (1970). *Proc. K. Ned. Akad. Wet*, **73**, 113–117 and 118–121
WALTERS, J. and OSBORNE, D.J. (1979). *Planta*, **146**, 309–317
WONG, C.H. and OSBORNE, D.J. (1978). *Planta*, **139**, 103–111
WRIGHT, M. and OSBORNE, D.J. (1974). *Planta*, **120**, 163–170

18

ETHYLENE, LATERAL BUD GROWTH AND INDOLE-3-ACETIC ACID TRANSPORT

J.R. HILLMAN
Botany Department, The University, Glasgow, UK
H.Y. YEANG
Rubber Research Institute of Malaysia, Kuala Lumpur, Malaysia
and V.J. FAIRHURST
Botany Department, The University, Glasgow, UK

Introduction

Phillips (1975) pointed out that the spatial and temporal organization of growth and differentiation in terrestrial, multicellular plants involves various forms of positional signalling. Of particular interest to plant physiologists is the fact that plant shape is dictated by the differential, but obviously coordinated, activity of meristems, notably those in the shoot and root apices and the lateral buds. The partial or complete inhibition of lateral (usually axillary) buds by an actively growing apical region, a phenomenon referred to as 'correlative inhibition', is an important form of correlative relationship in the shoots of higher plants. Awareness of the dominating role of the apex has influenced horticultural and agricultural practice, and raises fundamental questions about the factors regulating lateral growth.

All of the so-called 'plant hormones', or with due deference to etymology, 'plant growth substances', would appear to be prime candidates as signalling agents in growth (Wareing, 1977). Exogenous auxins, gibberellins, cytokinins, abscisic acid and ethylene, singly and in combination, affect bud growth. Whether their effects indicate the true role of the respective endogenous compounds in the intact plant is still a matter of conjecture. Without resource to unambiguous physicochemical methods of identification and quantification, coupled with knowledge of the biosynthesis, distribution and metabolism of these substances, much of the literature in this area of study must be regarded as somewhat preliminary in nature. Nevertheless, there are firm indications that indole-3-acetic acid (IAA), certain cytokinins and ethylene may play a crucial role in the control of lateral bud growth in herbaceous dicotyledonous plants (*see* Phillips, 1975; Hillman, 1984).

The first part of this chapter selectively reviews the relatively limited experimentation on the role of ethylene in the control of axillary bud growth, concentrating mainly on 19–29 day old plants of the dwarf French bean, *Phaseolus vulgaris* L. cv. Canadian Wonder. The special consideration given to this species merely reflects the fact that it is the most thoroughly investigated species with respect to the role of ethylene in bud growth. Even so, the results of these investigations are difficult to interpret and may well not apply to other species.

In view of the oft-cited involvement of IAA in correlative phenomena, the second part of the chapter is an examination of the relationship between ethylene effects and IAA transport in the potato, *Solanum tuberosum* L. cv. Arran Pilot.

Phaseolus vulgaris—an experimental system for correlative inhibition

THE FIRST TRIFOLIATE LEAF AXILLARY BUD

The first trifoliate leaf axillary bud from intact, 19–29 day old plants of the dwarf French bean, *Phaseolus vulgaris* L. cv. Canadian Wonder, shows slow, continuous extension growth at approximately $1-5\,\mu m\,h^{-1}$. Typically, this bud develops into an axillary shoot with a single internode bearing a trifoliate leaf and floral buds at the distal node. A second internode may sometimes develop. The internode, trifoliate leaf and floral structures are present in their rudimentary forms in the inhibited bud. Measurement of bud internode length rather than whole-bud dry weight provides a reasonable assessment of bud growth, since bud internode length and dry mass generally conform to the relationship:

$$\text{dry mass (g)} = 0.423 \times (\text{internode length in mm})^{1.35}$$
$$(r^2 = 0.948)$$

Following decapitation or other treatments to release these buds from correlative inhibition, the vigour of bud growth can be impeded by development of the lower primary-leaf axillary buds. Cultural conditions and variations between batches of plants must be carefully monitored.

SOURCE OF BUD INHIBITION IN CORRELATIVE INHIBITION

Selective removal of the apical meristem and leaf primordia in *P. vulgaris* reveals that these zones are not the source of bud inhibition (White *et al.*, 1975). Rather, it is the presence of the rapidly expanding trifoliate leaves, particularly the second and third leaves in 19–29 day old plants, that are responsible for inhibiting lateral growth. Physical restriction of the growth of the young trifoliate leaves is as effective in releasing lateral buds as decapitation (Hillman and Yeang, 1979), an effect previously noted in *Hevea* (*see* Leong, Leong and Yoon, 1976). Indeed, there are at least 12 types of non-surgical treatment which promote the growth of correlatively inhibited buds (*see* Hillman, 1984).

AXILLARY BUD GROWTH FOLLOWING SHOOT DECAPITATION

The time lag between removal of the source of correlative inhibition (the young leaves) and the subsequent detection of growth changes in the released bud, can be remarkably brief. Hall and Hillman (1975) observed enhanced outgrowth of the *entire* larger primary leaf axillary bud (in *P. vulgaris* the primary leaves are paired and the plant exhibits anisoclady) within 30 min of decapitation, and this growth was associated with the import of radioactively labelled photosynthetic assimilates from the subtending leaf. Measurements of internodal growth of the smaller first trifoliate leaf bud showed that sustained, accelerated growth following decapitation could be detected after a lag period of 3–5 h (Yeang and Hillman, 1981a). Anatomical analyses by Yeang and Hillman showed that cell extension could account for the total increase in internodal length one day after decapitation, but six days later, cell extension accounted for only 50% of internode elongation.

Rapid changes in bud growth and metabolism following decapitation have been detected in other species (e.g. *Pisum sativum*—Wardlaw and Mortimer, 1970; Nagao and Rubinstein, 1976: *Vicia faba*—Couot-Gastelier, 1978: *Glycine max*— Ali and Fletcher, 1970: *Cicer arietinum*—Guern and Usciati, 1972: *Bidens pilosus*— Kramer *et al.*, 1980).

It is unlikely that in *P. vulgaris* the primary control of bud development resides in the development or functionality of the phloem and xylem interconnections between the stem and the first trifoliate leaf axillary bud. Studies with decolorized basic fuchsin and radiolabelled sucrose indicate that functional xylem continuity and probably phloem continuity exist between the bud and stem at the 19–29 day stage (Yeang, 1980).

ETHYLENE PRODUCTION BY *PHASEOLUS*

The identities of ethylene and ethane in the gaseous extracts of *P. vulgaris* apical shoots have been established conclusively by a combined gas chromatography– mass spectrometric selective ion detection method (Yeang, Hillman and Math, 1980). Yeang and Hillman (1981b) noted that there were changes in the volume and composition of the gases recovered when the evacuation conditions during extraction were altered from 13.8 kPa for 2 min to 66.7 kPa for 3 min. These changes in volume and composition were different in the primary leaves and the second trifoliate leaves. Thus, estimates of *internal* ethylene and ethane concentrations (i.e. concentration in gases extracted under reduced pressure— 'vacuum' extraction) in plant tissues are influenced by the morphology of the tissue and the evacuation conditions. The internal ethylene concentrations in the primary, first trifoliate and second trifoliate leaves were correlated with their respective rates of ethylene *emanation* (i.e. release). Only trace emanation of ethane was recorded from leaf tissue.

Pertinent to the design of experiments is the timing of stress-induced ethylene production. Ethylene emanation by 5 mm sections (i.e. wounded tissue) of the fourth internode begins after a lag period of approximately 25 min with a peak output of around 60 mm^3 kg^{-1} h^{-1} at 90–100 min. Internodes from shoots enclosed for four days in an atmosphere depleted of ethylene (less than 0.005 mm^3 dm^{-3}) show a similar but less pronounced pattern. In sections of internodes pretreated for four days in an ethylene-enhanced atmosphere (*ca.* 0.5 mm^3 dm^{-3}), the wound ethylene released is similarly depressed but the peak occurs 50–70 min from the time of sectioning.

In contrast to internodal sections, the wound reaction in 4.7 mm diameter leaf discs punched from the second trifoliate leaf shows no consistent lag period and the surge in ethylene production may begin within 10 min of cutting. With leaf discs from control plants, the time to the peak rate of ethylene release (80–100 min from sectioning) is comparable to that for the fourth internode sections but the peak rate is relatively greater (*ca.* 130 mm^3 kg^{-1} h^{-1}).

In both leaf and internode sections, the rate of ethylene emanation declines slowly from the time of peak production, but does not return to prewounding rates even after 3 h. Only trace quantities of ethane are detected (*see also* Saltveit and Dilley, 1978).

Wound-induced ethylene in the internode sections does not appear to arise from water-stress effects although drying internodal and leaf tissues enhances ethylene release; nodal tissues are relatively unaffected. Pretreatment of internodal sections for 30 min with 0.7 kmol m^{-3} mannitol, which has a low water potential (ca. -19×10^{-5} Pa), does not increase wound ethylene emanation as compared with treatment with water. There is no diminution of wound ethylene production when water loss from the cut surfaces of the sections is reduced by covering with silicone grease. Cutting the sections under water gives only a small decrease, possibly resulting from the water-immersion treatment itself. In conclusion, wounding and water-stress are separate stimuli in respect of their effects on ethylene emanation.

Experiments on the role of ethylene in correlative inhibition must take into account the doses of ethylene applied, stress effects, long pretreatments and, not least, the relevance of isolated nodal segments bearing buds. Not only do these segments generate wound ethylene but also their buds have been relieved of correlative control. Therefore any experimental suppression of bud growth on these segments may operate via anomalous mechanisms.

Control of bud growth—ethylene in the apical region of the shoot

Were endogenous ethylene in the apical region of the shoot intimately involved in correlative inhibition, then treatments which affect the production, release, availability and effects of ethylene might be expected to modify bud growth.

EFFECTS OF PHYSICAL CONSTRICTION

Enclosure of the shoot within a sealed glass tube (14 mm diameter × 120 mm) above the third node was found to induce vigorous growth of the first trifoliate leaf axillary bud and frequently growth of the larger primary leaf axillary bud (Hillman and Yeang, 1979). With prolonged treatment (9–12 days) abscission of the terminal bud usually occurred, and in about half the instances the enclosed leaves senesced, but these developments occurred after axillary bud growth had been well established. Leaf expansion and internodal extension of the enclosed shoot were severely restricted. The axillary bud on the enclosed fourth node (second trifoliate leaf axillary bud) would usually sprout but further development was curtailed. The possibility of ethylene involvement in these symptoms arises from the observation that the concentration of ethylene in the sealed tubes during the first four days of treatment was calculated to be approximately 0.1 mm^3 dm^{-3}, but the tubes were not completely ethylene-tight and thus the figure is an underestimate.

When the shoot was enclosed above the second or fourth nodes, similar axillary bud development basipetal to the enclosure and inhibition of the enclosed shoot was observed. Thus, the physical constriction effect was common to buds other than that in the axil of the first trifoliate leaf.

A similar response of lateral bud development occurred when the experiments were repeated with the shoots enclosed in open-ended glass tubes or tubes of plastic mesh. However, the axillary buds beneath these ventilated tubes grew less vigorously than when the tubes were sealed, and there was no inhibition of internode growth and the shoot emerged from the upper, open end of the tube.

Additional evidence of ethylene involvement is provided by measurements of internal and emanated ethylene. The internal ethylene content of shoots enclosed in plastic mesh tubes was found to have increased 2.4-fold (from 0.34 to

0.82 mm^3 dm^{-3}) after treatment for three days. Enhancement of the rate of ethylene emanation from confined shoots was even more pronounced (8.28 mm^3 kg^{-1} h^{-1} compared with 1.81 mm^3 kg^{-1} h^{-1} in control shoots—a 4.6-fold increase).

EFFECTS OF ETHYLENE AND ETHEPHON

When apical shoots were enclosed in an atmosphere containing 0.5 mm^3 dm^{-3} ethylene, in glass vessels large enough not to constrict stem growth and leaf expansion, then rapid outgrowth of the axillary buds occurred beneath the enclosure (Hillman and Yeang, 1979). Bud growth remained inhibited in intact, untreated control plants, and in those controls where the vessel was either flushed with air or where endogenous ethylene emanated by the enclosed but unimpeded shoot was absorbed by mercuric perchlorate such that the ethylene concentration was maintained at less than 0.005 mm^3 dm^{-3}.

Extraction of the gases from apical shoots showed that all three treatments (ethylene enhancement, ethylene depletion and daily flushing with air) gave rise to elevated internal ethylene levels (Yeang and Hillman, 1981c). The rate of ethylene emanation, however, was not increased by any of the three treatments (*ca.* 3 mm^3 kg^{-1} h^{-1}).

Confirmation of the effect of ethylene on the shoot apical region with regard to bud growth was demonstrated by treatment with ethephon (Hillman and Yeang, 1979). Dilute aqueous solutions of ethephon (0.35–1.4 mol m^{-3}) painted on either the apical bud or expanding apical leaves induced the release of basipetally situated buds.

EFFECTS OF ETHYLENE INHIBITORS

The effects of the ethylene biosynthesis inhibitor aminoethoxyvinyl glycine (AVG) and silver nitrate, an inhibitor of ethylene action, on ethylene production and bud growth have been examined (Yeang and Hillman, 1981c). Treatments of the expanding second trifoliate leaf and terminal bud with 0.1 mol m^{-3} AVG or 0.5 mol m^{-3} silver nitrate were carried out at the beginning of the experiment and at day 2. By day 4 neither compound influenced the growth of the lateral buds or the treated shoot. Although these compounds did not affect bud growth they both altered levels of internal ethylene within the treated tissue. AVG reduced by about 50% both the internal ethylene content and rate of ethylene emanation, while Ag$^+$ had no significant effect on ethylene emanation but caused a doubling of the level of internal ethylene.

HYPOTHESIS ON THE RELATIONSHIP BETWEEN BUD GROWTH AND ENDOGENOUS ETHYLENE IN THE SHOOT APICAL REGION

Yeang and Hillman (1981c, 1984) expressed endogenous ethylene in the shoot apical region as (a) rate of ethylene emanation (release), (b) intercellular (free) ethylene, and (c) internal ethylene content (concentration of ethylene in gases recovered under reduced pressure). Intercellular ethylene is the non-bound, freely diffusible component, and may be enhanced exogenously by treating the tissue with ethylene or ethylene-generating chemicals, or endogenously. Internal ethylene is envisaged as incorporating intercellular ethylene together with that endogenous ethylene which can only be released under reduced pressure (possibly 'bound' or 'compartmentalized'; *see* Jerie, Zeroni and Hall, 1978).

Table 18.1 THE RELATIONSHIP BETWEEN LATERAL-BUD GROWTH AND THREE FORMS OF ENDOGENOUS ETHYLENE FROM THE APICAL PORTION OF THE SHOOT OF *PHASEOLUS VULGARIS*

Treatment to apical portion of the shoot	Endogenous ethylene		
	Internal	Emanated	Intercellular
Constriction	■	■	■
Ethylene-enhanced air[a]	■	▼	■
Ethylene-depleted air[b]	▲	□	□
Ventilation[c]	▲	□	□
Ethephon[d]	▨	▨	■
Ag$^{+\,e}$	▲	□	□
AVG[f]	□	□	□

Key: ▼ Lateral-bud growth enhanced
      ▲ Endogenous ethylene enhanced
      ▨ Endogenous ethylene not determined

Only in the case of intercellular ethylene is an increase consistently accompanied by an enhancement of lateral-bud growth, and vice versa (indicated by squares that are either completely filled or completely empty). [a], 0.5 mm^3 dm^{-3} ethylene; [b], ethylene absorbed by mercuric perchlorate; [c], vessel enclosing the shoot ventilated to preclude accumulation of endogenously released ethylene; [d], 0.7–1.4 mol m^{-3} ethephon; [e], 0.5 mol m^{-3} silver nitrate; [f], 0.1 mol m^{-3} AVG.
From Yeang and Hillman (1984) (Reproduced with permission of the publishers of *Physiologia Plantarum*)

The relationship between the forms of endogenous ethylene in the shoot apical region and axillary bud growth is presented in *Table 18.1*. Bud release cannot be correlated with enhanced ethylene emanation since bud growth occurs following ethylene treatment and ethylene emanation is not affected. Likewise, internal ethylene levels are not apparently critical because internal ethylene levels are enhanced in treatments which do not release dominance (e.g. ethylene-depleted air). On the other hand, the release of buds correlates with an increase in intercellular ethylene. Thus, the availability of freely diffusible ethylene in the apical region of the shoot may be related to the onset of axillary bud outgrowth.

Control of bud growth—ethylene in the region of lateral buds

In addition to having a possible role in the apical portion of the shoot, endogenous ethylene could influence bud growth at or adjacent to the axillary bud(*see* Abeles, 1973; Burg and Burg, 1968a,b; Blake, Reid and Rood, 1983).

EFFECTS OF ETHEPHON AND ETHYLENE INHIBITORS

Applications of ethephon over the range 0.7–1.4 mol m^{-3} painted on the axillary bud and adjacent tissue had no effect on stimulating the development of buds on intact plants, but inhibited the bud outgrowth of decapitated plants (Yeang and Hillman, 1982). Treatment with AVG (0.01 mol m^{-3}) or Ag$^+$ (0.1 to 1 mol m^{-3}) also had no effect on bud growth of axillary buds from intact plants. However higher concentrations (0.4 mol m^{-3} AVG or 4 mol m^{-3} Ag$^+$) inhibited the outgrowth of buds on decapitated plants. These results indicate that the restricted bud growth in intact plants is not due to a lack of free ethylene, and that the gas seems to play some role in the regulation of the growth of axillary buds released from dominance.

ENDOGENOUS ETHYLENE IN THE NODE AND INTERNODE

To determine if high ethylene levels at the nodal region of the stem were responsible for the inhibition of bud growth, Yeang and Hillman (1982) excised 6 mm sections of the third node (from which arises the first trifoliate leaf axillary bud) and 20 mm internodal sections immediately above and below the nodal section. The nodal section included the basal parts of the bud internode and first trifoliate leaf petiole. Internal ethylene and ethane and the rate of ethylene emanation were determined in intact control plants and in plants 2.5 and 5.5 h after decapitation, intervals selected to coincide with the time just before and just after bud growth can be detected. The precaution was taken of carrying out the analyses of endogenous ethylene within 27 min of excising the sections to prevent contamination of the samples with wound ethylene.

In intact plants there was no significant difference between the rates of ethylene emanation in the nodal and internodal sections. However, internal ethylene levels were generally *ca.* 25% higher in the nodal than in the internodal sections, a difference perhaps reflecting their differing morphology (*see* Yeang and Hillman, 1981b). Following stem decapitation, both ethylene emanation and internal ethylene levels in nodal and internodal sections declined, a decrease, then, which was not limited specifically to the node, but to the stem as a whole (*see also* Abeles and Rubinstein, 1964).

Unlike the internodes, the nodal sections contained high internal ethane levels ($5-8\,\text{mm}^3\,\text{dm}^{-3}$) and high rates of ethane emanation ($0.2-0.7\,\text{mm}^3\,\text{kg}^{-1}\,\text{h}^{-1}$).

IAA, ETHYLENE AND BUD GROWTH

The most widely cited evidence supporting the concept that apically synthesized IAA is involved in correlative inhibition is the observation that exogenous IAA will replace the shoot apex with respect to bud inhibition. Although the results of endogenous IAA determinations, the kinetics of bud release and the development of alternative theories point towards an, at best, indirect role for IAA (*see* Hillman, 1984), there is a possibility that the endogenous IAA may exert its effects on bud growth *via* ethylene.

Continuing their experiments on endogenous ethylene in the node and internode, Yeang and Hillman (1982) measured endogenous ethylene from nodal and internodal tissues of decapitated plants supplied with $0.02\,\text{cm}^3$ lanolin containing $10\,\text{mg}\,\text{g}^{-1}$ IAA (equivalent to approx. 200 µg) to the cut end of the third internode. Ethylene emanation from the internodal tissue adjacent to the site of IAA application was enhanced threefold compared with the comparable internode of the decapitated control. Nevertheless, in the nodal tissues, ethylene emanation and internal ethylene content were similar in the IAA-treated and corresponding untreated decapitated control plants. Such observations complement the report of Hall and Hillman (1975) where in a similar experimental system the translocation of radiolabelled IAA was poor and its metabolism extensive.

Because bud inhibition is maintained by IAA without affecting nodal ethylene levels, it can be concluded that IAA does not sustain the inhibition of bud growth in decapitated plants *via* IAA-induced ethylene.

It is commonplace for exogenous IAA applications to be associated with the inhibition of lateral bud outgrowth, yet $0.1\,\text{mol}\,\text{m}^{-3}$ IAA applied directly to the bud promotes extension growth over the first two days, with a growth increment close to that achieved by shoot decapitation (Yeang and Hillman, 1982). When combined with a low concentration of AVG ($0.01\,\text{mol}\,\text{m}^{-3}$) the increment attributable to IAA was annulled, despite the fact that the dosage of AVG applied did not affect the normal slow outgrowth of the bud on the intact plant or the enhanced outgrowth caused by shoot decapitation. Thus, IAA-induced bud growth appears to be different from the growth arising from shoot decapitation.

In a recent article, Blake, Reid and Rood carried out on *Pisum sativum*, experiments superficially similar to those of Yeang and Hillman. Contrary to the findings in *Phaseolus*, they found that ethylene release in the node plus lateral buds decreased upon decapitation, yet they did not report that they assayed ethylene from the nodes and internodes separately; thus the decrease may have reflected the general decrease occurring in the stem as a whole (similar to *P. vulgaris*). Blake and colleagues also supplied IAA to the cut stem stump of the decapitated plant and found that ethylene release in the node was increased, again an observation not in accord with the report by Yeang and Hillman. An explanation for this discrepancy could be that the semi-dwarf peas that were used are smaller than *P. vulgaris* and consequently any exogenous IAA applied to the stem stump has in pea a shorter transport path to the lateral bud. Moreover, although Yeang and Hillman applied approximately 200 µg IAA in lanolin to the stump compared with the 25 µg IAA in ethanol donated to the decapitated peas, the release of IAA from lanolin to the plant tissues is poor, yet ethanol readily evaporates depositing crystals of IAA on the stump.

Blake, Reid and Rood (1983) also employed ethylene antagonists (silver nitrate and canaline) to examine the role of ethylene in correlative inhibition. They applied the antagonist to decapitated plants bearing buds released from the supposed ethylene-induced inhibition. In their discussion, they suggest that the IAA-induced promotion of bud outgrowth on intact plants reported by Yeang and Hillman (1982) may be a swelling response perhaps attributable to IAA-induced ethylene. Yeang and Hillman, however, recorded bud internode *elongation*, and the increase in bud growth was 1.18 mm (*cf*. 0.18 mm cited by Blake, Reid and Rood, 1983, and *see* Yeang and Hillman, 1981a).

Conclusions

The release of dominated buds by ethylene applied to the whole shoot may occur in the continued presence of ethylene (*Gossypium*—Hall et al., 1957) or after an ethylene pulse treatment (*Petunia*—Burg, 1973; *Solanum*—Catchpole and Hillman, 1976). Ethephon treatment of whole shoots is also effective in stimulating bud growth (De Wilde, 1971; Lürssen, 1982; Nickell, 1982). These whole-shoot treatments with either ethylene gas or ethephon might act simply by suppressing the growth of the shoot apical region, eliminating physiologically rather than surgically its dominating role.

Ethylene appears to play an important role during normal bud outgrowth following shoot decapitation but on the intact plant restricted bud development is not due to a lack of freely diffusible ethylene.

In the absence of definitive experimentation on the transport of physiologically significant amounts of ethylene and ethylene precursors such as aminocyclopropanecarboxylic acid, it is premature to speculate about a direct, signalling role of ethylene—a true hormonal role—in bud growth regulation, but its seemingly poor transport (Zerino, Jerie and Hall, 1977; Jerie, Zeroni and Hall, 1978) may point towards it having an indirect role. Moreover, as the correlative signal has not been identified, nor its path and site of action in the bud characterized, there is clearly a requirement for further fundamental and applied research.

Ethylene and IAA transport

One of the most persistent concepts of integration in the shoots of higher plants is the view that apically synthesized IAA governs cellular development, the basipetally polarized transport of exogenous IAA reflecting the transport characteristics of endogenous IAA which are in turn inextricably linked to the morphological polarity of shoot tissue (but *see* Sheldrake, 1974). Further support for this concept comes from observations that endogenous IAA diffuses from the apices of *Zea* coleoptiles (Greenwood *et al.*, 1972) and *Phaseolus* shoots (White *et al.*, 1975), and exogenous IAA evokes a wide range of responses involving the division, enlargement and differentiation of cells (*see* Bandurski and Nonhebel, 1984). Gravity- and light-induced asymmetry in the distribution of IAA has been widely thought to account for tropic curvatures, although there are recognized to be deficiencies in regarding auxin redistribution as a universal reaction mechanism in tropisms (*see* Dennison, 1984 and Wilkins, 1984).

Ethylene has a potent effect on stem development, giving rise to the 'triple response' whereby stem elongation is inhibited, radial expansion occurs in the cells of the subapical region, and there is the loss of the normal response to light and gravity (*see* Abeles, 1973). Were IAA transport critical for growth correlation, then one way in which ethylene might act may be to modify longitudinal and/or lateral IAA transport. The second part of this review describes some experiments carried out on potato to investigate this.

SOLANUM TUBEROSUM—A SYSTEM FOR STUDYING IAA TRANSPORT

In recognizing the present technological limitations in establishing the transport characteristics of endogenous IAA, an attempt has been made to investigate the effects of exogenous ethylene on the transport of exogenous, radioactively labelled IAA. The plant species selected was an early variety of potato, *Solanum tuberosum* L. cv. Arran Pilot (Scotch-grown, foundation stock). This cultivar exhibits the triple response and radial swellings are also observed in the subapical regions of stolons; in appearance and anatomy, these structures are identical to normal tubers but do not contain starch detectable by iodine staining (Catchpole and Hillman, 1969). Furthermore, the symptoms of the coiled-sprout disorder in Arran Pilot can be attributed to ethylene (Catchpole and Hillman, 1975a,b,c, 1976).

Tubers were chitted at 15 °C and 30 g tuber pieces bearing a single short sprout were excised and dark-grown at 15 °C in vermiculite moistened with Long Ashton nutrient solution. Etiolated stems and stolons were selected for uniformity. Donations of $[1\text{-}^{14}C]$IAA and $[5\text{-}^{3}H]$IAA were made to subapical segments or whole organs.

LONGITUDINAL TRANSPORT

One technique used to measure IAA transport involves applying a known amount of labelled IAA at a concentration not exceeding 5 mmol m^{-3}, incorporated in a small 'donor' block of 1% agar, to one end of a segment which is in contact with, at the other end, a 'receiver' block of plain agar. After the appropriate transport period, the segment is cut into small pieces for radioassay by scintillation spectrometry following sample combustion. The radioactivity in the donor and receiver blocks, plus the atmosphere (for $^{14}CO_2$) is also checked. Chromatography of methanolic extracts of the segments is carried out in a parallel experiment to assess the extent of metabolism.

Over periods of up to 14 h and at temperatures between 15–25 °C, uptake of IAA-1-^{14}C from the donor block into apical stem segments of the first expanded internode behind the apex, reaches a maximum by 4–6 h, with the uptake from an apical donor more than twice the uptake from a basal donor. Transport of [1-^{14}C]IAA is strongly polarized in the basipetal direction—more IAA is transported from the apex to the base of a segment than from the base to the apex. The velocity of basipetal transport is approximately 4 mm h^{-1}. For segments 6 mm in length, the numbers of counts in the basipetal receivers increases with time to a maximum at 8 h and then declines. The decline is associated with resorption of radioactivity into the tissue and liberation of $^{14}CO_2$. For 8 mm segments, passage of [1-^{14}C]IAA into the receiver blocks almost stops after 6 h. This is confirmed by pretreating segments with unlabelled IAA donors for 6 h and then replacing them with donors containing radiolabelled IAA: few counts appear in the receiver block.

It might be expected that if the majority of the cells across the diameter of a stem segment contributed to the total transport of IAA, then the counts in a basipetal receiver would increase as the cross-sectional area of the segment is increased. If, instead, a ring of tissue were mainly responsible for IAA transport, then it might be expected that total transport would increase with increasing tissue circumference. For segments cut from batches of sprouts of varied size, no apparent correlation is found between radioactivity in basipetal receivers and values equivalent to either segment cross-sectional area or circumference. On the other hand, for segments cut from two positions on shoots of varied diameters, an average difference can often be noted between the transporting abilities of the two segment positions. Thus, it appears that the basipetal transporting abilities of segments varies with the position from which the segments were cut, and that innate differences between sprouts may be more important in determining the amount of transport than sprout size.

In segments 6–10 mm in length, there is no statistically significant effect of ethylene (1–10 mm^3 dm^{-3}) fumigation for 1–3 h prior to transport and for the 4 h transport period.

Similar results are obtained with subapical internodes of stolons with the exception that there is little difference in IAA uptake after apical and basal donations.

That chromatography of stolon- and stem-segment extracts on silica-gel thin layers and paper (solvent systems of propan-2-ol:ammonia:water, 10:1:1 and butan-1-ol:acetic acid: water, 25:5:11) reveals 95% of the radioactivity cochromatographing with authentic IAA, reinforces the finding that the polarly transported component from IAA donations to shoots is unmetabolized IAA (Nonhebel et al., 1984). Nevertheless, further analyses are required (see Nonhebel, Crozier and Hillman, 1983). Treatment of segments with 1–10 mm^3 dm^{-3} ethylene does not alter the chromatographic pattern.

Detailed photographic measurements of the subapical region of intact stems show a readily detectable inhibition of elongation by $1\,\text{mm}^3\,\text{dm}^{-3}$ ethylene after 20–35 min. However, isolated segments do not respond to ethylene (although they show gravitropic curvatures, as described below), and the failure to detect an effect of ethylene on IAA transport in stems and stolons may be expected.

LATERAL TRANSPORT

According to the Cholodny–Went theory, gravity-induced asymmetry in endogenous auxin can account for the gravitropic curvature seen in shoots (*see* Wilkins, 1984). Burg and Burg (1966) proposed that ethylene disturbs the normal gravitropic response by inhibiting the lateral movement (or possibly differential transport) of auxin. To investigate this hypothesis the potato plant provides an experimental system which has both negatively gravitropic shoots and diagravitropic stolons, the latter maintained in a more or less horizontal condition by the presence of an intact apex on the aerial plant.

After the initial ethylene-induced curvature, potato stems continue to grow, albeit slowly, in a horizontal direction. These treated stems return to the horizontal attitude if displaced, but no curvature is exhibited if they are grown on a Clinostat during treatment. Thus, ethylene causes stems to become stolon-like in growth habit though they will revert to negative gravitropism on removal of the ethylene.

Three methods are currently available to investigate the lateral distribution of exogenous IAA:

(1) various modifications of the agar-block technique,
(2) assessment of the distribution of radioactivity in the tissue by microautoradiography, and
(3) microdonation of high-specific-activity IAA by micropipettes.

All three methods can be combined.

Agar-block method

Using the asymmetrical donor arrangement it has not been possible to demonstrate significant lateral transport of IAA in potato stems and stolons. Segments, 7 or 8 mm in length, are excised from the subapical (responding) zones of sprouts and a half-cylinder, one-third of the length of the segment cut away from the apical end. A donor block containing radiolabelled IAA is applied to the protruding piece of tissue. After transport periods of 3–6 h the segments are split longitudinally into the longer (donor side) and shorter (receptor side) halves. Donor and receptor half cylinders are then divided transversely into three or two pieces respectively and the radioactivity assessed. When used, the receiver blocks are applied to the basal end of the segment and a section of razor blade used to both impale the segment slightly and separate the receiver block into portions adjacent to the donor and receptor sides. The segments are incubated in the vertical position, or horizontally with the protruding donor portion either on the upper or lower side. Segments from the region of maximum geotropic sensitivity and above and below this have been investigated over a period of five years of experimentation employing several thousand segments. No statistically significant effect of gravity on the distribution of radioactivity from ^{14}C-labelled and tritiated IAA has been observed, either in the tissues or receiver blocks when present. The amount of lateral transport of IAA occurring in segments is low—only about 5% (^{14}C) or 16% (^{3}H) of the total radioactivity in the tissues was obtained on the receptor side of the segments. Moreover, since only low levels of radioactivity were found in the receivers, it was the redistribution of radioactivity in the tissues which was evaluated, rather than the transported moiety, and this could mask any lateral redistribution.

No effect of ethylene at concentrations ranging between 1–10 mm^3 dm^{-3} was apparent on the distribution of radioactivity in the tissue regions or receiver blocks.

That these negative results were not simply attributable to excision of the segments desensitizing the tissue to gravitational stimuli was shown by the fact that the segments still exhibit gravitropic curvature.

Microautoradiography

Soluble-compound microautoradiography theoretically permits the distribution of radioactivity to be assessed at the tissue and cell level. Using the techniques of Bowen *et al.* (1972) and caveats of McWha and Hillman (1973), microautoradiographic assessments of IAA transport have been carried out using potato stems and stolons. Unfortunately, sections of stem embedded in liver bleach photographic film previously exposed to light, and the liver can contribute towards this negative chemography. While an alternative embedding medium, Tissue-Tek OCT compound, has proved to be more suitable in this respect, the basic technical difficulty lies with the oxidizing capacity of the tissue itself. Sections without embedding medium are completely able to bleach light-exposed film within 11 days at −15 °C. As the technique is *soluble-compound* microautoradiography, the sections are difficult to treat to remove the oxidizing agent(s). A reasonable density of silver grains can be obtained using radioisotopes of high specific activity, embedding in Tissue-Tek and a long exposure period of 38–40 days.

Sections, coated with photographic film emulsion, are viewed under dark-field illumination, and silver grains, produced by the ionizing radiation reducing the silver halide salts in the emulsion, are seen as small white dots. Larger white spots, caused by cell-wall refraction, foreign bodies or cellular inclusions can also be seen, usually as irregular aggregations.

Analysis of over 1400 micrographs has revealed that, following agar-block or microdonation of [5-^{3}H]IAA to the apical end of subapical stem and stolon segments, there is a high density of silver grains in the emulsion overlying the bicollateral vascular tissue. Relatively few grains can be seen overlying the pith, cortex and epidermis. It is not currently possible to distinguish whether the IAA is solely phloem- or xylem-conducted but no radioactivity is apparent in the cambium. Since the bleaching capacity of the tissue effectively reduces the sensitivity of the detection process, it is possible that radioactivity was conducted in the pith and cortex, though insufficient to mask the oxidizing capacity of the tissues. No uneven distribution of oxidizing capacity was found within sections.

No differences in the distribution of the silver-grain distribution has been observed in stem and stolon segments which have been held either vertically or horizontally for periods of 1–7 h, in the presence or absence of ethylene at 1 or 10 mm^3 dm^{-3}.

Microdonation

Lateral transport of IAA has been detected in intact *Zea* coleoptiles following gravitropic stimulation and microapplication of [5-^{3}H]IAA (Shaw, Gardner and

Wilkins, 1973). Using this technique with high specific activity [5-^{3}H]IAA, the tracer can be applied to intact stolons and stems. Since the apex remains largely undamaged, apart from a tiny epidermal wound caused by piercing with a glass micropipette not exceeding 150 μm in diameter (usually 100 μm), the natural IAA balance may not be significantly disturbed and the correlative relationships remain unaffected. Although in horizontal stems gravitropic curvature can be seen to commence within 20 min, neither segment orientation nor ethylene treatments for periods up to 6.5 h affected the distribution of radioactivity in the stem basipetal to the point of application of tritiated IAA, as judged by microautoradiography or scintillation spectrometry of portions of the stem. Ethylene at concentrations of 0.1 mm^3 dm^{-3} significantly inhibited the gravity response of horizontal shoots. The possibility of ethylene effecting graviperception by acting on the sedimentable starch grains in the endodermis, suggested by reports of ethylene influencing α-amylase activity (e.g. Jones, 1968), was discounted by microscopical examination of both stems and stolons.

The inability of ethylene to influence IAA transport following microapplication of tritiated IAA was also noted in horizontal and vertical (apex up or down) stolons over 6 h transport periods.

Conclusions

Reappraisal is needed of the role of IAA transport in the responses of potato shoots to the influence of gravity and the effects of ethylene. The dicotyledonous shoot and stolon apices are complex structurally, they respond rapidly to reorientation and are sensitive to ethylene. IAA in lanolin at a concentration of 10^{-3} mol m^{-3} will cause stem curvature when applied on one side of the subapical zone. Exogenous radiolabelled IAA has a marked basipetal polarity of transport in potato stems and stolons but neither gravity nor ethylene appear to affect the distribution of the exogenous IAA.

Several investigations have highlighted the deficiencies of the Cholodny-Went hypothesis as applied to dicotyledonous shoots (*see* Firn and Digby, 1980; Wilkins, 1984). The results obtained with potato sprouts and stolons add to the problems of specifying a mechanism of action of ethylene. If the results of this study on potato prove to be more widely applicable, then the correlative role of IAA needs to be examined, not at the level of monitoring the uptake, effects, transport and metabolism of the exogenous compound, but with regard to the sites of synthesis, paths of transport, target cells and turnover of the endogenous compound.

Acknowledgements

We thank the Rubber Research Institute of Malaysia for the provision of a scholarship and study leave to H.Y.Y., and the Potato Marketing Board for a research grant. Professor M.B. Wilkins is thanked for valuable advice during the course of our research studies.

References

ABELES, F.B. (1973). *Ethylene in Plant Biology*. Academic Press, New York and London
ABELES, F.B. and RUBINSTEIN, B. (1964). *Plant Physiology*, **39**, 963–969
ALI, A. and FLETCHER, R.A. (1970). *Canadian Journal of Botany*, **48**, 1989–1994
BANDURSKI, R. and NONHEBEL, H.M. (1984). In *Advanced Plant Physiology*, pp. 1–20. Ed. by M.B. Wilkins. Pitman, London
BLAKE, T.J., REID, D.M. and ROOD, S.B. (1983). *Physiologia Plantarum*, **59**, 481–487
BOWEN, M.R., WILKINS, M.B., CANE, A.R. and McCORQUODALE, I. (1972). *Planta*, **105**, 273–292
BURG, S.P. (1973). *Proceedings of the National Academy of Sciences of the USA*, **70**, 591–597
BURG, S.P. and BURG, E.A. (1966). *Proceedings of the National Academy of Sciences of the USA*, **55**, 262–269
BURG, S.P. and BURG, E.A. (1968a). *Plant Physiology*, **43**, 1069–1074
BURG, S.P. and BURG, E.A. (1968b). In *Biochemistry and Physiology of Plant Growth Substances*, pp. 1275–1294. Ed. by F. Wightman and G. Setterfield. Runge Press, Ottawa
CATCHPOLE, A.H. and HILLMAN, J.R. (1969). *Nature*, **223**, 1387
CATCHPOLE, A.H. and HILLMAN, J.R. (1975a). *Potato Research*, **18**, 282–289
CATCHPOLE, A.H. and HILLMAN, J.R. (1975b). *Potato Research*, **18**, 539–545
CATCHPOLE, A.H. and HILLMAN, J.R. (1975c). *Potato Research*, **18**, 597–607
CATCHPOLE, A.H. and HILLMAN, J.R. (1976). *Annals of Applied Biology*, **83**, 413–423
COUOT-GASTELIER, J. (1978). *Zeitschrift für Pflanzenphysiologie*, **89**, 189–206
DENNISON, D.S. (1984). In *Advanced Plant Physiology*, pp. 149–162. Ed. by M.B. Wilkins. Pitman, London
DE WILDE, R.C. (1971). *HortScience*, **6**, 364–370
FIRN, R.D. and DIGBY, J. (1980). *Annual Review of Plant Physiology*, **31**, 131–148
GREENWOOD, M.S., SHAW, S., HILLMAN, J.R., RITCHIE, A. and WILKINS, M.B. (1972). *Planta*, **108**, 179–183
GUERN, J. and USCIATI, M. (1972). In *Hormonal Regulation of Plant Growth and Development*, pp. 383–400. Ed. by H. Kaldewey and Y. Vardar. Verlag Chemie, Weinheim, Federal Republic of Germany
HALL, S.M. and HILLMAN, J.R. (1975). *Planta*, **123**, 137–143
HALL, W.C., TRUCHELUT, G.B., LEINWEBER, C.L. and HERRERO, F.A. (1957). *Physiologia Plantarum*, **10**, 306–317
HILLMAN, J.R. (1984). In *Advanced Plant Physiology*, pp. 127–148. Ed. by M.B. Wilkins. Pitman, London
HILLMAN, J.R. and YEANG, H.Y. (1979). *Journal of Experimental Botany*, **30**, 1079–1083
JERIE, P.H., ZERONI, M. and HALL, M.A. (1978). *Pesticide Science*, **9**, 162–168
JONES, R.L. (1968). *Plant Physiology*, **43**, 442–444
KRAMER, D., DESBIEZ, M.-O., GARREC, J.P., THELLIER, M., FOURCY, A. and BOSSY, J.P. (1980). *Journal of Experimental Botany*, **31**, 771–776
LEONG, W., LEONG, H.T. and YOON, P.K. (1976). *Some Branch Induction Methods for Young Buddings*. Rubber Research Institute of Malaysia, Kuala Lumpur, Malaysia
LÜRSSEN, K. (1982). In *Chemical Manipulation of Crop Growth and Development*, pp. 67–78. Ed. by J.S. McLaren. Butterworths, London
McWHA, J.A. and HILLMAN, J.R. (1973). *Stain Technology*, **48**, 136–138

NAGAO, M.A. and RUBINSTEIN, B. (1976). *Botanical Gazette*, **137**, 39–44
NICKELL, L.G. (1982). In *Chemical Manipulation of Crop Growth and Development*, pp. 167–189. Ed. by J.S. McLaren. Butterworths, London
NONHEBEL, H.M., CROZIER, A. and HILLMAN, J.R. (1983). *Physiologia Plantarum*, **57**, 129–134
NONHEBEL, H.M., HILLMAN, J.R., CROZIER, A. and WILKINS, M.B. (1984). *Journal of Experimental Botany*, in press
PHILLIPS, I.D.J. (1975). *Annual Review of Plant Physiology*, **26**, 341–367
SALTVEIT, M.E. and DILLEY, D.R. (1978). *Plant Physiology*, **61**, 447–450
SHAW, S., GARDNER, G. and WILKINS, M.B. (1973). *Planta*, **115**, 97–111
SHELDRAKE, A.R. (1974). *The New Phytologist*, **73**, 637–642
WARDLAW, I.F. and MORTIMER, D.C. (1970). *Canadian Journal of Botany*, **48**, 229–237
WAREING, P.F. (1977). In *Integration of Activity in the Higher Plants*, pp. 337–366. Ed. by D.H. Jennings. Cambridge University Press, Cambridge
WHITE, J.C., MEDLOW, G.C., HILLMAN, J.R. and WILKINS, M.B. (1975). *Journal of Experimental Botany*, **26**, 419–242
WILKINS, M.B. (1984). In *Advanced Plant Physiology*, pp. 163–185. Ed. by M.B. Wilkins. Pitman, London
YEANG, H.Y. (1980). PhD Thesis. University of Glasgow
YEANG, H.Y. and HILLMAN, J.R. (1981a). *Annals of Botany*, **48**, 25–32
YEANG, H.Y. and HILLMAN, J.R. (1981b). *Journal of Experimental Botany*, **32**, 381–394
YEANG, H.Y. and HILLMAN, J.R. (1981c). *Journal of Experimental Botany*, **32**, 395–404
YEANG, H.Y. and HILLMAN, J.R. (1982). *Journal of Experimental Botany*, **33**, 111–117
YEANG, H.Y. and HILLMAN, J.R. (1984). *Physiologia Plantarum*, in press
YEANG, H.Y., HILLMAN, J.R. and MATH, V.B. (1980). *Zeitschrift für Pflanzenphysiologie*, **99**, 379–382
ZERONI, M., JERIE, P.H. and HALL, M.A. (1977). *Planta*, **134**, 119–125

19

ETHYLENE AND PETIOLE DEVELOPMENT IN AMPHIBIOUS PLANTS

IRENE RIDGE
Biology Department, Open University, Walton Hall, Milton Keynes, UK

Introduction

As plant cells grow and mature it is not surprising to find that their response to a particular growth regulator changes. Inevitably, for example, the deposition of secondary wall material renders cells incapable of rapid-growth responses. But equally, differentiation into specialized cell types, as in abscission zones or guard cells, may render cells capable of totally new responses quite late in their development; this is implicit in the target cell approach described by Osborne (1982 and Chapter 17). The main theme of this chapter is the changing responses not of individual cells, but of a whole organ such as the leaf petiole in certain amphibious or wetland species. In these species the same cell types respond differently to raised levels of ethylene at different stages of petiole development. Furthermore, the levels of ethylene depend on environmental conditions, for instance whether or not the shoots are submerged, so that a differential response to ethylene during development reflects a differential response to the environment.

Effects of submergence on plant growth

It has been known since the last century that submerging the shoots of water plants causes organs to 'accommodate' (Arber, 1920, *see* Jackson, Chapter 20). The petioles and stems elongate until leaf laminae or flowers are restored to the water surface. This response is now known to occur because in these plants ethylene promotes elongation and ethylene levels rise in submerged organs, largely due to the slower diffusion of the gas through water and out of plant tissues (Musgrave, Jackson and Ling, 1972). Much is also known about the mechanism of the response, at least for the petioles of almost fully grown leaves. In this case ethylene-induced elongation involves cell elongation only, and is a rapid-growth phenomenon requiring the presence of auxin (Cookson and Osborne, 1978; Walters and Osborne, 1979; Malone and Ridge, 1983).

Of necessity studies on the growth of water plants have used petiole segments or detached petioles from leaves that were at least half-grown and often fully developed. However, when investigating petiole responses to submergence or ethylene using whole plants from a wide range of species, we have consistently

Table 19.1 EFFECT OF GROWTH CONDITIONS ON PETIOLE LENGTH OF *RANUNCULUS REPENS*

Growth condition	Final length (mm)				
	Leaf 1 (%)	Leaf 2 (%)	Leaf 3 (%)	Leaf 4	New leaves
Moist air	45.5 (17)	56.0 (40)	66.3 (89)	87.4	102.4
Air + 10 µl l^{-1} ethylene	49.5 (32)	57.8 (66)	[a]94.0 (205)	[b]142.0	[b]231.7
Submerged in 40 cm of deionized water	47.5 (30)	53.0 (33)	[a]76.4 (125)	[b]149.8	[b]212.7

Notes
Values are means, $n = 5$ (leaves 1–4) and 10–14 (new leaves). Values in brackets show the percentage increase in petiole length above initial length.
Values differ from air controls at [a]$P \leq 0.05$ or [b]$P \leq 0.01$, Student's t test

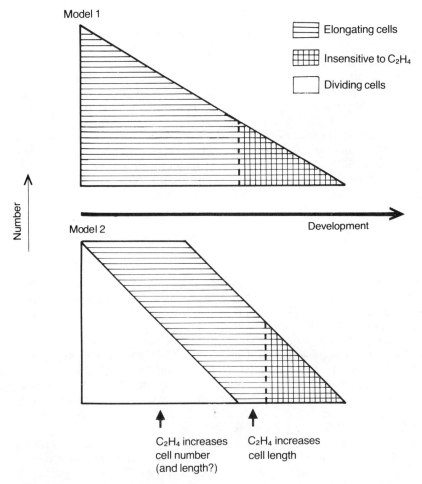

Figure 19.1 Two models which account for the greater enhancement of growth by ethylene in young petioles

found that leaves which were very young at the start of treatment, or developed during treatment, showed a disproportionately large elongation response (Ridge and Amarasinghe, 1984). The data in *Table 19.1* illustrate this for young plants of *Ranunculus repens* (Creeping buttercup) which were submerged or exposed to $10\,\mu l\,l^{-1}$ ethylene in air at 20 °C at the four-leaf stage for five weeks. Only the younger leaves responded strongly and older leaves, although still capable of elongation, showed little (leaf 3) or no (leaves 1 and 2) growth promotion compared to controls maintained in moist air (80% RH). In several species, for example *Ranunculus lingua* L., *Oenanthe crocata* L. and *Geum rivale* L., it is only the new leaves which develop during submergence or ethylene treatment that show any growth promotion at all.

This high sensitivity of petioles from young leaves could be explained by either of two models of ethylene action. The first model assumes that ethylene solely promotes cell elongation, envisaging that more cells respond to the gas from younger than older petioles, and that cells become increasingly less sensitive to ethylene during the later stages of petiole growth. The second model assumes that ethylene can promote both cell division and elongation depending on the developmental age of the petiole. The primary effect in young petioles would be to promote cell division and also possibly cell elongation. In older petioles where growth occurs mainly through cell elongation, ethylene either increases cell length or has no effect. These models are summarized in *Figure 19.1*.

One early report provided evidence suggesting that model 2 might be closest to reality. Funke and Bartels (1937) showed that accommodation in petioles of *Limnanthemum nymphoides* (=*Nymphoides peltata*) could not be explained solely on the basis of greater cell elongation and involved an increase in cell number. I have carried out further experiments with *Nymphoides*, a species which responds very rapidly to ethylene (Malone and Ridge, 1983), and with *R. repens*, which shows a smaller, slower response. The results support model 2 and suggest not only that ethylene has differential effects on cell number and cell size at different stages of petiole development but also that growth patterns may be altered irreversibly in petioles submerged very early in their development.

Accommodation and responses to ethylene in *Ranunculus repens*

Ranunculus repens L. is a widespread perennial weed which commonly grows in damp or seasonally flooded grassland. Leaves arise from creeping stolons and

Table 19.2 LEAF MEASUREMENTS FOR *RANUNCULUS REPENS* GROWING AT THREE DIFFERENT WATER DEPTH SITES IN ONE LOCATION

Site (water depth)	Petiole (a)	Length (mm) Basal sheath (b)	Total (a+b)	Ratio b:(a+b) (%)
(1) Bank of stream (0 mm)	28.2 ± 1.1	9.8 ± 0.9	38.0 ± 1.7	25.7 ± 1.5
(2) Still water (70–75 mm)	66.2 ± 1.6	24.4 ± 1.5	89.4 ± 2.5	27.5 ± 1.1
(3) Still water (150–210 mm)	99.5 ± 6.1	53.3 ± 7.1	153.3 ± 12.2	34.1 ± 2.7

Petiole lengths relate to the distance between the basal sheath and the first pair of leaflets. At each site, all mature leaves showing no signs of damage or senescence were measured; at site 3, upper leaflets of all such leaves had reached the surface of the water.
Values are means ± s.e. (n = 10–21)

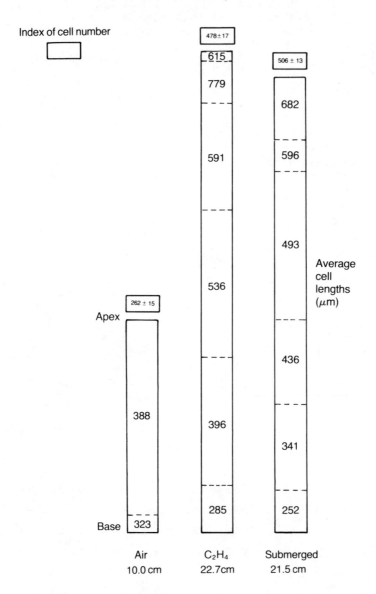

Figure 19.2 Total length, epidermal cell lengths and index of total epidermal cell number in petioles of *Ranunculus repens* which developed entirely in moist air, air containing 10 μl l^{-1} ethylene or submerged in 25 cm of distilled water. Whole plants with five leaves were treated in a controlled environment chamber

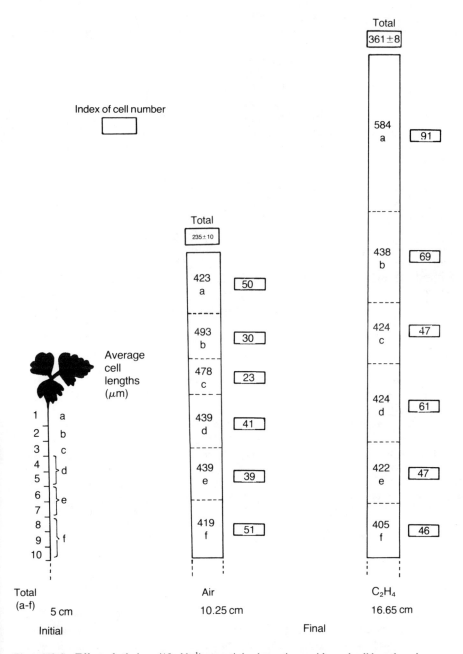

Figure 19.3 Effect of ethylene (10 μl l^{-1}) on petiole elongation, epidermal cell length and indices of epidermal cell number in marked zones of older petioles of *Ranunculus repens*. Whole plants were treated in a controlled environment chamber. Initial length of the petioles monitored were 8.7 cm (air controls) and 9.1 cm (ethylene treated) and final lengths were 14 cm and 21.6 cm respectively. Ten × 5 mm distal zones were marked at the start of the experiment

observations under different depths of flooding, show how effectively petiole length is adjusted to water level (*Table 19.2*). These data also show that the basal sheath may contribute significantly more to total petiole length in deeper water but the sheath was excluded from all further measurements described in this chapter. Given the elongation response of petioles when whole plants of *R. repens* were submerged or exposed to ethylene (*Table 19.1*), the first question asked was whether the much greater length of petioles developing entirely under water or in an ethylene-enriched atmosphere resulted from increased cell lengths, increased cell number, or both. An index of epidermal cell number provided a reliable measure of total cell number (Ridge and Amarasinghe, 1984) and the data in *Figure 19.2* show clearly a large increase in cell number with both treatments. This increase in number could account for 65% of the additional elongation in ethylene and 80% of the additional growth when submerged, assuming cell number to be the only parameter that changed. There was no increase in cell length relative to controls in air until about halfway through development, so that only cells at the distal ends of petioles were significantly longer.

These data for petioles developing entirely in the presence of ethylene appear to support model 2 but a further prediction of this model is that, during the late phase of petiole growth when elongation occurs mainly through cell expansion, the main effect of ethylene (if any) should be to increase cell length. In *R. repens*, petioles which have reached 25–30% of their final length in air contain 85% of the final cell number so that the last 70–75% of elongation is accomplished largely through cell expansion. We therefore monitored cell number and cell lengths in petioles which were half-grown when exposed to ethylene and allowed to complete their growth. Surprisingly, the growth promotion in ethylene still occurred mainly through an increase in cell number; this was evident in a distal zone of the petiole which was initially 15 mm long (*Figure 19.3*). Cell length was increased by ethylene only in the most distal zone, initially 5 mm long, and cells below this zone, even though capable of elongating, were not affected.

I interpret these data for *R. repens* in the following way. Early in petiole development ethylene causes a doubling in the rate of petiole elongation (Ridge, unpublished) which is achieved by an increase in cell number with no increase in cell lengths relative to air controls (*Figure 19.2*). A faster rate of cell elongation coupled with shorter cell cycles could explain this observation. As petiole development proceeds, a distal zone arises where cell lengths are increased by ethylene (*Figures 19.2* and *19.3*). This zone must be defined by its position on the organ rather than by the age of individual cells within it, since cell division may occur proximally but cells do not then attain longer lengths in ethylene (*Figure 19.3*).

Accommodation and responses to ethylene in *Nymphoides peltata*

Nymphoides peltata S.G. Gmelin (O. Kuntze), the fringed waterlily, usually occurs as a floating-leaved aquatic but is truly amphibious and grows readily on damp soil in air. Leaves arise from creeping rhizomes and petioles are erect when plants grow in air. When air-grown plants were submerged, very large accommodation responses were found, especially in the younger petioles (*Figure 19.4*). This response could not be explained solely in terms of accumulated ethylene and the additional factor acting on submerged petioles was shown to be tension caused by the buoyant nature of the leaf (Ridge and Amarasinghe, 1984) In responding to

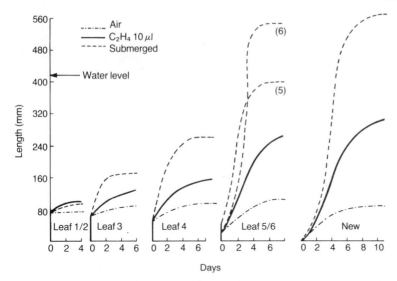

Figure 19.4 Effects of ethylene (10 µl l^{-1}) and submergence (to 42 cm) on the growth of petioles of different ages in air-grown *Nymphoides peltata*. Whole plants ($n = 5$ per treatment) with five or six leaves were treated in a controlled environment chamber, and petiole length measured daily. Results for the youngest leaf (5 or 6) were pooled except for submerged plants

this buoyant pull, *Nymphoides* differs from *Ranunculus repens* but behaves exactly like the amphibious fern, *Regnellidium diphyllum* (Musgrave and Walters, 1974).

Paralleling the experiment with *R. repens* (*Figure 19.2*), the relative contributions of cell division and elongation were determined for the growth response of the youngest petiole (leaf 6, *Figure 19.4*) from *Nymphoides* plants. In both the ethylene and submergence treatments substantial increases in cell number occurred (*Figure 19.5*). In ethylene this accounted for 64% of the additional elongation relative to air controls, although cells were longer in all parts of the petiole and especially the distal 2 cm. In submerged plants the effect of buoyant tension during elongation through the water column (42 cm) was to cause a doubling of cell length relative to petioles in ethylene-enriched air. Cell lengths declined markedly once laminae reached the water surface but petiole elongation continued for some time.

In older petioles of air-grown plants, ethylene and submergence both promoted elongation mainly through increases in cell lengths (*Figure 19.6*). There was some increase in cell number in the distal zones but this accounted for only 13–14% of the additional growth. Cell lengths were again longer in submerged petioles than in ethylene-treated plants, presumably due to the influence of buoyant tension.

Are there irreversible effects of ethylene on development?

When hormones are applied to plants, any effects on growth usually stop as soon as the hormone is removed. In *Nymphoides peltata*, however, there is some evidence that exposure of tissues to ethylene by submergence very early in petiole development (and I have assumed that raised ethylene levels is the most significant consequence of submergence) causes changes in growth that are at least partly

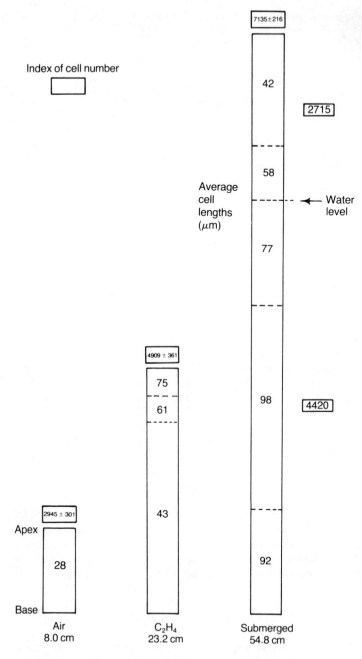

Figure 19.5 Effects of ethylene (10 μl l^{-1}) and submergence on petiole elongation, epidermal cell lengths and indices of epidermal cell number for the youngest leaf of air-grown *Nymphoides*. Initial petiole length was 18–22 mm with an average epidermal cell length of 21.2 ± 1.7 μm and index of epidermal cell number of 1240 ± 21

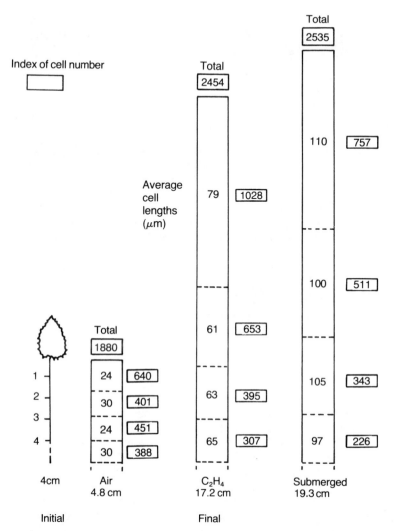

Figure 19.6 Effects of ethylene (10 μl l^{-1}) and submergence on epidermal cell length and indices of cell number for marked 1 cm zones on leaf 4 of air-grown *Nymphoides* plants

irreversible. Detailed discussion of this will be published elsewhere but three pieces of evidence are described briefly here. Firstly, when leaves developed entirely or from an early age in submerged conditions, rapid elongation did not stop immediately after reaching the water surface (*Figure 19.4*). This 'carryover' growth occurred in the absence of buoyant tension or of raised ethylene levels since both petioles and lamina were now floating at the surface. Nevertheless, the index of cell number continued to increase in the floating part of the petiole (*Figure 19.5*) and the average cell length exceeded 40 μm. This is considerably longer than the cell lengths attained in wholly air-grown petioles, which were consistently between 20 and 30 μm (*Figures 19.5* and *19.6*) and assuming that these cells were fully turgid (a not unreasonable assumption, given that plants grew in water-saturated compost and an atmosphere of 80% RH and cells attained the same lengths in detached, floating petioles), the difference in cell lengths must reflect differences in osmotic potential, wall extensibility or both.

A second piece of evidence relates to a physiological difference between the distal part of mature, floating petioles that had stopped growing and petioles of a comparable age on wholly air-grown plants. The former were much more sensitive to ethylene plus applied tension and this treatment (or resubmergence of the lamina) caused elongation at rates of 8–10 cm day^{-1} compared with only 1–2 cm day^{-1} in the wholly air-grown petioles.

Finally, the capacity for cell division seems to be irreversibly increased if petioles begin development under water. The final index of cell number in air-grown petioles was relatively constant at between 2000 and 3000. But if a young leaf elongating through the water column was excised and held vertically with the lamina in air, petiole cell number continued to increase up to an index value characteristic of submerged leaves (6000 to 7000, cf. *Figure 19.5*). Unless these excised petioles were under tension, average cell length declined in the distal region.

Concluding remarks

Ethylene promotes elongation in petioles of amphibious or flooding-tolerant plants in two distinct ways. Early in development it causes an increase in cell number which may (*Nymphoides*) or may not (*Ranunculus repens*) be accompanied by an increase in cell length; this effect causes the large elongation response of young petioles and is probably the only effect of ethylene in species with a very delayed response to flooding. Later in petiole development a distal zone may arise within which cells attain considerably greater lengths in the presence of ethylene; this effect is involved in the rapid growth responses to ethylene or submergence, and in at least two species (*Nymphoides* and *Regnellidium diphyllum*), buoyant tension acts synergistically to give even longer cells.

This 'first level' explanation provides only a framework within which to interpret the varying responses of different species to ethylene and submergence and to explore further the mechanism of ethylene action on petiole growth.

Three questions about ethylene action seem in greatest need of attention:

(1) What determines where on the petiole, and when during development, cells arise in which ethylene promotes much greater elongation? Is it related to the levels of other hormones, such as auxin; the osmotic pressure and turgor of the cells; or to the presence of new kinds of receptors?
(2) How does ethylene cause an increase in cell number? Is it simply a consequence of faster cell elongation resulting in a shortening of the cell cycle; or is there also an increase in the number of cycles per cell or of the number of cells able to divide; and is the presence of other hormones necessary, as it is for ethylene-promoted elongation?
(3) Is there, as preliminary evidence suggests, a critical, early period of petiole development during which exposure to ethylene causes irreversible or long-lasting changes in growth and physiology?

The answers to these questions are of considerable importance for a better understanding of how ethylene may control development and could give useful information about the wider subject of developmental control and the interactions between plants and their environment.

Acknowledgement

I wish to thank Ivan Amarasinghe for skilled technical support.

References

ARBER, A. (1920). *Water Plants: a study of aquatic angiosperms.* Cambridge University Press, Cambridge
COOKSON, C. and OSBORNE, D.J. (1978). *Planta,* **144**, 39–47
FUNKE, G.L. and BARTELS, P.M. (1937). *Biol Jaarbuek,* **4**, 316–344
MALONE, M. and RIDGE, I. (1983). *Planta,* **157**, 71–73
MUSGRAVE, A., JACKSON, M.B. and LING, E. (1972). *Nature (New Biology),* **238**, 93–96
MUSGRAVE, A. and WALTERS, J. (1974). *Planta,* **121**, 51–56
OSBORNE, D.J. (1982). In *Plant Growth Substances 1982,* pp. 279–290. Edited by P.F. Wareing. Academic Press, London
RIDGE, I. and AMARASINGHE, I. (1984). *Plant Growth Regulation,* **2**, 235–249
WALTERS, J. and OSBORNE, D.J. (1979). *Planta,* **146**, 309–317

20

ETHYLENE AND THE RESPONSES OF PLANTS TO EXCESS WATER IN THEIR ENVIRONMENT—A REVIEW*

MICHAEL B. JACKSON
Agricultural and Food Research Council, Letcombe Laboratory, Letcombe Regis, Wantage, Oxfordshire, UK

Introduction

HISTORICAL

The first suggestions of ethylene involvement in the responses of plants to overwet conditions are perhaps to be found in the work of Neilson Jones (1935) and Turkova (1944). Neilson Jones, working at Bedford College, London, found an organic soil containing gas with biological properties similar to those of ethylene. He concluded that '...seedlings in some soils may be influenced by the presence in the soil atmosphere of varying amounts of ethylene...'. Nine years later, Turkova (1944) reported that wet soil in the Alma-Ata region of the Soviet Union caused pronounced leaf epinasty (downward curling and growth) in potatoes and tomatoes. She described the phenomenon as 'strikingly similar to the usual habitus of plants that were under the influence of ethylene...'. In making their conclusions both authors drew upon earlier American work published in volume seven of Contributions from Boyce Thompson Institute (1935) that describes many now well-known effects of ethylene on plants. Among these was evidence for the growth-promoting activity of ethylene. This challenge to the then pre-eminence of auxin as *the* plant growth hormone may have been at least partly responsible for adverse criticism of ethylene work at the Boyce Thompson Institute in Went and Thimann's influential book about auxins entitled 'Phytohormones' (Went and Thimann, 1937). Subsequent research has shown many of Went and Thimann's criticisms to be misplaced. Furthermore an alleged arithmetic error by Boyce Thompson workers of 'ten times' that is discussed at length in Went and Thimann's book (p. 192) is not really justified. There is certainly a mistake of this size in the second step of a four-part calculation deriving the minimum physiologically active concentration of ethylene in water or living tissue (Crocker, Hitchcock and Zimmerman, 1935, p. 239), but the final calculated figure of 1 part in $65\,800 \times 10^6$ w/w is arithmetically acceptable. The 'error' was therefore that of the printer and could not have been employed by Crocker, Hitchcock and Zimmerman otherwise the outcome of their calculation would accordingly have been incorrect.

---

*In recognition of the pioneering contribution to this subject by Professor Makoto Kawase, of Wooster, Ohio, USA, who died on 2nd August 1983 after a long illness.

242 Ethylene and plant response to excess water in their environment

This adverse commentary on work at the Boyce Thompson Institute was a set-back for establishing a hormonal role for ethylene in plant growth and development. Consequently, little if any further work on links between ethylene and soil or water effects on plants was published until 1969 when Smith and Russell (1969) reported the presence of the gas in wet soils at concentrations known to affect plants in laboratory tests. During the intervening years Kramer (1951) and his colleague W.T. Jackson (1956) mentioned the possibility of a connection between the effects of excess water and those of ethylene. Their remarks were based on the old results of Crocker et al. and Turkova rather than on any new research findings.

PHYSICAL CHEMISTRY

A similarity between the appearance of plants treated with ethylene or waterlogged is not the only basis for suspecting a causal relationship between the two. Since most if not all plant parts produce the gas it is obvious that situations such as waterlogging or submergence which slow the rate of gaseous diffusion between the plant and its environment will also increase internal ethylene concentrations. Ethylene-sensitive tissues will then respond accordingly.

The relevant facts are that although ethylene is moderately soluble in water at growing temperatures (*Figure 20.1*a), the diffusion coefficient of ethylene is about 10 000 times less in water than in air (*Figure 20.1b*). So whenever a significant proportion of the plant's surface area becomes water covered, metabolically-generated gases such as ethylene become trapped within source tissue. The amount will be some function of the rate at which ethylene is produced by the tissue and the depth of the water covering. The use of a simple diffusion equation shows these

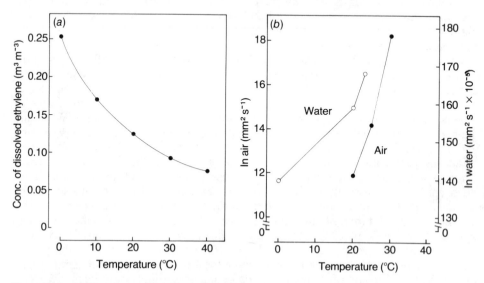

Figure 20.1 (a) Experimentally derived relationship between the solubility of pure ethylene in water at various temperatures. The value for 0 °C is that quoted by Porritt (1951). (b) Diffusivity of ethylene in air and water at different temperatures—based on Burg and Burg (1965); Elliott and Watts (1972); Huq and Wood (1968)

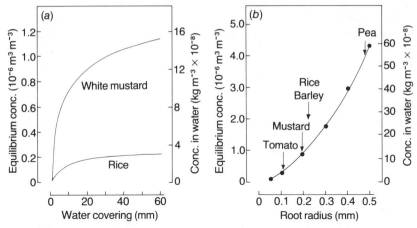

Figure 20.2 Effects of (a) increasing the depth of water covering, and (b) increasing root thickness on the accumulation of ethylene in submerged roots. In (a) the effect of water depth is shown for roots of two species with contrasting rates of ethylene production (white mustard — 1.8×10^{-12} m³ kg⁻¹ s⁻¹; rice — 0.4×10^{-12} m³ kg⁻¹ s⁻¹). In (b) the effect under 20 mm of water of root thickness at one rate of ethylene production (1.8×10^{-12} m³ kg⁻¹ s⁻¹) is shown. Radii of typical seedling roots of various species are also given. See Konings and Jackson (1979) for the diffusion equation used to calculate these results. (Note — 1 nl g⁻¹ h⁻¹ ≡ 0.28×10^{-12} m³ kg s⁻¹)

effects in *Figure 20.2a* for seedling roots of two species with contrasting rates of ethylene production. Faster ethylene production and deeper water inevitably mean more accumulated gas.

There is also an interaction with temperature. While the solubility of ethylene in water decreases at warmer temperatures, the converse is true for the coefficient of diffusion (*Figure 20.1*). By analogy with studies of oxygen (P.S. Blackwell, Letcombe Laboratory, personal communication) the flux of ethylene out of roots and into water will therefore increase with temperature, assuming all other things remain unchanged. This potential for faster depletion by diffusion as temperatures increase is however likely to be offset by an increased rate of ethylene biosynthesis within the roots. A further consideration is the thickness of the tissue. Again it is convenient to consider seedling roots. Thick roots generating ethylene at a similar rate to thin roots on a fresh weight or length basis inevitably will form more of the gas for each unit of surface area available for radial diffusive loss. This is a consequence of their smaller surface:volume ratio. Thick roots are therefore likely to accumulate more ethylene at similar depths of water covering and similar rates of ethylene production per unit length or weight (*Figure 20.2b*).

The situation becomes more complex when account is taken also of the link between oxygen supply and ethylene production. A covering of water not only entraps gases such as ethylene but is also as effective in preventing oxygen from diffusing to respiring tissue. (Coefficient of diffusion at 20 °C for oxygen in air is 20.1 mm² s⁻¹, and in water 21×10^{-4} mm² s⁻¹ (Armstrong, 1979).) Decreases in oxygen availability in turn can influence the rate of ethylene production. In some tissues such as mungbean hypocotyls (*Figure 20.3a*; Imaseki, Watanabe and Odawara, 1977) ethylene evolution declines in oxygen concentrations below 15 kPa (Air = 20.8 kPa). But in roots of barley (*Hordeum vulgare* L. 'Midas') and also maize (*Zea mays* L. 'LG 11') ethylene production is promoted at about 3 and 5 kPa

(*Figure 20.3b*) although in the complete absence of oxygen almost no ethylene is emanated. The arresting effect of anoxia is thought to be a consequence of a dependence upon oxygen by the final step in an ethylene biosynthetic pathway involving the conversion of 1-aminocyclopropane-1-carboxylic acid (ACC) to ethylene gas (Adams and Yang, 1979; Yang *et al.*, Chapter 2). The biochemical basis for the stimulating effects of partial oxygen deficiency (hypoxia) is unknown, but it does not appear to involve this final step (author's unpublished result).

The solubility and diffusivity of ethylene and also of its immediate precursor ACC are relevant to interpreting ethylene-like effects of waterlogging or submergence that can be detected some distance from inundated parts. For example, ethylene around whole root systems of tomato plants in excess of $2 \times 10^{-6} \text{ m}^3 \text{ m}^{-3}$ (2 ppm, v/v) can stimulate leaf epinasty several centimetres above, a result of

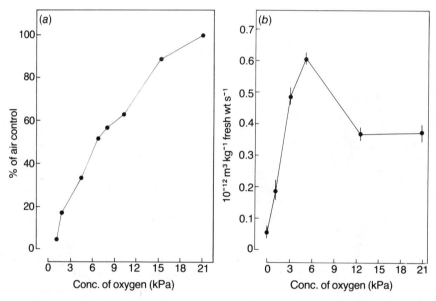

Figure 20.3 Effect of partial pressures of oxygen at and below that of air (20.8 kPa) on the rate of ethylene production. (a) Mungbean hypocotyl (redrawn from Imaseki, Watanabe and Odawara, 1977). (b) Roots of barley (redrawn from Jackson *et al.*, 1984)

ethylene diffusing up the plant through interconnected gas-filled space (Jackson and Campbell, 1975). Since the stems of plants are also porous radially (although usually less so than in the longitudinal direction) much ethylene is lost *en route* (Zeroni, Jerie and Hall, 1977). It has yet to be shown that ethylene can be transported as a gas dissolved in the transpiration stream moving in xylem vessels. The amount is unlikely to be physiologically significant since at best only 11–13% of the ethylene in a gas phase can be stripped out by an equal volume of water at 25 °C (*Figure 20.1a*, Porritt, 1951; McAuliffe, 1966). However, water readily dissolves ACC which has been measured in the xylem sap of waterlogged tomato plants at concentrations ($>10^{-3}$ mol m^{-3}) sufficient to increase ethylene production in recipient aerial shoots of transpiring plants (Bradford and Yang, 1980).

Submergence and shoot growth

RICE COLEOPTILES

Takahashi (1905) and Nagai (1916) provide two early reports that elongation by the coleoptile of germinating rice seed (*Oryza sativa* L.) is stimulated in partial pressures of oxygen less than that of air. The response of coleoptiles to reduced external oxygen (about 3 kPa is optimal in the absence of photosynthetic oxygen) was for many years held to be primarily responsible for explaining why coleoptiles of some types of rice (notably lowland japonica) extend faster under water than in the air (e.g. Nagao and Ohwaki, 1953; Yamada, 1954; Ranson and Parija, 1955). An inhibition of the oxidative destruction of growth-promoting auxin was a favoured hypothesis (Yamada, 1954; Wada, 1961) for explaining the effect, which may promote successful seedling establishment in wet conditions, e.g. when rice is 'direct-seeded' rather than transplanted to save labour costs (Turner, Chen and McCauley, 1981) or to protect seedlings from frost (Takahashi, 1978). Only when the coleoptile elongates sufficiently to break the water surface can it act as a 'snorkel' (Kordan, 1974) allowing oxygen to diffuse internally to the roots and the first leaves which would otherwise fail to extend (Turner, Chen and McCauley, 1981).

A major shortcoming of most of the above studies is their failure to separate the effect of low oxygen from that of entrapped metabolic gases such as ethylene and carbon dioxide. Ohwaki (1967) had noted that the elongation of coleoptile sections was faster in 'still' rather than agitated water and that 3 kPa oxygen was a more effective stimulator of elongation in water than in air. Takahashi *et al.* (1975) also

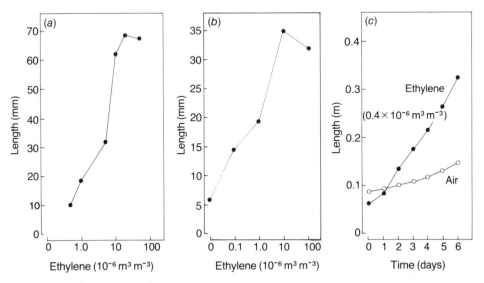

Figure 20.4 Ethylene as a stimulator of elongation in rice. (a) Effect on coleoptile length of increasing concentrations of ethylene flowing for 8 days over dark, germinating rice seed (redrawn from Ku *et al.*, 1970). (b) Effect on mesocotyl length after 8 days in sealed flasks in the dark after heat treatment to sensitize the tissue (redrawn from Suge, 1972). (c) Effect of 0.4×10^{-6} m^3 m^{-3} ethylene applied to non-submerged plants on total internode length of deep water rice (redrawn from Metraux and Kende, 1983)

showed that shoot length under water greatly exceeded that obtained in gaseous phases containing less than 20.8 kPa oxygen. These discrepancies were largely explained by Ku et al. (1970) who found ethylene to be a powerful promoter of coleoptile extension. The effect was dramatic. A doubling of ethylene concentration from $0.5 \times 10^{-6}\,m^3\,m^{-3}$ stimulated growth by 66% (*Figure 20.4a*). This was the first published demonstration of ethylene-promoted elongation in plants, although many examples of stimulated lateral expansion by ethylene were well-known at that time. Ku et al. also found that if ethylene enrichment ($10 \times 10^{-6}\,m^3\,m^{-3}$) was combined with oxygen deficiency (4 kPa), coleoptile extension was faster than when either treatment was given separately. Furthermore, carbon dioxide also promoted elongation, while removing carbon dioxide and ethylene from the gaseous environment by chemical trapping inhibited elongation by over 50%, when measured in sealed containers over five days. It is reasonable to conclude that faster elongation caused by submersion in stagnant water is a result of the combined influence of diminished oxygen supply, and the accumulation of growth-stimulating concentrations of ethylene and carbon dioxide. Each component appears to act independently (Raskin and Kende, 1983). Unfortunately, the time course measurements of Atwell, Waters and Greenway (1982) under cleverly varied conditions of oxygenation in flowing systems, contradict these conclusions. Atwell, Waters and Greenway show small concentrations of oxygen to have almost *no* stimulatory or inhibitory effect on elongation. They state (but without supporting evidence) that carbon dioxide was also inactive and thus ascribe the submergence response almost entirely to ethylene. This view was supported by finding that $10^{-3}\,mol\,m^{-3}$ silver nitrate (an inhibitor of ethylene action (Beyer, 1976)) almost entirely eliminated the stimulation of extension caused by submergence. Unfortunately Atwell, Waters and Greenway did not actually test silver nitrate for its ability to offset the effect of ethylene on the coleoptile, nor eliminate the possibility that it may have acted non-specifically as a toxin. New research to clarify matters could profitably take account of the large differences in the response of various cultivars or ecotypes (e.g. Suge, 1971a; Takahashi et al., 1975; Atwell, Waters and Greenway, 1982), the effects of illumination (e.g. Suge, Katsura and Inanda, 1971) and the importance shown by Atwell and colleagues of separating effects on the early cell division stage from those on the later cell extension phase of coleoptile growth. Only the latter is affected by ethylene while the former can be much inhibited by oxygen deficiency. A shortcoming of all previous work is the absence of direct measurements of carbon dioxide, oxygen and ethylene in submerged seedlings. At present we must rely on deductions from gassing and enclosing treatments that mimic submergence.

The mechanism by which ethylene stimulates coleoptile extension is unknown. Submergence and small concentrations of oxygen can increase cell wall extensibility (Furuya et al., 1972; Zara and Masuda, 1979). Ethylene has also been found to have this effect while stimulating rachis elongation by the water fern *Regnellidium diphyllum* (Cookson and Osborne, 1979). The response is probably mediated by cell-wall acidification (Malone and Ridge, 1983). Zara and Masuda (1979) measured surprisingly high osmotic potentials (i.e. dilute cell sap) in submerged coleoptiles (*cf.* Atwell, Waters and Greenway, 1982). But control of extension may nevertheless reside in small changes of osmotic potential since cell walls remain highly extensible even when growth finally ceases with age. It can be calculated from Zara and Masuda's results that the minimum turgor required for growth (threshold turgor) is extremely small under submerged conditions.

Most research suggests that auxin is unnecessary for ethylene action in the rice coleoptile. For example, elongation can be stimulated in the presence of the anti-auxin *p*-chlorophenoxyisobutyric acid or in the absence of the seed or the coleoptile tip (putative sources of auxin) (Furuya, Masuda and Yamamoto, 1972; Katsura and Suge, 1979). However, Imaseki and Pjon (1970) believe that auxin is essential, finding no stimulation of elongation by ethylene in apical coleoptile segments unless auxin is also added. In a more convincing study Ishizawa and Esashi (1983) found that exogenous ethylene *can* stimulate elongation in excised coleoptiles, even when de-tipped (auxin deficient). Gibberellins may be required for maximum responsiveness to ethylene (Suge, 1974).

The rice coleoptile is not unique in its response to submergence. Other allied species likely to possess a similar physiology include the so-called wild rice, *Zizania aquatica* (Campiranon and Koukkari, 1977), and barnyard grass, *Echinochloa crus-galli* (Civico and Moody, 1979; Kennedy *et al.*, 1980) which is a weed of rice fields.

RICE MESOCOTYL

Much less information is available concerning the growth of this normally short stem-like tissue situated between the coleoptile and the seed. From two papers (Nakayama, 1942; Takahashi, 1978 (cited by Takahashi)) the excessively wet conditions that stimulate coleoptile elongation seem to *suppress* mesocotyl growth. In wet soil or vermiculite, mesocotyl elongation is accelerated by increasing moisture up to about 100% of dry weight before growth declines again at the greater moisture contents favouring coleoptile extension (*Figure 20.5*). Ethylene is

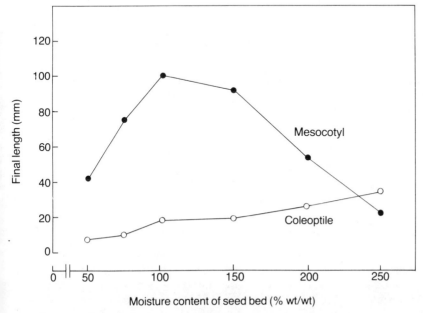

Figure 20.5 Elongation over 7 days by mesocotyls and coleoptiles of germinating rice seedlings sown 10 mm deep in vermiculite of increasing water content (cultivar Surjamki — Indica ecotype). Redrawn from Takahashi (1978)

probably involved in controlling mesocotyl growth under these partially saturated conditions. Takahashi (1978) found the gas in vermiculite that was wetted sufficiently to ensure rapid extension of the mesocotyl, and depletion of this ethylene with a mercuric perchlorate trap decreased growth from 112 mm to 13 mm over seven days. Carbon dioxide was also shown to be involved in a similar depletion experiment using a KOH trap. Although Takahashi was unable to stimulate mesocotyl growth with exogenous ethylene, Suge (1971b, 1972) gives clear evidence of ethylene-stimulated extension. Japonica rice was found to be less sensitive to ethylene than Indica rice, but a heat treatment (40 °C) rectified this shortcoming (*Figure 20.4b*, Suge, 1972). Japonica rice that is not heat treated may lack abscisic acid, a hormone that can act synergically with ethylene in promoting mesocotyl elongation (Takahashi, 1973). Measurements of endogenous ethylene and studies with inhibitors of ethylene synthesis and action could usefully be applied to establishing more firmly a role for ethylene in mesocotyl growth in wet conditions.

A partial explanation for the differential effect of submergence on mesocotyl and coleoptile may lie in the tolerance of the coleoptile to oxygen deficiency and to anoxia. The mesocotyl may not possess this tolerance and thus like the roots of rice fail to grow when oxygen is depleted, even in the presence of concentrations of ethylene that would otherwise promote elongation.

DEEP WATER RICE

An outstanding feature of this ecotype is the ability of established plants (but not seedlings or flowering plants) to elongate quickly under water on a scale about one order of magnitude greater than that previously described for coleoptiles and mesocotyls. Deep water rice can survive and even yield grain in several metres of water, an achievement at least partly due to the fast underwater growth that maintains some of the foliage above the rising flood water (Vergara, Jackson and De Datta, 1976). Growth is located at internodes and in the leaf-bases that surround them. Internode lengths of up to 60 cm have been recorded (Vergara, Jackson and De Datta, 1976). Métraux and Kende (1983) found that fast growing, submerged plants contain up to $1 \times 10^{-6} \, m^3 \, m^{-3}$ of ethylene in the lumen of the shoots. When non-submerged plants were supplied with less than half this concentration, internode growth accelerated to rates similar to those of submerged plants (*Figure 20.4c*). A role for endogenous ethylene in promoting elongation in submerged conditions was supported further by experiments in which inhibitors of ethylene biosynthesis (aminoethoxyvinylglycine (AVG) and amino-oxyacetic acid (AOA)) also prevented the submergence response by the internodes. The effects of these inhibitors could be reversed by supplying ACC to restore ethylene production in the rice shoot. How far ethylene may also stimulate growth of the surrounding leaf sheaths is not clear. Métraux and Kende also found that submergence not only causes endogenous ethylene to increase simply as a result of entrapment but it also promotes faster biosynthesis. How this is brought about is unknown. It may be a response to the small oxygen concentrations (7–8 kPa) measured in the water used for submersion. It is known already that roots of barley and maize produce more ethylene when partially depleted of oxygen (*Figure 20.2b*; Jackson, 1982; Jackson *et al.*, 1984).

Tolerance of deep-water conditions is also associated with the formation of a larger diameter lumen (central cavity) and numerous gas-filled sacs in the cortex (Roy, 1972; Datta and Banerji, 1974). Possibly ethylene may also effect these changes in submerged deep water rice since ethylene can enhance gas-space formation in sunflower stems (Kawase, 1979) and also maize roots (Drew, Jackson and Giffard, 1979). The likely decrease in resistance to the longitudinal diffusion of oxygen within plants containing extensive gas-space can be expected to help aerate roots positioned several metres below the water surface in anaerobic soil. Without an adequate internal oxygen supply by this pathway the resultant poor root growth would cause instability and wilting (Anonymous, 1979) and depress nutrient uptake. Deep water rice is especially vulnerable to uprooting in windy conditions (Vergara, Jackson and De Datta, 1976).

STEM AND PETIOLE EXTENSION IN SPECIES OTHER THAN RICE

A range of aquatic and semi-aquatic dicotyledonous plants, and a semi-aquatic fern have been shown to respond to submergence and ethylene in a broadly similar way to that described for rice. In some instances it is the stem that responds, in others the petiole. The effect appears to be of acclimatic significance, helping to maintain leaves and flowers above water level where photosynthesis and pollination can proceed effectively.

Callitriche spp. (Starworts) inhabit ponds or slow-moving streams. The plants comprise a floating rosette of leaves separated by short internodes (less than 1 mm long) connected to a root system in the substratum by a series of submerged, elongated internodes. Each internode can be several centimetres long. If a floating rosette becomes submerged, extension of its short internodes is stimulated sequentially (oldest first) until the apex again reaches the surface. McComb (1965) found no evidence that anoxia or changes in incident light intensity or wavelength caused the fast underwater growth. However, coating the leaves with 'Vaseline' grease did simulate submergence suggesting that interference in gaseous or vapour diffusion is involved. McComb concluded that inhibited transpiration, coupled with the action of gibberellic acid, promoted underwater growth.

Increasing evidence that ethylene could promote rather than inhibit elongation (e.g. rice coleoptile (Ku *et al.*, 1970); fern gamotophyte (Miller, Sweet and Miller, 1970); and fruit-stalk of the squirting cucumber (Jackson, Morrow and Osborne, 1972)) prompted Musgrave, Jackson and Ling (1972) to investigate the involvement of ethylene in the underwater growth of *Callitriche*. They found concentrations between 0.01 and $1.0 \times 10^{-6} \, m^3 \, m^{-3}$ stimulated internode extension to rates approaching those of submerged plants and with a similar response time of 30 min or less. The effects were reversible by floating submerged rosettes or withdrawing ethylene or by applying silver nitrate, an ethylene antagonist (Jackson, 1982). The concentration of endogenous ethylene within submerged plants exceeded that required to promote extension (Musgrave, Jackson and Ling, 1972). Thus ethylene gas (in concert with gibberellins), entrapped within submerged stems is the likely cause of the submergence response in *Callitriche*. In contrast to similar work with rice, the possible involvement of changes in oxygen and carbon dioxide partial pressures has not been examined. Accumulated carbon dioxide and partial oxygen deficiency may exercise some slight *restriction* on underwater growth in *Callitriche*. Certainly the growth response to a combination of ethylene (applied as ethephon)

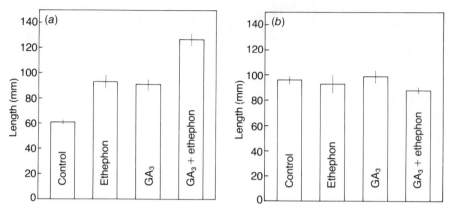

Figure 20.6 Effect on extension growth of (a) floating or (b) submerged rosettes of *Callitriche platycarpa* (Kutz) treated with ethephon, 10^{-3} kg m^{-3} or gibberellic acid, 0.01 mol m^{-3}, separately or together for 2 days in 1 litre volumes of nutrient solution. (Means of four replicates ± s.e.)

and gibberellic acid is more effective on floating than on submerged plants (*Figure 20.6*) suggesting that submergence sets an upper limit on the activity of these hormones.

Studies of other species have established further points of detail. The petiole of *Ranunculus sceleratus* responds to ethylene and submergence in a similar manner to the stem of *Callitriche* (Musgrave and Walters, 1973). Indole acetic acid also promotes elongation and independently of any additional ethylene production that growth-promoting applications of IAA can bring about (Cookson and Osborne, 1978). These workers also found that an inhibitor of ethylene production (1,2,amino-4-(2'-aminoethoxy)-trans butenoic acid) largely prevented fast underwater growth, an effect at least partially reversible by applying exogenous ethylene. Thus ethylene and not IAA appears to control the rate of underwater growth. However, a certain minimal endogenous concentration of endogenous IAA may be needed for ethylene activity in submerged petioles, since application of naptalam, an inhibitor of auxin transport or action blocks the response (Horton and Samarakoon, 1982). In contrast to the preceding examples petiole growth seems not to be primarily a consequence of submergence of the elongating part itself but rather a response to submergence of adjacent leaf-blade tissue. The petiole will elongate rapidly even in *air* if the leaf-blade alone is under water (i.e. a detached leaf turned upside-down, Horton and Samarakoon, 1979). The mechanism of communication between leaf-blade and petiole is unknown, but could conceivably involve internal diffusion of ethylene or movement of a precursor of the gas such as ACC. It may be possible that elongation occurs where leaf-blade attaches to the petiole (*see* Ridge, Chapter 19). Thus the distance for any internally transported message to travel may be minimal. A related phenomenon in this genus is the change in shape by leaves developing under water (heterophylly). Ethylene is not involved but other hormones may play some part although available evidence is preliminary (Young and Horton, 1983).

Other dicotyledonous species with similar but less intensively researched characteristics to *Callitriche* and *Ranunculus* include *Hydrocharis morsus - ranae* L. a floating-rosette plant (petiole responsive, Cookson and Osborne, 1979) *Nymphoides peltata*, the aquatic fringed waterlily (Malone and Ridge, 1983), *Sagittaria pymaea*, the perennial aquatic arrowhead and *Potamogeton distinctus*, a deep water pondweed and perennial weed of rice fields (shoot responsive, Suge and Kusanagi, 1975). Carbon dioxide also has some promotive effect in the last two species and acts synergistically with ethylene in a manner reminiscent of rice coleoptiles. Ridge and Amarsinghe (1981) state that *Ranunculus repens*, *Plantago major* (rat-tail plantain) and 16 other species, none of them grasses, also respond similarly to ethylene and submergence.

To the already diverse taxonomic spectrum of flowering plants in which ethylene regulates their response to submergence can be added the Brazilian water fern *Regnellidium diphyllum* (Lindm). The upright, frond-stalk (the rachis), originating from a submerged, horizontal rhizome gives the depth response. The fern has many features in common with *Callitriche* but exhibits a particular need for longitudinal tension arising from the buoyancy of the hollow fronds before submergence can effect faster extension (Musgrave and Walters, 1974). Ethylene can even be promotive in air but again activity is fully expressed only if the rachis is stretched a little, for example by suspending it above a small weight (3 g). Under these conditions response time can be as short as 10 min. Musgrave and Walters (1974) deduced that ethylene action involves weakening of the cell walls. This was subsequently confirmed by Cookson and Osborne (1978, 1979). Ethylene activity seems to require IAA supplied from the frond. If this is excised, IAA must be given generously in the bathing medium. Cell wall loosening by ethylene could be a consequence of acidification following accelerated proton extrusion as shown in ethylene-treated petioles of *Nymphoides peltata* (Malone and Ridge, 1983).

Questions remain concerning firstly, the involvement of small partial pressures of oxygen and increased partial pressures of carbon dioxide in the depth accommodating response of this group of plants; secondly, the possibility that submergence may increase ethylene synthesis and thirdly, the significance of fast underwater growth for survival and competitive advantage. This latter point is often assumed but experimental testing is now becoming possible with the availability of chemical means for suppressing ethylene formation and action.

Submergence and root extension

Roots of different species in well aerated conditions elongate faster when exposed to small concentrations of ethylene, but more slowly when these are increased above a certain value (*Figure 20.7*). The optimum concentration differs between species and this may be related to different rates of endogenous ethylene production (Konings and Jackson, 1979). For example white mustard roots produce ethylene at a fast rate (*Figure 20.7*) and consequently very little additional ethylene (0.02–0.05×10^{-6} m^3 m^{-3}) is required before root extension is retarded. However, rice roots produce much less ethylene (*Figure 20.7*), and more exogenous ethylene (1×10^{-6} m^3 m^{-3} is needed therefore to supplement this before inhibition is achieved. Indeed, at smaller concentrations (e.g. 0.05×10^{-6} m^3 m^{-3}) a considerable stimulation of extension (approximately 18%) can be obtained (*Figure 20.7*; Smith and Robertson, 1971; Konings and Jackson, 1979). It has been calculated that the slow rate of ethylene production of rice roots will restrict the extent of ethylene accumulation in roots under water to growth-promoting concentrations (Konings and Jackson, 1979). This may be of acclimatic significance for rice. It seems especially relevant to nodal or adventitious roots of deep water

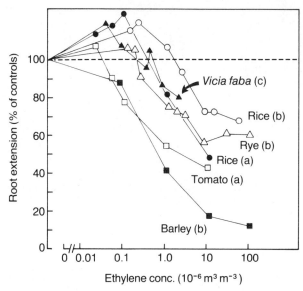

Figure 20.7 Effect of exogenous ethylene on extension by seedling roots of various species. Based on results of Konings and Jackson, 1979(a); Smith and Robertson, 1971(b); Kays, Nicklow and Simons, 1974(c)

rice that emerge into the flood water above soil level (Yamaguchi, 1976). Such roots are not subjected to the dominating influence of oxygen deficiency and toxins to which soil-bound roots may be exposed and thus may be able to respond fully to ethylene.

Shortcomings in the evidence presently available include the lack of any direct demonstration that submergence hastens root extension in rice, although a slowing of extension has been reported in submerged but presumably well-oxygenated roots of species with faster rates of ethylene production (e.g. cress—Larqué-Saavedra, Wilkins and Wain, 1975; mustard—Haberkorn and Sievers, 1977; MacDonald and Gordon, 1978). The effect of submergence on increasing the ethylene content of roots needs to be monitored directly. So far this effect has only been calculated and not assayed.

Soil waterlogging and responses of roots

When about 90% of interconnected soil pores become water-filled rather than gas-filled (Cannell, 1977) roots and aerobic soil micro-organisms begin to asphyxiate. Gaseous products of their metabolism accumulate while the entry of atmospheric oxygen to replace that used in respiration is denied. Later when reserves of oxygen dissolved in the soil water become exhausted toxic substances accumulate such as partly reduced intermediates of carbohydrate metabolism (e.g. ethanol) and chemically reduced alternatives to oxygen for accepting electrons from microbial respiration (Cannell and Jackson, 1981; Jackson, 1983). Ethylene is most likely to play a significant part in the responses of roots at or near the soil surface where some oxygen from the air may be available for growth reactions to ethylene and for ethylene biosynthesis. Some effects of the gas have the appearance of improving the ability of roots and thus the whole plant to survive the anaerobic and toxin forming conditions typical of prolonged waterlogging.

AERENCHYMA DEVELOPMENT

Armstrong (1979) has reviewed the early work of Van Raalte, Conway, Lang and others demonstrating the close association between the survival of roots in anoxic surroundings and the existence of an internal, longitudinal gas-filled space (aerenchyma) connecting with above-ground parts. Not only is such gas-space thought to allow oxygenation of the roots and some dispersal of carbon dioxide, but oxygen leaking radially from these conduits oxidizes chemically reduced toxins in the waterlogged root environment, rendering them less harmful. In many plants gas-space results from the collapse and death of cells, often in species-specific patterns (Armstrong, 1979; Smirnoff and Crawford, 1983). The resulting hollow roots will use less oxygen while retaining their potential for absorbing inorganic nutrient (Drew *et al.*, 1980). The dying cells may also provide nutrition for their surviving neighbours.

In some species poor aeration *stimulates* the development of gas-space within the root cortex (maize—Norris, 1913; wheat—Trought and Drew, 1980; rice—Katayama, 1961). The reason for this has been examined thoroughly in adventitious roots of maize. McPherson (1939) believed that total oxygen depletion (anoxia) kills the cells to form aerenchyma but recent tests have shown anoxia actually inhibits its formation (Drew, Jackson and Giffard, 1979). An alternative theory implicates ethylene, a view substantiated by the following evidence. In stagnant but *not* completely anaerobic solution cultures, aerenchyma formation is stimulated in growing roots in association with an increase in endogenous ethylene concentration. Applying similar concentrations to well-aerated roots also stimulates aerenchyma but silver ions at non-toxic concentrations block aerenchyma development in both ethylene-treated roots and in roots growing in stagnant conditions (Drew, Jackson and Giffard, 1979; Drew *et al.*, 1981; Konings, 1982). Treatment with anoxia to prevent ethylene synthesis also arrests aerenchyma formation. Konings (1982) succeeded in inhibiting aerenchyma formation by supplying cobalt chloride, a likely inhibitor of ACC breakdown to ethylene, and by supplying amino-oxyacetic acid (AOA), a substance that can block ACC production from SAM. Unfortunately in these experiments ethylene or ACC were not added together with AOA or AVG to test whether the effect of the inhibitors was specifically to inhibit ethylene formation.

The additional ethylene found in roots in poorly aerated surroundings results not only from entrapment, but also from a boost to ethylene synthesis caused by oxygen concentrations intermediate between those of solutions vigorously bubbled with a stream of air (20.8 kPa) and anoxia (Jackson, 1982). Most ethylene is formed under conditions of 3–5 kPa oxygen, concentrations that are also the most effective in stimulating gas-space development (Jackson *et al.*, in preparation). The biosynthetic pathway for ethylene production in maize roots has not been established. Preliminary testing based on the assumption that it involves methionine disassembly via s-adenosylmethionine (SAM) and ACC, show that 3–5 kPa oxygen does not stimulate the conversion of ACC to ethylene. Its action must therefore be located earlier in the pathway.

254 *Ethylene and plant response to excess water in their environment*

Interestingly, a deficiency of nitrate (Konings and Verschuren, 1980) and also phosphate (Drew and Saker, 1983) encourages gas-space formation in maize roots. The possible interactive relationships between the effects of nutrient deficiency, oxygen deficiency and those of ethylene are therefore of interest. Oxygen deficiency and ethylene can inhibit phosphorus uptake (e.g. Jackson, Drew and Giffard, 1981; Jackson, 1984), but the influence of nutrient deficiency on ethylene production or action remains to be examined. An excess of phosphorus can slow ethylene formation, in carrot roots (Chalutz, Mattoo and Fuchs, 1980) suggesting that deficiency brought about by oxygen shortage may achieve the opposite result and promote aerenchyma formation by way of increasing the ethylene content of maize roots.

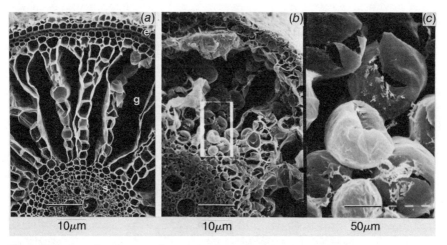

Figure 20.8 Scanning electron micrographs of cross-sections of rice roots (*Oryza sativa* L. 'IR3') after freeze-drying. (a) Root tissue about 6 days old. (b) Younger tissue showing (c) detail of degeneration of cortical cells. (Gas space 43.3% of cortex.) From left to right horizontal scale bars represent 10 μm, 10 μm and 50 μm respectively. (Photographs by S.F. Young, AFRC Letcombe Laboratory)

The process of cell dissolution that ethylene stimulates during gas-space formation has not been investigated in detail. One of the first degenerative events is loss of tonoplast integrity and turgor followed by inward collapse as adjacent expanding cells exert pressure on their weakening neighbours (Campbell and Drew, 1983). These effects are reminiscent of ethylene-promoted abscission (Leopold, 1967; Wright and Osborne, 1974) and cavitation in petioles (Linkins, Lewis and Palmer, 1973), flower stalks (Wong and Osborne, 1978) and stems (Aloni and Rosenshtein, 1982). Extra cellulase-enzyme activity is found in most of these degenerating tissues. The enzyme may therefore be active in degrading cortical cells of maize roots. Kawase (1981b) makes this proposal in a recent review.

Aerenchyma in the adventitious roots of rice is even more extensive than in maize. *Figure 20.8a* shows the internal structure of six day old root tissue—the cortex is 46% gas-space. Younger tissue still undergoing cell expansion and breakdown is shown in *Figure 20.8b*. The extent of any involvement of ethylene is at present unclear. Unlike maize, rapid aerenchyma formation does not require an external stimulus such as stagnant conditions, since even in briskly bubbled nutrient solutions, gas-space is evident in one day old root tissue (Jackson and Fenning, unpublished). Imposing stagnant conditions has only a small accelerating effect, and applying ethylene in the form of 'ethephon' or as $5 \times 10^{-6} \, m^3 \, m^{-3}$ ethylene gas also gives only a slight promotion of aerenchyma (Jackson and Fenning, unpublished). These results suggest that gas-space forms at close to maximal rate in well aerated rice roots. If ethylene is involved it must be the small amount of endogenous hormone present under these conditions. Experiments with inhibitors of ethylene synthesis or action have however given equivocal results. Silver nitrate at non-toxic concentration inhibits aerenchyma development, but weakly. Only when sufficient silver nitrate is supplied almost to halt root extension is the formation of gas-space completely arrested. This is presumably a toxicity effect and not related to the action of small amounts of silver ions that can block ethylene action (Beyer, 1976). The slow rate of ethylene production by roots of rice poses problems for studies with presumptive inhibitors of ethylene biosynthesis such as cobalt chloride and AVG. Their application separately over a wide concentration range has failed to inhibit aerenchyma formation. When administered together, considerable depression of ethylene production has been achieved but with no associated diminution of aerenchyma (Jackson and Fenning, unpublished). Subject to further research it is tentatively concluded that a major role for ethylene in controlling aerenchyma formation in rice roots is unlikely.

Ethylene activity in the many other species of aquatic and semi-aquatic species that form aerenchyma (for examples see Armstrong, 1979; Smirnoff and Crawford, 1983) remains to be examined. It is possible that the gas will prove to be more important in non-aquatic species that form extensive aerenchyma only under poorly aerated conditions. Perversely, *Bouteloua gracilis* (H.B.K.) Lag. a grass from western North America forms aerenchyma for reasons unconnected with tolerance of wet conditions. In this plant a watery environment or small oxygen concentrations actually *decrease* cortical disintegration (Beckel, 1956).

MISCELLANEOUS EFFECTS ON ROOTS

Because roots are so likely to experience wet conditions and thus an elevated ethylene content it is appropriate to identify some additional effects of ethylene on roots that also occur in the presence of excess water. Nodulation by species of *Rhizobium* and the rate of nitrogen fixation by existing nodules are severely inhibited by $>0.4 \times 10^{-6} \, m^3 \, m^{-3}$ ethylene (Grobbelaar, Clark and Hough, 1970, 1971; Goodlass and Smith, 1979). This sensitivity may contribute to the well established susceptibility of legume crops to wetness (e.g. Minchin and Pate, 1975). Ethylene can also cause root swelling, a phenomenon also seen by Eavis (1972) in pea roots growing in oxygen-deficient (0.16–0.08 kPa) soil. The orientation of root growth with respect to gravity can be changed from positive to diageotropism or negative geotropism by flooding (e.g. tomato—Guhman, 1924; sunflower—Wample and Reid, 1978; *Ludwigia peploides*—Ellmore, 1981) and also by ethylene (tobacco—Zimmerman and Hitchcock, 1935, Zimmerman and Wilcoxon, 1933; maize—Bucher and Pilet, 1982). It is conceivable that such redirecting of growth may enable roots to extend upwards, towards better aerated regions closer to the soil surface. The effect may be that termed 'aerotropism' by Molisch (1884) and demonstrated more recently by Wiersum (1967).

Environmental soil ethylene

Hitherto the discussion has concentrated on the significance of ethylene produced endogenously by plants growing in wet conditions. But there is evidence that soil also forms ethylene. Smith and Russell (1969) found the equivalent of $20 \times 10^{-6} \, m^3 \, m^{-3}$ of ethylene dissolved in the water of flooded field soil and rather more if wet soil was sealed in containers for seven days. However, despite subsequent confirmation that abnormally high concentrations of ethylene do occur in wet soil (e.g. Leyshon and Sheard, 1978; Yoshida and Suzuki, 1975) the source of the ethylene and its significance for plants remains uncertain.

SOURCES OF ETHYLENE IN WATERLOGGED SOIL

Using results from a closely monitored laboratory study of soil suspensions, Smith and Restall (1971) concluded that temperature-dependent microbial breakdown of soil organic matter stimulated by anaerobic or nearly anaerobic (1 kPa oxygen) conditions is responsible for much of the ethylene to be found in waterlogged soil. Intermediates in the anaerobic breakdown of carbohydrates are thought to be the substrates for this ethylene (Goodlass and Smith, 1978) that is probably generated by anaerobic bacteria (Smith, 1976; Primrose, 1976). Evidence from Abeles *et al.* (1971), Cornforth (1975) and Smith, Bremner and Tabatabi (1973) shows that other soil micro-organisms (probably Gram-positive mycobacteria (De Bont, 1975)) can *degrade* ethylene, but only in the presence of oxygen. Therefore, it may reasonably be concluded that waterlogging favours ethylene accumulation in soil because firstly, microbial production is increased by anaerobiosis; secondly, vigorous microbial degradation is prevented by anaerobiosis; and thirdly, the balance is entrapped within the soil by the presence of water in the interconnected pore-space.

Less easily incorporated into the above scheme is the work of Lynch and co-workers (Lynch, 1972; Lynch and Harper, 1974a; Lynch, 1975) identifying *aerobic* fungi such as *Mucor hiemalis* as copious ethylene generators in soil containing readily degradable organic matter. An oxygen requiring pathway with methionine as the precursor seems to be involved. At first glance this work does not help to explain the presence of ethylene in waterlogged soils that are anaerobic. Certainly it does not readily explain how anaerobic soils *generate* ethylene. But some of the ethylene found in such soil was possibly formed by aerobic fungi such as *M. hiemalis* (and possibly aerobic bacteria (Pazout and Wurst, 1977)) in the early stages of waterlogging before the supply of dissolved oxygen became exhausted. The ethylene thus synthesized would be retained by the sealing-in effect of the water and carried over into the subsequent anoxic conditions that suppress microbial degradation. The action of aerobic, ethylene-producing micro-organisms may also explain reports of considerable quantities of ethylene in wet field soil that is not entirely free of oxygen (e.g. Campbell and Moreau, 1979; Smith and Dowdell, 1974). In laboratory experiments too, time course measurements of ethylene and oxygen show increases in ethylene to occur before oxygen is entirely depleted in the soil (e.g. Jackson, 1979a).

Lynch suggests three possible alternatives to the above for explaining how aerobic soil fungi such as *M. hiemalis* may enrich over-wet anaerobic soil with ethylene. The first derives from observations that slowly growing fungi produce the most ethylene (Lynch and Harper, 1974b). In small oxygen concentrations growth rates are also slow and these authors speculate that *a priori* ethylene production should increase. Unfortunately conflicting laboratory tests showed that a restricted oxygen supply to the fungus decreased ethylene production rather than stimulated it, although growth was severely retarded under these conditions (Lynch and Harper, 1974a). It is also proposed by Lynch and Harper that in partially anaerobic soil, zones of anaerobiosis (e.g. sites rich in degrading organic matter) may accumulate ethylene precursors such as methionine because no oxygen is available to complete the pathway to ethylene. The precursors may then diffuse or be transported (mobilized) out to better aerated zones where aerobic micro-organisms such as *M. hiemalis* would be able to complete the conversion to ethylene (Drew and Lynch, 1980). There is however no experimental evidence of these processes taking place. Their third proposal is that *extracellular* ethylene production may be possible from the products of fungal autolysis that anaerobiosis can induce; such production possibly taking place without need of oxygen (Lynch, 1974). The evidence however lacks any demonstration that lysis brought about by anaerobic conditions actually releases the components required to boost extracellular ethylene production. Instead, the phenomenon seems to be associated specifically with senescence and fungal degeneration caused by ageing, rather than by a lack of oxygen.

SIGNIFICANCE OF SOIL ETHYLENE

The principal unresolved question here is the extent to which roots exposed to a depletion of oxygen and enrichment with carbon dioxide caused by waterlogging the soil can respond to elevated concentrations of ethylene. Only in the case of aerenchyma formation is it reasonably certain that ethylene is active in the presence of small oxygen partial pressures (e.g. 5 kPa). But, even here the potentially interfering effect of carbon dioxide on this ethylene response has yet to be investigated. Other work suggests that ethylene inhibition of extension may be enfeebled by oxygen deficiency. Cornforth and Stevens (1973) found no effect of $1–20 \times 10^{-6} m^3 m^{-3}$ ethylene on barley roots in 5 kPa of oxygen. Similarly, the growth-promoting effect of 3,5-diiodo-4-hydroxybenzoic acid (DIHB) an inhibitor of ethylene action is diminished in barley roots by small oxygen partial pressures (Jackson *et al.*, 1984), suggesting little involvement for ethylene in controlling extension under these conditions. Furthermore, Cornforth and Stevens' work shows inhibition of root extension by ethylene to be suppressed by carbon dioxide. More detailed examination of the interactive effects of ethylene with oxygen and carbon dioxide is needed before the significance of soil ethylene (and endogenous ethylene) for the growth of roots, nodulation, nitrogen fixation, etc. in over-wet soil can be fully evaluated.

Because ethylene has some mobility within plants it is conceivable that soil ethylene may enter the shoot system by internal diffusion. Tissue at the base of the aerial shoot and at the soil–air interface is likely to be well oxygenated and thus potentially responsive to this ethylene. There is certainly a zone of rapidly changing morphology at this site. Stem and lenticel hypertrophy (swelling) in association with cavitation resembling aerenchyma (Mishra, 1973; Kawase, 1974, 1979; Wample and Reid, 1975; Kawase and Whitmoyer, 1980) are typical features of this region of waterlogged sunflower and tomato and other species. Applying ethylene can mimic these symptoms (Wallace, 1926; Kawase, 1981a) and the gas is present in

Figure 20.9 (a) Adventitious root system on sunflower plants waterlogged for 14 days. (b) Effect of removing the adventitious roots daily during 17 days of continuous waterlogging. Treatments from left to right are control, waterlogged, waterlogged with adventitious roots removed by daily excision. Plants 22 days old when treatment commenced (Jackson and Palmer, unpublished)

greater concentrations in hypertrophic tissue of flooded plants (Kawase, 1972a, 1974; Wample and Reid, 1979). Involvement of the gas is therefore probable and some of the ethylene could come from sources within the soil.

Adventitious root formation at the stem-base of flooded plants is also thought by Kawase (1974) and Wample and Reid (1979) to be under ethylene control but Jackson and Palmer (1981) found $5-10 \times 10^{-6}\,m^3\,m^{-3}$ ethylene applied to the stem base (hypocotyl) of sunflowers induces only hypertrophy and not root formation. But in two week old maize plants, ethylene can bring forward in time the outgrowth of *preformed* adventitious roots (also in willow—Kawase, 1972b) through the leaf bases that surround the stem (Jackson, Drew and Giffard, 1981). Their subsequent elongation is however retarded.

Hypertrophy and stem aerenchyma are likely to increase the chances of plant survival by improving internal aeration of the newly forming adventitious roots that may themselves be aerenchymatous. These roots (*Figure 20.9a*) appear to replace those afflicted by anoxia at depth in the flooded soil. In sunflower, removal of these new surface roots severely depresses the growth of waterlogged plants and increases foliar senescence. With the roots left intact the plants survive and succeed in flowering, although overall size is smaller (*Figure 20.9b*).

Soil waterlogging and the responses of shoots

Soil waterlogging can enrich shoot tissue with ethylene for considerable distances above the soil surface, including the apex (El-Beltagy and Hall, 1974; Smith and Jackson, 1974; Jackson and Campbell, 1975). The effect is not attributable principally to ethylene entering from the soil, since it can be reproduced in tomato by exposing the roots of plants to anaerobic or near anaerobic nutrient solution (Jackson and Campbell, 1976; Bradford and Dilley, 1978). The presence of anaerobic roots is required, and split-root and root excision experiments suggested that such roots provide a precursor of ethylene biosynthesis that passes up the plant in the xylem to the aerial shoot. (Jackson and Campbell, 1975, 1976; Jackson, Gales and Campbell, 1978; Jackson, 1980). Here readily available oxygen would permit its conversion to ethylene. In this way the capacity of the root system for producing the gas could be transferred to the shoots. The proposed precursor was identified in xylem sap by Bradford and Yang (1980) as ACC. These authors showed the flux of ACC from roots to shoot is sufficient to explain the faster ethylene production in the aerial parts of flooded tomatoes. They also deduced that in addition to ensuring its accumulation by blocking conversion to ethylene at an oxygen-requiring step, flooding also stimulates ACC biosynthesis in the roots. As would be expected if this mechanism operates, applying inhibitors of ACC formation such as AVG or AOA largely prevents root anaerobiosis from stimulating ethylene production in shoot tissues (Bradford, Hsiao and Yang, 1982).

Physical wounding (cutting) of tomato petioles stimulates ethylene production (Jackson, 1976). The effect is probably a result of enhancing ACC formation from SAM (Bradford and Yang, 1980). Ethylene production by petioles in response to wounding is enhanced if plants are first waterlogged for 48 h (Jackson, Gales and Campbell, 1978). This suggests either that wounding can stimulate ethylene formation from the additional ACC provided by the flooded roots or that flooding enhances the potential for shoot tissue to form ACC when wounded.

Increased ethylene content of the shoot systems of waterlogged plants has been linked causally by several authors to various symptoms of flooding injury such as slow shoot growth, leaf senescence, leaf and flower abscission (El-Beltagy and Hall, 1974; Dodds, Smith and Hall, 1983), epinasty (*see* references in Introduction), reorientation of shoot growth by diageotropic tomatoes (Jackson, 1979b; Jackson and Campbell, 1976). Only in the latter two examples is the link other than the simplest correlative evidence of 'joint occurrence'. For epinasty the supporting case is particularly strong. Increases in ACC and ethylene in the leaves of flooded plants precede or coincide with the development of epinasty while the amount of additional ethylene is demonstrably sufficient to promote epinasty. When inhibitors of ethylene action such as carbon dioxide, DIHB, silver nitrate or benzothiadiazole are applied, epinasty in flooded plants is inhibited (Jackson and Campbell, 1976; Bradford and Dilley, 1978; Wilkins, Alejar and Wilkins, 1978; Bradford and Yang, 1980). The extent to which changes in other hormones associated with flooding (e.g. a decrease in gibberellins and cytokinins) influence sensitivity to ethylene (*see* Jackson and Drew, 1984 for discussion) is somewhat uncertain. Foliar applications of those hormones inhibit epinasty in tomato plants treated with ethylene (Jackson and Campbell, 1979) or with their roots in stagnant nutrient solution (Selman and Sandanam, 1972) suggesting that deficiencies in gibberellins and cytokinins may sensitize the petioles to ethylene.

The action of endogenous abscisic acid in closing the stomata of flooded plants (Jackson, Hall and Kowalewska, 1983) may help retain ethylene within the leaves.

An effect of flooding that merits further study is that of stolon growth of the strawberry clover (*Trifolium fragiferum*). This species is naturally prostrate in well-drained conditions, but when flooded it rapidly grows upwards (Bendixen and Peterson, 1962). The response is thought to aid survival. Applying ethylene to well-aerated, prostrate plants reproduces the effect of flooding within 24 h (Hansen and Bendixen, 1974).

Conclusions

The assembled evidence probably comprises the most extensively researched and convincing example of hormone mediation in plant responses to an environmental change. The inherent strong physiological activity of ethylene, its slow diffusivity in water and the susceptibility of its production to changes in oxygen supply are properties exploited by a wide variety of plant species to sense they are submerged or waterlogged and to respond quickly in ways that often appear conducive to survival.

Experimental examination of ethylene action in responses to excess water have benefited from several physical properties of the gas and from technical and biochemical advances not always available in work with the other, non-gaseous hormones. These advantages include, simple extraction techniques (e.g. diffusion or mass flow under vacuum), extremely sensitive identification and detection methods that approach the sensitivity of the plant itself (flame ionization gas chromatography), the possibility of making many well replicated measurements over short periods of time, the ease of applying ethylene in small continuous doses for pharmacological studies, the availability of chemical inhibitors of ethylene action (e.g. silver ions), a well understood biosynthetic pathway, and a selection of chemicals that inhibit biosynthesis with a physiologically useful degree of specificity. With these tools considerable progress has been made while further consolidation, together with new examples of ethylene involvement in the effects of environmental water on growth and development can be expected. The subject could usefully be extended by critical experimental testing of the assumed and likely adaptive or acclimatic significance of ethylene mediated responses to excess water.

References

ABELES, F.B., CRAKER, L.E., FORRENCE, L.E. and LEATHER, G.R. (1971). *Science*, **173**, 914–915
ADAMS, D.O. and YANG, S.F. (1979). *Proceedings of the National Academy of Science USA*, **76**, 170–174
ALONI, B. and ROSENSHTEIN, G. (1982). *Physiologia Plantarum*, **56**, 513–517
ANONYMOUS (1979). *International Rice Research Institute Annual Report for 1978*. IRRI, Los Baños, The Philippines
ARMSTRONG, W. (1971). *Physiologia Plantarum*, **25**, 192–197
ARMSTRONG, W. (1979). *Advances in Botanical Research*, **7**, 225–332
ATWELL, B.J., WATERS, I. and GREENWAY, H. (1982). *Journal of Experimental Botany*, **33**, 1030–1044
AVADHANI, P.N., GREENWAY, H., LEFROY, R. and PRIOR, L. (1978). *Australian Journal of Plant Physiology*, **5**, 15–25
BECKEL, D.K.B. (1956). *The New Phytologist*, **55**, 183–191
BENDIXEN, L.E. and PETERSON, M.L. (1962). *Crop Science*, **2**, 223–228
BEYER, E. Jr. (1976). *HortScience*, **11**, 195–196

BRADFORD, K.J. (1981). *HortScience*, **16**, 25–30
BRADFORD, K.J. and DILLEY, D.R. (1978). *Plant Physiology*, **61**, 506–509
BRADFORD, K.J., HSIAO, T.C. and YANG, S.F. (1982). *Plant Physiology*, **70**, 1503–1507
BRADFORD, K.J. and YANG, S.F. (1980). *Plant Physiology*, **65**, 322–326
BUCHER, D. and PILET, P-E. (1982). *Physiologia Plantarum*, **55**, 1–4
BURG, S.P. and BURG, E.A. (1965). *Physiologia Plantarum*, **18**, 870–884
CAMPBELL, R. and DREW, M.C. (1983). *Planta*, **157**, 350–357
CAMPBELL, R.B. and MOREAU, R.A. (1979). *American Potato Journal*, **56**, 199–210
CAMPIRANON, S. and KOUKKARI, L. (1977). *Physiologia Plantarum*, **41**, 293–297
CANNELL, R.Q. (1977). *Advances in Applied Biology*, **2**, 1–86
CANNELL, R.Q. and JACKSON, M.B. (1981). In *Modifying the Root Environment to Reduce Crop Stress*, pp. 141–192. Ed. by G.F. Arkin and H.M. Taylor. American Society of Agricultural Engineers, St. Joseph, USA
CHALUTZ, E., MATTOO, A.K. and FUCHS, Y. (1980). *Plant, Cell and Environment*, **3**, 349–356
CIVICO, R.S.A. and MOODY, K. (1979). *Philippines Journal of Weed Science*, **6**, 41–49
COOKSON, C. and OSBORNE, D.J. (1978). *Planta*, **144**, 39–47
COOKSON, C. and OSBORNE, D.J. (1979). *Planta*, **146**, 303–307
CORNFORTH, I.S. (1975). *Plant and Soil*, **42**, 85–96
CORNFORTH, I.S. and STEVENS, R.J. (1973). *Plant and Soil*, **38**, 581–587
CROCKER, W., HITCHCOCK, A.E. and ZIMMERMAN, P.W. (1935). *Contributions from Boyce Thompson Institute*, **7**, 231–248
DATTA, S.K. and BANERJI, B. (1974). *Phytomorphology*, **24**, 164–174
DEBONT, J.A.M. (1975). *Annals of Applied Biology*, **81**, 119–121
DODDS, J.H., SMITH, A.R. and HALL, M.A. (1983). *Plant Growth Regulation*, **1**, 203–207
DREW, M.C., CHAMEL, A., GARREC, J.P. and FOURCY, A. (1980). *Plant Physiology*, **65**, 506–511
DREW, M.C., JACKSON, M.B. and GIFFARD, S. (1979). *Planta*, **147**, 83–88
DREW, M.C., JACKSON, M.B., GIFFARD, S.C. and CAMPBELL, R. (1981). *Planta*, **153**, 217–224
DREW, M.C. and LYNCH, J.M. (1980). *Annual Review of Phytopathology*, **18**, 37–66
DREW, M.C. and SAKER, L.R. (1982). *Agricultural Research Council Letcombe Laboratory Annual Report for 1982*, pp. 43–44. Agricultural Research Council, London
EAVIS, B.W. (1972). *Plant and Soil*, **37**, 151–158
EL-BELTAGY, A.S. and HALL, M.A. (1974). *The New Phytologist*, **73**, 47–60
ELLIOTT, R.W. and WATTS, H. (1972). *Canadian Journal of Chemistry*, **50**, 31–34
ELLMORE, G.S. (1981). *American Journal of Botany*, **68**, 557–568
FURUYA, M., MASUDA, Y. and YAMAMOTO, R. (1972). *Development, Growth and Differentiation*, **4**, 95–105
GOODLASS, G. and SMITH, K.A. (1978). *Soil Biology and Biochemistry*, **10**, 201–205
GOODLASS, G. and SMITH, K.A. (1979). *Plant and Soil*, **51**, 387–395
GROBBELAAR, N., CLARKE, B. and HOUGH, M.C. (1970). *Agroplantae*, **2**, 81–82
GROBBELAAR, N., CLARKE, B. and HOUGH, M.C. (1971). *Plant and Soil, Special Volume*, 215–223
GUHMAN, H. (1924). *Ohio Journal of Science*, **24**, 199–208
HABERKORN, H.-R. and SIEVERS, A. (1977). *Naturwissenschaften*, **64**, 639–640
HANSEN, D.J. and BENDIXEN, L.E. (1974). *Plant Physiology*, **53**, 80–82
HORTON, R.F. and SAMARAKOON, A. (1979). In *The Tenth International Conference*

on *Plant Growth Regulators—Abstracts*, p. 47. International Plant Growth Substances Association: Madison, USA
HORTON, R.F. and SAMARAKOON, A.B. (1982). *Aquatic Botany*, **13**, 97–104
HUQ, A. and WOOD, T. (1968). *Journal of Chemical and Engineering Data*, **13**, 256–259
IMASEKI, H. and PJON, C. (1970). *Plant and Cell Physiology*, **11**, 827–829
IMASEKI, H., WATANABE, A. and ODAWARA, S. (1977). *Plant and Cell Physiology*, **18**, 577–586
ISHIZAWA, K. and ESASHI, Y. (1983). *Journal of Experimental Botany*, **34**, 74–82
JACKSON, M.B. (1976). *Planta*, **129**, 273–274
JACKSON, M.B. (1979a). *Journal of the Science of Food and Agriculture*, **30**, 143–152
JACKSON, M.B. (1979b). *Physiologia Plantarum*, **46**, 347–351
JACKSON, M.B. (1980). *Acta Horticulturae*, **98**, 61–78
JACKSON, M.B. (1982). In *Plant Growth Substances 1982*, pp. 291–301. Ed. by P.F. Wareing. Academic Press, London
JACKSON, M.B. (1983). *Aspects of Applied Biology*, **4**, 99–116
JACKSON, M.B. (1984). In *Mechanism of Assimilate Distribution and Plant Growth Regulators*, pp. 158–182. Ed. by J. Kralovic: Slovak Society of Agriculture, Food and Forestry of the Slovak Academy of Sciences, Piestany, Czechoslovakia
JACKSON, M.B. and CAMPBELL, D.J. (1975). *Annals of Applied Biology*, **81**, 102–105
JACKSON, M.B. and CAMPBELL, D.J. (1975). *The New Phytologist*, **74**, 397–406
JACKSON, M.B. and CAMPBELL, D.J. (1976). *The New Phytologist*, **76**, 21–29
JACKSON, M.B. and CAMPBELL, D.J. (1979). *The New Phytologist*, **82**, 331–340
JACKSON, M.B., DOBSON, C.M., HERMAN, B. and MERRYWEATHER, A. (1984). *Plant Growth Regulation*, **2**, 251–262
JACKSON, M.B. and DREW, M.C. (1984). In *Flooding and Plant Growth, Physiology Ecology Series*, pp. 47–128. Ed. by T.T. Kozlowski. Academic Press, New York
JACKSON, M.B., DREW, M.C. and GIFFARD, S.C. (1981). *Physiologia Plantarum*, **52**, 23–28
JACKSON, M.B., GALES, K. and CAMPBELL, D.J. (1978). *Journal of Experimental Botany*, **29**, 183–193
JACKSON, M.B., HALL, K.A. and KOWALEWSKA, A.K.B. (1983). *Journal of the Science of Food and Agriculture*, **34**, 944–945
JACKSON, M.B., MORROW, I.B. and OSBORNE, D.J. (1972). *Canadian Journal of Botany*, **50**, 1465–1471
JACKSON, M.B. and PALMER, J.H. (1981). *Plant Physiology*, **67**, 58 (abstract)
JACKSON, W.T. (1956). *American Journal of Botany*, **43**, 637–639
JOHN, C.D. (1977). *Plant and Soil*, **47**, 269–274
KATAYAMA, T. (1961). *Proceedings of the Crop Science Society of Japan*, **29**, 229–233 (English summary)
KATSURA, N. and SUGE, H. (1979). *Plant and Cell Physiology*, **20**, 1147–1150
KAWASE, M. (1972a). *Journal of the American Society of Horticultural Science*, **97**, 584–588
KAWASE, M. (1972b). *Proceedings of the Plant Propagation Society*, **22**, 360–367
KAWASE, M. (1974). *Physiologia Plantarum*, **31**, 29–38
KAWASE, M. (1979). *American Journal of Botany*, **66**, 183–190
KAWASE, M. (1981a). *American Journal of Botany*, **68**, 651–658
KAWASE, M. (1981b). *HortScience*, **16**, 30–34
KAWASE, M. and WHITMOYER, R.E. (1980). *American Journal of Botany*, **67**, 18–22

KAYS, S.J., NICKLOW, C.W. and SIMONS, D.H. (1974). *Plant and Soil*, **40**, 565–571
KENNEDY, R.A., BARRETT, S.C.H., ZEE, D.V. and RUMPHO, M.E. (1980). *Plant Cell and Environment*, **3**, 243–248
KONINGS, H. (1982). *Physiologia Plantarum*, **54**, 119–124
KONINGS, H. and JACKSON, M.B. (1979). *Zeitschrift für Pflanzenphysiologie*, **92**, 385–397
KONINGS, H. and VERSCHUREN, G. (1980). *Physiologia Plantarum*, **49**, 265–270
KORDAN, H.A. (1974). *Journal of Applied Ecology*, **11**, 685–690
KRAMER, P.J. (1951). *Plant Physiology*, **26**, 722–736
KU, H.S., SUGE, H., RAPPAPORT, L. and PRATT, H.K. (1970). *Planta*, **90**, 333–339
LARQUÉ-SAAVEDRA, A., WILKINS, H. and WAIN, R.L. (1975). *Planta*, **126**, 269–272
LEOPOLD, A.C. (1967). *Symposium of the Society for Experimental Biology*, **21**, 507–516
LEYSHON, A.J. and SHEARD, R.W. (1978). *Canadian Journal of Soil Science*, **58**, 347–355
LINKINS, A.E., LEWIS, N. and PALMER, R.L. (1973). *Plant Physiology*, **52**, 554–560
LYNCH, J.M. (1972). *Nature*, **240**, 45–46
LYNCH, J.M. (1974). *Journal of General Microbiology*, **83**, 407–411
LYNCH, J.M. (1975). *Agricultural Research Council Letcombe Laboratory Annual Report for 1974*, pp. 88–95. Agricultural Research Council, London
LYNCH, J.M. and HARPER, S.H.T. (1974a). *Journal of General Microbiology*, **80**, 187–195
LYNCH, J.M. and HARPER, S.H.T. (1974b). *Journal of General Microbiology*, **85**, 91–96
MACDONALD, I.R. and GORDON, D.C. (1978). *Plant, Cell and Environment*, **1**, 313–316
MALONE, M. and RIDGE, I. (1983). *Planta*, **157**, 71–73
McAULIFFE, C. (1966). *The Journal of Physical Chemistry*, **70**, 1267–1275 (Table II)
McCOMB, A.J. (1965). *Annals of Botany*, **29**, 445–459
McPHERSON, D.C. (1939). *The New Phytologist*, **38**, 190–202
MÉTRAUX, J-P. and KENDE, H. (1983). *Plant Physiology*, **72**, 441–446
MILLER, P.M., SWEET, H.C. and MILLER, J.H. (1970). *American Journal of Botany*, **57**, 212–217
MINCHIN, F.R. and PATE, J.S. (1975). *Journal of Experimental Botany*, **26**, 60–69
MISHRA, B.N. (1973). *Indian Scientific Congress Association Proceedings*, **60**, 389–390
MOLISCH, H. (1884). *Berichte der Deutschen Botanischen Gesellschaft*, **2**, 160–169
MUSGRAVE, A., JACKSON, M.B. and LING, E. (1972). *Nature New Biology*, **238**, 93–96
MUSGRAVE, A. and WALTERS, J. (1973). *The New Phytologist*, **72**, 783–789
MUSGRAVE, A. and WALTERS, J. (1974). *Planta*, **121**, 51–56
NAGAI, I. (1916). *Journal of the College of Agriculture, Tokyo Imperial University*, **3**, 109–158
NAGAO, M. and OHWAKI, Y. (1953). *Science Report Tôhoku University, Series (IV) (Biology)*, **20**, 54–71
NEILSON JONES, W. (1935). *Nature*, **136**, 554
NORRIS, F. DE LA, M. (1913). *Proceedings of the Bristol Naturalists Society*, **4**, 134–136
OHWAKI, Y. (1967). *Science Report of the Tôhoku University* Series (IV) (Biology), **33**, 1–5
ORUNRISEM, S.V. (1976). *Journal of General Microbiology*, **97**, 343–346

PAZOUT, J. and WURST, M. (1977). *Folia Microbiologia*, **22**, 458–459
PORRITT, S.W. (1951). *Scientific Agriculture*, **31**, 99–112
RANSON, S.L. and PARIJA, B. (1955). *Journal of Experimental Botany*, **6**, 80–93
RASKIN, I. and KENDE, H. (1983). *Journal of Plant Growth Regulation*, **2**, 193–203
RIDGE, I. and AMARSINGHE, I. (1981). In *Responses of Plants to Environmental Stress and Their Mediation by Plant Growth Substances*, pp. 13–14. Ed. by M.B. Jackson and M.C. Drew. Association of Applied Biologists—British Plant Growth Regulator Group, Wantage, UK (Abstract)
ROY, J.M. (1972). *Riso (Milan)*, **21**, 157–160
SELMAN, I.W. and SANDANAM, S. (1972). *Annals of Botany*, **36**, 837–848
SMIRNOFF, N. and CRAWFORD, R.M.M. (1983). *Annals of Botany*, **51**, 237–249
SMITH, A.M. (1976). *Soil Biology and Biochemistry*, **8**, 293–298
SMITH, K.A., BREMNER, J.M. and TABATABI, M.A. (1973). *Soil Science*, **116**, 313–319
SMITH, K.A. and DOWDELL, R.J. (1974). *Journal of Soil Science*, **25**, 217–230
SMITH, K.A. and JACKSON, M.B. (1974). *Agricultural Research Council Letcombe Laboratory Annual Report for 1973*, pp. 60–75. Agricultural Research Council, London
SMITH, K.A. and RESTALL, S.W.F. (1971). *Journal of Soil Science*, **22**, 430–443
SMITH, K.A. and ROBERTSON, P.D. (1971). *Nature*, **234**, 148–149
SMITH, K.A. and RUSSELL, R.S. (1969). *Nature*, **222**, 769–771
SUGE, H. (1971a). *Japanese Journal of Soil and Manure Science*, **40**, 127–131
SUGE, H. (1971b). *Plant and Cell Physiology*, **12**, 831–837
SUGE, H. (1972). *Plant and Cell Physiology*, **13**, 401–405
SUGE, H. (1974). *Proceedings of the Crop Science Society of Japan*, **43**, 83–87
SUGE, H., KATSURA, N. and INANDA, K. (1971). *Planta*, **101**, 365–368
SUGE, H. and KUSANAGI, T. (1975). *Plant and Cell Physiology*, **16**, 65–72
TAKAHASHI, K. (1973). *Planta*, **109**, 363–364
TAKAHASHI, N. (1978). *Australian Journal of Plant Physiology*, **5**, 511–517
TAKAHASHI, T. (1905). *Bulletin of the College of Agriculture, Imperial University, Tokyo*, **6**, 439–442
TAKAHASHI, H., PONGSROYPACH, C., GUNTHARAROM, S. and SASIPRAPA, V. (1975). *Japanese Agricultural Research Quarterly*, **9**, 73–75
TROUGHT, M.C.T. and DREW, M.C. (1980). *Journal of Experimental Botany*, **31**, 1573–1585
TURKOVA, N.S. (1944). *Comptes Rendus (Doklady) de l'Academie des Sciences de l'URSS*, **42**, 87–90
TURNER, F.T., CHEN, C-C. and McCAULEY, G.N. (1981). *Agronomy Journal*, **73**, 566–570
VERGARA, B.S., JACKSON, B. and DE DATTA, S.K. (1976). In *Rice and Climate*, pp. 301–319. The International Rice Research Institute, Los Baños, The Philippines
WADA, S. (1961). *Science Report of the Tôhoku University Series (IV) (Biology)*, **27**, 237–249
WALLACE, R.H. (1926). *Bulletin of the Torrey Botanical Club*, **53**, 385–402
WAMPLE, R.L. and REID, D.M. (1975). *Planta*, **127**, 262–270
WAMPLE, R.L. and REID, D.M. (1978). *Physiologia Plantarum*, **44**, 351–358
WAMPLE, R.L. and REID, D.M. (1979). *Physiologia Plantarum*, **45**, 219–226
WENT, F.W. and THIMANN, K.V. (1937). *Phytohormones*, pp. 181–182; 191–192. Macmillan, New York
WIERSUM, L.K. (1967). *Naturwissenschaften*, **54**, 203–204
WILKINS, H., ALEJAR, A.A. and WILKINS, S.M. (1978). In *Opportunities for Chemical*

Plant Growth Regulation, Monograph 21, pp. 83–94. British Crop Protection Council, Croydon
WONG, C.H. and OSBORNE, D.J. (1978). *Planta*, **139**, 103–111
WRIGHT, M. and OSBORNE, D.J. (1974). *Planta*, **120**, 163–170
YAMADA, N. (1954). *Plant Physiology*, **29**, 92–96
YAMAGUCHI, T. (1976). *Japanese Journal of Tropicl Agriculture*, **20**, 33–34
YOSHIDA, T. and SUZUKI, M. (1975). *Soil Science and Plant Nutrition*, **21**, 129–135
YOUNG, J.P. and HORTON, R.F. (1983). In *Proceedings of the Canadian Society of Plant Physiologists*, p. 25. CSPP Waterloo, Canada (Abstract)
ZARA, I. and MASUDA, Y. (1979). *Plant and Cell Physiology*, **20**, 1117–1124
ZERONI, M., JERIE, P.H. and HALL, M.A. (1977). *Planta*, **134**, 119–125
ZIMMERMAN, P.W. and HITCHCOCK, F. (1935). *Contributions from Boyce Thompson Institute*, **7**, 209–229
ZIMMERMAN, P.W. and WILCOXON, F. (1933). *Contributions from Boyce Thompson Institute*, **5**, 351–369

21
ETHYLENE AND FOLIAR SENESCENCE

JEREMY A. ROBERTS
Physiology and Environmental Science, University of Nottingham Faculty of Agriculture
GREGORY A. TUCKER
Applied Biochemistry and Food Science, University of Nottingham Faculty of Agriculture
MARTIN J. MAUNDERS
Physiology and Environmental Science, University of Nottingham Faculty of Agriculture

Introduction

Senescence may be considered as the sequence of metabolic events which culminate in cell death. This developmental phenomenon may be restricted to individual cells within a tissue, for example during the final stages of xylem differentiation or alternatively it may occur synchronously in all the cells comprising a plant organ (flower, fruit, leaf) or even, as in monocarpic plants, a whole organism (Woolhouse, 1983). Various plant growth substances have been implicated in the regulation of senescence. The role of ethylene in the ripening and senescence of fruit (Grierson *et al.*, Chapter 14; Knee, Chapter 24) and the senescence of flowers (Stead, Chapter 7; Manning, Chapter 8; Nichols and Frost, Chapter 28) has been discussed previously in this volume. The aim of this chapter is to review the role of ethylene in foliar senescence.

Types of foliar senescence

Simon (1967) describes three types of naturally occurring senescence processes in leaves. Firstly, progressive or sequential senescence where the oldest leaves senesce first. Secondly, synchronous senescence where all the leaves senesce at the same time, and finally overall senescence, characteristic of monocarpic species, where flowering is followed by overall death of the plant. It is likely that these various types of senescence result from different physiological and biochemical processes occurring within the leaf (Simon, 1967).

Many experimental systems use either detached leaves or leaf discs to study senescence. In general detachment of the leaf results in accelerated senescence and in tissues such as oat (*Avena sativa*) leaves the accompanying biochemical changes although temporarily advanced mimic those which occur during natural senescence (Thimann, Tetley and Van Thanh, 1974). However, in other cases detachment results in marked changes in the senescence pattern of the leaves (Simon, 1967), for example apple (*Pyrus malus*) leaf discs show a reversal in the sequence of chlorophyll and protein loss compared to attached leaves (Spencer and Titus, 1973). Experimental conditions may also influence foliar senescence. A number of workers have shown that detached leaves senesce faster in the dark than in the light

(Simon, 1967; Thimann, Tetley and Krivak, 1977) and there are differences in their respective metabolisms (Malik, 1982).

In view of the variety of naturally occurring senescence processes which exist and the experimental procedures adopted to study them it is difficult to relate results from any one experimental system to foliar senescence *in situ*. However, although there are undoubtedly many different patterns of foliar senescence it is possible to draw some general conclusions concerning the physiological and biochemical changes involved.

Physiological and biochemical changes during foliar senescence

The metabolic changes accompanying leaf senescence have been reviewed comprehensively (Beevers, 1976; Thimann, 1980; Thomas and Stoddart, 1980; Woolhouse, 1967, 1982, 1983) and will only be outlined briefly here.

The two major biochemical changes consistently observed are extensive proteolysis and chlorophyll loss. In general proteolysis precedes chlorophyll loss, for example in detached oat leaves proteolysis commences 6 h after detachment and chlorophyll degradation occurs 18 h later (Tetley and Thimann, 1974). The period of proteolysis is accompanied by an increase in free amino acids (Malik, 1982; Wang, Cheng and Kao, 1982) which in attached leaves are transported out of the leaf (Thimann, Tetley and Van Thanh, 1974; Beevers, 1976). In detached leaves or leaf discs the accumulation of amino acids may promote senescence (Malik and Thimann, 1980). Accompanying proteolysis and yellowing is a decrease in the nucleic acid content of the cells with loss of RNA occurring faster than loss of DNA (Wollgiehn, 1967; Dyer and Osborne, 1971; Harris *et al.*, 1982; Dhillon and Miksche, 1981). Also solubilization of cell-wall polysaccharides occurs (Matile, 1974) however the free reducing sugars generated do not appear to be transported out of the leaf (Thimann, Tetley and Van Thanh, 1974).

Early in senescence there is a marked decline in photosynthetic activity of the leaf tissue (Woolhouse, 1967). This is associated with the loss of certain key enzymes of the photosynthetic carbon reduction cycle, notably the large subunit of ribulose bisphosphate carboxylase, as well as several thylakoid membrane proteins (Woolhouse, 1982). There are conflicting reports on how respiration rates change during senescence. Several early reports simply demonstrated a gradual decline in respiration during senescence (*see* Rhodes, 1980a). Woolhouse (1967) using *Perilla fructescens* showed that respiration remained constant during the early stages of senescence then went through a 'climacteric-like' rise prior to a marked decline. In contrast Tetley and Thimann (1974) have shown that in detached oat leaves respiration increased at the same time as loss of chlorophyll occurred. Mitochondria remain structurally intact until senescence is well advanced and it is possible that respiration may have an important role. Indeed Satler and Thimann (1983) have demonstrated that experimental conditions which delay the respiratory climacteric in oat leaves also delay senescence.

Although senescence is characterized by extensive degradation of nucleic acid and protein, synthesis of RNA (Osborne, 1967; Wollgiehn, 1967) and protein (Shiboaka and Thimann, 1970) continues. Many new enzyme activities, especially proteases, RNAses and other hydrolases, appear during senescence (Lauriere, 1983). Thus the process may be regulated at either the transcriptional or translational level or perhaps both.

In general leaf senescence is insensitive to the application of the transcriptional inhibitors actinomycin-D and rifampicin (Thomas and Stoddart, 1977) and also to chloramphenicol which inhibits translation on 70S ribosomes (Kao, 1978). However a number of workers have shown that senescence is delayed by the application of either cycloheximide (Kao, 1978) or 1-(4-methyl-2,6-dinitroanilino)-N-methylpropionamide (Thomas, 1976). Both of these chemicals specifically inhibit translation on 80S ribosomes. From the results of studies such as these the consensus of opinion is that translational control in the cytoplasm plays a significant role in the regulation of foliar senescence. However, Yu and Kao (1981) have shown that whilst cycloheximide has a pronounced delaying effect on the senescence of soybean (*Glycine max*) leaves, actinomycin-D, rifampicin and chloramphenicol also have a small effect. Thus the involvement of transcriptional control or protein synthesis in chloroplasts or mitochondria cannot be excluded.

It is generally considered that the primary genetic control of senescence resides in the nuclear genome (Thomas and Stoddart, 1980). However the nature of this control is unknown. Further understanding of the process may be facilitated by the use of mutants such as those of *Festuca pratensis* which whilst retaining chlorophyll and some thylakoid proteins still undergo extensive proteolysis during 'senescence' (Thomas and Stoddart, 1975; Thomas, 1982, 1983).

Factors affecting the rate of senescence

Several environmental factors can influence foliar senescence (Woolhouse, 1967; Thomas and Stoddart, 1980). Light delays senescence of excised leaf tissues but may promote the process in attached leaves (Maunders and Brown, 1983). Exposure to extremes of temperature can also initiate the onset of yellowing. Other stress factors for instance drought, nutrient deficiency and invasion by pathogens can similarly promote chlorophyll loss. Thus leaf senescence *in situ* may be induced or accelerated by environmental factors including competition for space, light and nutrients or may be genetically programmed (Thomas and Stoddart, 1980).

At the cellular level senescence is probably controlled by plant growth regulators. Richmond and Lang (1957) showed that the application of kinetin to detached leaves of *Xanthium* effectively delayed senescence. It is now known that cytokinins can delay senescence in a wide variety of plants (Osborne, 1967; Thimann, 1980) and that auxins can behave similarly in some species (Osborne, 1967). Gibberellic acids have been shown to delay senescence in only a very few species (Thimann, 1980).

El-Antably, Waring and Hillman (1967), during an investigation into the effects of abscisic acid (ABA) on plants, found that it accelerated leaf senescence. This observation has been repeated on many species (for a review *see* Thimann, 1980). It is not, however, clear what role, if any, ABA may play in inducing senescence since conflicting reports on ABA levels in senescing leaves have appeared in the literature. Even-Chen and Itai (1975) demonstrated a sharp rise in ABA levels during the early stages of tobacco (*Nicotinia tabaccum*) leaf ageing, followed by a decline. The initial increase was prevented by the application of kinetin. Samet and Sinclair (1980) demonstrated increasing levels of ABA during the senescence of soybean leaves. However, the increase occurred after total soluble protein and chlorophyll had started to decline. Thus ABA may not initiate the senescent processes but may act to control their subsequent rates. Gepstein and Thimann

(1980) have demonstrated a correlation between ABA level and rate of senescence in oat leaves under a variety of experimental conditions.

These plant growth regulators are not the only endogenous factors capable of influencing foliar senescence. Several amino acids, especially serine, can promote senescence, and this effect is antagonized by arginine (Martin and Thimann, 1972). Also polyamines, diamines such as spermine and putrescine, and calcium ions are effective antagonists of the process (Altman, 1982; Ferguson, Watkins and Harman, 1983).

Role of ethylene in leaf senescence

It is evident from the preceding discussion that a number of different types of foliar senescence may be recognized and that these can be influenced by a variety of factors. The aim of this section of our review is to examine the role of ethylene in leaf senescence and to critically appraise the evidence that the gas is 'a dominant hormone in the regulation of leaf senescence' (Ferguson, Watkins and Harman, 1983).

Implication of a role for ethylene in senescence has originated from a number of sources. Firstly, incubation of plant tissue such as leaves or stems in the presence of ethylene hastens loss of chlorophyll (Mack, 1927; Steffens, Alphin and Ford, 1970; Phan and Hsu, 1975; Nilsen and Hodges, 1983). The efficacy of the ethylene treatment is most pronounced using excised tissue maintained under continuous light (Gepstein and Thimann, 1981) and may also be dependent upon the state of maturity of the tissue (Brady, Scott and Munns, 1974). Treatment of excised leaf tissue with 1-aminocyclopropane carboxylic acid (ACC) (a biosynthetic precursor of ethylene) will similarly promote chlorophyll breakdown (Gepstein and Thimann, 1981; Kao and Yang, 1983).

The second piece of evidence that has been used to invoke a role for ethylene in senescence is the demonstration of an ethylene climacteric during the course of this developmental event. Such a rise in endogenous ethylene has been observed in both freshly detached senescing leaves (Aharoni, Lieberman and Sisler, 1979) and mature green leaf tissue excised and induced to senesce off the plant (McGlasson, Poovaiah and Dostal, 1975; Aharoni, Lieberman and Sisler, 1979; Roberts and Osborne, 1981). In the latter instance the rise in ethylene production can occur within 24 h of leaf removal and may be related to a wound phenomenon associated with tissue excision (Kao and Yang, 1983). The evidence supporting ethylene as an initiator of senescence might be strengthened if there was a close correlation between the timing of the rise in ethylene production and the onset of leaf senescence but as yet this is lacking. Indeed, a number of workers have documented the appearance of the ethylene climacteric as an event subsequent to the first visible signs of senescence (Even-Chen, Atsmon and Itai, 1978; Aharoni, Lieberman and Sisler, 1979). The increasing number of reports correlating the ethylene climacteric with the period of most rapid chlorophyll loss (Aharoni, Lieberman and Sisler, 1979; Gepstein and Thimann, 1981; Ferguson, Watkins and Harman, 1983) may however implicate the gas as a regulator of the rate of senescence.

The origin of the rise in ethylene production associated with leaf senescence is presently unknown. The most likely source is via a stimulation of the ACC mediated biosynthetic pathway (*see* Yang *et al.*, Chapter 2) and indeed Kao and

Yang (1983) have demonstrated that levels of ACC increase within 24 h of excision-induced senescence of rice leaves. However the speed of this response suggests that the rise in ACC might be wound-induced rather than related to the senescence event *per se*. A stimulation of the ethylene biosynthetic pathway could arise either as a consequence of an increase in activity of enzymes associated with the pathway, such as ACC synthase (Saltveit and Dilley, 1979; Boller and Kende, 1980) or as a consequence of the release of cofactors of ethylene biosynthesis. A cofactor release mechanism for the regulation of ethylene production was originally proposed by Osborne, Jackson and Milborrow (1972). If cofactors of ethylene are compartmentalized within a membrane-bound organelle, it would follow that any event modifying membrane integrity or permeability would affect cofactor release and hence ethylene production. Since changes in phase properties and microviscosity of plant membrane lipids have been reported to occur during the senescence of a variety of tissues (Borochov *et al.*, 1978; Barber and Thompson, 1980; Lees and Thompson, 1980; Dhindsa, Plumb-Dhindra and Thorpe, 1981), such changes might precipitate the rapid upsurge in ethylene production associated with the latter stages of tissue senescence.

An alternative source of ethylene might be that released from compartmentation within the tissue. Sequestration of the gas has been reported to occur in a number of different tissues (Jerie *et al.*, 1978; Jerie, Shaari and Hall, 1979) and the capacity of the tissue to compartmentalize ethylene has been found to vary during development (Jerie *et al.*, 1978).

The use of correlative data such as that described above to infer a role for ethylene in leaf senescence has severe limitations. Such correlations are restricted by the technical difficulties associated with ethylene determinations and by our present lack of knowledge of the primary event(s) subsequent to the senescence programme being triggered. Also ethylene is only one of a family of regulatory molecules which can influence senescence. A decline in the level of an ethylene antagonist or a modification in tissue sensitivity by an alternative mechanism (Trewavas, 1981) might result in ethylene initiating the sequence of events culminating in senescence in the absence of a detectable change in the level of the regulator.

The advent of chemicals which can specifically block ethylene biosynthesis (e.g. aminoethoxyvinylglycine—AVG) or ethylene action (e.g. Ag^+) have provided the tools to examine the role of ethylene in developmental events more critically. Such an approach to study leaf senescence has been recently employed by a number of workers. Aharoni and Lieberman (1979) found that treatment of tobacco leaf discs with AVG markedly retarded both the loss of chlorophyll and the respiratory climacteric. However, as the authors point out the senescence-retarding property of AVG was not completely nullified by exogenous ethylene indicating that the effects of the chemical were not limited to an inhibition of the biosynthetic pathway of the gas. Senescence of excised oat leaves is similarly retarded by AVG, with the chemical being most effective at inhibiting chlorophyll loss of tissue maintained in continuous light (Gepstein and Thimann, 1981). Under these conditions AVG did not modify the time of onset of senescence but rather the rate at which it progressed suggesting that ethylene was not the initiator of the process. In this study the specificity of the inhibitor was not tested.

Ag^+ is a potent inhibitor of ethylene action in plants (Beyer, 1976a and b). It has been demonstrated to effectively inhibit chlorophyll loss from dark incubated tobacco leaf discs (Aharoni and Lieberman, 1979) and detached rice leaves maintained under either light or dark conditions (Kao and Yang, 1983). No time course data are presented by either of these groups and so it is not possible to determine at what stage in the senescence sequence ethylene might act. Information on the time course of Ag^+ on senescence of excised oat leaves has been published by Gepstein and Thimann (1981). These workers found that Ag^+ blocked chlorophyll loss from oat tissue in the light but was ineffective if the tissue was held in the dark. If the premise that Ag^+ specifically inhibits ethylene action is correct then the results of Gepstein and Thimann (1981) implicate a role for ethylene in the senescence of light maintained but not dark maintained oat leaf tissue! The data highlight once more the difficulties of drawing conclusions about the regulation of developmental phenomena *in situ* solely from laboratory based studies.

Conclusion

The involvement of ethylene in the senescence of flowers and the ripening of fruit has been clearly demonstrated both *in situ* and under experimental conditions following removal from the plant. An increase in ethylene production is one of the earliest detectable events in many cases of flower (Nichols and Frost, Chapter 28) and fruit (Grierson *et al.*, Chapter 14) senescence, and the gas has been implicated in both the initiation and acceleration of the developmental process. The physiological and biochemical changes occurring during flower senescence (Halevy and Mayak, 1979, 1981; Mayak and Halevy, 1980) and fruit ripening (Sacher, 1973; Coombe, 1976; Rhodes, 1980b) are well documented. Although there are some obvious similarities with these processes and the changes that occur during foliar senescence, there are also marked differences. Since the developmental processes are different the demonstration of a role for ethylene during flower or fruit senescence need not imply a similar role for the gas in foliar senescence. Furthermore since leaf senescence itself occurs by several distinct mechanisms (Simon, 1967) a role for ethylene in one experimental system need not infer a similar role for foliar senescence in general.

It has been shown that detached leaves or leaf discs from a variety of species exhibit a rise in ethylene production during senescence under experimental conditions. This could imply that ethylene is in some way involved with the senescence programme. However, whether or not a similar rise in ethylene production occurs during leaf senescence on the plant is unclear. In contrast to flowers and fruit the increase in ethylene production during foliar senescence appears to occur after the onset of the senescence programme when a decline in protein and chlorophyll has already commenced. Thus on this evidence alone a role for ethylene in the initiation of foliar senescence seems unlikely. However, it is possible that a change in tissue sensitivity to ethylene might be involved in the initiation process. Evidence is accumulating to invoke a role for ethylene in the acceleration of foliar senescence. However, the majority of data comes from experiments using detached leaves which again may not accurately reflect the situation in the whole plant. In this respect, and before any firm conclusions on the role of ethylene in foliar senescence can be made, much more work is required on the process as it occurs in attached leaves.

The precise molecular mechanisms by which ethylene exerts its influence on senescence are unknown even for fruit ripening where the importance of ethylene has been clearly established. Two other contributors to this volume, namely Grierson *et al.* (Chapter 14) and Tucker *et al.* (Chapter 15), present evidence that fruit ripening involves changes in the mRNA composition of the tissue and hence control of gene expression at the level of transcription. Thus in fruit at least ethylene could act to initiate or accelerate these changes. Foliar senescence is generally presumed to be under strict genetic control (Thomas and Stoddart, 1980) and, despite evidence from experimental studies using inhibitors of transcription and translation, may also involve changes in mRNA composition. Two recent studies have demonstrated such changes in mRNA populations during the senescence of excised wheat leaves (Watanabe and Imaseki, 1982) and soybean cotyledons (Skadsen and Cherry, 1983). Clearly however more work on the transcriptional and translational controls acting during foliar senescence *in situ* is required before adequate comparisons with fruit senescence may be made. It is interesting to note that Sexton *et al.* (Chapter 16) present evidence for transcriptional changes occurring during leaf abscission. This is again a process in which ethylene has been clearly implicated and which in many instances is closely linked with the latter stages of foliar senescence. One experimental approach that would advance our knowledge of senescence, would be to compare the transcriptional and translational changes which occur during *in situ* and ethylene promoted senescence and abscission of various organs within a single plant species.

References

AHARONI, N.M. and LIEBERMAN, M. (1979). *Plant Physiology*, **64**, 801–804
AHARONI, N.M., LIEBERMAN, M. and SISLER, H.D. (1979). *Plant Physiology*, **64**, 796–800
ALTMAN, A. (1982). *Physiologia plantarum*, **54**, 189–193
BARBER, R.F. and THOMPSON, J.E. (1980). *Journal of Experimental Botany*, **31**, 1305–1313
BEEVERS, L. (1976). In *Plant Biochemistry*, pp. 771–794. 3rd Edn. Ed. by J. Bonner and J.E. Varner. Academic Press, London
BEYER, E.M., Jr (1976a). *HortScience*, **11**, 195–196
BEYER, E.M., Jr (1976b). *Plant Physiology*, **58**, 268–271
BOLLER, T. and KENDE, H. (1980). *Nature*, **286**, 259–260
BOROCHOV, A., HALEVY, A.H., BOROCHOV, H. and SHINITZKY, M. (1978). *Plant Physiology*, **61**, 812–815
BRADY, C.J., SCOTT, N.S. and MUNNS, R. (1974). In *Mechanisms of regulation of plant growth*, pp. 403–409. Bulletin 12 of the Royal Society of New Zealand. Ed. by R.L. Bieleska, A.R. Ferguson and M.M. Cresswell.
COOMBE, B.G. (1976). *Annual Review of Plant Physiology*, **27**, 207–228
DHILLON, S.S. and MIKSCHE, J.P. (1981). *Physiologia Plantarum*, **51**, 291–298
DHINDSA, R.S., PLUMB-DHINDSA, P. and THORPE, T.A. (1981). *Journal of Experimental Botany*, **32**, 93–101
DYER, T.A. and OSBORNE, D.J. (1971). *Journal of Experimental Botany*, **22**, 552–560
EL-ANTABLY, H.M.M., WARING, P.F. and HILLMAN, J. (1967). *Planta*, **73**, 74–90
EVEN-CHEN, Z.C. and ITAI, C. (1975). *Plant Physiology*, **34**, 97–100
EVEN-CHEN, Z.C., ATSMON, D. and ITAI, C. (1978). *Physiologia Plantarum*, **44**, 377–382

FERGUSON, I.B., WATKINS, C.B. and HARMAN, J.R. (1983). *Plant Physiology*, **71**, 182–186
GEPSTEIN, S. and THIMANN, K.V. (1980). *Proceedings of the National Academy of Sciences, USA*, **77**, 2050–2053
GEPSTEIN, S. and THIMANN, K.V. (1981). *Plant Physiology*, **68**, 349–354
HALEVY, A.H. and MAYAK, S. (1979). *Horticultural Reviews*, **1**, 204–236
HALEVY, A.H. and MAYAK, S. (1981). *Horticultural Reviews*, **3**, 54–143
HARRIS, J.B., SCHAEFER, V.G., DHILLON, S.S. and MIKSCHE, J.P. (1982). *Plant Cell Physiology*, **23**, 1267–1273
JERIE, P.H., SHAARI, A.R., ZERONI, M. and HALL, M.A. (1978). *New Phytologist*, **81**, 499–504
JERIE, P.H., SHAARI, A.R. and HALL, M.A. (1979). *Planta*, **144**, 503–507
KAO, C.H. (1978). *Proceedings of the National Science Council*, **2**, 391–398
KAO, C.H. and YANG, S.F. (1983). *Plant Physiology*, **73**, 881–885
LAURIERE, C. (1983). *Physiologie Végétale*, **21**, 1159–1177
LEES, G.L. and THOMPSON, J.E. (1980). *Physiologica Plantarum*, **49**, 215–221
McGLASSON, W.B., POOVAIAH, B.W. and DOSTAL, H.C. (1975). *Plant Physiology*, **56**, 547–549
MACK, W.B. (1927). *Plant Physiology*, **2**, 103
MALIK, N.S.A. (1982). *Plant Cell Physiology*, **23**, 49–57
MALIK, N.S.A. and THIMANN, K.V. (1980). *Plant Physiology*, **65**, 855–858
MARTIN, C. and THIMANN, K.V. (1972). *Plant Physiology*, **50**, 432–437
MATILE, P. (1974). *Experimentia*, **30**, 98–99
MAUNDERS, M.J. and BROWN, S.B. (1983). *Planta*, **158**, 309–311
MAYAK, S. and HALEVY, A.H. (1980). In *Senescence in Plants*, pp. 131–156. Ed. by K.V. Thimann. CRC Press Inc.
NILSEN, K.W. and HODGES, C.F. (1983). *Plant Physiology*, **71**, 96–101
OSBORNE, D.J. (1967). *Symposium of the Society of Experimental Biology*, **21**, 179–213
OSBORNE, D.J., JACKSON, M.B. and MILBORROW, B.V. (1972). *Nature (New Biol.)*, **240**, 98–101
PHAN, C.T. and HSU, H. (1975). *Physiologie Végétale*, **13**, 427–434
RHODES, M.J.C. (1980a). In *The Biochemistry of Plants, a Comprehensive Treatise*, pp. 419–462. Vol. 2. Metabolism and Respiration. Ed. by D.D. Davies. Academic Press, London
RHODES, M.J.C. (1980b). In *Senescence in Plants*, pp. 157–205. Ed. by K.V. Thimann. CRC Press Inc.
RICHMOND, A.E. and LANG, A. (1957). *Science*, **125**, 650–651
ROBERTS, J.A. and OSBORNE, D.J. (1981). *Journal of Experimental Botany*, **32**, 875–887
SACHER, J.A. (1973). *Annual Review of Plant Physiology*, **24**, 197–224
SALTVEIT, Jr. M.E. and DILLEY, D.R. (1979). *Plant Physiology*, **64**, 417–420
SAMET, J.S. and SINCLAIR, T.R. (1980). *Plant Physiology*, **66**, 1164–1168
SATLER, S.O. and THIMANN, K.V. (1983). *Plant Physiology*, **72**, 540–546
SHIBOAKA, H. and THIMANN, K.V. (1970). *Plant Physiology*, **46**, 212–220
SIMON, E.W. (1967). *Symposium of the Society of Experimental Biology*, **21**, 215–230
SKADSEN, R.W. and CHERRY, J.H. (1983). *Plant Physiology*, **71**, 861–868
SPENCER, P.W. and TITUS, J.S. (1973). *Plant Physiology*, **51**, 89–92
STEFFENS, G.L., ALPHIN, J.G. and FORD, Z.T. (1970). *Beitr. Tabak. Forsch.*, **5**, 262–265

TETLEY, R.M. and THIMANN, K.V. (1974). *Plant Physiology*, **54**, 294–303
THIMANN, K.V. (1980). In *Senescence in Plants*, pp. 85–115. Ed. by K.V. Thimann. CRC Press Inc.
THIMANN, K.V., TETLEY, R.M. and KRIVAK, B.M. (1977). *Plant Physiology*, **59**, 448–454
THIMANN, K.V., TETLEY, R.M. and VAN THANH, T. (1974). *Plant Physiology*, **54**, 859–862
THOMAS, H. (1976). *Plant Science Letters*, **6**, 369–377
THOMAS, H. (1982). *Planta*, **154**, 212–218
THOMAS, H. (1983). *Photosynthetica*, **17**, 506–514
THOMAS, H. and STODDART, J.L. (1975). *Plant Physiology*, **56**, 438–441
THOMAS, H. and STODDART, J.L. (1977). *Annals of Applied Biology*, **85**, 461–463
THOMAS, H. and STODDART, J.L. (1980). *Annual Review of Plant Physiology*, **31**, 83–111
TREWAVAS, A.J. (1981). *Plant Cell Environment*, **4**, 203–228
WANG, C.Y., CHENG, S.H. and KAO, C.H. (1982). *Plant Physiology*, **69**, 1348–1349
WATANABE, A. and IMASEKI, H. (1982). *Plant Cell Physiology*, **23**, 489–497
WOLLGIEHN, R. (1967). *Symposium of the Society of Experimental Biology*, **21**, 231–266
WOOLHOUSE, H.W. (1967). *Symposium of the Society of Experimental Biology*, **21**, 179–214
WOOLHOUSE, H.W. (1982). In *The Molecular Biology of Plant Development*, pp. 256–281. Ed. by H. Smith and D. Grierson. Blackwell, Oxford
WOOLHOUSE, H.W. (1983). In *Post-harvest Physiology and Crop Protection*, pp. 1–43. Ed. by M. Lieberman. Plenum Press,
YU, S.M. and KAO, C.H. (1981). *Physiologia Plantarum*, **52**, 207–210

22
ETHYLENE AS AN AIR POLLUTANT

DAVID M. REID and KEVIN WATSON
Plant Physiology Research Group, Biology Department, University of Calgary, Alberta, Canada

Introduction

Ethylene appears to be a ubiquitous constituent of air (Altshuller and Bellar, 1963; Abeles, Forrence and Leather, 1971; Abeles and Heggestad, 1973; Harbourn and McCambly, 1973; Guicherit, 1975). It is produced as a by-product of man's activities (Altschuller and Bellar, 1963; Gerakis, Versoglu and Sfakiotakis, 1978; Fluckiger *et al.*, 1979; Abeles, Chapter 1), and influences both plant growth and development (Abeles, 1973). Thus it is reasonable to consider it as an air pollutant. However, it is an unusual pollutant in that it is an endogenous plant growth regulator, produced normally in small quantities and in larger quantities as a result of stress (Abeles, 1973; Kimmerer and Kozlowski, 1982).

Few reports dealing with measurements of the ethylene content of air examine the possible impact of such levels on plant growth. In this chapter we wish to follow up the pioneering work of Abeles, Forrence and Leather (1971) and Abeles and Heggestad (1973), and attempt to answer the following three questions. Firstly, what are the levels of ethylene found in our area? Secondly, if plants are constantly fumigated with such levels does this affect their growth or response to other environmental stresses? Finally, are ambient ethylene levels sufficiently high to interfere with laboratory experiments?

Determination of ethylene levels

Ethylene concentrations were measured using the gas chromatographic system described by Fabijan, Taylor and Reid (1981). Low ethylene levels (below $5\,\text{nl}\,\text{l}^{-1}$) were determined by first trapping the gas from a 50 ml air sample on a cooled ($-80\,^\circ\text{C}$) pre-column of Poropak S, and then by heating this pre-column to $100\,^\circ\text{C}$. The ethylene released was then passed through a normal gas chromatography column (Stinson and Spencer, 1969; DeGreef, DeProft and DeWinter, 1976; Eastewell, Bassi and Spencer, 1978). All air samples were stored in Teflon air bags.

This Study was part of a larger investigation funded by: The Alberta Environmental Research Trust, Alberta Gas Ethylene Co., Union Carbide Canada, Dow Chemical, Alberta Environment and NSERC (Canada).

Treatment of plant material

The species used were oat (*Avena sativa* L. cv. Random), rapeseed or Canola (*Brassica campestris* L. cv. Candle) and sunflower (*Helianthus annuus* cv. Russian). Canola and oat plants were grown in plastic pots in granite grit in sealed Plexiglass chambers (1.7 m high, 1.0 m wide, 0.7 m deep). Sunflowers were grown in similar but smaller chambers. Each pot was watered and fertilized with half-strength Hoagland's solution with a hydroponic system controlled by time clocks. Temperatures were held at 25 °C day and 14 °C night with the relative humidity maintained between 50 and 60%. The chambers received light from banks of VHO Grolux wide spectrum and Grolux wide spectrum fluorescent tubes (320 µmol m^{-2} s^{-1} PAR, 660 nm/730 nm = 4.5, 16 h photoperiod). The level of ethylene in the air supply to each chamber was reduced to below 0.2 nl l^{-1} by passing the air through a Purafil (KMnO$_4$) filter. Before entering a chamber, pure ethylene was readded to the air flow in a mixing compartment, to produce the appropriate concentrations of the hydrocarbon. Each growth chamber was provided with a fan to ensure complete mixing. The fan motor was outside the chamber. The flow rate of the gases ensured that the entire volume of air in a chamber was changed twice each hour. The ethylene and carbon dioxide levels of each chamber were monitored regularly by gas chromatography and infrared gas analysis respectively. When appropriate, carbon dioxide was added to the gas flow to ensure that a low level of this gas was not a limiting factor in plant growth. Carbon dioxide levels were maintained at normal ambient concentrations (i.e. 330 µl l^{-1}).

Stomatal frequencies were estimated using the technique of Sampson (1961), and stomatal resistance with a Licor LI 700 transient porometer.

Data was analysed by ANOVA and means compared using the Newman-Keuls Test.

Ambient ethylene levels

Although there are a number of papers in which measurements of the quantities of ethylene in the air have been made (e.g. Abeles and Heggestad, 1973; Nassan and Goldbach, 1979) it is clear that there are considerable variations in level depending upon such variables as time of day and year and location. Therefore it was necessary to check the quantities of ethylene that were present in our area.

Figure 22.1 shows the ethylene levels of an unpolluted natural forest area. Ethylene was present in variable amounts with the peak concentration probably related to the rapid midday metabolic activities of plants in the area.

On the windward side of the city ethylene concentration was higher (*Figure 22.2*) than in the forest and there was frequently a substantial rise in levels of the gas towards the end of the day. This usually occurred shortly after the peak rush-hour traffic, and was generally higher in the winter than in the summer. The variable nature of ethylene concentrations in the air is clear from the data in *Table 22.1*. All these samples were taken at exactly the same time of day, at the same location (university campus) and with the wind blowing from the same quadrant (N.W.—S.W.). *Figure 22.3* shows the approximate locations of the university (site M) in relation to the industrial area (site N) and the direction of the prevailing wind. The ethylene levels found at the university campus are higher than the forest samples and this is probably the result of automobile pollution (Menzies and Shumate,

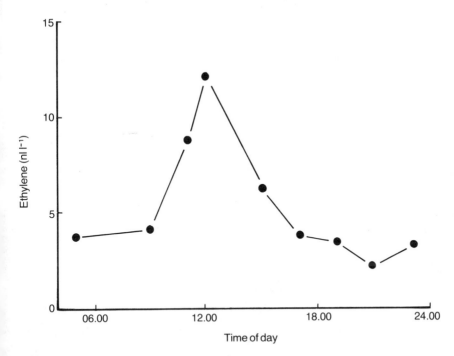

Figure 22.1 Concentration of ethylene (nl l^{-1}) in a subalpine coniferous forest 72 km N.W. of Calgary (200 m West of the unoccupied North Ghost River Campground, summer 1981). No traffic, industrial or combustion activity nearby. The wind was from the westerly direction. Each value is the mean of three samples

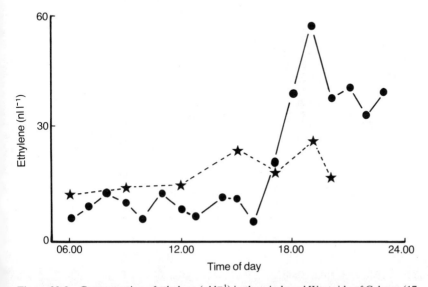

Figure 22.2 Concentration of ethylene (nl l^{-1}) in the windward West side of Calgary (17 Avenue and 51 Street N.W.) (----) measurements taken 29–30 June 1981; or (———) February 1984. The wind was from the westerly direction on both occasions. Each value is the mean of three samples

Table 22.1 ETHYLENE LEVEL (nl l^{-1}) 2 m ABOVE GROUND LEVEL, 100 m UPWIND OF LIGHTLY USED ROAD, 400 m FROM NEAREST BUILDING ON WEST EDGE OF UNIVERSITY OF CALGARY CAMPUS

Date	Ethylene (nl l^{-1})
27 Jan	98 ± 6.8 (inversion layer)
10 Feb	17 ± 0.18
13	26
16	22
17	29 ± 0.78 (55 in downtown area)
18	12
12 May	27 ± 0.38
13	20

All samples taken at 0.800h in 1982. Values are means of two different samples, or three samples ± s.e. of means

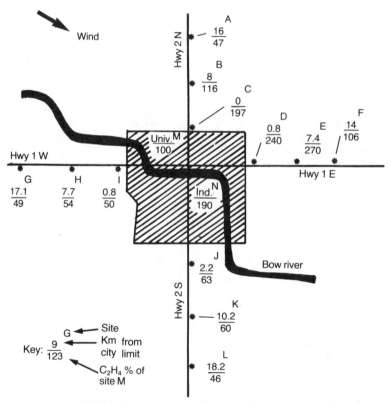

Figure 22.3 Ethylene concentrations expressed as per cent of control value (University campus, site M). Each value is the mean of three samples, all taken between 09.00 and 10.00 h on four consecutive days (16–19 February 1981) with very similar weather conditions and wind direction. All samples (except N) taken 200 m to the windward side of the road in undisturbed fescue prairie or grazing grassland, and well away from industrial or combustion activities or grazing animals. Mean ethylene concentration at site M over the four days = 22.3 ± s.e. 7.43 nl l^{-1}

1976; Flückiger et al., 1979; Nassan and Goldbach, 1979). On a few days we have measured levels of ethylene approaching $100\,nl\,l^{-1}$ (e.g. 27 Jan. in *Table 22.1*). However, these have always been when there was an atmospheric inversion layer trapping polluted air in the Bow River Valley. On such occasions brown photochemical smog was clearly visible hanging over the city.

Invariably we have found higher concentration of ethylene as one neared the city centre. This prompted a survey of ethylene levels in different locations. The data in *Figure 22.3* are typical of the normal range of values of the gas in this area. Generally samples taken at sites in the direction of the prevailing wind through the city centre and industrial area (north and east of site N) showed increasing ethylene levels reaching values of from three to five times those found in rural areas (sites A, G or L). Sites D, E and F are located in areas of fairly intensive agriculture.

The results of this study therefore clearly show that concentrations of ethylene in the air around the Calgary area can range from 3 to $100\,nl\,l^{-1}$, and the level of the gas depends upon meteorological, spatial and temporal factors.

Effects of ethylene on plant growth

A series of studies were carried out to assess the effect of constant fumigation of low levels of ethylene on plant growth and development. An initial experiment with oats, lasting three weeks, was set up in order to test the uniformity of the fumigation chambers. If any of the chambers had differed in performance from the others, this would invalidate the conclusions that could be drawn regarding the effects of ethylene. Measurements of stem height, leaf number, fresh and dry weights (roots and shoots) showed that there were no significant differences between the growth in any of the chambers in the presence of an ambient air atmosphere.

The chambers were then set up with oats and Canola plants grown in a constant supply of oxygen but with supplemented ethylene levels (56, 126, 341 and $831\,nl\,l^{-1}$) for 23 days. Canola plants grown at these concentrations showed a difference in the numbers of leaves they bore, but at $126\,nl\,l^{-1}$ and above shoot dry weight and leaf area were reduced (*Table 22.2*). In contrast, the tillering, shoot dry weight and total numbers of leaves of oat plants were all increased after ethylene treatment (*Table 22.2*).

In a long-term experiment with a lower range of ethylene concentrations both species exhibited similar effects to those described above after three weeks, but by harvest the picture differed (*Table 22.3*). By this time the higher concentrations of ethylene brought about a substantial reduction in seed yield (150 and $600\,nl\,l^{-1}$) and shoot dry weight ($600\,nl\,l^{-1}$) in Canola plants. Interestingly, there was a marked stimulation compared to the control of shoot dry weight and seed weight/number after fumigation by $10\,nl\,l^{-1}$ ethylene. With oats the initial ethylene-stimulating effects disappeared after prolonged fumigation and no difference in tillering was observed and a significant lowering in numbers of florets was also apparent. Thus these results indicate that a constant supply of ethylene at the levels found in Calgary ($3-100\,nl\,l^{-1}$), can either inhibit or promote aspects of growth and development of oats and Canola. The extent of the effect is dependent upon the experimental conditions.

The effects of low levels of ethylene on the stimulation of various processes (initial tillering in oats (*Table 22.2*) and seed yield in Canola (*Table 22.3*)) are of

282 Ethylene as an air pollutant

Table 22.2 EFFECTS OF A RANGE OF ETHYLENE CONCENTRATIONS ON THE GROWTH OF CANOLA AND OAT PLANTS AFTER 23 DAYS CONSTANT FUMIGATION

Ethylene (nl l^{-1})	Mean no. leaves	Canola Mean DW shoot (g)	Mean leaf area per plant (cm^2)
0	8.1a	2.25a	698a
56	9.0a	2.32a	721a
126	9.3a	1.59b	505b
341	9.8a	1.23b	294c
831	8.9a	0.84b	223c

Ethylene (nl l^{-1})	Mean no. leaves	Oats Mean DW shoot (g)	Mean no. tillers
0	10.0a	0.62a	2.7a
56	12.9a	0.94b	3.1a
126	16.6a	1.95b	4.6b
341	17.9b	1.00b	4.1b
831	18.9b	0.96b	4.4b

Means in any one column followed by the same letter are not statistically significantly different at the 5% level

Table 22.3 EFFECTS OF VARIOUS CONCENTRATIONS OF ETHYLENE ON THE GROWTH OF CANOLA AFTER 87 DAYS AND OATS AFTER 100 DAYS CONSTANT FUMIGATION

Ethylene (nl l^{-1})	Mean DW shoots	Canola Weight seeds/plant (g)	No. seed/plant
0	381a	23.1a	1008a
10	495b	56.0b	2902b
35	400a	20.5a	1381a
150	360a	8.3c	477c
600	216c	0.2c	41c

Ethylene (nl l^{-1})	Oats Mean no. tillers	Mean no. florets/plant
0	4.4a	464a
7	3.8a	360b
35	4.0a	267c
70	4.4a	41d
150	4.8a	2d

Means in any one column followed by the same letter are not statistically significantly different at the 5% level

interest in that ethylene is commonly considered, at least at higher concentrations, to be an inhibitor. This of course depends on one's perspective. For instance, although the gas reduces stem elongation it can promote stem width (Sargent, Attack and Osborne, 1973). At low concentrations it will also promote elongation of rice roots (Konings and Jackson, 1979). It has also been shown to stimulate epicotyl/hypocotyl hook opening (Goeschl, Pralt and Bonner, 1967; Kang et al., 1967) and stem lengthening in hydrophytes (Ku et al., 1969). Lieberman and Kunishi (1972), Zobel and Roberts (1974), Zobel (1978) and Huxter, Reid and Thorpe (1979) have suggested that small quantities of ethylene might be required for normal growth and development. Perhaps a biphasic growth curve of initial promotion by low levels followed by inhibition at higher, is more common than is generally supposed. It might be that the lower and stimulatory end of this curve can only be seen when one uses a true control with no ethylene. The use of antagonists of ethylene action and/or synthesis (Beyer, 1976; Fabijan, Taylor and Reid, 1981) or the examination of the responses of certain mutants (Zobel, 1973) might also reveal this normally hidden promotory effect of ethylene.

Interaction with drought

We observed that plants grown in the absence of ethylene were more susceptible to wilting than the ethylene treated plants. We thus decided to see if low levels of ethylene altered the response of sunflower seedlings to drought stress. In three experiments sunflower seedlings were grown in the presence of 0, 15, 30 or 60 nl l^{-1} ethylene. Seedlings were planted in the small chambers which were held at a

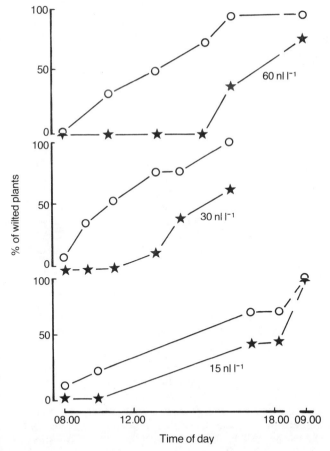

Figure 22.4 The effect of cessation of watering on sunflower seedlings that had been constantly exposed to various levels of ethylene for 11 days. Watering ceased at 12.00 h on the previous day

particular ethylene concentration for 11 days. The water supply was then turned off (at 12.00 h) and the plants were observed for signs of wilting. A plant was determined to have wilted when a primary leaf showed any drooping of the lamina on either side of the mid rib. Wilting first occurred on the following morning (*Figure 22.4*) and in all cases ethylene-treated plants wilted later than did the controls.

In an attempt to discover the basis of this increased drought tolerance we measured the stomatal resistance of both leaf surfaces. Pallas and Kays (1982) found that 1000 nl l^{-1} ethylene decreased leaf conductance in peanut. However, we saw no significant differences in stomatal resistance of adaxial or abaxial leaf surfaces after ethylene treatments. Measurements of leaf area, stomatal number and shoot/root dry weight ratio did show differences (*Table 22.4*). For instance, plants exposed to 60 nl l^{-1} ethylene had smaller leaves and cotyledons, fewer

Table 22.4 THE EFFECTS OF ETHYLENE ON THE GROWTH OF SUNFLOWER SEEDLINGS AFTER 11 DAYS' CONSTANT FUMIGATION

	Zero ethylene			Ethylene (60 nl l^{-1})		
Leaf position	*Cotyledon*	*1st*	*2nd*	*Cotyledon*	*1st*	*2nd*
Leaf area (mm)	669a	1933b	549a	296c	987d	216c
Total number of stomatas per leaf (×10)	476a	313b	—	450a	160c	—
Mean shoot dry wt (mg)		223a			135b	
Mean root dry wt (mg)		50a			38b	
Shoot dry wt/Root dry wt		4.46a			3.55b	

Means in any one horizontal column followed by the same letter are not statistically significantly different at the 5% level

stomata on the primary leaves and reduced shoot/root ratio compared to controls. We have consistently found this to be the case in other experiments. Funke *et al.* (1938) reported that levels of ethylene affected stomatal frequency in sunflower cotyledons, and Abeles and Heggested (1973) showed reductions in leaf area in other species with ethylene concentrations similar to ours. We feel that the increased ability of the ethylene treated plants to withstand drought could be related to these alterations in leaf area, numbers of stomata, and the alteration in relative sizes of shoots to roots.

The lack of effect of ethylene on numbers of stomata in cotyledons may be because the stomata in these organs had differentiated prior to ethylene treatment.

Conclusions

The air in the Calgary region always seems to contain ethylene and depending upon meteorological conditions, industrial activities, automobile traffic, and time of day, the concentration of ethylene in the air may vary between 3 and almost 100 nl l^{-1}. *Avena sativa* (oat) and *Brassica campestris* (Canola) plants were grown for varying periods in chambers at different constant low levels of ethylene. After three weeks in ethylene atmospheres of above 56 nl l^{-1}, oat plants showed significant increases in tillering, total leaf area and dry weight compared to controls. After 100 days these stimulatory effects disappeared and instead 7 nl l^{-1} ethylene and above caused substantial reductions in the number of florets. With

Canola short-term fumigations with 56 nl l^{-1} ethylene or less produced little effect, however, at seed harvest 7 nl l^{-1} significantly stimulated seed production while it took 150 nl l^{-1} and above to lower seed production compared to controls.

Helianthus annuus seedlings were grown for 11 days in 0, 15, 30 or 60 nl l^{-1} ethylene. All concentrations reduced the rate at which the plants wilted in response to withdrawal of water. The increase in drought tolerance may be related to the ethylene-induced reduction in leaf area, numbers of stomata and decreased shoot/root ratio.

We conclude that the level of ethylene found normally in the air around Calgary might be sufficient to affect certain aspects of plant growth and development in our experimental plants.

References

ABELES, F.B. (1973). *Ethylene in Plant Biology*. Academic Press, New York and London

ABELES, F.B. (1982). *Agriculture and Forestry Culture* (of the University of Alberta Edmonton, Alberta, Canada). Vol. 5, No. 1, pp. 4–12

ABELES, F.B., FORRENCE, L.E. and LEATHER, G.R. (1971). Ethylene air pollution. Effects of ambient levels of ethylene on the glucanase content of bean leaves. *Plant Physiology*, **48**, 504–505

ABELES, F.G. and HEGGESTAD, H.E. (1973). *Journal of the Air Pollution Control Association*, **23**, 517–521

ALTSCHULLER, A.P. and BELLAR, T.A. (1963). *Journal of the Air Pollution Control Association*, **13**, 81–87

BEYER, E.M. Jr (1976). *Plant Physiology*, **58**, 208–271

DEGREEF, J., DEPROFT, M. and DEWINTER, F. (1976). *Analytical Chemistry*, **48**, 38–41

EASTWELL, K.C., BASSI, P.W. and SPENCER, M.E. (1978). *Plant Physiology*, **60**, 723–726

FABIJAN, D., TAYLOR, J.S. and REID, D.M. (1981). *Physiologia Plantarum*, **53**, 589–597

FLÜCKIGER, W., OERTLI, J.J., FLUCKIGER-KELLER, H. and BRAUN, S. (1979). *Environmental Pollution*, pp. 171–176. Applied Science, London

FUNKE, G.L., DECOEYER, F., DEDECKER, A. and MATON, J. (1938). *Biol Jaabuck*, **5**, 335

GERAKIS, P.A., VERSOGLU, D.S. and SFAKIOTAKIS, E. (1978). *Bulletin of Environmental Contamination and Toxicology*, **20**, 657–661

GOESCHL, J.D., PRATT, H.K. and BONNER, B.A. (1967). *Plant Physiology*, **42**, 1077–1080

GUICHERIT, R. (1975). *Staub-Reinhalt Luft*, **35** (No. 3), 89–95

HARBOURN, C.L. and McCAMBLY, T. (1973). *Proceedings of the Third International Air Congress, Dusseldorf (FRG)*, 38–41

HUXTER, T.J., REID, D.M. and THORPE, T.A. (1979). *Physiology of the Plant*, **4**, 374–380

KANG, B.G., YOCUM, C.S., BURG, S.P. and RAY, P.M. (1967). *Science*, **156**, 958–959

KIMMERER, T.W. and KOZLOWSKI, T.T. (1982). *Plant Physiology*, **69**, 840–847

KONINGS, H. and JACKSON, M.B. (1979). *Zeitschrift Pflanzenphysiologie*, **92**, 385–397

KU, H.S., SUGE, H., RAPPAPORT, L. and PRATT, H.K. (1970). *Planta (Berlin)*, **90**, 333–339

LIEBERMAN, M. and KUNISHI, A.T. (1972). In *Plant Growth Substances 1970*, pp. 549–560. Ed. by D.J. Carr. Springer-Verlag, New York
MENZIES, R.T. and SHUMATE, M.S. (1976). *International Conference on Environmental Sensing and Assessment*, Vol. 2, pp. 1–3. Inst. Elect. Electron Engineers, New York
NASSAN, J. and GOLDBACH, J. (1979). *International Journal of Environmental Analytical Chemistry*, **6**, 145–159
PALLAS, J.E. and KAYS, S.J. (1982). *Plant Physiology*, **70**, 598–601
SAMPSON, J. (1961). *Nature (London)*, **191**, 932–933
SARGENT, J.A., ATTACK, A.V. and OSBORNE, D.J. (1973). *Planta (Berlin)*, **109**, 185–192
STINSON, R.A. and SPENCER, M. (1969). *Plant Physiology*, **44**, 1217–1226
ZOBEL, R.W. (1973). *Plant Physiology*, **52**, 385–389
ZOBEL, R.W. (1978). *Canadian Journal of Botany*, **56**, 987–990
ZOBEL, R.W. and ROBERTS, L.W. (1974). *Canadian Journal of Botany*, **52**, 735–741

23
SOURCES OF ETHYLENE OF HORTICULTURAL SIGNIFICANCE

FRED B. ABELES
US Department of Agriculture, Agricultural Research Service, Appalachian Fruit Research Station, Kearneysville, West Virginia, USA

Introduction

The major losses caused by ethylene are associated with high value crops such as greenhouse crops, cut flowers, fresh fruit and vegetables and indoor ornamentals. Because of this, it is important to have an understanding of the sources, action and sinks of the gas. In addition, it is appropriate to develop strategies for controlling emissions, scrubbing the air and minimizing physiological effects on plants. The major sources of ethylene are emissions from internal combustion engines, fires, and plants themselves. Soil is both a source and sink for ethylene, but under conditions of flooding, physiologically active levels may appear in the root zone. The reaction of ethylene with ozone and ultraviolet light represents additional sinks in the atmosphere. Practical controls to minimize ethylene damage are limited to devices which minimize hydrocarbon production by cars. At the present time, scrubbers designed to reduce ethylene levels in the air of greenhouses and storage facilities are effective but costly. Other strategies and techniques used in minimizing adverse economic effects of ethylene include the storage of produce at low temperatures, controlled atmospheres, and the use of compounds which prevent ethylene action or production.

Ethylene as an air pollutant

The initial reports of the effect of ethylene on plants were associated with its action as an air pollutant. As reviewed earlier (Abeles, 1973), reports as early as 1864 described the defoliating effects of leaking illuminating gas on street trees. Smoke was subsequently shown to cause other physiological effects such as the promotion of ripening, initiation of flowering in pineapples, and altering the sex expression of curcurbits. The report by Burg and Stolwijk (1959) that ethylene could be readily measured by gas chromatography set the stage for the rapid growth in what is currently known about ethylene as an air pollutant. The role of ethylene as an air pollutant has been reviewed earlier (Abeles, 1973, 1982; Arnts and Meeks, 1981). Four phenomena play a role in the action of ethylene as either an air pollutant or a hormone. They are: multiple effects of the gas, dose response curves, exposure durations, and interaction with stress induced ethylene.

MULTIPLE EFFECTS

Ethylene has many effects on plant growth and development. A list of the most important of these include effects on dormancy, growth, epinasty, hook opening, induction of rooting and root hairs, hypertrophy, exudation, flowering, gravitational responses, effects on motor cells, disease resistance, stress responses, flower senescence, ripening, and abscission. In an early study of ethylene as an air pollutant, Heck and coworkers (Heck, Pires and Hall, 1961; Heck and Pires, 1962) described a number of these effects.

DOSE RESPONSES

Give or take an order of magnitude, the typical dose response curve of ethylene causing the above effects is: theshold $\simeq 10\,nl\,l^{-1}$, 1/2 maximal effect $\simeq 100\,nl\,l^{-1}$, and 90% maximum effect $\simeq 1000\,nl\,l^{-1}$. Concentrations higher than this rarely have 'toxic' effects. The dose response curve for ethylene is unlike that for typical toxic air pollutants where a low concentration promotes or stimulates a process followed by the effect of toxic higher concentrations (Arndt-Schulz Law; Tingey, 1980). Associated with the dose response curve is the experimentally useful knowledge that hydrocarbon analogues of ethylene, such as propylene, can be used as a diagnostic test for ethylene effects, and that CO_2 and silver ions are often used as competitive inhibitors of ethylene action.

DURATION OF EXPOSURE

Some physiological processes, such as elongation, respond rapidly (in the order of minutes) to ethylene, while others, like senescence and abscission, require hours or days of fumigation in order to have a detectable effect. The time required for ethylene to have an effect depends upon the plant's inherent genetic make up, stage of development, and internal levels of juvenility (auxin) hormones. Unlike the dose response curve, there are no generalities in this aspect of ethylene physiology and the biology of individual systems varies from plant to plant.

STRESS ETHYLENE

Under field conditions, plants are exposed to two sources of ethylene. Internal ethylene, normal or stress induced, and that produced by an external source. A variety of perturbations such as temperature extremes, wind, grazing, moisture stress, physical damage, insects, and disease can cause a plant to produce stress ethylene. This fact is a key to understanding the potential effects of air pollution. Levels of ethylene capable of causing responses in a laboratory experiment may not have significance in the field where plants are already responding to stress ethylene. For example, Jaffe (1980) has shown that stress ethylene plays a role in the physiology of plants exposed to mechanical stimulation. Mechanical perturbations caused by wind were also reported to alter the response of plants to SO_2 (Ashenden and Mansfield, 1977). In addition, roots are exposed to an edaphic gas phase which can contain ethylene produced by soil inhabiting fungi.

Unlike studies with ozone or SO_2 pollution, any evaluation of ethylene effects has to take into account the fact that a certain amount of ethylene is always present in the plant.

Economic losses due to ethylene air pollution

Two situations lead to economic loss as a result of ethylene air pollution. The first is the effect of man-made ethylene in predominantly urban areas, and the second, the 'pollution', due to ethylene released by produce in storage. This latter ethylene production is considered in detail by Schouten in Chapter 29.

A table summarizing reports of ethylene damage to plant material has been presented earlier (Abeles, 1973). For the most part, the dollar evaluation of losses is relatively small when compared to major air pollutants such as ozone and SO_2. For example, the 1962 crop of orchids in the Bay Area of California was one million blooms. Of this number, 10% were unsaleable because of ethylene damage. At an average value of $1 per bloom the loss was $100 000 (James, 1963).

The task of assigning a specific degree of loss to ethylene is difficult because ethylene levels in urban areas are normally so low that plants do not demonstrate obvious symptoms of ethylene damage such as epinasty. In addition, since all plants in an area are exposed to the same low concentration of gas, it is difficult to find unfumigated plants to serve as controls. Abeles and Heggestad (1973) grew plants in chambers equipped with potassium permanganate filters and compared the growth of these plants with those grown in ambient air. In all cases it could be observed that ambient air resulted in reduced vegetative growth and fruit yield. However, it is important to recall that these plants were grown under optimal conditions and that ethylene was the only stress factor in the system. Under normal field conditions, the plants would be expected to produce stress ethylene generated by wind movement, drought, temperature extremes, insect, disease and oxidant air pollution.

Sources of ethylene

AUTOMOBILE EXHAUST

Automobiles represent the major source of urban ethylene pollution. Automobile exhaust consists of 18% CO_2, 6% CO, 0.4% NO and 0.6% hydrocarbons (Barth, 1970). The major hydrocarbon fractions are acetylene, 27%; ethylene, 20%; propene, 8%; butane, 7%; and toluene, 5%. The remaining 33% is made up of a large number of compounds of which 44 have been identified (Altshuller, 1968). Effluents from transportation accounts for 70% of the hydrocarbons produced by man and various estimates place this annual production in the order of $1 \times$ to 30×10^6 MT (metric tons) per year for the USA (Altshuller, 1968; Barth, 1970). Some perspective for this figure is provided by the report that global production of carbon has been estimated to be 55×10^9 MT year^{-1} (Idso, 1983). Although many believe that a rise in CO_2 levels will eventually result in warming of the earth's atmosphere, Idso (1983) has indicated that actual data suggest the opposite. Earlier we estimated that 13.5×10^6 MT of ethylene were produced as a pollutant in the USA in 1966 (Abeles *et al.*, 1971). For the sake of comparison, ethylene production in

1981 by industry was equivalent to 15×10^6 MT and its commercial value was $7 billion (Anon., 1981). Since the USA is about 8×10^8 hectares in area, the rate of ethylene production per unit surface area is equivalent to 17 kg ha^{-1} per year. However, automobiles are localized in urban areas with the result that ethylene levels accumulate in cities. Detroit, Washington DC, and San Francisco were reported to have ethylene levels of 120, 93, and 76 kg ha^{-1} per year, respectively (Barth, 1970). Lonneman (1977) observed hydrocarbon levels at five different US cities between the hours of 6 and 9 a.m. The amounts of ethylene expressed as nl l^{-1} were: Wilmington, Ohio, 10.4; Bayonne, New Jersey, 14; New York City, 18; St Louis, Missouri, 16; and Los Angeles, 14. Another study of urban air pollution indicated that ethylene levels ranged from 700 nl l^{-1} in central Washington DC to 39 nl l^{-1} near Beltsville, Maryland, located on the peripheral beltway. In rural areas, ethylene levels were usually lower than 5 nl l^{-1} (Abeles and Heggestad, 1973). As yet, reports of ethylene levels adjacent to a major roadway have not appeared. However, in an analogous study Flückiger et al. (1979) reported that CO_2 at the median was 4000 nl l^{-1}; at the shoulder, 2800 and 200 m distant, 80 nl l^{-1}. Assuming ethylene levels were 2% that observed for CO_2 ethylene levels adjacent to some roadways might be as high as 80 nl l^{-1}. Automobiles have also been shown to be the major source of hydrocarbons at airports (Clark et al., 1983).

Ethylene production by automobiles is influenced by the type of engine and emission control devices. Carey and Cohen (1980) compared the hydrocarbon emissions from light duty gasoline engines equipped with emission control devices. The ethylene emission rate (mg km^{-1}) for a standard non-catalytic engine was 90 compared to 66 and 17 for lean burn and a catalyst-equipped vehicle, respectively. The effect of catalyst-equipped vehicles on atmospheric hydrocarbon levels has been reported for the period of 1972 to 1976. During that time emission of hydrocarbons from automobiles in the USA dropped from 30×10^6 MT to 28×10^6 MT (EPA, 1977).

BURNING VEGETATION

Yamate (1974) estimated that 53 million hectares of vegetation were burned in the USA. Of this total, 38% were forest wild fires; 25% were managed burnings and 37% were agricultural fires. The total amount of vegetation burned was 80×10^6 MT. The average, normal field emission factors (kg MT^{-1}) for various gases were CO_2, 1.24; water, 500; CO, 130; hydrocarbons, 12; and NO, 2 (Geomet, 1978). Boubel, Darley and Schuck (1969) reported that burning straw produced 3 kg of hydrocarbon gases per MT, of which 27% was ethylene. Darley et al. (1966) performed similar studies with wood chips and brush. The yield of ethylene per MT varied from 0.1 kg for the wood chips to 2.7 kg for the green brush. From these studies, the conversion factor for ethylene production from fires is approximately 1 kg MT^{-1} (0.1%). A different estimate of 1.5% ethylene yield was obtained by Feldstein et al. (1963). Using the data provided by Yamate (1974) the total tonnage of ethylene from fires in the USA is 80×10^3 MT. Since fires are point sources, this may result in the production of physiologically significant levels of ethylene within a smoke plume. Pineapple and mango growers have taken advantage of this method of ethylene production to accelerate the development of female flowers (Rodriquez, 1932). Levels of ethylene in smoke plumes varied from 4 to 129 μl l^{-1} (McElroy, 1960; Westberg, Sexton and Flyckt, 1981).

INDUSTRIAL

Hall *et al.* (1957) reported that cotton grown in the vicinity of a polyethylene plant in Texas showed symptoms similar to cotton gassed with ethylene. The ethylene content of the air near the plant was $3000\,\text{nl}\,l^{-1}$ and returned to ambient levels 6.5 km away. Most of the vegetation damage recorded occurred within a 3.2 km radius of the plant. Fukuchi and Yamamoto (1969) reported premature abscission of mandarin fruit in the vicinity of a gas works. Reports of ethylene levels near petroleum refineries indicate maximum levels of $118\,\text{nl}\,l^{-1}$ (Arnts and Meeks, 1981) and $353\,\text{nl}\,l^{-1}$ (Mohan Rao *et al.*, 1983).

OTHER EXOGENOUS SOURCES

Hanan (1973) and Ashenden, Mansfield and Harrison (1977) reported that kerosene heaters used in greenhouses are a significant source of ethylene. Ethylene and other gases can also be a problem in growth chambers (Tibbitts *et al.*, 1977). Ethylene from buffers (Forsyth and Eaves, 1969) and rubber stoppers (Jacobsen and McGlasson, 1970) may also contaminate physiological experiments.

PLANTS

Earlier, we (Abeles *et al.*, 1971) estimated that plants produced 2×10^4 MT of ethylene per year in the USA. For the most part, plants produce little ethylene, though as discussed earlier, stress can increase the rate of production. In some cases, the damage caused by the stress may in fact be due to the ethylene produced, since many air pollutants have been shown to increase ethylene production (Abeles and Abeles, 1972; Bressan *et al.*, 1979; Craker, Fillatti and Grant, 1982; Hogsett, Raba and Tingey, 1981; Stan, Schicker and Kassner, 1981; Tingey, 1980; Tingey, Standley and Field, 1976). A number of workers (Craker and Fillatti, 1982; Craker, Fillatti and Grant, 1982; Simon, Decoteau and Craker, 1983; Tingey, 1980; Tingey, Standley and Field, 1976) have shown that stress ethylene can be used to measure the effect of air pollution stress on plants. The mode of action of stress agents seems to be localized at the same point at which indole-acetic acid is thought to act—namely, the enhancement of the synthesis of ACC synthase (Stan, Schicker and Kassner, 1981).

SOIL

Soil acts as both an ethylene sink and source. Fungi are the primary producers and actinomycetes the primary consumers. Following the original observations by Smith and Russell (Smith and Russell, 1969), a large number of workers have shown that soil is a source of ethylene (Campbell and Moreau, 1979; Considine, Flynn and Patching, 1977; Dowdell *et al.*, 1972; Goodlass and Smith, 1978; Hunt, Campbell and Moreau, 1980; Kawase, 1976; Lindberg, Granhall and Hall, 1979; Lynch, 1975, 1983; Rovira and Vendrell, 1972; Smith, 1973; Smith and Cook, 1974; Smith and Dowdell, 1974; Smith and Restall, 1971; Yoshida and Suzuki, 1975). However, it is important to note that a number of workers have shown that air-drying soils selectively inhibits or kills ethylene consumers leaving ethylene producers intact (Cornforth, 1975; Goodlass and Smith, 1978; Rovira and Vendrell, 1972; Yoshida and Suzuki, 1975).

Field studies with intact soil have shown that ethylene can accumulate in the gas phase. Dowdell et al. (1972) measured ethylene from waterlogged clay soils in winter. The highest levels, $5-10\,\mu l\,l^{-1}$, were found at depths of 30–90 cm. Levels were greatest in the winter and lowest in the spring and then rose again in the summer. Smith and Dowdell (1974) found ethylene levels rose to $10\,\mu l\,l^{-1}$ at a soil depth of 15 cm when oxygen levels fell to 2%. At higher oxygen levels, 19%, little ethylene accumulated in the gas phase. Campbell and Moreau (1979) demonstrated that ethylene levels increased in soils following flooding. However, even though the soil remained flooded, ethylene levels returned to zero suggesting that ethylene consumers eventually dominated the system. In a study of ethylene levels at various soil depths, they found little or no ethylene at depths of 30 cm, a maximum at 10 cm and again no ethylene near the soil surface.

For the most part, the ethylene producers are probably fungi. Ilag and Curtis (1968) compared the ability of 228 species of fungi to produce ethylene. They observed that approximately 25% of the organisms they examined produced the gas, and that the most effective producers were *Aspergillus* and *Penicillium*.

A number of workers have assayed the ability of bacteria to produce ethylene. Swanson, Wilkins and Kennedy (1979) assayed 20 bacteria for the ability to produce ethylene. Both *Pseudomonas solanacearum* and *P. pisi* produced ethylene when grown on potato dextrose broth. Fourteen other species of bacteria also produced ethylene, but did so only after methionine was added to their growth medium. This effect may be due to the production of an extracellular ethylene generating system via a mechanism described earlier by Mapson and Wardale (1970).

Levels of ethylene in the air

The level of ethylene in rural areas varies from nondetectable to $5\,nl\,l^{-1}$. Levels of ethylene in a canopy of cotton plants were found to vary from undetectable at the upper portion to $80\,nl\,l^{-1}$ in the middle of the canopy on a calm night (Heilman, Meredith and Gonzales, 1971). In contrast, ethylene levels in metropolitan areas can increase to physiologically significant levels. Various studies have reported that ethylene levels in urban air were approximately $100\,nl\,l^{-1}$ (Altshuller, 1968; Barth, 1970; Clayton and Platt, 1967; Gordon, Mayrsohn and Ingels, 1968; Stephens and Burleson, 1967). A high value for Pasadena of $500\,nl\,l^{-1}$ was recorded by Scott et al. (1957). In our own work in Washington DC, we reported levels of ethylene equal to $700\,nl\,l^{-1}$ in a downtown area to $100\,nl\,l^{-1}$ in the surrounding suburbs (Abeles and Heggestad, 1973). Ethylene levels remain constant for some distance above street levels. Values of $45\,nl\,l^{-1}$ at the top of a five-storey building and $51\,nl\,l^{-1}$ at the second floor have been reported (Stephens and Burleson, 1967). A number of workers have reported that ethylene levels vary as a function of traffic flow, with maximums at 8–9 a.m. and again 5–6 p.m. (Altshuller and Bellar, 1963; Gordon, Mayrsohn and Ingels, 1968). Hassek, James and Sciaroni (1969) reported that inversion layers could increase the night-time levels of ethylene. James (1963) gathered crop loss data from a number of orchid growers in the San Francisco area and found that the greatest losses occurred in the winter. Yearly trends in November showed an increase from approximately 30% losses in 1952 to 40%

losses in 1962. During this ten-year period, crop losses steadily increased until the number of growers either stopped producing plants or went into another form of business.

Degradation of atmospheric ethylene

Atmospheric levels of ethylene in rural areas remain low in spite of the fact that the gas is produced constantly by fires and internal combustion engines. At the present time, three mechanisms for ethylene degradation are known. Ozone reacts rapidly with ethylene to form formaldehyde and formic acid (Leighton, 1961; Lindberg, Granhall and Berg, 1979). Other products formed include water, CO_2, CO and formic acid. Ultraviolet irradiation also reduces the levels of ethylene and other alkenes and alkanes (Scott, Wills and Patterson, 1971; Stephens, Darley and Burleson, 1967). Abeles *et al.* (1971) demonstrated that soil samples were able to remove ethylene and other auto exhaust hydrocarbons from the air. Other investigators have also reported that soils act as ethylene sinks. Smith, Bremner and Tabatabai (1973) reported that soils are an important sink for ethylene, SO_2, H_2S, CO and acetylene. Both groups of workers demonstrated that sterilization prevented gas uptake, suggesting that soil microbes were involved in gas consumption. Cornforth (1975) concluded that ethylene was microbially decomposed in aerobic soils 50 times faster than it was being produced.

In 1956, Davis, Chase and Raymond (1956) isolated an ethane utilizing actinomycete from soil. The organism, subsequently named *Mycobacterium paraffinicum*, was reported to oxidize ethylene as well as other hydrocarbons. Subsequent work by De Bont (De Bont, 1976; De Bont *et al.*, 1983) resulted in the isolation and characterization of the organism from a variety of soils.

Control of ethylene pollution

EMISSION REGULATORS

Since automobiles produce 90% of atmospheric ethylene it seems reasonable that the easiest way to control air pollution is to limit hydrocarbon production by using catalytic filters. As discussed earlier, Carey and Cohen (1980) reported that catalysts reduced ethylene production by 80%. Other filters have been used to reduce ethylene levels in greenhouses and fruit storage areas. Eastwell, Bassi and Spencer (1978) described a variety of filters capable of removing ethylene from air streams. Other workers have also described the use of various ethylene scrubbers (Lang and Tibbitts, 1983; Sorensen and Nobe, 1972). Colbert (1952) assessed the ability of ozone to reduce ethylene from storage atmospheres. His studies show that it might be feasible to generate ozone, mixing it with the storage atmosphere, and then decomposing the excess ozone by passing it through a layer of charcoal. Scott, Wills and Patterson (1971) evaluated the possibility of using ultraviolet lamps as an ozone generating source and reacting the ozone thus produced with the ethylene in a storage atmosphere. We have evaluated the feasibility (unpublished results) of using *M. paraffinicum* as a biological scrubber for ethylene in fruit storage atmospheres. This organism is capable of reacting with ethylene under storage conditions—namely low temperature, low oxygen and high CO_2 levels. However, the relatively low affinity of this organism for ethylene and the fact that it releases ethylene oxide when air is bubbled through the medium (De Bont *et al.*, 1983) may preclude the use of this organism as a biological ethylene scrubber for fruit storage atmospheres.

References

ABELES, F.B. (1973). *Ethylene in Plant Biology*. Academic Press, New York
ABELES, F.B. (1982). *Agriculture and Forestry Bulletin of the University of Alberta*, **5**, 3-12
ABELES, F.B. and ABELES, A.L. (1972). *Plant Physiology*, **50**, 496-498
ABELES, F.B., CRAKER, L.E., FORRENCE, L.E. and LEATHER, G.R. (1971). *Science*, **173**, 914-916
ABELES, F.B. and HEGGESTAD, H.E. (1973). *Journal of the Air Pollution Control Association*, **23**, 517-521
ALTSHULLER, A.P. (1968). *Advances in Chromatography*, **5**, 229-262
ALTSHULLER, A.P. and BELLAR,T.A. (1963). *Journal of the Air Pollution Control Association*, **13**, 81-87
ANONYMOUS (1981). *Chemical & Engineering News*, **59**, 12
ARNTS, R.R. and MEEKS, S.A. (1981). *Atmospheric Environment*, **15**, 1643-1651
ASHENDEN, T.W. and MANSFIELD, T.A. (1977). *Journal of Experimental Botany*, **28**, 729-735
ASHENDEN, T.W., MANSFIELD, T.A. and HARRISON, R.M. (1977). *Environmental Pollution*, **14**, 93-100
BARTH, D.S. (1970). *National Air Pollution Control Administration*, Publication Number AP-64. US Department of Health, Education and Welfare, Washington DC
BOUBEL, R.W., DARLEY, E.F. and SCHUCK, E.A. (1969). *Journal of the Air Pollution Control Association*, **19**, 497-500
BRESSAN, R.A., LeCUREUX, L., WILSON, L.G. and FILNER, P. (1979). *Plant Physiology*, **63**, 924-930
BURG, S.P. and STOLWIJK, J.A.J. (1959). *Journal of Biochemical and Microbiological Technology and Engineering*, **1**, 245-259
CAMPBELL, R.B. and MOREAU, R.A. (1979). *American Potato Journal*, **56**, 199-210
CAREY, P. and COHEN, J. (1980). *Technical Report* CTAB/PA/80-5 USEPA 2565 Plymonth Road, Ann Arbor, Michigan 48105
CLARK, A.I., McINTYRE, A.E., PERRY, R. and LESTER, J.N. (1983). *Environmental Pollution, (B)*, **6**, 245-261
CLAYTON, G.D. and PLATT, T.S. (1967). *American Industrial Hygiene Association Journal*, **28**, 151-160
COLBERT, J.W. (1952). *Refrigeration Engineering*, **60**, 265-267, 306
CONSIDINE, P.J., FLYNN, N. and PATCHING, J.W. (1977). *Applied Environmental Biology*, **33**, 977-979
CORNFORTH, I.S. (1975). *Plant Soil*, **42**, 85-96
CRAKER, L.E. and FILLATTI, J.J. (1982). *Environmental Pollution, (A)*, **28**, 265-272
CRAKER, L.E., FILLATTI, J.J. and GRANT, L. (1982). *Atmospheric Environment*, **16**, 371-374
DARLEY, E.F., BURLESON, F.R., MATEER, E.H., MIDDLETON, J.T. and OSTERLI, V.P. (1966). *Journal of the Air Pollution Control Association*, **16**, 685-690

DAVIS, J.B., CHASE, H.H. and RAYMOND, R.L. (1956). *Applied Microbiology*, **4**, 310–315
DE BONT, J.A.M. (1976). *Antonie Van Leeuwenhoek*, **72**, 59–71
DE BONT, J.A.M., VAN GINKEL, C.G., TRAMPER, J. and LUYBEN, K.Ch.A.M. (1983). *Enzyme Microbiology Technology*, **5**, 55–59
DOWDELL, R.J., SMITH, K.A., CRESS, R. and RESTALL, R.F. (1972). *Soil Biology & Biochemistry*, **4**, 325–331
EPA (1977). *National Air Quality and Emissions Trends Reports*, 1976. Publication Number EPA-450/1-77-002. USEPA. Research Triangle Park, North Carolina 27711
EASTWELL, K.C., BASSI, P.K. and SPENCER, M.E. (1978). *Plant Physiology*, **63**, 723–726
FELSTEIN, M., DUCKWORTH, S., WOHLERS, H.C. and LINSKY, B. (1963). *Journal of the Air Pollution Control Association*, **13**, 542–545, 565
FLÜCKIGER, W., OERTLI, J.J., FLÜCKIGER-KELLER, H. and BRAUN, S. (1979). *Environmental Pollution*, **20**, 171–176
FORSYTH, F.R. and EAVES, C.A. (1969). *Physiologia Plantarum*, **22**, 1055–1058
FUKUCHI, T. and YAMAMOTO, T. (1969). *Kogai to Taisaku (Journal for Pollution Control)*, **5**, 17–23
GEOMET INC. (1978). *EPA Report* No. 910/9-78-052. NTIS PB-290-472
GOODLASS, G. and SMITH, K.A. (1978). *Soil Biology & Biochemistry*, **10**, 193–199
GOODLASS, G. and SMITH, K.A. (1978). *Soil Biology & Biochemistry*, **10**, 201–205
GORDON, R.J., MAYRSOHN, H. and INGELS, R.M. (1968). *Environmental Science & Technology*, **2**, 1117–1120
HALL, W.C., TRUCHELUT, G.B., WEINWEBER, C.L. and HERRERO, F.A. (1957). *Physiologia Plantarum*, **10**, 306–317
HANAN, J.J. (1973). *HortScience*, **8**, 23–24
HASEK, R.F., JAMES, H.A. and SCIARONI, R.H. (1969). *Florists Review, 144(3721)*, **21**, 65–68, 79–82
HECK, W.W. and PIRES, E.G. (1962). *Texas Agricultural Experiment Station Bulletin*, MP-613
HECK, W.W., PIRES, E.G. and HALL, W.C. (1961). *Journal of the Air Pollution Control Association*, **11**, 549–556
HEILMAN, M.D., MEREDITH, F.I. and GONZALES, C.L. (1971). *Crop Science*, **11**, 25–27
HOGSETT, W.E., RABA, R.M. and TINGEY, D.T. (1981). *Physiologia Plantarum*, **53**, 307–314
HUNT, P.G., CAMPBELL, R.B. and MOREAU, R.A. (1980). *Soil Science*, **129**, 22–27
IDSO, S.B. (1983). *Journal of Environmental Quality*, **12**, 159–163
ILAG, C. and CURTIS, R.W. (1968). *Science*, **159**, 1357–1358
JACOBSEN, J.V. and McGLASSON, W.B. (1970). *Plant Physiology*, **45**, 631
JAFFE, M.J. (1980). *BioScience*, **30**, 239–243
JAMES, H.A. (1963). *Bay Area Air Pollution Control District, Information Bulletin*, 8–63. 939, Ellis Street, San Francisco, California 94109
KAWASE, M. (1976). *Physiologia Plantarum*, **36**, 236–241
LANG, S.P. and TIBBITTS, T.W. (1983). *HortScience*, **18**, 179–180
LEIGHTON, P.A. (1961). *Photochemistry of Air Pollution*, Academic Press, New York
LINDBERG, T., GRANHALL, U. and BERG, B. (1979). *Soil Biology & Biochemistry*, **11**, 637–643
LONNEMAN, W.A. (1977). *EPA Report* Number 600/3-77-011a. NTIS PB-264-232

LYNCH, J.M. (1975). *Nature (London)*, **256**, 576–577
LYNCH, J.M. (1983). *Plant Soil*, **70**, 415–420
MAPSON, L.W. and WARDALE, D.A. (1970). *Phytochemistry*, **10**, 29–39
McELROY, J.J. (1960). *California Agriculture*, **14**, 3
MOHAN RAO, A.M., NETRAVALKAR, A.J., ARORA, P.K. and VOHRA, K.G. (1982). *Atmospheric Environment*, **17**, 1093–1097
RODRIQUEZ, A.B. (1932). *Journal of the Department Agriculture, Puerto Rico*, **26**, 5–18
ROVIRA, A. and VENDRELL, D.M. (1972). *Soil Biology & Biochemistry*, **4**, 63–67
SCOTT, K.J., WILLS, R.B.H. and PATTERSON, B.D. (1971). *Journal of the Science of Food and Agriculture*, **22**, 496–497
SCOTT, W.E., STEPHENS, E.R., HANST, P.C. and DOERR, R.C. (1957). *Proceedings, American Petroleum Institute*, **37**, 171–183
SIMON, J.E., DECOTEAU, D.R. and CRAKER, L.E. (1983). *Environmental Technology Letters*, **4**, 157–162
SMITH, A.M. (1973). *Nature (London)*, **246**, 311–313
SMITH, A.M. and COOK, R.J. (1974). *Nature (London)*, **252**, 703–705
SMITH, K.A., BREMNER, J.M. and TABATABAI, M.A. (1973). *Soil Science*, **116**, 313–319
SMITH, K.A. and DOWDELL, R.J. (1974). *Journal of Soil Science*, **25**, 217–230
SMITH, K.A. and RESTALL, S.W.F. (1971). *Journal of Soil Science*, **22**, 430–443
SMITH, K.A. and RUSSELL, R.S. (1969). *Nature (London)*, **222**, 769–771
SORENSEN, L.L.C. and NOBE, K. (1972). *Environmental Science and Technology*, **6**, 239–242
STAN, H.-J., SCHICKER, S. and KASSNER, H. (1981). *Atmospheric Environment*, **15**, 391–395
STEPHENS, E.R. and BURLESON, F.R. (1967). *Journal of the Air Pollution Control Association*, **17**, 147–153
STEPHENS, E.R., DARLEY, E.F. and BURLESON, F.R. (1967). *American Petroleum Institute Division Refining Meeting*. Los Angeles, California
SWANSON, B.T., WILKINS, H.F. and KENNEDY, B.W. (1979). *Plant Soil*, **51**, 19–26
TIBBITTS, T.W., McFARLANE, J.C., KRIZEK, D.T., BERRY, W.L., HAMMER, P.A., HODGSON, R.H. and LANGHANS, R.W. (1977). *HortScience*, **12**, 310–311
TINGEY, D.T. (1980). *HortScience*, **15**, 630–633
TINGEY, D.T., STANDLEY, C. and FIELD, R.W. (1976). *Atmospheric Environment*, **10**, 969–974
WESTBERG, H., SEXTON, K. and FLYCKT, D. (1981). *Journal of the Air Pollution Control Association*, **31**, 661–667
YAMATE, G. (1974). *EPA Report* No 450/3-74-062. NTIS PB-238-766
YOSHIDA, T. and SUZUKI, M. (1975). *Soil Science and Plant Nutrition*, **21**, 129

24

EVALUATING THE PRACTICAL SIGNIFICANCE OF ETHYLENE IN FRUIT STORAGE

MICHAEL KNEE
East Malling Research Station, East Malling, Maidstone, Kent, UK

Introduction

Mention of ethylene to a plant physiologist will almost certainly evoke thoughts of its role in the ripening of climacteric fruits. Equally the first thought about the practical benefits of controlling ethylene responses is likely to focus on such fruits. The use of ethylene to promote ripening is more or less established for several fruits (*see* Proctor and Caygill, Chapter 25). However the removal of ethylene to prevent induction of ripening has taken half a century to become a commercial reality for a few fruits. Progress was retarded by technical and conceptual problems; understanding the nature and the resolution of these problems should help in assessing the role of ethylene in post-harvest changes in a wider range of crops. With this wider understanding the scope for practical control of ethylene mediated processes will become clear. This can be seen in a wider context as a case study in the exploitation of our understanding of the hormonal control of plant developmental processes.

History

The discoveries that many ripening fruits produce ethylene and that this gas could itself elicit ripening were quickly followed by suggestions that ethylene accumulation during storage was deleterious. Thus ripe and unripe apples should not be stored together (Kidd and West, 1938) and removal of ethylene from the store atmosphere could prolong storage life (Smock, 1943). However doubt was soon cast on the practical significance of ethylene in refrigerated fruit stores. Primitive methods of analysis revealed that ethylene levels were often above $10\,\mu l\,l^{-1}$ in refrigerated apple stores and above $100\,\mu l\,l^{-1}$ in modified atmosphere stores; yet the fruit survived in reasonable condition (Fidler, 1960) and (partial) removal of ethylene made little or no difference (Fidler, 1949). Gerhardt and Siegelman (1955) found that the firmness and colour of unripe apples were unaffected by storage with ripe fruit.

Early fruit physiologists placed great emphasis on respiratory rates as indicators of the overall metabolic state of the tissue. The demonstration that ethylene would elicit a typical climacteric rise in respiration in apples at 12.5 °C or avocados at 10 °C but produced no response at 3.3 °C in apples or 5 °C in avocados (Fidler, 1960; Biale, 1960a) seemed impressive. This should not have been surprising since it had long been known that the respiration of untreated apples held at 3–4 °C does not show a climacteric rise in response to endogenous ethylene (Kidd and West, 1930). However it seemed to Fidler and Biale that ethylene was physiologically inactive at 5 °C or below. When they observed that in cherimoya and feijoa (Biale, Young and Olmsted, 1954) and in Newton Wonder apples (Fidler and North, 1971) the climacteric rise occurred before measurable ethylene production these authors reached the conclusion that ethylene might not be a natural regulator of ripening at any temperature. These observations can now be partly explained by the low sensitivity of chemical and early chromatographic methods used for ethylene measurement. There are better substantiated examples of temporal discrepancies between the rises in respiration and ethylene production (Biale and Young, 1981). These examples are troublesome but not overwhelming; there are many reasons why the respiration rate of a plant organ can change and respiration is not tightly coupled to ripening; increases in respiration and in ethylene production are *both* symptoms of ripening rather than primary regulatory events (Pratt and Goeschl, 1968). Nevertheless ease of accurate measurement makes these processes useful (and non-destructive) experimental markers of ripening. Much of our understanding of fruit ripening relies on their measurement, but it has to be remembered that they are secondary consequences of underlying regulatory changes and, hence, that they give *indirect* evidence of those changes.

Retardation of fruit ripening by inhibitors of ethylene synthesis (Bangerth, 1978) ethylene action (Janes and Frenkel, 1978) and by hypobaric storage (Burg and Burg, 1966) have confirmed the natural role of ethylene in regulation of ripening in climacteric fruits. It is equally clear that rising ethylene production is not the basis of regulation; ethylene is acting at a time when its concentration is low and static so that the ability to respond to ethylene is seen as the basis of regulation (McGlasson, Wade and Adato, 1978). Changes in the responsiveness of tissues to plant growth regulators rather than the levels of these compounds *per se* may be a general feature of growth regulation and developmental change in plants (Trewavas, 1982). The role of ethylene in the senescence of non-climacteric fruit and vegetables is unclear. They all produce and are more or less responsive to the gas; thus there may be contexts in which control of ethylene may have practical benefits with these commodities.

The practical findings of those early workers, who were sceptical about the importance of ethylene, are not in doubt. The difficulty of preventing ethylene action in some contexts may prove insurmountable; the inevitable conclusion will be that, in those contexts, ethylene cannot be manipulated to practical advantage and the technology of ethylene removal will be irrelevant. That is to say that ethylene is too active rather than that it is inactive, which was the wider conclusion drawn by Fidler, Biale and others. The remainder of this chapter will consider how understanding the physiology of ethylene involvement in fruit ripening can guide an approach to practical control. This will entail a review of information on the production of ethylene, sensitivity to the gas, and how these change during fruit development and can be affected by environmental and chemical treatments. This might serve as a model for extension to a wider range of crops. Since the earlier workers did not have sensitive gas chromatographs or insights into the biochemistry of ethylene synthesis and action; they were forced to adopt an empirical approach. Their conclusions were rational at the time, although they may now seem erroneous. But there is no reason to continue their experimental approach or their pattern of thinking today.

Production of ethylene

Ripening climacteric fruit produce ethylene and this gas will induce ripening. It is customary to think of one fruit setting off the ripening of another. However if each fruit is individually supplied with a flow of ethylene free air each fruit will ripen in isolation. This autonomous ripening sets a natural limit on any attempt to control ripening by removal of ethylene in bulk storage of fruit. As the efficiency of ethylene removal is increased towards infinity, that is towards maintenance of zero ethylene concentration, so each fruit will tend to approach autonomy. Instead of ripening being synchronized by the earliest fruit, the natural distribution of ripening times in the population will be revealed.

Although there may be no ethylene in the surrounding atmosphere, at the cellular level each isolated fruit is continuously exposed to ethylene. Fruits are bulky organs with diffusion barriers, mainly in the skin (Burg and Burg, 1965). At any rate of production ethylene is released from the fruit in a way which obeys Fick's law. In its simplest form this can be stated as

$$D = a\,(C\text{ in} - C\text{ out})$$

where D = rate of diffusion out of the fruit
C in = internal concentration
C out = external concentration
a = a factor which depends on the diffusivity of the gas, the porosity of the barrier and its surface area.

In a flowing system under equilibrium D is equal to the rate of production and it can readily be seen that even if C out is zero then

$$C\text{ in} = D/a$$

that is to say there is a theoretical lower limit for internal ethylene for a particular fruit and scaling up an ethylene removal system could never eliminate it.

The ethylene concentrations reported in the intercellular space of preclimacteric fruits range from 0.01 to 0.9 μl l^{-1} with many values around 0.05 μl l^{-1} (Biale and Young, 1981). Such levels are maintained over the final days or weeks of the fruit's development on the plant or after harvest and up to the point where a rapid rise marks the onset of ripening. A prevalent idea has been that because the fruit remains in a condition of apparent physiological stasis these low levels of ethylene must be ineffective. Peacock (1972) treated bananas with 0.44 μl l^{-1} ethylene for four days after harvest and observed no immediate response. However treated fruit ripened on average 26.4 days after harvest whereas untreated fruit ripened at 38.9 days. Thus the ethylene treatment was effective although the effect was delayed. As will be shown below there is no evidence that ethylene ceases to be effective in promoting ripening below a particular concentration so the fruit should also respond to its endogenous ethylene. Peacock (1972) concluded that the low levels of ethylene in preclimacteric fruit must cause it to advance towards the climacteric.

Table 24.1 RATES OF PRODUCTION OF ETHYLENE BY CLIMACTERIC FRUIT

Fruit	Rate of production ($\mu l\ kg^{-1}\ h^{-1}$)		Temperature (°C)	Reference
	Preclimacteric	Climacteric		
Apple	0.1	100	12	Reid, Rhodes and Hulme (1973)
Apricot	0.03	0.4	20	Reid (1975)
Kiwi-fruit	<0.1	50	20	Pratt and Reid (1974)
Melon	0.1–0.2	20–50	20	Pratt and Goeschl (1968)
Peach	0.1–0.2	50–100	20	Looney, McGlasson and Coombe (1974)
Pear	0.5–0.6	10–30	20	Wang, Mellenthin and Hansen (1972)
Tomato	0.2		20	Hoffman and Yang (1980)
Tomato		120	20	Lyons and Pratt (1964)

The cellular basis of this developmental process and its regulation by ethylene are unclear. What is clear is that when internal ethylene concentrations are around $0.05\ \mu l\ l^{-1}$ and rates of production are about $0.02\ \mu l\ kg^{-1}\ h^{-1}$, this is the time to be controlling ethylene accumulation and preventing its action. After the onset of the climacteric, ethylene production will rise to up to $100\ \mu l\ kg^{-1}\ h^{-1}$ depending on the fruit (*Table 24.1*). Even if it were thought that removal of this ethylene could arrest the ripening processes, once rapid ethylene production had started removal would have become a physical impossibility. The minimum internal concentration that could be achieved would be from 3 to $500\ \mu l\ l^{-1}$; all concentrations in this range are highly physiologically active.

Sensitivity to ethylene

Earlier workers might have found it hard to believe that concentrations of ethylene below the limits of detection of their analytical methods were nevertheless physiologically active upon fruit. It was known for example that concentrations below $1\ \mu l\ l^{-1}$ would cause abnormal growth in etiolated peas or epinasty in tomato plants (Fidler, 1960) but there was no theoretical basis which would have led them to predict that all parts of all plants were equally sensitive. It was Burg and Burg (1967) who provided this theoretical base after analysis of the concentration–activity relationship for various plant responses. Their use of an analogy with enzyme–substrate binding kinetics can be questioned on theoretical grounds; it can also be empirically awkward as a Lineweaver–Burk analysis of concentration and response data for ethylene will often produce a curvilinear plot. The semilog plot adopted by Goeschl and Kays (1975) overcomes these objections, is more readily interpretable and still allows interpolation of a concentration for 50% of maximum response ($C_{1/2}$). In their experiments Goeschl and Kays (1975) found that various responses in different plant species were half-maximal at concentrations ranging from 0.095 to $1.35\ \mu l\ l^{-1}$.

A similar treatment of literature values for ethylene effects on respiration of various fruit is shown in *Figure 24.1*. The $C_{1/2}$ for the rate of induction of the climacteric by continuous ethylene treatments is between 0.1 and $0.5\ \mu l\ l^{-1}$ for avocado, banana, Cantaloupe melons and pears. Less complete data for softening of Kiwi fruit (Harris, 1981), induction of the climacteric and rapid ethylene synthesis in apple (Harkett *et al.*, 1971) and softening of water melons (Risse and Hatton, 1982) all suggest a similar sensitivity.

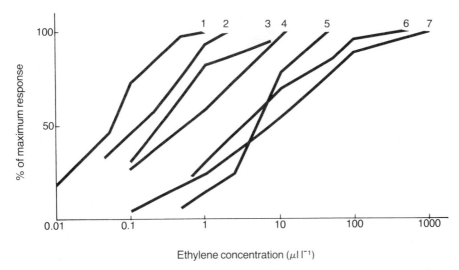

Figure 24.1 Concentration dependency of responses to ethylene in various fruit. (1) Induction of ripening in Dwarf Cavendish banana (Liu, 1976). (2) Induction of ethylene synthesis in Anjou pear (Wang, Mellenthin and Hansen, 1972); induction of the climacteric in Canteloupe melon (McGlasson and Pratt, 1964) is very similar. (3) Induction of climacteric in Gros Michel banana (Biale and Young, 1981). (4) Induction of climacteric in avocado pear (Biale, 1960b). (5) Stimulation of respiration in immature Honeydew melon (Pratt and Goeschl, 1968). (6) Stimulation of respiration in Navel Orange (Biale, 1960a). (7) Stimulation of respiration in lemons (Biale and Young, 1981)

Ethylene can stimulate respiration in non-climacteric fruit. However, unlike climacteric fruits the rate declines when the ethylene is withdrawn (Reid and Pratt, 1972) and high concentrations are required to elicit a maximum response (*Figure 24.1*). However, the limited data available suggest that the $C_{1/2}$ for chlorophyll destruction in the non-climacteric citrus fruit is around $0.1\,\mu l\,l^{-1}$ (Wheaton and Stewart, 1973). Interestingly immature Anjou pears behave like citrus in that respiration rate rises with ethylene concentration and before endogenous ethylene synthesis is stimulated (Wang, Mellenthin and Hansen, 1972). Fruit softening in the Anjou pear is promoted by ethylene concentrations as low as $0.05\,\mu l\,l^{-1}$ although the data do not permit calculation of $C_{1/2}$. Similarly in immature Honeydew melons ethylene stimulates respiration but not endogenous ethylene synthesis and $C_{1/2}$ is approximately $6\,\mu l\,l^{-1}$. When this fruit is more mature low concentrations of ethylene stimulate both endogenous ethylene production and the climacteric rise in respiration (Pratt and Goeschl, 1968). Goeschl and Kays (1975) report a $C_{1/2}$ of $0.2\,\mu l\,l^{-1}$ for the induction of softening in this fruit.

In mature climacteric fruits ethylene production is itself stimulated at the same time as respiration by ethylene treatment. So the rate of respiration cannot be related to the concentration of ethylene applied (McGlasson, Wade and Adato, 1978).

These apparently conflicting observations can be rationalized (*Figure 24.2*) if it is supposed that ethylene acts at a high affinity site to initiate softening and chlorophyll degradation and that it also acts at a low affinity site to stimulate respiration. In mature climacteric fruits ethylene synthesis is triggered by a high affinity response; this ensures that, in most fruit of this type, there is enough endogenous ethylene to saturate the low affinity site and stimulate maximal rates of respiration. An exception seems to be the mango where ethylene treatment during the climacteric can cause a further respiratory rise (Burg and Burg, 1961).

The sensitivity of plant tissues to ethylene is influenced by their endogenous content. As noted by Goeschl and Kays (1975) the $C_{1/2}$ is displaced upwards by an amount equal to the internal concentration; that is the tissue becomes less sensitive to external ethylene. Variations in internal ethylene probably account for much of the variation in apparent sensitivity in *Figure 24.1*. The high affinity site may be uniform in all fruit and other plant tissues with a true $C_{1/2}$ at $0.1\,\mu l\,l^{-1}$ or less.

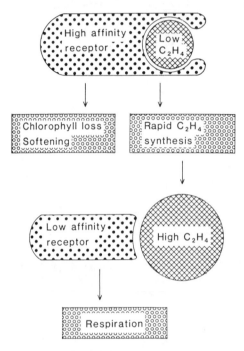

Figure 24.2 Possible involvement of different ethylene receptors in ripening responses of fruit. Rapid ethylene synthesis occurs only in mature climacteric fruit

Much effort has been expended in determining 'threshold' concentrations of ethylene for particular fruits. If the argument developed above is valid this is misconceived. Theoretically there is no lower limit at which ethylene becomes inactive. Detection of a response to applied ethylene will depend upon its concentration relative to endogenous ethylene and the precision of the experiment. A further misconception arises from the observation that fruit can become more responsive to applied ethylene during development (Harkett *et al.*, 1971; Wang, Mellenthin and Hansen, 1972; Lyons and Pratt, 1964); this is often described as 'increasing sensitivity' but there is no evidence that the $C_{1/2}$ decreases. What is observed is an earlier response to a particular concentration as the fruit advances towards autonomous ripening. Trewavas (1982) cites this as evidence of the regulation of ripening via ethylene receptors. It is surely to be expected that if fruit is naturally close to ripening then ethylene treatment will result in an earlier response than in immature fruit where the necessary preparations have not been

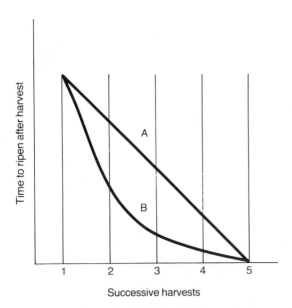

Figure 24.3 Theoretical response of fruit to endogenous and applied ethylene at different stages in development, assuming increasing sensitivity. A, ripening of control fruit (endogenous ethylene); B, ripening in presence of a certain concentration of applied ethylene

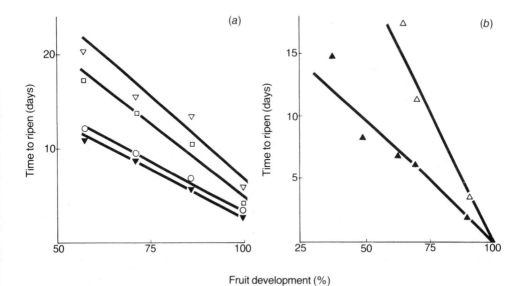

Figure 24.4 Observed response of fruit to ethylene at different stages of development. (a) Time taken to reach firmness of 4.5 kg in Anjou pear harvested at different times and treated with ethylene at 0.2 µl l^{-1} ($\triangledown$), 0.5 µl l^{-1} ($\square$), 1 µl l^{-1} ($\bigcirc$) and 2 µl l^{-1} ($\blacktriangledown$). (Data of Wang, Mellenthin and Hansen, 1972). (b) Time to onset of lycopene formation in 'VC243-20' tomatoes harvested at different times and treated with ethylene at 1000 µl l^{-1} ($\blacktriangle$) or untreated ($\triangle$) (Data of Lyons and Pratt, 1964)

made. Increased concentration or sensitivity of receptors would be expected to cause an acceleration of progress to ripening in response to ethylene; the plot of days to ripen in response to ethylene should be curvilinear (*Figure 24.3*). Actual examples of such data show rectilinear plots (*Figure 24.4*) and hence imply constant concentration and sensitivity of receptors through at least the latter half of fruit development. Fruits which are less than half developed may not ripen if detached from the plant either autonomously or if treated with ethylene (McGlasson, Wade and Adato, 1978). Whether this is because they lack ethylene receptors or are blocked at some other stage in the ripening response has not been determined. There are also fruits whose ripening is delayed or which will not ripen if left attached to the plant. The best example is avocado, for which there is some evidence that the response to ethylene is blocked while the fruit is on the bush; it remains relatively unresponsive for at least 24 h after harvest, particularly when immature (Adato and Gazit, 1974).

It seems that the time of initiation of ripening in climacteric fruit cannot be accounted for by changes in ethylene synthesis or sensitivity. All that can be said is that the fruits contain low concentrations of ethylene, and from a certain stage of development, or in some instances after removal from the parent plant, the fruit becomes responsive to ethylene so that it advances towards the climacteric. The rate of advance could be influenced by manipulating ethylene concentrations or sensitivity. Furthermore the absence of a rapid rise in ethylene synthesis in non-climacteric fruit does not exclude ethylene from a role in regulation of ripening in such fruit.

The problem of practical control

All fruit (and indeed probably all higher plant crops) produce ethylene. For preclimacteric and non-climacteric fruit the production rate is of the order of

Table 24.2 RATES OF PRODUCTION OF ETHYLENE BY NON-CLIMACTERIC FRUIT

Fruit	Rate of production ($\mu l\, kg^{-1}\, h^{-1}$)	Temperature (°C)	Reference
Blueberry	0.04–0.05	—	Frenkel (1972)
Cranberry	0.02–0.04	21	Forsyth and Hall (1969)
Cucumber	0.02–0.16	—	Saltveit and McFeeters (1980)
Orange	0.02–0.06	—	Burg and Burg (1961)
Pineapple	0.01–0.3	20	Dull, Young and Biale (1967)
Strawberry	0.1	20	Knee, Sargent and Osborne (1977)

$0.02\,\mu l\, l^{-1}\, h^{-1}$ (*Tables 24.1* and *24.2*) and internal concentrations are around $0.05\,\mu l\, l^{-1}$ (Biale and Young, 1981). Presumably in less bulky, leafy vegetables and flowers internal concentrations are lower. If there is a finite concentration then physiological activity must be assumed, although the consequences may not be commercially significant. If there are significant effects, control of ethylene may be worthwhile; this control can include control of accumulation, control of synthesis and control of action. For climacteric fruit these measures will only be effective if they are applied in the preclimacteric state.

Control of ethylene accumulation

In much storage and post-harvest handling air circulation around crops is restricted. Problems of ethylene contamination from other produce or pollution can arise (*see* Abeles Chapter 23) but even in the absence of these the produce is liable to experience enhanced ethylene levels. A simple calculation shows that if a crop producing $0.01\,\mu l\,kg^{-1}\,h^{-1}$ occupies one-third of the volume of a sealed container or room, in 24 h $0.12\,\mu l\,l^{-1}$ will accumulate. Of course, in practice, there is usually some ventilation and this is the simplest control measure. For example ventilation can prevent ethylene damage to tulip bulbs in store (de Munk, 1972).

Ventilation is worthwhile only if ethylene free air is available, and may not be feasible when low temperatures or modified atmospheres have to be maintained. Product generated atmospheres, enriched with carbon dioxide or depleted of oxygen, are almost hermetically sealed. This leads to higher accumulation of ethylene than in air storage for both fruit (Knee, 1980) and vegetable crops (Toivonen *et al.*, 1982). A continuous flow of externally generated atmosphere would be an expensive control measure on the large scale and some other means of ethylene removal is necessary. This can involve oxidation with potassium permanganate (Smock, 1943), ozone (Fidler and North, 1969) or active oxygen (Scott, Wills and Patterson, 1971) or catalysts (Dilley, 1982; Dover, 1983) or adsorption by charcoal (Stoll, Hansen and Datwyler, 1974) or brominated charcoal (Fidler, 1949) (*see* Dover, Chapter 31, Liu, Chapter 32 and Blanpied, Chapter 33).

For reasons explained above ethylene removal alone will not prevent the response of the crop to endogenous ethylene and there are relatively few instances of untreated crops in normal air storage responding positively to this treatment. The maintenance of lemon quality by ethylene removal during storage is an exception (Wild, McGlasson and Lee, 1976). Usually suppression of ethylene synthesis or interference with ethylene action is necessary. But equally the effects of these treatments are not absolute; residual production or activity can easily negate their effects unless supplemented by ethylene removal.

The only way to overcome the limitation imposed by the diffusion process is to increase the rate of diffusion. This can be done by reducing the density of the storage atmosphere. Substitution of hydrogen or helium for nitrogen is a theoretical possibility but, in practice, reduction of pressure has been adopted (Burg and Burg, 1966). The effect of lower ethylene levels is often compounded by the concomitant reduction in the partial pressure of oxygen and in many experiments it is not possible to separate the consequences of the dual treatment. Exceptionally, supply of ethylene under hypobaric conditions has shown the extent of involvement of ethylene in the fruit response (Burg and Burg, 1966; Bangerth, 1975). After many years of commercial development it seems unlikely that hypobaric storage will be widely adopted.

Effects of temperature

Ethylene production by post-climacteric apple slices varies directly with temperature; Burg and Thimann (1959) reported a Q_{10} of 2.8 from 10 to 25 °C whereas Mattoo *et al.* (1977) reported a Q_{10} of 1.74 from 10 to 30 °C and 2.70 from 0 to 10 °C. The latter authors observed a similar transition in the relationship of ethylene synthesis to temperature in post-climacteric tomato slices; they accounted for the transition by suggesting that at least part of the pathway of ethylene biosynthesis occurs in a lipid domain whose fluidity decreases sharply at the transition temperature. Since the discovery of ACC as an intermediate in ethylene biosynthesis it has been postulated that the enzyme converting ACC to ethylene is membrane localized (Yang, 1981). The relevance of such short-term observations of response of plant tissue to a range of temperatures depends on whether the effects are manifested in the long term with intact plant organs. Unpublished observations by the author reveal a Q_{10} of about 2 for ethylene production by whole apples in both the pre- and post-climacteric states (*Figure 24.5*). An Arrhenius plot of the data does not reveal an obvious transition temperature. Since response is related to log concentration a 10 °C reduction in the temperature of a crop would not be expected to change substantially the response to endogenous ethylene.

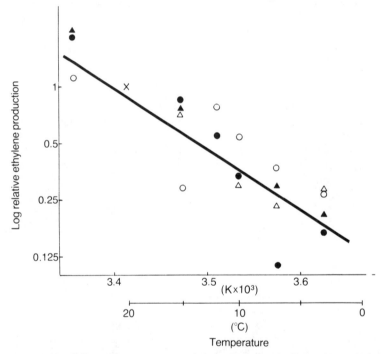

Figure 24.5 Effect of temperature on ethylene production by fruits of two varieties of apples before and after the climacteric rise in respiration. The mean rates for individual apples are expressed relative to the rate at 20°C (×). Preclimacteric Cox apples (○) had a rate at 20°C of 0.04 µl kg^{-1} h^{-1}; post-climacteric Cox (●), 100 µl kg^{-1} h^{-1}; preclimacteric Golden Delicious (△) 0.026 µl kg^{-1} h^{-1}; post-climacteric Golden Delicious (▲) 240 µl kg^{-1} h^{-1}. The regression line is shown

Ethylene synthesis may be enhanced by low temperature stress of plants (Wright, 1974), although ethylene is not thought to have a causal role in development of low temperature injury. Many varieties of pear and some apple varieties held at low but non-injurious, temperatures show accelerated ethylene production earlier than fruit at higher temperatures (Looney, 1972; Knee et al., 1983). This implies that the fruit responds more quickly to endogenous ethylene at low temperature. This could occur because ethylene itself is more active at low temperatures or because of an increase in receptors. The affinity of an enzyme for its substrate can increase or decrease with temperature, depending upon the activation energy of the formation of the enzyme-substrate complex (Gibson, 1953). There is no definitive information on the activity of ethylene in relation to temperature although some data on degreening of citrus fruits suggest that $C_{1/2}$ decreases with temperature (Wheaton and Stewart, 1973). It is not known whether other fruits respond to temperature as do the pomes or whether ethylene might be involved in breaking of dormancy of various plant parts by chilling treatments. Obviously the response in pears and some apple varieties compounds the difficulty of controlling ethylene action in refrigerated storage; it may account, to a large extent, for the negative results obtained by early workers.

There is no sound evidence that ethylene becomes inactive below threshold temperature, as thought previously, although comparatively few authors have thought it worthwhile to demonstrate the action of ethylene below 5°C. Harris (1981) showed that ethylene at $0.1\,\mu l\,l^{-1}$ promotes softening of kiwi fruit at 0°C. In apples at 3°C endogenous ethylene rises to high levels five to ten days after harvest and, as ripening processes are slow at this temperature, a difference of initiation time of the same order is about the least measurable effect. Ethylene treatment at 3°C does cause slightly earlier softening of apples (*Table 24.3*) although effects on

Table 24.3 EFFECT OF ETHYLENE ON FLESH FIRMNESS OF APPLES STORED AT 3.3°C

	Apple variety				
	James Grieve			Tydeman's Late Orange	
Added C_2H_4 ($\mu l\,l^{-1}$)	Mean firmness (kg 10–56 days) a	b	Added C_2H_4 ($\mu l\,l^{-1}$)	Mean firmness (kg 11–49 days) a	b
0	3.86	3.86	0	5.30	5.30
1.2	3.75	} 3.76	0.8	5.29	} 5.27
9.8	3.77		5.4	5.25	
300	3.80	} 3.75	410	5.18	} 5.13
1900	3.70		1400	5.07	
(LSD)	(0.12)	(0.10)		(0.18)	(0.14)

The least significant differences at $P=0.05$ (LSD) apply for comparisons between ethylene treatments ($n = 10$) and controls ($n = 20$) in columns a or any comparison on columns b

chlorophyll loss were not detected (Knee, 1976). The high concentrations required to promote the maximum effect are explicable as a displacement of the response-concentration relationship by high endogenous ethylene levels (Goeschl and Kays, 1975). These results confirm the practical conclusions of Fidler (1960) and Gerhardt and Siegelman (1955) that there is unlikely to be significant commercial loss from exposure of unripe apples to ethylene in refrigerated air storage.

Carbon dioxide and oxygen effects

Oxygen is required for the conversion of ACC to ethylene (Adams and Yang, 1979). Results obtained with apple slices by Burg and Thimann (1959) suggest that the affinity of the oxidase involved is similar to that of the respiratory oxidase. The affinity of cytochrome oxidase for O_2 is so high that inhibition is unlikely under commercial low oxygen storage conditions (Knee, 1980). However ethylene synthesis by intact post-climacteric apples is inhibited by reduction of oxygen from 21% to 2% (Knee, 1980). Whether a similar inhibition would occur in preclimacteric fruit is questionable.

Burg and Burg (1967) examined the interaction of oxygen and carbon dioxide with ethylene inhibition of pea epicotyl elongation. Raising the carbon dioxide and lowering the oxygen concentration increased the concentration of ethylene required for half maximum response (*Table 24.4*). This suggests that carbon dioxide is an inhibitor, and oxygen a cooperative effector of ethylene action. The fragmentary data available for fruit suggest that similar interactions are operative. Thus Harris (1981) showed that 6% CO_2 delays softening of kiwi fruit and ethylene at $0.1\,\mu l\,l^{-1}$ reversed this effect. Liu (1976) found that bananas responded about equally to $0.1\,\mu l\,l^{-1}$ of ethylene in air or $0.5\,\mu l\,l^{-1}$ of ethylene in 4% oxygen.

Table 24.4 SENSITIVITY OF PEA SEEDLINGS TO ETHYLENE IN DIFFERENT ATMOSPHERES ESTIMATED FROM DATA OF BURG AND BURG (1967)

Atmosphere		$C_{1/2}$
% O_2	% CO_2	($\mu l\,l^{-1}$)
0.7	0	0.6
2.2	0	0.3
18	0	0.14
18	1.8	0.3
18	7.1	0.6

$C_{1/2}$ is the concentration at which half maximum inhibition of pea stem elongation occurs

If fruit are held in a flowing stream of a high carbon dioxide or low oxygen mixture ethylene accumulation is prevented and the onset of ripening delayed. Mapson and Robinson (1966) ascribed the delay in low oxygen atmospheres to an effect on ethylene synthesis. If the argument developed above from Peacocks's (1972) observations is correct it is more likely that the atmosphere lowers the sensitivity of the fruit to its endogenous ethylene and therefore delays a ripening response.

The effects of modified atmospheres on fruit ripening seem to be fully reversible by ethylene in kiwi fruit (Harris, 1981) and banana (Mapson and Robinson, 1966; Quazi and Freebairn, 1970). So the commercial benefit of storage of these fruit in such atmospheres will only be realized if low ethylene concentrations are maintained. In apples the effects of low oxygen atmospheres are not fully reversible by ethylene (Knee, 1976). This appears to be because oxygen is required in the metabolic processes underlying ripening changes such as chlorophyll loss (Knee, 1980; Knee, Hatfield and Ratnayake, 1979) and softening (Knee, 1982). Thus low oxygen storage of apples is commercially successful even though ethylene levels of several hundred $\mu l\,l^{-1}$ accumulate in the stores. There have been many investigations of the consequences of removal of that ethylene with apparently conflicting results. However those authors who demonstrate a control of ethylene such that minimal levels ($<0.1\,\mu l\,l^{-1}$) were maintained for a substantial part of the storage period (>50 days) also showed that apples remained firmer than when ethylene was allowed to accumulate (e.g. Forsyth, Eaves and Lightfoot, 1969; Liu, 1979; Knee, 1975).

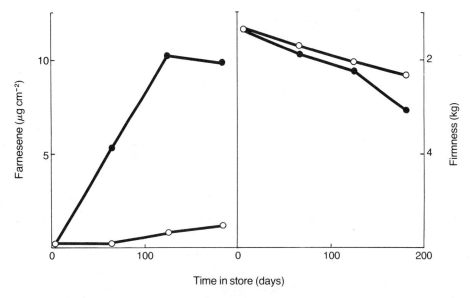

Figure 24.6 Effect of ethylene (1100 µl l^{-1}) on farnesene synthesis and softening of Bramley apples stored in 5% CO_2:3% O_2 at 3.5°C. ●—● Ethylene; ○—○ control. (Data from Knee and Hatfield, 1981)

There are aspects of metabolism in apples that are relatively unaffected by modified atmosphere storage but nevertheless stimulated by ethylene. When Bramley's seedling apples are held in a flow of 5% carbon dioxide, 3% oxygen, low endogenous ethylene levels are maintained for over 200 days. Ethylene treatment under these conditions promotes a rapid rise in concentration of α-farnesene in the surface wax but the promotion of softening is much more gradual (*Figure 24.6*). It is not known whether synthesis of farnesene is specifically promoted or whether there is a general enhancement of terpenoid metabolism. Farnesene is of interest because it has been implicated in the aetiology of a storage disorder of apples, superficial scald (Huelin and Coggiola, 1970). When Bramley apples are stored in the usual product-generated atmosphere ethylene levels soon build up and trigger farnesene synthesis; later in storage this compound is involved in autoxidation reactions leading to death of cells in the fruit peel. Continuous ethylene removal is an alternative to the usual control of the disorder by application of anti-oxidants to the fruit surface, and has been shown to be effective on a commercial scale (*see* Dover, Chapter 31).

Whenever a crop is held in a product generated atmosphere, ethylene will accumulate to the µl l^{-1} level even if rates of production are very low. It is too early to say what effects the interaction of ethylene, carbon dioxide and oxygen will cause in many crops. Certainly there are crops such as cabbage which, when stored in modified atmospheres, manifest little of their normal response to ethylene (Toivonen *et al.*, 1982).

Chemical treatments

As understanding of the natural regulation of ethylene synthesis and action grows this should suggest opportunities for manipulating these processes in crops. The discovery of rhizobitoxine and its analogues as inhibitors of ACC synthase presented one such opportunity. The analogue aminoethoxyvinylglycine (AVG) inhibits ethylene synthesis and, consequently, ripening in apples (Bangerth, 1978) and pears (Romani *et al.*, 1982). Unfortunately ethylene synthesis is much less inhibited at lower temperatures (Mattoo *et al.*, 1977) and this may be one reason for the disappointing results of the use of the compound on fruit prior to refrigerated storage. Alternatively, failure to ensure low background ethylene levels during storage could have vitiated the AVG effect in some experiments. Autio and Bramlage (1982) showed that, provided ethylene is removed during storage, AVG treated apples can remain preclimacteric at low temperature for up to 200 days. AVG is an inhibitor of many pyridoxal phosphate enzymes and its approval for food use seems unlikely. This consideration may have led to the decision not to develop it as an agrochemical.

Similarly silver ion as an antagonist of ethylene action (Beyer, 1976) is unlikely to be permitted for use on food crops. However its successful use on cut flowers could obviate all other ethylene control technology for these (*see* Nichols, Chapter 28).

Daminozide is an agrochemical in current use which interferes with ethylene induction of ripening in pome fruit (Looney, 1968). It has no effect on ACC synthesis (Walsh and Solomos, 1981); and daminozide and its breakdown product, dimethyl hydrazine, do not inhibit ethylene production by apple slices (*Table 24.5*). Preclimacteric ethylene synthesis is slower in daminozide treated whole apples but the inhibitor is usually thought to act at some point in gibberellin biosynthesis (Lawrence, 1984). This would imply that the effects of daminozide on ethylene synthesis are indirect and effects at the level of ethylene action cannot be excluded. An indirect mechanism is supported by the fact that daminozide promotes ripening in stone fruit (Looney, McGlasson and Coombe, 1974).

A useful feature of daminozide is that the delay of rapid ethylene synthesis is longer at lower temperatures; the chemical can restore the normal physiological response to temperature (*Table 24.6*). Consequently it can make possible control of ethylene accumulation in storage for apples that would otherwise not be amenable

Table 24.5 EFFECT OF DAMINOZIDE AND DIMETHYL HYDRAZINE ON ETHYLENE PRODUCTION BY APPLE FRUIT CORTICAL DISCS

Concentration (mM)	Ethylene production ($nl\ g^{-1}\ h^{-1}$) in presence of			
	Daminozide		Dimethyl hydrazine	
	Exp. 1	Exp. 2	Exp. 3	Exp. 4
0	25	18	23	19
2	18	19	26	19
10	18	15	21	16
50	20	18	18	11

Compounds were supplied in 0.5 M glycerol, 0.001 M $CaCl_2$ solution during 4 h incubation at 25 °C

Table 24.6 EFFECT OF DAMINOZIDE SPRAY ($0.85\ g\ l^{-1}$) ON TIMING OF RISE IN ETHYLENE PRODUCTION BY INDIVIDUAL APPLES IN RELATION TO STORAGE TEMPERATURE

Treatment	Control		Daminozide	
Storage temperature (°C)	7	25	7	25
Mean time to reach $0.1\ \mu l\ kg^{-1}\ h^{-1}$ (day)	7.3	7.8	14.8	7.5

(LSD = 1.8 days when $P = 0.05$)

to control; the main benefit of dual treatment with daminozide and ethylene removal is a retention of fruit firmness which neither treatment alone can provide (Liu, 1979, Chapter 32). There are side effects of daminozide which will limit its general acceptability for use on apples; many growers would not accept the loss of yield resulting from its growth inhibitory effects and in some varieties the chemical proves injurious to fruit stored for long periods (Sharples, 1971).

Genetic control

There are varieties of apple which are much more amenable to control of ethylene accumulation during storage than others (*Table 24.7*). Whether this results from variation in endogenous ethylene synthesis or readiness to respond remains to be seen. Varietal differences need to be considered when evaluating the practical value of ethylene control technology for particular crops. So far ethylene synthesis and response have not been selection criteria in breeding programmes for any crop.

Table 24.7 DELAY IN RISE OF ETHYLENE PRODUCTION FOR DIFFERENT VARIETIES OF APPLE STORED IN 2% OXYGEN 98% NITROGEN AT 3.5°C WITH CONTINUOUS ETHYLENE REMOVAL

Variety	Time (days) to reach $0.1\ \mu l\ l^{-1}\ C_2H_4$	Variety	Time (days) to reach $0.1\ \mu l\ l^{-1}\ C_2H_4$
Bramley	>150	Jonagold	49
Cox[a]	36	Jonathan	51
Cox[b]	48	Jupiter	29
Crispin	>150	Kent	63
Gloster 69	>150	McIntosh	52
Golden Delicious	37	Spartan	20
Ida red	38	T3131	29

[a,b]Cox samples from different orchards

The non-ripening mutants of tomato are unresponsive to ethylene (Tigchelaar, McGlasson and Buescher, 1978) and the homozygous lines have no commercial value. However the heterozygotes produce fruit with an intrinsically long shelf-life and it would be interesting to know whether this results, to any extent, from blockages in ethylene synthesis or action.

Prospects

Experimental control of ethylene has prevented deterioration of a number of fruit: apples (Forsyth, Eaves and Lightfoot, 1969; Liu, 1979; Knee and Hatfield, 1981); avocados (Hatton and Reeder, 1972; Oudit and Scott, 1973); bananas (Scott and Gandanegara, 1974; Esquerra, Mendoza and Pantastico, 1978); lemons (Wild, McGlasson and Lee, 1976); pears (Scott and Wills, 1974); plantains (Hernandez, 1973) and tomatoes (Geeson, Brown and Guaraldi, 1983). Ethylene decreases the post-harvest life or acceptability of other crops including cabbage (Pendergrass *et al.*, 1976); carrots (Carlton, Peterson and Tolbert, 1961); cucumber (Poenicke *et al.*, 1977); kiwi fruit (Harris, 1981); lettuce (Morris *et al.*, 1978); citrus fruit (Hatton and Reeder, 1972); strawberries (El-Kazzaz, Sommer and Fortlage, 1983) and water melons (Risse and Hatton, 1982). Only for apples is there a large scale semicommercial system for ethylene removal during storage. Ethylene oxidants are used on a small scale in short-term holding and distribution of other crops on a rather haphazard basis. The full extent of applicability of ethylene removal has yet to be seen. It may be useful for crops not so far considered; commercial application for some of those named may not prove to be advantageous or feasible. Although the climacteric fruit may seem the obvious subjects for further work, their inherent tendency to develop rapid ethylene synthesis presents a formidable problem. Non-climacteric fruit and vegetables may prove much more tractable commodities. For reasons explained above removal of ethylene may not be sufficient to prevent its deteriorative effects. Means of controlling ethylene levels in the internal gas space or the activity of ethylene at its acceptor site may be necessary. At present high carbon dioxide and low oxygen atmospheres are the only widely applicable methods of manipulating ethylene responses. The potential for chemical and genetic control is becoming apparent as understanding of the biochemistry of ethylene synthesis and action increases.

Much of the experimentation on ethylene control has been empirical; a commodity has been stored with an ethylene removal agent and the visible consequences recorded. From such information it is equally difficult to account for failures and to repeat successes. To progress biologists must define the production rate, internal concentration and sensitivity of the crop, in relation to stage of development and under the temperature and atmosphere conditions which the crop will experience post-harvest. With this information the engineer can apply the chemistry and physics necessary for ethylene control. It will be necessary to monitor the effectiveness of ethylene removal by regular analysis.

Finally, although this chapter has been mainly concerned with deleterious effects of ethylene and their control, the useful effects should be acknowledged. Research on tomato storage has led to the novel idea that there can be an optimum ethylene level for a commodity in storage (Geeson, Brown and Guaraldi, 1983). If tomatoes are kept too long with minimal ethylene they fail to ripen normally; if the level is too high they ripen prematurely. Perhaps 'management of ethylene levels' is a more appropriate concept than 'ethylene removal'.

Acknowledgements

Thanks are due to Stephen Hatfield for helpful criticism of concepts and their presentation.

References

ADAMS, D.O. and YANG, S.F. (1979). *Proceedings of the National Academy of Sciences*, **76**, 170–174
ADATO, I. and GAZIT, S. (1974). *Plant Physiology*, **53**, 899–902
AUTIO, W.R. and BRAMLAGE, W.J. (1982). *Journal of the American Society for Horticultural Science*, **107**, 1074–1077
BANGERTH, F. (1975). In *Facteurs et Regulation de la Maturation des Fruits*, pp. 183–187, CNRS, Paris

BANGERTH, F. (1978). *Journal of the American Society for Horticultural Science*, **103**, 401–404
BEYER, E. (1976). *HortScience*, **11**, 195–196
BIALE, J.B. (1960a). In *Encyclopedia of Plant Physiology*, XII (2), pp. 536–592. Ed. by W. Ruhland. Springer Verlag, Berlin
BIALE, J.B. (1960b). *Advances in Food Research*, **10**, 293–354
BIALE, J.B. and YOUNG, R.E. (1981). In *Recent Advances in the Biochemistry of Fruits and Vegetables*, pp. 1–39, Ed. by J. Friend and M.J.C. Rhodes. Academic Press, London
BIALE, J.B., YOUNG, R.E. and OLMSTEAD, A.J. (1954). *Plant Physiology*, **29**, 168–174
BURG, S.P. and BURG, E.A. (1961). *Plant Physiology*, **37**, 179–189
BURG, S.P. and BURG, E.A. (1965). *Physiologia Plantarum*, **18**, 870–874
BURG, S.P. and BURG, E.A. (1966). *Science*, **153**, 314–315
BURG, S.P. and BURG, E.A. (1967). *Plant Physiology*, **42**, 144–152
BURG, S.P. and THIMANN, K.V. (1959). *Proceedings of the National Academy of Sciences*, **45**, 335–344
CARLTON, B.C., PETERSON, L.E. and TOLBERT, N.E. (1961). *Plant Physiology*, **36**, 550–552
DE MUNK, W.I. (1972). *Netherlands Journal of Plant Pathology*, **78**, 168–178
DILLEY, D.R. (1982). *Michigan State Horticultural Society 110th Annual Report*, pp. 132–145
DOVER, C.J. (1983). *Proceedings of 16th International Congress of Refrigeration Commission C2*, pp. 165–170
DULL, G.C., YOUNG, R.E. and BIALE, J.B. (1967). *Physiologia Plantarum*, **20**, 1059–1065
EL-KAZZAZ, M.K., SOMMER, N.F. and FORTLAGE, R.J. (1983). *Phytopathology*, **73**, 282–285
ESGUERRA, E.B., MENDOZA, D.B. and PANTASTICO, E.B. (1978). *Philippine Journal of Science*, **107**, 23–31
FIDLER, J.C. (1949). *Journal of Horticultural Science*, **25**, 81–110
FIDLER, J.C. (1960). In *Encyclopedia of Plant Physiology*, XII (2), pp. 347–359. Ed. by W. Ruhland. Springer Verlag, Berlin
FIDLER, J.C. and NORTH, C.J. (1969). *Journal of the Science of Food and Agriculture*, **20**, 521–526
FIDLER, J.C. and NORTH, C.J. (1971). *Journal of Horticultural Science*, **46**, 23–243
FORSYTH, F.R., EAVES, C.A. and LIGHTFOOT, H.J. (1969). *Canadian Journal of Plant Science*, **49**, 567–572
FORSYTH, F.R. and HALL, I.V. (1969). *Naturaliste Canadien*, **96**, 257–259
FRENKEL, C. (1972). *Plant Physiology*, **49**, 757–763
GEESON, J.D., BROWNE, K.M. and GUARALDI, F. (1983). *Agricultural Research Council Food Research Institute Biennial Report*, 1981 and 1982, p. 81
GERHARDT, F. and SIEGELMAN, H.W. (1955). *Agricultural and Food Chemistry*, **3**, 428–433
GIBSON, K.D. (1953). *Biochimica et Biophysica Acta*, **10**, 211–229
GOESCHL, J.D. and KAYS, S.J. (1975). *Plant Physiology*, **55**, 670–677
HARKETT, P.J., HULME, A.C., RHODES, M.J.C. and WOOLTORTON, L.S.C. (1971). *Journal of Food Technology*, **6**, 39–45
HARRIS, S. (1981). *Orchardist of New Zealand*, **54**, 105
HATTON, T.T. and REEDER, W.F. (1972). *Journal of the American Society for Horticultural Science*, **97**, 339–341

HERNANDEZ, I. (1973). *Journal of Agriculture of University of Puerto Rico*, **57**, 100–106
HOFFMAN, N.E. and YANG, S.F. (1980). *Journal of the American Society for Horticultural Science*, **105**, 492–495
HUELIN, F.E. and COGGIOLA, I.M. (1970). *Journal of the Science of Food and Agriculture*, **21**, 44–48
JANES, H.W. and FRENKEL, C. (1978). *Journal of the American Society for Horticultural Science*, **103**, 394–397
KIDD, F. and WEST, C. (1930). *Proceedings of the Royal Society B*, **106**, 93–109
KIDD, F. and WEST, C. (1938). *Journal of Pomology and Horticultural Science*, **16**, 274–279
KNEE, M. (1975). In *Facteurs et Regulation de la Maturation des Fruits*, pp. 341–345. CNRS, Paris
KNEE, M. (1976). *Journal of the Science of Food and Agriculture*, **27**, 383–392
KNEE, M. (1980). *Annals of Applied Biology*, **96**, 243–253
KNEE, M. (1982). *Journal of Experimental Botany*, **33**, 1263–1269
KNEE, M. and HATFIELD, S.G.S. (1981). *Annals of Applied Biology*, **98**, 157–165
KNEE, M., HATFIELD, S.G.S. and RATNAYAKE, M. (1979). *Journal of Experimental Botany*, **30**, 1013–1020
KNEE, M., LOONEY, N.E., HATFIELD, S.G.S. and SMITH, S.M. (1983). *Journal of Experimental Botany*, **34**, 1207–1212
KNEE, M., SARGENT, J.A. and OSBORNE, D.J. (1977). *Journal of Experimental Botany*, **28**, 377–396
LAWRENCE, D.K. (1984). In *Biosynthesis and Metabolism of Plant Hormones*, Society for Experimental Biology, Seminar Series Volume. Ed. by A. Crozier and J.R. Hillman. Cambridge University Press, Cambridge (in press)
LIU, F.W. (1976). *Journal of the American Society for Horticultural Science*, **101**, 222–224
LIU, F.W. (1979). *Journal of the American Society for Horticultural Science*, **104**, 599–601
LOONEY, N.E. (1968). *Plant Physiology*, **43**, 1133–1137
LOONEY, N.E. (1972). *Journal of the American Society for Horticultural Science*, **97**, 81–83
LOONEY, N.E., McGLASSON, W.B. and COOMBE, B.G. (1974). *Australian Journal of Plant Physiology*, **1**, 77–86
LYONS, J.M. and PRATT, H.K. (1964). *Proceedings of the American Society for Horticultural Science*, **84**, 491–500
MAPSON, L.W. and ROBINSON, J.E. (1966). *Journal of Food Technology*, **1**, 215–225
MATTOO, A.K., BAKER, J.E., CHALUTZ, E. and LIEBERMAN, M. (1977). *Plant and Cell Physiology*, **18**, 715–719
McGLASSON, W.B. and PRATT, H.K. (1964). *Plant Physiology*, **39**, 120–127
McGLASSON, W.B., WADE, N.L. and ADATO, I. (1978). In *Phytohormones and Related Compounds—A Comprehensive Treatise*, Volume 2, pp. 447–493. Ed. by D.S. Letham, P.B. Goodwin and T.J.V. Higgins. Elsevier/North Holland Biomedical Press, Amsterdam
MORRIS, L.L., KADER, A.., KLAUSTERMEYER, J.A. and CHEYNEY, C.C. (1978). *California Agriculture*, **32**, 14–15
OUDIT, D.S. and SCOTT, K.J. (1973). *Tropical Agriculture*, **50**, 241–243
PEACOCK, B.C. (1972). *Queensland Journal of Agricultural and Animal Science*, **29**, 137–145

PENDERGRASS, A., ISENBERG, F.M.R., HOWELL, L.L. and CARROLL, J.E. (1979). *Canadian Journal of Plant Science*, **56**, 319–324
POENICKE, E.F., KAYS, S.J., SMITTLE, D.A. and WILLIAMSON, R.E. (1977). *Journal of the American Society for Horticultural Science*, **102**, 303–306
PRATT, H.K. and GOESCHL, J.D. (1968). In *Biochemistry and Physiology of Plant Growth Substances*, pp. 1295–1302. Ed. by F. Whiteman and G. Setterfield
PRATT, H.K. and REID, M.S. (1974). *Journal of the Science of Food and Agriculture*, **25**, 747–757
QUAZI, M.H. and FREEBAIRN, H.T. (1970). *Botanical Gazette*, **131**, 5–14
REID, M.S. (1975). In *Facteurs et Regulation de la Maturation des Fruits*, pp. 177–182. CNRS, Paris
REID, M.S. and PRATT, H.K. (1972). *Plant Physiology*, **49**, 252–255
REID, M.S., RHODES, M.J.C. and HULME, A.C. (1973). *Journal of the Science of Food and Agriculture*, **24**, 971–979
RISSE, L.A. and HATTON, T.T. (1982). *HortScience*, **17**, 946–948
ROMANI, R., PUSCHMANN, R., FINCH, J. and BEUTEL, J. (1982). *HortScience*, **17**, 214–215
SALTVEIT, M.E. and McFEETERS, R.F. (1980). *Plant Physiology*, **66**, 1019–1023
SCOTT, K.J. and GANDANEGARA, S. (1974). *Tropical Agriculture*, **51**, 23–26
SCOTT, K.J. and WILLS, R.B.H. (1974). *Australian Journal of Experimental Agricultural and Animal Husbandry*, **14**, 266–268
SCOTT, K.J., WILLS, R.B.H. and PATTERSON, B.D. (1971). *Journal of the Science of Food and Agriculture*, **22**, 496–497
SHARPLES, R.O. (1971). *Annual Report of East Malling Research Station for 1970*, p. 66
SMOCK, R.M. (1943). *Bulletin 799 Cornell University Agricultural Experiment Station*. Ithaca, New York
STOLL, K., HAUSER, F. and DATWYLER, D. (1974). In *Facteurs et Regulation de la Maturation des Fruits*, pp. 81–85. CNRS, Paris
TIGCHELAAR, E.C., McGLASSON, W.B. and BEUSCHER, R.W. (1978). *HortScience*, **13**, 508–512
TOIVONEN, P., WALSH, J., LOUGHEED, E.C. and MURR, D.P. (1982). In *Proceedings of the Third National Controlled Atmosphere Research Conference*, pp. 299–305. Ed. by D.G. Richardson and M. Meheriuk. Oregon State University School of Agriculture. Timber Press, Beaverton, Oregon
TREWAVAS, A.J. (1982). *Physiologia Plantarum*, **55**, 60–72
WALSH, C.S. and SOLOMOS, T. (1981). *HortScience*, **16**, 431
WANG, C.Y., MELLENTHIN, W.M. and HANSEN, E. (1972). *Journal of the American Society for Horticultural Science*, **97**, 9–12
WHEATON, R.A. and STEWART, I. (1973). *Journal of the American Society for Horticultural Science*, **98**, 337–340
WILD, B.L., McGLASSON, W.B. and LEE, T.H. (1976). *HortScience*, **11**, 114–115
WRIGHT, M. (1974). *Planta*, **120**, 63–69
YANG, S.F. (1981). In *Recent Advances in the Biochemistry of Fruits and Vegetables*, pp. 89–106. Ed. by J. Friend and M.J.C. Rhodes. Academic Press, London

25
ETHYLENE IN COMMERCIAL POST-HARVEST HANDLING OF TROPICAL FRUIT

F.J. PROCTOR and J.C. CAYGILL
Tropical Development and Research Institute, Gray's Inn Road, London, UK

Introduction

In the past decade there has been a significant growth in international trade in tropical and subtropical fruit, particularly in avocado, mango and pineapple. Such fruit satisfy a demand for variety or indeed luxury in affluent countries or regions where they cannot be grown. The trade is illustrated in *Table 25.1*, which shows the growth of imports of selected tropical and subtropical fruit into the European Community, with particular reference to the UK. The international trade in fruit such as banana and citrus has been well established for several decades and these fruit are imported in large volumes in contrast to the lesser known tropical and subtropical fruit.

For those fruit with an established market a considerable body of research data and of practical commercial experience enables these commodities to be handled, packaged, transported and distributed to ensure that good quality produce is delivered to the consumer at least cost. The purpose of this review is to consider the

Table 25.1 IMPORTS OF MAJOR TROPICAL AND SUBTROPICAL FRUIT INTO THE EUROPEAN COMMUNITY WITH SPECIFIC REFERENCE TO IMPORTS INTO UK (TONNES)

Commodity	1977		1982	
	Total European Community	UK	Total European Community	UK
Avocado	28 099	4222	57 001	9310
Banana	1 982 146	305 129	1 875 597	327 876
Citrus:				
Grapefruit	430 291	102 252	460 002	97 776
Lemon	317 226	38 826	355 831	50 073
Oranges	2 085 261	336 919	1 900 786	315 066
Mandarins, Clementines, Wilkings, others	660 008	59 916	777 467	110 813
Mango, Mangosteen, Guava	3473	1126	10 419	4292
Papaya	NA	NA	863	300
Pineapple	66 268	4813	100 025	15 809

Source: Analytical Tables of Foreign Trade NIMEXE.
NA: not available

limitations of knowledge in the context of commercial needs for both the established tropical and subtropical commodities and those which only relatively recently have entered international trade in any significant volume.

Ripening and the respiratory climacteric

Fruit may be classified on the basis of their respiratory rate between growth and senescence as climacteric or non-climacteric (Biale and Young, 1981) (*Table 25.2*). Mature climacteric fruit, after harvest, normally show a decline in respiration to a minimum rate followed by a rapid rise to a maximum rate, the respiratory 'climacteric' peak. If climacteric fruit of appropriate maturity are exposed for a

Table 25.2 SELECTED TROPICAL CLIMACTERIC AND NON-CLIMACTERIC FRUIT

Common name	Scientific name
Climacteric	
Avocado	*Persea americana*
Banana	*Musa sapientum*
Breadfruit	*Artocarpus altilis*
Cherimoya	*Annona cherimola*
Guava	*Psidium guajava*
Mango	*Mangifera indica*
Papaya	*Carica papaya*
Passion fruit	*Passiflora edulis*
Soursop	*Annona muricata*
Non-climacteric	
Cashew apple	*Anacardium occidentale*
Grapefruit	*Citrus paradisi*
Lemon	*Citrus lemonia*
Lychee	*Litchi chinensis*
Orange	*Citrus sinensis*
Pineapple	*Ananas comosus*
Tamarillo	*Cyphomandra betacea*

After Biale and Young (1981)

sufficient time to concentrations of ethylene higher than a threshold minimum, irreversible ripening is triggered. Non-climacteric fruit, in contrast, do not show any climacteric burst of respiration. Exposure to ethylene produces a small rise in respiratory rate proportional to the ambient concentration of ethylene, but fruit return to their resting respiratory rate when the ethylene is removed. A similar response is observed, in mature fruit, for other ripening parameters such as degreening.

While such laboratory studies may assist the plant physiologist to classify the post-harvest behaviour of the fruit, of far greater significance to the consumer are the other changes which occur during ripening. These include softening and changes in internal colour (e.g. the synthesis of carotenoids in mango pulp), external colour (e.g. chlorophyll breakdown unmasking carotenoids in banana peel, Marriott, 1980), in internal composition (e.g. breakdown of starch to sucrose and reducing sugars and a fall in acid content of mango, Bhatnagar and Subramanyam, 1973) and in particular the development of the characteristic

Table 25.3 RIPENING CHANGES OF COMMERCIAL SIGNIFICANCE

Chloroplast/chlorophyll breakdown
Starch hydrolysis
Loss of organic acids
Pectin solubilization/softening
Release of ethylene
Carotenoid/anthocyanin formation
Increased respiration
Synthesis of flavour compounds

flavour of the fruit. These changes, like the burst of respiration in climacteric fruit, occur after the exposure of mature unripe fruit to ethylene (Biale and Young, 1981). Indeed, release of ethylene by climacteric fruit is one of the characteristic aspects of their ripening. *Table 25.3* summarizes some of the more important changes, which are of commercial significance.

The relative importance of these changes during ripening in different fruit, and the relative rates at which they occur may differ between cultivars and with maturity, temperature and possibly other factors. One difficulty experienced in reviewing experimental studies for selected fruits and relating these to commercially feasible procedures is that not all factors have been or can be accurately controlled. In considering the effect of environmental factors, observers tend to rely on a few parameters such as visible changes (e.g. in peel colour) or easily measured chemical changes (e.g. increased reducing sugars or decreased acidity). Changes in firmness or texture, though of obvious interest to the consumer, are less well understood and accordingly more difficult to quantify while flavour changes which are of prime interest to the consumer but highly subjective, are often assumed to mirror the physicochemical changes. It is of considerable concern, therefore, to establish fully and for each cultivar or species, the effect on the rate of each ripening change of manipulating the temperature, or ethylene concentration before or during ripening.

The importance of fruit maturity is referred to throughout this review. One aspect of fruit maturity is the development of the capacity to ripen. In any particular species or cultivar the response to ethylene is influenced not only by the maturity of the fruit but also by the relative concentrations of other plant growth regulators, for example gibberellic acid, and by the relative concentrations of minerals. Calcium chloride treatment of avocados inhibited respiration, delayed the onset of the climacteric and depressed the peak of ethylene production (Tingwa and Young, 1984); such effects were not, however, observed in banana (Wills, Tirmazi and Scott, 1982). The literature on the effects of ethylene on fruit ripening contains much less information concerning other compounds which must also be involved in regulating metabolic processes, and this too requires more attention.

Finally, one factor which is generally of great significance in handling tropical or subtropical fruit is their susceptibility to chilling injury. Exposure to temperatures below a critical threshold value results in subsequent failure to ripen normally.

Prevention and control of ripening in banana

The consumer in many countries expects to receive a ripe, undamaged banana fruit (*Musa* AAA) and in the retail outlet usually selects fruit with an unblemished yellow skin. Trade practices to supply such fruit are based on the following factors. Firstly unripe fruit are much easier to transport without mechanical damage than ripe fruit. Secondly ripening can be delayed by reducing the temperature, which reduces the respiratory rate, the rate of water loss and in general the rate of microbial attack. However, since bananas suffer chilling injury most internationally traded cultivars cannot be stored below about 13 °C. Thirdly ripening is triggered by exposure to ethylene. Hence the system which has evolved is based on transporting unripe fruit at the lowest safe temperature, holding in a buffer store near retail terminal markets until required, initiating ripening with ethylene and distributing fruit so that they are just becoming ripe on the day of sale.

Bananas possess certain characteristics which differentiate them from most other fruit. Bananas are produced on a single shoot (other shoots having been removed) which emerges from the pseudostem formed from a closely packed leaf sheath, and the fruit develop parthenocarpically from the proximal flowers which are female. The fruit in one bunch are thus of known age, and, in cultivars under commercial cultivation, receive a similar supply of photosynthate (Simmonds, 1966). Avocados, mangoes and papaya, in contrast, are tree crops, producing fruit from flowers which open at different times in the season, and are so distributed about the branches that the supply of nutrients to the developing fruit may be unequal.

Most export bananas today belong to the rather narrow germplasm base of the Cavendish group of clones. They are classified as *Musa* AAA, being triploids with contributions from a few *Musa acuminata* genotypes. Cooking bananas and plantains are classified as *Musa* AAB, having contributions from *Musa balbisiana* genotypes (Simmonds, 1966). The Banana Breeding Research Scheme in the West Indies has developed tetraploid (*Musa* AAAA) clones. These tetraploid clones are resistant to Panama disease and Sigatoka disease, the former disease almost eliminated the Gros Michel (*Musa* AAA) cultivars exported prior to 1939. There are significant differences between clones, some of which are noted below. Bananas are harvested by cutting the bunch either when the fingers have filled to a specified girth or 'grade' measured with calipers, or when they have reached a specified age. If bananas are allowed to grow to full maturity their subsequent preclimacteric period, or 'green life' before spontaneous ripening occurs, is reduced. Neither method is totally precise (Marriott, 1980), though green life is more closely correlated with physiological age than with grade at harvest (Montoya *et al.*, 1984). Very highly significant differences in preclimacteric period were observed between clones when 450 bunches from 29 clones were studied. For tetraploid fruit this varied between 8 and 39 days and appeared to be related in part to the diploid parents (Marriott *et al.*, 1979).

To allow an adequate time for shipping and to provide some buffer stocks to match supplies to market demand, a preclimacteric life of approximately 20 days at 13.5–14 °C is required for the transatlantic trade (New and Marriott, 1974). Fruit with insufficient preclimacteric life may ripen prematurely and produce ethylene, thus triggering the ripening of neighbouring fruit. After harvest hands are removed, washed, treated with fungicide, packed in boxes with polyethylene liners and shipped at 13.5–14 °C to ripening rooms where they can be held until ripening is initiated. For a recent review of handling practices *see* Marriott and Lancaster (1983).

Ripening of banana fruit involves a number of changes, which must be synchronized to produce fruit suited as ideally as practicable to the consumers' tastes. These changes include:

(1) the degreening of the skin, which is of great importance since the consumer judges fruit by appearance;
(2) the development of a characteristic banana flavour, which taste panels report to be the prime factor in organoleptic acceptability of banana fruit from different clones (Baldry, Coursey and Howard, 1981);
(3) softening; and
(4) conversion of starch to sugar.

Ripening can be triggered by exposure to ethylene or certain similar gases such as acetylene (Thompson and Seymour, 1982) or propylene. The ripening of bananas has been initiated by exposure to hydrocarbons from a range of sources, such as burning leaves in India, coal in Egypt, coal gas in Japan and storage with granadillas in South Africa. Exposure to acetylene released from wetted calcium carbide has been used for many years, though it is often reported that the resultant ripe fruit is less satisfactory to the consumer than if it had been ripened by ethylene. It should be noted, however, that it is often used at tropical ambient temperature, and some of the criticisms may arise from the effects of high ambient temperatures on the various processes occurring during ripening. Above a critical temperature, for instance 30 °C, the fruit may soften and the starch hydrolyse to sugar, but the peel does not degreen, thereby producing a 'green ripe' fruit. Ripening of almost all the internationally traded bananas, today, is initiated by exposure to ethylene.

Freshly harvested mature bananas are less sensitive than stored fruit to ethylene (Liu, 1976; Peacock, 1972), possibly because another plant growth substance reduces either the sensitivity or the response to ethylene. Marriott (1980) discusses the evidence for the influence of gibberellic acids, auxins and kinetin. Considerable clonal differences in preclimacteric period and in response to ethylene have been noted. For example, the preclimacteric period of two tetraploid clones (crosses between Highgate, a dwarf mutant of Gros Michel and synthetic diploids) was 30–40% less than for Valery fruit (Cavendish type) at an equivalent stage of fruit development. Furthermore, the pulp firmness of preclimacteric tetraploid fruit was 20–30% less than that of Valery fruit and these differences persisted during ripening. The softening response to applied ethylene was 15 h earlier in fruit of tetraploid clones than in Valery, but respiratory patterns, colour development and starch to sugar conversion were similar (New and Marriott, 1974). Unlike Valery fruit, ripe tetraploid fruit did not develop senescent spotting, and shelf-life was terminated by rapid deterioration of peel strength and severe 'finger drop'.

Once triggered, the fruit ripen, even if the exogenous ethylene source is removed. This is due to the ability of ripening bananas to produce endogenous ethylene. For example, triploid fruit (clone Valery) showed a peak rate of ethylene production of $0.11\,\mu l\,kg^{-1}\,h^{-1}$ at 14 °C and $2.1\,\mu l\,kg^{-1}\,h^{-1}$ at 25 °C, whereas one tetraploid clone, which had a shorter green life, produced ethylene at $0.38\,\mu l\,kg^{-1}\,h^{-1}$ at 14 °C and $6.73\,\mu l\,kg^{-1}\,h^{-1}$ at 25 °C (New and Marriott, in preparation). The quantity of ethylene required to trigger ripening depends on the cultivar, the physiological age at harvest, the time from harvest, the temperature and the duration of the exposure.

Commercial ripening schedules have been published (Anon., 1964) and generally recommend raising the pulp temperature to 18 °C, whilst maintaining high relative humidity. The rooms are then sealed and ethylene gas mixed with nitrogen metered to give an ethylene concentration of $1\,ml\,l^{-1}$. After 24 h the room is ventilated, and the fruit are cooled to pulp temperatures of 15–17 °C, with a

Table 25.4 POST-HARVEST CONSIDERATIONS RELEVANT TO CLIMACTERIC TROPICAL FRUIT

Harvesting	Select maturity for particular cultivar
	Select undamaged fruit
	Avoid damage
Transit	Identify lowest safe temperature for cultivar
	Minimize ethylene concentration
	Minimize moisture loss
Ripening	Select temperature giving good appearance/flavour
	Expose to sufficient ethylene for adequate time to trigger
Distribution	Distribute triggered or ripened fruit
	Avoid damage

relative humidity of around 80%. Ripening is slower, the lower the temperature, which gives the storage operator some control over the date on which the fruit will reach a predetermined colour stage; fruit of more advanced colour being distributed in cooler weather.

Considerable skill in accurate temperature control is required, and the management of fruit ripening still relies heavily on the ripener's experience of variation due to differing fruit origins, finger size and seasonal factors. *Table 25.4* summarizes the main post-harvest considerations for bananas which are also relevant to other climacteric tropical fruit.

Improvement of visual appeal in citrus fruits

Consumers show a preference for citrus fruits showing an evenly uniform, bright colour. The most attractive fruit is grown in regions of the world with a dry climate and where the night temperatures are below 13 °C during fruit maturation. For Valencia oranges studies have shown that the best coloured fruit are produced under 20 °C daytime air temperature, 7 °C night-time air and 12 °C soil temperatures (Young and Erickson, 1961). Fruit grown under humid climatic conditions may not have a uniform colour particularly early in the season; such fruit reaches eating maturity without the development of full peel colour. Consumers in developed market economies associate immaturity with this green coloration.

The peel colour of citrus fruits is dependent upon the chlorophyll and various carotenoid pigments present within the flavedo. The chlorophyll can partly or entirely obscure the carotenoid pigments; where chlorophyll is not present the carotenoids determine the external colour. Ethylene can act in two ways, firstly through its effect on the biosynthesis of the carotenoid, β-citraurin (Stewart and Wheaton, 1972) and secondly, through causing the degradation of chlorophyll. The formation of β-citraurin is sensitive to temperature, and is inhibited above 30 °C. At decreasing temperatures of 25, 20, 15 °C a decreasing amount of ethylene is required to produce maximum colour. However, to achieve maximum colour an exposure period of up to 12 days is required. This method of enhancing the colour of citrus fruits has, therefore, serious disadvantages for the future storage and marketing of the fruit.

Ethylene, however, also promotes chlorophyll breakdown, and this action of ethylene is exploited commercially to improve fruit colour. Chlorophyll degradation depends on the fruit, the ethylene concentration, temperature and time (Barmore, 1975a). Recommendations for commercial post-harvest citrus degreening, which vary between the major citrus producing countries, have been developed. In Florida the recommended conditions are 27.8–29.5 °C, 90–96% relative humidity, 1–5 µl l^{-1} ethylene, and one air change per hour (McCormack and Wardowski, 1977), to achieve a degreening time of 72 h for early season fruit; this reduces in time as the season progresses. In California, the recommendations are ethylene 5–10 µl l^{-1}, temperature 20–25 °C, relative humidity 90%, air circulation one room volume per minute and ventilation 1–2 air changes per hour or sufficient to maintain 1 ml l^{-1} carbon dioxide.

Two citrus degreening methods have been developed. One is the 'shot' method in which ethylene is introduced at intervals into a gas-tight room, with controlled air ventilation to avoid carbon dioxide build up. Ethylene is introduced every 6–8 hours over a period of 24–72 hours. The other method, known as the 'trickle' method allows a continuous addition of ethylene into a room with regular air changes. It is this latter method which is now most commonly commercially practised.

The physiological basis for continuous ethylene treatment is identified in work by Barmore (1975a) who showed that 8–12 hours of ethylene treatment was required before increased chlorophyllase activity was observed. Differences in the rate and extent of chlorophyll degradation were found between citrus cultivars when tested under standard conditions of a 15 h ethylene treatment followed by three or four days at normal temperature (Kitagawa, Kawada and Tarutani, 1978). The resistance of gas diffusion into and out of the fruit appeared to be one of the factors influencing this.

While it has been reported that the process of degreening is reduced at 1 ml l^{-1} carbon dioxide and inhibited at 10 ml l^{-1} carbon dioxide (Grierson and Newhall, 1960), studies in Israel indicate that adequate ventilation in the degreening chamber is required to maintain oxygen level rather than remove carbon dioxide and that for lemons, levels of 50 ml l^{-1} carbon dioxide did not cause an inhibition in the development of peel colour (Cohen, 1973; 1977).

The most desirable way to improve citrus colour would be for the process to take place on the tree prior to harvest or during storage or transport after post-harvest treatment. Trials of pre-harvest treatments, namely spray programmes using ethephon (Young, Jahn and Smoot, 1974) have not been encouraging in terms of commercial application.

Trials of post-harvest treatment with ethephon have been reported. For lemons a delay in the degreening of ethephon treated fruit was reported following waxing (Fuchs and Cohen, 1969). However, for orange and grapefruit, degreening was not inhibited by washing and waxing prior to ethephon treatment (Grierson, Ismail and Obenbacher, 1972). This method of degreening, however, has not been commercially adopted.

Degreening of citrus with ethylene after harvest has several commercial disadvantages. For instance, there is a need for specialist degreening rooms. There may be a delay of several days before full colour is attained, and post-harvest waxing of citrus fruit can take place. There is a requirement for more complex regimens of application of post-harvest fungicide treatments due to delays between harvest and full post-harvest treatment. Finally there is a potential increase in development and/or sensitivity to peel injuries and post-harvest disease.

Adoption of technology for other tropical fruits

AVOCADO

The main cultivars entering international trade are Fuerte, Hass and Ettinger, although the tropical hybrids such as Lula, Booth are increasing in importance.

Minimum maturity specifications established by exporting countries have been defined in order to ensure that early season fruit ripens normally and reaches an acceptable eating quality. These specifications differ by cultivar and by region of production and have been reviewed (Lewis, 1978). Identification of horticultural maturity is difficult in the case of avocado as maturation is not accompanied by significant changes in external appearance. Recent studies in California (Lee et al., 1983) indicate that further definition of the existing maturity criteria are justified on the basis of consumer response to ripened fruit originating from different areas within the State.

Fruit imported into the UK are, on occasion, immature. The reasons for this are difficulties in implementing maturity assessment at farm and packhouse level, the lack of suitable criteria for assessment of maturity, and inadequate enforcement of the existing specifications. In some major producer regions, dependent on long-distance sea transportation to enter international trade, there also exists a need to determine the end of season high risk picking and shipping period to ensure adequate storage life after harvest (Durand, 1981).

Mature avocado fruit ripen in 6–12 days at 20 °C, the ripening time being dependent on physiological maturity. While the climacteric patterns of avocado fruit harvested at various stages of development are similar, the preclimacteric period decreases as the fruit matures (Zauberman and Schiffman-Nadel, 1972). The climacteric peak rate and the peak rate of ethylene production also increase as the fruit matures (Eaks, 1980). The treatment of mature avocado fruit with exogenous ethylene after harvest hastens fruit ripening (Eaks, 1966). The effect of the time lapse between harvest and the start of the ethylene treatment should be taken into consideration as some workers indicate that ethylene applied immediately after harvest does not accelerate ripening (Gazit and Blumenfeld, 1970). Hastening of fruit ripening was more marked following ethylene treatment which began 48 h after harvest than immediately following harvest (Adato and Gazit, 1974). From a commercial operator's viewpoint the adverse effects of low levels of ethylene during low temperature storage on fruit ripening, and post-climacteric fruit quality should not be overlooked (Zauberman and Fuchs, 1973). The deleterious effects of insufficient oxygen and excessive carbon dioxide on the post-harvest life of avocados have been the subject of some study (Spalding and Marousky, 1981). The carbon dioxide level should probably not exceed 20 ml l^{-1} at the end of the treatment; however, precise data are not available.

The optimum ripening conditions for each major variety have to some extent been defined. Rousseau (1981) in South Africa recommended a 24 h treatment with ethylene at 1 ml l^{-1}, at 16 °C, followed by two to four days at 16–18 °C for Fuerte and 18–20 °C for Hass and Ettinger. Optimal ripening temperatures for Florida avocados were reported as 15–16 °C (Hatton, Reeder and Campbell, 1965a).

The trend for supply of quality goods to the consumer and increasing consumer awareness has encouraged the relatively recent adoption of supply of 'triggered' and 'ready to eat' avocado fruit by a limited number of retail groups and hotel and restaurant chains. This involves approximately 10% of the total fruit marketed in the UK.

The most commonly adopted practice used almost exclusively for varieties Hass and Fuerte is to bring the product temperature from the storage temperature of 4–5 °C to the ripening temperature of 20 °C in 8–10 h, treat with a single introduction of ethylene at $0.8–1$ ml l^{-1}, and seal the room for approximately 12 h, before ventilation. 'Stubborn' fruit are subjected to a further ethylene treatment depending on requirements. 'Triggered' fruit are removed from the ripening room soon after the 12-h treatment and 'ready to eat' fruit, with an expected retail shelf-life of two to three days are removed after two to four days at 20 °C. The time in the ripening room is therefore influenced by the response of the fruit and the stage required for removal. For 'ready to eat' fruit consideration is also given to ease of handling. Generally, Hass is removed in a more advanced stage of ripening than Fuerte, the latter being sensitive to handling damage when ripe is therefore usually removed at the 'just give' stage. The fruit is rapidly air-cooled to bring the product temperature down to the storage and distribution temperature of 0–1 °C.

Facilities used by UK commercial fruit ripeners are converted cold rooms or rooms originally designed for banana ripening. With the relatively small volume of fruit involved, avocado is often ripened in the same rooms and at the same time as other fruit requiring ripening, e.g. pears, plums, peaches. Adjustments to the ethylene level or temperature depending on specific batch or cultivar needs are not facilitated by these circumstances.

A number of technical problems face the commercial fruit ripener. The main problem being the lack of ability to predict the ripening time for a given batch of fruit combined with the variability in response to ripening of fruit within a single batch. The problems of delivery of immature fruit present obvious difficulties. The fruit distributor has largely overcome the lack of predictability through operating a stock holding system of fruit prior to and after ripening. This is greatly facilitated by the reduced sensitivity of post-climacteric avocados to chilling injury; the cultivars Fuerte and Hass can be held for a number of weeks post-climacteric at 1–2 °C (Kosiyachinda and Young, 1976). This, however, requires that additional storage space be made available.

The variability of response to ripening reflects the inherent product characteristics, but this may also be influenced by the post-harvest history of the product. For example, harvest to UK arrival times may be three to four days from USA, seven days from Israel and up to 21 days from South Africa. Avocado fruit are also highly susceptible to a range of physiological disorders and diseases which, influenced by both pre- and post-harvest environment may only become apparent when the fruit is ripe. The industry may therefore not wish to ripen high risk fruit or a fruit with an unknown history, preferring perhaps to pass the potential problem onto the consumer.

Changes in post-harvest practices at the producer/packhouse level have also influenced fruit ripening. The adoption of the practice of waxing, for the cultivar Fuerte led in the early years to greater variation in response of fruit to ethylene treatment. However, the advantages of extended shelf-life combined with improved wax application techniques have largely overcome this early difficulty. Difficulties now arise when mixed batches of waxed and unwaxed fruit are received.

Package design, internal fitments and palletization influence commercial fruit ripening. In particular they may affect the efficient warming up of fruit prior to ripening and the subsequent need for rapid cooling to the temperature required for retail distribution.

The growth of the scale of operation of supplying ripened avocado fruit to the consumer is limited, not only by the technical constraints listed above, but also by the following commercial considerations:

(1) The general lack of a 'closed circuit' marketing system involving the close commitment and involvement of the producer, exporter, importer, distributor/fruit ripening room operator and retail outlet/chain.
(2) The increase in product cost incurred by fruit ripening.
(3) The increased sensitivity of ripe avocado to mechanical damage and hence potential increase in wastage at retail level.
(4) Greater risk incurred to the retailer/retail outlet in marketing produce with limited shelf-life.

MANGO

The consumer requires a mango showing a bright fully developed skin colour, a uniformly softened flesh and a fruit with full flavour development. Mango fruit ripened on the tree develop such characteristics. However, such fruit, whether fully or partially ripe at the time of harvest, display an unacceptably short market and shelf-life. Mango fruit are therefore usually picked and shipped to markets in the mature firm condition.

Mango is a short season crop from any single source of supply. It is therefore of critical importance to suppliers that quality produce of optimum ripeness is placed at the retail level in order to market these significant volumes effectively over a short time period. Within the importing countries of the European Community (EC), fruit, usually transported by air, arrives in conditions ranging from immature to mature, and unripe, fully ripe or over-ripe. Retail buyers respond to this by inspection and purchase on a shipment by shipment basis, generally seeking out mature fruit just showing signs of ripening. Such an *ad hoc* approach is not satisfactory for the future expansion of this industry, as no effective control of supply to retail and consumer levels can be achieved. The import of mature firm fruit, achieved through attention to harvest maturity and control of transport conditions, followed by controlled ripening at distribution level, offers one theoretical alternative.

Recommended optimum post-harvest ripening temperatures vary between cultivars from different regions. Work in India (Thomas, 1975) has concluded that storage temperatures below 25 °C adversely affect the development of typical aroma, flavour and carotenoid formation of Alphonso mango during ripening. Ethylene was not used in these trials to stimulate ripening. Trials conducted by Lakshminarayana, Subbiah Shetty and Krishnaprasad (1975) using a post-harvest treatment with 2-chloroethylphosphoric acid (ethephon) at $500 \,\mu l \, l^{-1}$ and $1000 \,\mu l \, l^{-1}$ in hot (54 °C ± 1 °C for 5 min) or cold (24–28 °C for 5 min) water with subsequent holding conditions at 24–28 °C indicated that accelerated fruit ripening and improved surface colour was achieved using the ethephon in hot water compared with untreated control and cold water treatments. Ripening treatments for mature Florida mangoes are recommended (Hatton, Reeder and Campbell, 1965b) as

21–24 °C, however at 15.5–19 °C the brightest and most attractive skin colour developed. Fruit ripened at 15.5 and 19 °C was usually tart and required a further two to three days at 21 or 24 °C to attain a fuller flavour. At 26.7 °C some Florida varieties developed a mottled skin, although others, e.g. Kent and Keitt, did not. In general, the average time to softening decreased with increasing ripening temperature within the range of 15.5–26.7 °C and ranged from 4 to 20 days depending on variety.

Subsequent work on Florida mangoes led to recommendations to use $5-10\,\mu l\,l^{-1}$ ethylene for 24–48 h at 30 °C with a high relative humidity to achieve ripening. Lower temperatures can be used, but the rate of ripening is reduced (Barmore, 1974), for instance seven days to full ripening compared with four days using the cultivar Tommy Atkins. While uniformity of ripening was achieved colour development was limited to development of yellow colour associated with chlorophyll degradation. The development of red colour in the peel was not affected.

Successful trials have been conducted in Florida (Barmore, 1975b) to trigger ripening of mangoes prior to shipment to both US and external markets. The recommended conditions for Florida cultivars are treatment of mature firm fruit with $10-20\,\mu l\,l^{-1}$ ethylene at 21 °C for 12–24 h under high relative humidity of 92–95%. The duration of treatment varies with the time of the season, earlier season fruit requiring a longer time period. A temperature of 21 °C was selected to achieve a balance between the rate of colour development and ripening. Colour development continued following ethylene exposure when the fruit were subsequently shipped at 15.6 °C. Additional benefits from the use of ethylene before shipping were a reduction of the incidence of internal breakdown in fruit, in particular with the cultivar Tommy Atkins and a reduction in the rate of development of anthracnose spotting. Mango fruit are commercially ripened in Israel to enable picking to start early in the season, to regulate the flow of fruit onto the market and to improve the uniformity of fruit colour (Fuchs et al., 1975). Recommended conditions are $100\,\mu l\,l^{-1}$ ethylene for 48 h at 25 °C and 90% relative humidity.

Information on changes which occur during mango fruit development is however, inadequate. There are a lack of criteria for identification of different stages of maturity, and information on biochemical changes occurring during fruit ripening with related developments of flavour and skin colour is limited. Guidelines for optimum post-harvest conditions for fruit ripening of the major varieties through their production season do not therefore exist. From a commercial viewpoint, factors limiting the marketing and distribution of ripened fruit are similar to those identified for avocados. Very small quantities of mango are commercially ripened in the UK. Problems of mottled and irregular ripening have been reported subsequent to ethylene treatment.

PAPAYA

While commercial trade in papaya between Hawaii and the mainland USA is significant, imports into the EC markets remain small at below 1000 tonnes in 1982 (*Table 25.1*). Relatively sophisticated post-harvest procedures have been developed for the US industry; current practice involves a combined hot water and fumigation treatment for the control of fruit flies and post-harvest decay (Akamine, 1976). However, one of the major problems facing papaya fruit marketing still remains the identification of optimum harvest maturity to ensure adequate fruit ripening to good eating quality. Most research work has been conducted on the Hawaiian solo type papaya. Hawaii specifies a minimum total soluble solids content of 11.5%, identified commercially as fruit showing at least 6% surface coloration at the blossom end region (Akamine and Goo, 1971). Changes in the carbohydrate content and composition during fruit development have been examined (Chan and Tang, 1979) and could be used to establish a biochemical index of maturity. The study did, however, identify a ten-day difference between summer and winter grown fruit in Hawaii in achieving the increase in total sugars and sucrose.

Akamine and Goo (1977) indicated a relationship between ethylene and the triggering of the onset of the climacteric although few studies have been conducted to examine this relationship. Rodriguez, Guadalupe and Iguina de George (1974) successfully ripened local Puerto Rican cultivars at 25 °C, 85–95% relative humidity and $1 \, ml \, l^{-1}$ ethylene. The fruit ripened in six to seven days. This work also emphasized the importance of determination of the optimum harvest maturity to ensure good eating quality.

A significant proportion of fruit imported into the EC markets is harvested immature; such fruit does not therefore subsequently ripen or ripens to a poor eating quality and is often highly susceptible to development of post-harvest diseases. A small number of fruit distributors treat papaya with ethylene under the conditions given earlier for avocados; however, the problems of immature fruit and susceptibility to disease render this a risky operation to both the fruit store operator and the retail distributor. Until solutions to these problems and to problems associated with retail management of delicate ripe fruit have been found, the market opportunities in the EC for this fruit will remain somewhat restricted.

PINEAPPLES

On the basis of examination of respiratory pattern of pineapple from termination of flowering to senescence of the fruit, the pineapple has been classified as non-climacteric (Dull, Young and Biale, 1967). Fruit treated with up to $1 \, ml \, l^{-1}$ ethylene at the start of ripening showed no change in respiration or chemical composition which could be interpreted as affecting the ripening processes.

Four distinct stages of development of the pineapple fruit have been defined on a chemical basis (Gortner, Dull and Krauss, 1967). Ripening of the pineapple could be considered as the terminal period of maturation during which the fruit attains the most desirable quality; one aspect of this concerns fruit shell colour which generally develops with an unmasking of the carotenoid pigments through a decline in chlorophyll content.

Pineapples are treated with ethylene releasing compounds for the purposes of flower induction and fruit coloration. The application of ethephon just prior to harvest leads to faster and more even orange colour development in the fruits. From trials conducted in the Ivory Coast it was shown that the best results were obtained when the application coincided with the theoretical date of harvest, if applied too early an adverse effect on taste was reported (Crochon *et al.*, 1981).

Conclusions

Bananas and citrus are two fruits which have achieved large volumes in international trade. This has resulted in significant research and development work in the post-harvest field of which much has been applied to commercial practice. This has been facilitated in the case of banana by the integrated structure of the international banana industry. The extent to which even this industry will, however, respond to changes within the industry, e.g. the possible introduction of different cultivars of bananas, which require significantly different post-harvest management to existing commercial cultivars, remains to be seen. Similarly, the full extent of adoption of research knowledge has yet to be achieved, e.g. achievement of maximum yield, optimum use of infrastructure such as ripening room facilities.

For all other tropical and subtropical fruits an integrated approach to commercial fruit marketing, embracing both technical and commercial components, has yet to emerge. In general, research work on these fruits has concentrated on a narrow approach or the investigation of a physiological principle rather than the examination and identification of a strategy for commercial fruit post-harvest handling. This results in the fruit industry being confronted with fragmentary information which cannot always be put to good commercial use.

To achieve improved tropical fruit management two aspects must be examined concurrently. Firstly, the commercial structure of international trade from producer through to importer/distributor requires full understanding and appropriate arrangements made to enable the implementation of improved technologies, and secondly the need exists to identify areas of deficiency in research and development work in this field.

Control of exposure of fruit to ethylene is one major factor influencing successful fruit marketing. It is suggested that the current information concerning cultivars of climacteric tropical and subtropical fruits showing commercial potential is generally inadequate, and attention should be given to the following:

(1) Criteria, particularly criteria suitable for field use at harvest, of maturity and the effect of maturity on ethylene production and sensitivity.
(2) Systematic studies of the rate of ethylene production at various temperatures and the corresponding studies of their sensitivity to ethylene.
(3) Examination of the effects of storage environment including modified atmosphere, and time on ethylene production and subsequent fruit ripening.
(4) Examination of the effects of other compounds, e.g. growth regulators or mineral salts on ethylene production and fruit ripening.
(5) Studies to define the optimum condition for the handling of ripened fruit.

For non-climacteric fruit, a better fundamental understanding of maturation and ripening is required. Biale and Young (1981) concluded their review on respiration and ripening in fruits by remarking on the number of investigations on climacteric as opposed to non-climacteric and on the response to ethylene as a means of distinguishing the two types. They suggested that at present it was not clear what investigations of non-climacteric fruit were most likely to be productive in terms of better control of fruit quality.

If the industry were to adopt commercial ripening of climacteric subtropical and tropical fruits it would require commitment and close management at all stages of the industry, together with capital investment. Current progress within the EC is in response to the growing consumer and trade realization that through sound produce management and the adoption of optimum post-harvest handling procedures a supply of consistent quality fruit can be offered at the retail level. Further progress depends on an expansion of applied research in which research workers and industry in producing and consuming countries collaborate.

References

ADATO, I. and GAZIT, S. (1974). *Plant Physiology*, **53** (6), 899–902
AKAMINE, E.K. (1976). *Acta Horticulturae*, **57**, 151–161
AKAMINE, E.K. and GOO, T. (1971). *HortScience*, **6** (6), 567–568
AKAMINE, E.K. and GOO, T. (1977). *Hawaii Agricultural Experiment Station*, Technical Bulletin No. 93, 12 pp
ANON. (1964). *Banana Ripening Manual*, United Fruit Sales Corporation, Boston
BALDRY, J., COURSEY, D.G. and HOWARD, G.E. (1981). *Tropical Science*, **23**, 33–66
BARMORE, C.R. (1974). *Proceedings of the Florida State Horticultural Society*, **87**, 331–334
BARMORE, C.R. (1975a). *HortScience*, **10**, 595–596
BARMORE, C.R. (1975b). *Proceedings of the Florida State Horticultural Society*, **88**, 469–471
BHATNAGAR, H.C. and SUBRAMANYAM, H. (1973). *Indian Food Packer*, **27** (4), 33–52
BIALE, J.B. and YOUNG, R.E. (1981). In *Recent Advances in the Biochemistry of Fruit and Vegetables*, pp. 1–39. Ed. by J. Friend and M.J.C. Rhodes. Academic Press, London
CHAN, H.T. and TANG, C.S. (1979). In *Tropical Foods: Chemistry and Nutrition*, Vol. 1, 33–53. Ed. by G.E. Inglett and G. Charalambous. Academic Press, New York, London
COHEN, E. (1973). In *First International Citrus Congress*, **3**, 297–301
COHEN, E. (1977). *Proceedings of the International Society of Citriculture*, **1**, 215–219
CROCHON, M., TISSEAU, R., TEISSON, C. and HUET, R. (1981). *Fruits*, **36** (7–8), 409–415
DULL, G.G., YOUNG, R.E. and BIALE, J.B. (1967). *Physiologia Plantarum*, **20**, 1059–1065
DURAND, B.J. (1981). *South African Avocado Growers' Association Yearbook*, **4**, 39–41
EAKS, I.L. (1966). *California Avocado Society Yearbook*, 128–133
EAKS, I.L. (1980). *Journal of the American Society for Horticultural Science*, **105** (5), 744–747
FUCHS, Y. and COHEN, A. (1969). *Journal of the American Society for Horticultural Science*, **94**, 617–618
FUCHS, Y., ZAUBERMAN, G., YANKO, U. and HOMSKY, S. (1975). *Tropical Science*, **17** (4), 211–216
GAZIT, S. and BLUMENFELD, A. (1979). *Journal of the American Society for Horticultural Science*, **95** (2), 229–231
GORTNER, W.A., DULL, G.G. and KRAUSS, G.G. (1967). *Proceedings of the American Society for Horticultural Science*, **2**, 141–144
GRIERSON, W. and NEWHALL, W.F. (1960). *University of Florida, Agricultural Experimental Station Bulletin*, No. 620, 80pp
GRIERSON, W., ISMAIL, F.H. and OBENBACHER, M.F. (1972). *Journal of the American Society for Horticultural Science*, **97**, 541–544
HATTON, T.T., REEDER, W.F. and CAMPBELL, C.W. (1965a). Ripening and storage of Florida avocados. *United States Department of Agriculture, Agricultural Research Service Marketing Research Report*, No. 697, 13 pp
HATTON, T.T., REEDER, W.F. and CAMPBELL, C.W. (1965b). Ripening and storage of

Florida mangoes. *United States Department of Agriculture, Agricultural Research Service, Marketing Research Report*, No. 725, 9 pp

KITAGAWA, H., KAWADA, K. and TARUTANI, T. (1978). *Journal of the American Society for Horticultural Science*, **103** (1), 113–115

KOSIYACHINDA, S. and YOUNG, R.E. (1976). *Journal of the American Society for Horticultural Science*, **101** (6), 665–667

LAKSHMINARAYANA, S., SUBBIAH SHETTY, M. and KRISHNAPRASAD, C.A. (1975). *Tropical Science*, **17** (2), 95–101

LEE, S.K., YOUNG, R.E., SCHIFFMAN, P.M. and COGGINS, C.W. (1983). *Journal of the American Society for Horticultural Science*, **108**, (3), 390–394

LEWIS, C.E. (1978). *Journal of the Science of Food and Agriculture*, **29**, 857–866

LIU, F.W. (1976). *Journal of the American Society for Horticultural Science*, **101**, 222–224

McCORMACK, A.A. and WARDOWSKI, W.F. (1977). *Proceedings of the International Society of Citriculture*, **1**, 211–215

MARRIOTT, J. (1980). *CRC Critical Review in Food Science and Nutrition*, **13**(1), 41–88

MARRIOTT, J., NEW, S., DIXON, E.A. and MARTIN, K.J. (1979). *Annals of Applied Biology*, **93**, 91–100

MARRIOTT, J. and LANCASTER, P.A. (1983). In *Handbook of Tropical Foods*, pp. 85–143. Ed. by H.T. Chan, Jnr. Marcel Dekker Inc., New York and Basel

MONTOYA, J., MARRIOTT, J., QUIMI, V.H. and CAYGILL, J.C. (1984). *Fruits*, **39**, (5), 293–296

NEW, S. and MARRIOTT, J. (1974). *Annals of Applied Biology*, **78**, 193–204

PEACOCK, B.C. (1972). *Queensland Journal of Agriculture and Animal Science*, **29**, 137–145

STATISTICAL OFFICE OF THE EUROPEAN COMMUNITIES (1977). *Analytical Tables of Foreign Trade, NIMEXE*. Office for Official Publications of the European Communities, Luxemburg

STATISTICAL OFFICE OF THE EUROPEAN COMMUNITIES (1982). *Analytical Tables of Foreign Trade, NIMEXE*. Office for Official Publications of the European Communities, Luxemburg

RODRIGUEZ, A.J., GUADALUPE, R. and IGUINA, L.M. DE GEORGE (1974). *Journal of Agriculture of the University of Puerto Rico*, **58** (2), 184–196

ROUSSEAU, G.G. (1981). *South African Avocado Growers' Association Yearbook*, **4**, 36–37

SIMMONDS, N.W. (1966). *Bananas*, 2nd ed. Longman, London and New York

SPALDING, D.H. and MAROUSKY, F.J. (1981). *Proceedings of the Florida State Horticultural Society*, **94**, 299–301

STEWART, I. and WHEATON, T.A. (1972). *Journal of Agriculture and Food Chemistry*, **20**, 448–449

THOMAS, P. (1975). *Journal of Food Science*, **40** (4), 704–706

THOMPSON, A.K. and SEYMOUR, G.B. (1982). *Annals of Applied Biology*, **101**, 407–410

TINGWA, P.O. and YOUNG, R.E. (1974). *Journal of the American Society for Horticultural Science*, **99** (6), 540–542

WILLS, R.B.H., TIRMAZI, S.I.H. and SCOTT, K.J. (1982). *Journal of Horticultural Science*, **57**, 431–435

YOUNG, L.B. and ERICKSON, L.C. (1961). *Proceedings of the American Society for Horticultural Science*, **78**, 197–200

YOUNG, R.H., JAHN, O.L. and SMOOT, J.J. (1974). *Proceedings of the Florida State Horticultural Society*, **87**, 24–28

ZAUBERMAN, G. and FUCHS, Y. (1973). *Journal of the American Society for Horticultural Science*, **98** (5), 477–480

ZAUBERMAN, G. and SCHIFFMAN-NADEL, M. (1972). *Journal of the American Society for Horticultural Science*, **97** (3), 313–315

26
RESPIRATION AND ETHYLENE PRODUCTION IN POST-HARVEST SOURSOP FRUIT (Annona muricata L.)

J. BRUINSMA
Department of Plant Physiology, Agricultural University, Wageningen, The Netherlands
and
R.E. PAULL
Department of Botany, University of Hawaii at Manoa, Honolulu, Hawaii, USA

Introduction

The ripening of fruit showing a climacteric rise in respiration is triggered by endogenously produced ethylene. It can also be advanced by exogenously applied ethylene. If the endogenous, autocatalytic ethylene production is genetically or chemically inhibited, ripening processes including the climacteric are prevented unless ethylene is added. Climacteric fruit can, therefore, be defined as fruit able to autocatalytically produce ethylene which in turn induces the other ripening phenomena including the climacteric respiratory rise (Bruinsma, 1983; Solomos, 1983).

Annonaceous fruit seem to behave differently. Shortly after picking, their respiration starts to increase irregularly, days before the onset of autocatalytic ethylene evolution. This irregularity of the respiratory curve has been ascribed to the composite nature of the primitive fruit, its many ovaries being at different stages of development (Biale and Barcus, 1970; Paull, 1982). However, this leaves the regularity of the later occurring ethylene peak unexplained. In order to investigate the relationship between respiration and ripening of post-harvest Annona fruit, respiration and ethylene evolution were analysed with soursop fruit (*A. muricata* L.).

Post-harvest gas exchange of the whole fruit

Whole fruit were individually enclosed in glass jars and production of carbon dioxide and ethylene monitored every hour. A typical result is presented in *Figure 26.1*. Shortly after harvest the CO_2 production rose to a first peak around the second day after harvest. The ethylene evolution at this stage was still at a constant low level, varying from fruit to fruit from a few nanolitres to at most $3\,\mu l\,kg^{-1}\,h^{-1}$. It rose three to five days after harvest, concomitant with a further increase in the respiration rate. The autocatalytic ethylene production suddenly dropped after five to six days whereas respiration increased further. In this final stage the cell membranes collapsed and the decay of the fruit was accompanied by fungal development.

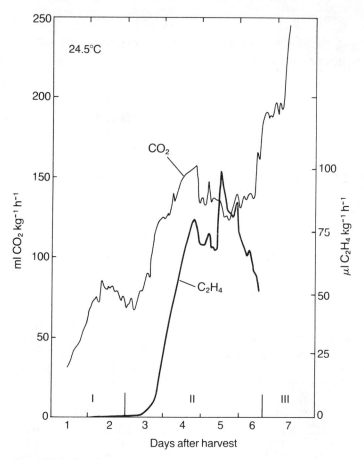

Figure 26.1 Post-harvest evolution of carbon dioxide and ethylene from soursop fruit

The typical ripening phenomena coincided with the autocatalytic ethylene evolution, that is the skin colour changed from green to dark brown, the water potential of the tissue decreased, and volatiles were produced. When the data for CO_2 and ethylene evolution at the beginning of this period (day 2 to 4) were plotted against each other, it was often observed that a certain level of ethylene production was surpassed before the respiration rate increased sharply. This phenomenon may also be observed in the ripening of other climacteric fruit (Sawamura, Knegt and Bruinsma, 1978).

Respiration of fruit tissue discs

The respiration of 1 mm thick discs of parenchymous tissue, 10 mm in diameter, was determined from the oxygen uptake by 10 discs in a Warburg flask using a Gilson respirometer. To provide an isotonic medium, of a similar pH and cation composition as the fruit press sap, the bathing fluid contained 630 mosmol sorbitol, 3 mM $CaCl_2$, 5 mM $MgCl_2$ and 50 mM KH_2PO_4, at pH 4.0. When respiratory

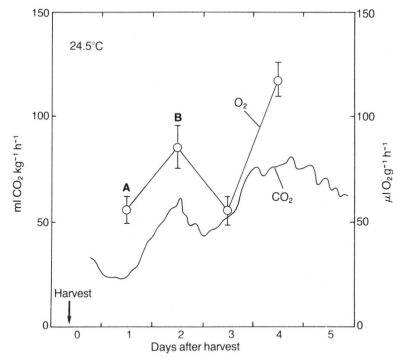

Figure 26.2 Respiration of a whole fruit (CO_2 production) and of discs from that fruit (O_2 uptake), both at 24.5 °C

substrates were used, the sorbitol was partially or wholly replaced by iso-osmotic amounts of carbohydrates or carboxylates, at the same pH.

Freshly prepared discs followed the same respiratory pattern as the whole fruit, but at a higher level (*Figure 26.2*), probably owing to a wound effect (Theologis and Laties, 1978). This similarity in respiration disproves the view that the irregular respiratory curve of the whole fruit is caused by subsequent ripening of the different ovaries. On the contrary, the curve is a characteristic of the fruit flesh as a whole. Ripening was confined to the period of autocatalytic ethylene production and this period was preceded by a preclimacteric rise in respiration.

The nature of this preclimacteric rise was investigated in more detail. Analysis of fruit tissue had shown the presence of sucrose, glucose, fructose and carboxylates, particularly malate, as possible substrates for respiration (Paull, Deputy and Chen, 1983). Substitution of sorbitol by these substrates only weakly promoted respiration, except for malate (*Table 26.1*). Also when respiration was uncoupled by

Table 26.1 STIMULATION OF DISC RESPIRATION BY SUBSTRATES AND BY THE UNCOUPLER, DNP

	Respiration ($\mu l\ O_2\ g^{-1}\ h^{-1}$)	
	− DNP	+20 μM DNP
Sorbitol	87 ± 6	142 ± 10
Sucrose	96 ± 2	135 ± 6
Glucose	104 ± 11	148 ± 9
Fructose	96 ± 1	145 ± 4
Malate	139 ± 11	179 ± 16

Table 26.2 EFFECTS ON DISC RESPIRATION IN 0.15 M MALATE OF INHIBITORS OF THE CYTOCHROME PATHWAY, KCN AND NaN$_3$, AND OF THE ALTERNATE PATHWAY, SALICYLHYDROXAMIC ACID (SHAM)

	Respiration (μl O_2 g^{-1} h^{-1})	
	Exp. 1	Exp. 2
Control	116 ± 6	114 ± 2
0.4 mM KCN	132 ± 4	—
4 mM SHAM	91 ± 7	—
0.4 mM KCN + 4 mM SHAM	6 ± 2	—
10 mM NaN$_3$	—	7 ± 2
1 mM NaN$_3$	—	28 ± 8
1 mM NaN$_3$ + 4 mM SHAM	—	4 ± 1

2,4-dinitrophenol (DNP), malate still increased its rate considerably. Since other members of the tricarboxylic-acid cycle exerted similar effects it can be concluded that tissue respiration was limited by phosphorylation, on the one hand, and by mitochondrial substrates, on the other hand. Limitation of mitochondrial electron flow by both substrates and phosphorylation is usually found with non-isolated mitochondria (Solomos, 1983).

That tissue respiration is almost completely mitochondrial, i.e. without a substantial contribution from any oxidative processes in the cytoplasm, was shown by the effects of inhibitors of the cytochrome and alternate pathways (*Table 26.2*). Of the latter, salicylhydroxamic acid (SHAM) had little effect when present alone. Of the former, NaN$_3$ inhibited respiration considerably, whilst KCN actually stimulated oxygen uptake. However, after 24 h the KCN stimulated respiration had vanished (data not shown), and when KCN was added, together with SHAM, respiration was immediately reduced to a very low level. These results show that nearly all the oxygen was consumed in mitochondrial respiration, and that electron flow can be shifted from the cytochrome to the alternate pathway.

What then causes the oxygen consumption to increase during the preclimacteric rise in *Figure 26.2*? In experiments without (gas-absorbing) bathing fluid, the respiratory quotient for tissue respiration was demonstrated to remain fairly constant throughout the post-harvest period: RQ = 0.94 ± 0.07, the variation being largely between fruits and not in time. To test whether the capacity of the mitochondrial apparatus increased during the preclimacteric rise, respiration was maximally stimulated with substrates and uncoupler (*Table 26.3*). Replacing sorbitol completely by succinate rendered the mitochondria very sensitive to DNP, showing the same reduction of oxygen uptake as caused by an overdose of DNP; some carboxylates, e.g. acetate and glutarate, turned out to be highly inhibitory in

Table 26.3 STIMULATION OF RESPIRATION OF FRESH DISCS BEFORE AND AT THE PEAK OF THE PRECLIMACTERIC RISE

	Respiration (μl O_2 g^{-1} h^{-1})	
	Stage A	Stage B (Figure 26.2)
Sorbitol − DNP	52 ± 7	75 ± 6
+ DNP	127 ± 6	129 ± 10
Succinate − DNP	129 ± 7	120 ± 11
+ DNP	7 ± 0	6 ± 1
Ascorbate − DNP	145 ± 6	152 ± 6
+ DNP	153 ± 2	125 ± 4

Table 26.4 HPLC-ANALYSIS OF ENDOGENOUS SUBSTRATES IN DISCS BEFORE AND AT THE PEAK OF THE PRECLIMACTERIC RISE

	mM g^{-1} fresh wt	
	Stage A	Stage B (Figure 26.2)
Malic acid	14.0 ± 1.6	38.0 ± 5.5
Citric acid	8.4 ± 1.0	12.3 ± 1.4
Sucrose	52.2 ± 4.6	65.7 ± 4.3
Glucose	116 ± 5	120 ± 5
Fructose	116 ± 4	121 ± 4

themselves. However, the data show stimulations to the same levels of respiration irrespective of the stage of the fruit at which the discs were sampled.

Since the capacity of the mitochondrial apparatus is unchanged, the higher respiration rate might be due to an increased level of substrate. In order to investigate this, samples taken simultaneously with the discs used in the experiment shown in *Table 26.3*, were analysed for their contents of sugars and carboxylates (*Table 26.4*). Indeed, the endogenous levels of substrates increased during the preclimacteric rise, in particular the content of rate-limiting carboxylates was found to double.

Effect of harvesting on preclimacteric respiration

The increase in substrates and, thereby, of the preclimacteric rise in respiration, was most probably caused by the picking of the fruit. This could not be demonstrated directly by the comparison of respiration of post-harvest fruit with that of fruit still on the tree, since the soursop grove was located on another island. However, although it was very difficult to identify fruit on the tree which were about to ripen, the preclimacteric rise invariably occurred about two days after harvest. It has been shown that mature green tomatoes, kept under low oxygen and high carbon dioxide partial pressures immediately after harvest so that their ripening is prevented, will initiate starch breakdown and changes in metabolism of organic acids and sugars without any change in ethylene production. Only upon transfer to ambient atmosphere do polygalacturonase activity and pigment changes occur simultaneously with autocatalytic ethylene evolution (Goodenough *et al.*, 1982; Jeffery *et al.*, 1984). Apparently, such a separation between post-harvest effects and ripening phenomena occurs naturally in Annona fruit.

Picking may interrupt the supply from the vegetative parts of the plant of a labile inhibitor of ethylene action, so that shortly after harvest ethylene may become active even at the low level prevailing in the freshly harvested fruit. Removal from the tree also terminates the supply of assimilates from the tree as a substrate for fruit respiration, and the fruit has to shift to its own store of starch which rapidly decreases following harvest (Paull, Deputy and Chen, 1983). It is possible that this shift is causally related to the disappearance of the ethylene inhibitor. Anyway, glycolytic breakdown of starch followed by carboxylation of phosphoenolpyruvate into malate may well start at a rate that surpasses the substrate demand. Such a temporary overshooting of starch degradation has been observed in green bananas (McGlasson and Wills, 1972) and may result in a temporary rise in substrate-limited respiration prior to the climacteric rise proper. The dissimilation of starch through malic acid into carbon dioxide results in a respiratory quotient of unity according to:

$$\text{starch} \rightarrow \text{P-enolpyruvate} + CO_2 \rightarrow \text{malate} + 3O_2 \rightarrow 4CO_2 + 3H_2O$$

Conclusion

Respiration of a post-harvest Annona fruit can be considered to occur in three phases (*Figure 26.1*). Stage I is a preclimacteric peak, probably due to the removal from the tree and is caused by a temporary overproduction of rate-limiting substrate. During this period there may also be a reduction in the levels of an inhibitor of ethylene action. This could then lead to the induction of stage II, which can be considered as the respiratory rise of a climacteric fruit induced by and concomitant with autocatalytic ethylene production and the other ripening phenomena. The climacteric ripening of this primitive fruit is unusual in that after a few days the membranes collapse, leading to the rapid deterioration of the fruit in stage III.

A more detailed report will be published elsewhere (Bruinsma and Paull, 1984).

References

BIALE, J.B. and BARCUS, D.E. (1970). *Tropical Science*, **12**, 93–104
BRUINSMA, J. (1983). In *Post-Harvest Physiology and Crop Preservation*, pp. 141–163. Ed. by M. Lieberman. Plenum Press, New York, London
BRUINSMA, J. and PAULL, R.E. (1984). *Plant Physiology*, **76**, 131–138
GOODENOUGH, P.W., TUCKER, G.A., GRIERSON, D. and THOMAS,T. (1982). *Phytochemistry*, **21**, 281–284
JEFFERY, D., SMITH, C., GOODENOUGH, P., PROSSER, I. and GRIERSON, D. (1984). *Plant Physiology*, **74**, 32–38
McGLASSON, W.B. and WILLS, R.B.H. (1972). *Australian Journal of Biological Science*, **25**, 35–42
PAULL, R.E. (1982). *Journal of the American Society of Horticultural Sciences*, **107**, 582–585
PAULL, R.E., DEPUTY, J. and CHEN, N.J. (1983). *Journal of the American Society of Horticultural Sciences*, **108**, 931–934
SAWAMURA, M., KNEGT, E. and BRUINSMA, J. (1978). *Plant Cell Physiology*, **19**, 1061–1069
SOLOMOS, T. (1983). In *Post-Harvest Physiology and Crop Preservation*, pp. 61–98. Ed. by M. Lieberman. Plenum Press, New York, London
THEOLOGIS, A. and LATIES, G.G. (1978). *Plant Physiology*, **62**, 249–255

27

THE EFFECT OF HEAVY METAL IONS ON TOMATO RIPENING

GRAEME E. HOBSON, ROYSTON NICHOLS and CAROL E. FROST
Glasshouse Crops Research Institute, Littlehampton, West Sussex, UK

Introduction

It is well recognized that silver is a powerful ethylene antagonist in plant tissues, and in the form of the thiosulphate complex (AgTS) is capable of being translocated more readily than the nitrate (*see* Veen, 1983). AgTS has been widely used to inhibit such ethylene-promoted processes as the irreversible wilting of cut flowers and petal drop in intact plants (Veen, 1983). Although silver ions strongly inhibit fruit ripening at relatively low concentrations (Saltveit, Bradford and Dilley, 1978; Hobson, Harman and Nichols, 1984), this treatment has had only a limited application in the study of fruit physiology (Lis, Kwakkenbos and Veen, 1984).

Previous studies on the effects of heavy metal ions on fruit ripening have involved the use of tissue discs (Saltveit, Bradford and Dilley, 1978). This chapter describes a technique to study the effect, in intact fruit, of heavy metal ions on ripening, CO_2 production and ethylene production of tomato fruit.

Infiltration of intact fruit with silver thiosulphate

Mature green tomatoes (cv. Sonatine) were removed, with as much of the peduncle attached to the calyx as possible, from trusses bearing one or more ripening fruit. Ethylene evolution from the unripe fruit was measured by gas–liquid chromatography immediately after picking, and if the rate was less than $1 \, nl \, g^{-1} \, h^{-1}$ the fruit were judged to be preclimacteric and were used for experimental purposes. Immediately before infiltration with solutions, the peduncle of each fruit was broken at the abscission point, a silicone-rubber tube with a small narrow-bore plastic container attached was pushed over the end of the peduncle and 0.5 ml test solution placed in the container. For the metal ions investigated, solutions of salts ($AgNO_3$, $CdCl_2$, cupric acetate, $Pb(NO_3)_2$ or $Co(NO_3)_2$), were mixed with $Na_2S_2O_3$ in a molar ratio of 1:4. As a mercuric–thiosulphate mixture was unstable, mercuric acetate was used alone. Control solutions contained only appropriate amounts of $Na_2S_2O_3$ except for experiments with mercury salts where Na acetate was used instead. Most of the solutions applied were absorbed by the fruit within 24 h. Ethylene and CO_2 production was measured daily on individual fruit for at least five days after treatment. Damage to the tissues was assessed after ripening the fruit at 20 °C. All experiments were repeated at least twice.

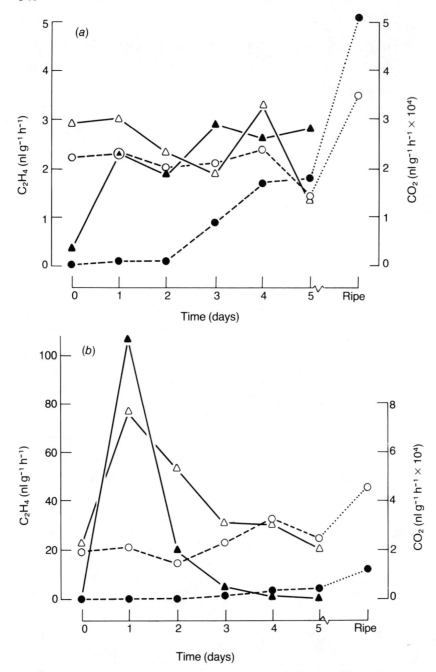

Figure 27.1 The response of mature green tomato fruit to infiltration with AgTS. Fruit were treated with (a) 1 μmol AgTS or (b) 10 μmol AgTS, and ethylene (▲) and CO_2 (△) production monitored. Control fruit were treated with sodium thiosulphate, and ethylene (●) and CO_2 (○) monitored. Control fruit ripened normally. Peak ethylene (●) and CO_2 (○) production by control fruit during the climacteric are shown at the end of the dotted line

Effect of silver thiosulphate on tomato ripening

Previous results have shown that infiltration of 1 µmol of AgTS is sufficient to cause a proportion of the locule walls to fail to change colour as the rest of the walls ripened more or less normally (Hobson, Harman and Nichols, 1984). As indicated in *Figure 27.1a* a similar infiltration using 2 mM AgTS led to an immediate rise in ethylene evolution which persisted during the five days over which readings were taken. Carbon dioxide production by the treated fruit showed a less clearcut effect. There are grounds for suggesting that the initial surge in ethylene is a wound response, but the picture was confused by ethylene production from those parts of the fruit which the silver did not reach and which were beginning to ripen. Control fruit subjected only to equivalent amounts of $Na_2S_2O_3$ invariably ripened quite uniformly. As indicated in *Figure 27.1a*, typical peak values for ethylene evolution during ripening were 5 nl $g^{-1} h^{-1}$ and for CO_2 3.5 nl $g^{-1} h^{-1} \times 10^4$.

In order to confirm these responses of unripe tomatoes to silver, further experiments were carried out using 20 mM AgTS. As shown in *Figure 27.1b*, both ethylene and CO_2 production increased rapidly after treatment, only to fall again by the second day. A typical climacteric pattern for ethylene production and respiration was seen in the control fruit, which ripened normally.

Effects of other metal ions on tomato ripening

The effects of silver and the ions of Cu, Cd, Hg, Co and Pb are summarized in *Table 27.1*. Fruit infiltrated with cadmium showed a similar but more limited response compared to silver treated fruit and the part of the fruit near to the point of attachment of the calyx failed to ripen. The ions of copper, lead, mercury and cobalt had progressively smaller effects on ethylene and CO_2 production, and fruit treated with such salts seemed to ripen normally. After cutting the fruit, a certain amount of tissue damage was apparent, as is indicated in *Table 27.1*.

Results from several studies involving exposure of plant material to various metal ions have underlined the relative effectiveness of silver in counteracting the normal response to exogenous and endogenous ethylene. Beyer (1976) tested a wide range of salts using peas, cotton and orchids as experimental material, and found that silver was uniquely effective in its anti-ethylene properties. Saltveit, Bradford and

Table 27.1 SOME EFFECTS OF INFILTRATION WITH HEAVY METAL IONS (10 µmol USUALLY AS THE THIOSULPHATE COMPLEX) ON TOMATO FRUIT RIPENING (EACH FIGURE IS THE MEAN OF RESULTS FROM THREE FRUIT)

Chemical infiltrated (0.5 ml 20 mM metal salt- 80 mM $Na_2S_2O_3$)	Ethylene evolution 24 h after treatment (nl $g^{-1} h^{-1}$)	CO_2 evolution 24 h after treatment (nl $g^{-1} h^{-1} \times 10^4$)	Extent of tissue damage	Extent of inhibition of ripening
Control (NaTS)	0.5	2.2	None	None
Silver (AgTS)	107	7.8	Slight	Pericarp
Cadmium (CdTS)	39	5.1	Extensive	Calyx area
Copper (CuTS)	8.3	4.2	Moderate	None
Lead (PbTS)	4.9	4.0	Slight	None
Mercury (Hg acetate)	1.1	1.8	Moderate	None
Cobalt (CoTS)	0.1	2.2	None	None

Dilley (1978) exposed discs of unripe apple, banana and tomato fruit tissue to another set of salts (including silver) over a range of concentrations. Silver ions were clearly the most effective in inhibiting ethylene synthesis and fruit ripening.

Conclusion

These results demonstrate that silver as the thiosulphate complex at a tissue concentration between 0.03–0.3 mM causes the production of 'wound' or 'stress' ethylene, presumably by activation of the normal metabolic pathway for ethylene synthesis from methionine. Nevertheless, continuing synthesis does not take place, fruits fail to ripen and as such do not initiate the autocatalytic ethylene production associated with the climacteric.

The site of action of silver is still largely unknown. Silver appears to move apoplastically in the tissue (data not shown), and we have no evidence that the ions enter the cytoplasm of the conducting tissue. Hence if silver is blocking autocatalytic ethylene production the mechanism that normally controls this in fruit such as the tomato may be in close proximity to the plasmalemma. Alternatively silver may preferentially bind to a site responsive to ethylene, replace another metal in a binding complex, or interfere with the rate of ethylene metabolism (*see* Hall *et al.*, 1982 and Chapter 10). These are at present open questions. In mutant tomatoes, a failure by the fruit to ripen does not appear to be due to a lack of ethylene-binding sites nor to an inhibition of ethylene binding (Sisler, 1982). There is, of course, no reason why there should be only one cause for a fruit to fail to ripen, but there are a number of similarities between mutant and silver-treated tissue. A detailed explanation for the action of silver on tomato ripening must await further experimental information concerning this interesting phenomenon.

References

BEYER, E. Jr. (1976). *Plant Physiology*, **58**, 268–271
HALL, M.A., EVANS, D.E., SMITH, A.R., TAYLOR, J.E. and AL-MUTAWA, M.M.A. (1982). In *Growth Regulators in Plant Senescence*, pp. 103–111. Ed. by M.B. Jackson, B. Grout and I.A. Mackenzie. British Plant Growth Regulator Group, Monograph No. 8. Wessex Press, Wantage
HOBSON, G.E., HARMAN, J.E. and NICHOLS, R. (1984). In *Ethylene—Biochemical, Physiological and Applied Aspects*, pp. 281–290. Ed. by Y. Fuchs and E. Chalutz. Martinus Nijhoff/Dr W. Junk, Amsterdam. In press
LIS, E.K., KWAKKENBOS, A.A.M. and VEEN, H. (1984). *Plant Science Letters*, **33**, 1–6
SALTVEIT, M.E., BRADFORD, K.J. and DILLEY, D.R. (1978). *Journal of the American Society for Horticultural Science*, **103**, 472–475
SISLER, E.C. (1982). *Journal of Plant Growth Regulation*, **1**, 219–226
VEEN, H. (1983). *Scientia Horticulturae*, **20**, 211–224

POST-HARVEST EFFECTS OF ETHYLENE ON ORNAMENTAL PLANTS

R. NICHOLS and CAROL E. FROST
Glasshouse Crops Research Institute, Littlehampton, West Sussex, UK

Introduction

Since the recognition of ethylene as the phytotoxic component of illuminating gases, there have been numerous investigations to quantify the response of flowers to ethylene derived from a variety of sources. The need for such studies was further rationalized as more sensitive analytical methods for quantifying ethylene were devised. Further impetus was given as potentially hazardous sources of ethylene were identified such as ripening fruits, decaying vegetation and emissions from internal combustion engines.

The important point is that ethylene is a gas and therefore its production from one source may affect sensitive organisms by aerial diffusion and accumulation in the atmosphere. It appears that flowers, or certain species of them, are particularly sensitive. The reason is that ethylene is also one of the complex of naturally occurring growth substances which control development, maturation and senescence of plant tissues and flowers are no exception. Exposure of a flower to an external source of the gas is analogous to presenting it with one of its own hormones. The flower does not distinguish between ethylene (endogenous) produced during its normal metabolism and that from an outside (exogenous) source. The exogenous source may be, and frequently is, contaminated with other gases such as oxides of nitrogen and sulphur dioxide derived from industrial pollution (Abeles, 1973); these gases may also contribute to the flower and plant response. However, if the principal contaminant is ethylene and this is present in sufficient concentration, the flower will respond in a characteristic way. The morphological and physiological responses of the flower in such circumstances often resemble those which are seen during the normal pattern of senescence and which are instigated by increases in endogenous ethylene. The effect of the exogenous ethylene is to accelerate the symptoms and thus to impair the shelf-life and keeping quality of the produce.

Physiological responses of flowers to ethylene

In the carnation, exposure of mature flowers to about $1 \mu l\, l^{-1}$ ethylene for 24 h results in curling (inrolling) of petals one or two days later; this is usually

Figure 28.1 Effect of ethylene on carnation flowers. Cut flowers were photographed after 5 days in air (left) or after 4 days following a 24 h exposure to 1 µl l^{-1} ethylene (right). The ethylene treatment caused wilting of petals, enlarged receptacle and swelling of the ovary

irreversible and petals shrivel and dry. The petal curling is accompanied or followed by visible swelling of the ovary if the cut stem is placed in water (*Figure 28.1*). The accelerated ovary swelling and irreversible petal wilting identifies the response of the flower to ethylene. The response of the flower to exogenous ethylene appears to be similar to that of a carnation which has been pollinated (*Figure 28.2*) and the reason appears to be that germinating pollen induces formation of ethylene in the flower parts (Nichols *et al.*, 1983). The ovary also swells after pollination so that one effect of exogenous ethylene is to mimic the effects of a natural phenomenon (pollination) in which ethylene has a physiological role.

It is not known for certain if pollination promotes ethylene production in all flowers but it has been reported to occur in a number of other genera (Burg and Dijkman, 1967; Hall and Forsyth, 1967). It is also significant that pollens from a range of species contain 1-aminocyclopropane-1-carboxylic acid (ACC) (Whitehead, Fujino and Reid, 1983), the precursor of ethylene in higher plants (Adams and Yang, 1979; Lürssen, Naumann and Schroeder, 1979). Ethylene does not evoke all the responses caused by pollination; other growth substances are clearly involved (Strauss and Arditti, 1982) particularly with respect to seed set and further growth of the fruit. It is pertinent that growth substances such as auxins or ethylene analogues, such as propylene, will promote accelerated wilting of carnations, possibly by inducing the endogenous ethylene-forming systems. We have found that 0.2–0.5 µmol of indole-acetic acid is about the minimum amount which will accelerate wilting of a carnation flower under our experimental conditions.

Factors affecting sensitivity of flowers to ethylene

GENUS

When considering factors which may affect the sensitivity of flowers to ethylene a level of the gas must be used which is relevant to commercial practice. Ethylene under commercial conditions may arise from a variety of sources (Abeles Chapter 23, Schouten Chapter 29) and the concentration encountered may vary. Smith, Meigh and Parker (1964) reported ethylene levels of about $0.06\,\mu l\,l^{-1}$ in commercial boxes of narcissus flowers, and Harkema and Woltering (1981) used the gas at $3.0\,\mu l\,l^{-1}$ in experimental trials on a range of species. It follows that it is difficult to make generalizations about the levels of ethylene encountered under commercial conditions. However, there is good evidence that flowers differ in their sensitivity to ethylene. Much of the work on cut flowers in relation to ethylene sensitivity has been done with the inflorescence at the current marketing stage of development. Flowers, as with fruit, seem to be less susceptible to exogenous ethylene when they are immature and so it is difficult to make comparisons when the physiological stages of development at which flowers are marketed are very different.

FLOWER MATURITY

Buds of carnation are more tolerant of ethylene than open flowers (Barden and Hanan, 1972; Camprubi and Nichols, 1978), although the former may require bud-opening solution (e.g. sugar plus a germicide) to achieve full opening and a reasonable vase-life. The current tendency to pick and transport flowers in bud has probably reduced their inherent susceptibility to ethylene but it has also encouraged denser packing and may exacerbate problems associated with temperature.

The observations relating to ethylene sensitivity and flower maturity have been made with genera which do not shed their flower parts, although it probably also applies to those which do. In the latter instance ethylene affects the activation of specialized layers of cells which are responsible for shedding flowers or flower parts.

ENVIRONMENT

Temperature

The reaction of flowers to exogenous ethylene and their production of endogenous ethylene are both a function of temperature. Carnations, for example, have been shown to be less sensitive to ethylene at low temperatures (Smith, Parker and Freeman, 1966; Barden and Hanan, 1972). Nonetheless at these temperatures they are still responsive to ethylene and this is important during long-term cool storage. Clearly, however, other metabolic activities continue during cool storage and these in turn will affect the physiology of the tissue and its ethylene response (*see* review Halevy and Mayak, 1981), particularly on transfer to ambient temperature.

Oxygen and carbon dioxide

Flowers can tolerate the low levels of oxygen required to inhibit ethylene production without apparent injury. In practice reduced levels of oxygen in a closed market container, generated as a result of the respiration of the flower, will be invariably accompanied by increased levels of carbon dioxide. Carbon dioxide counteracts ethylene and has been shown to mitigate the effects of ethylene injury to carnations in conditions simulating those which might be found in commercial practice (Smith and Parker, 1966). Concentrations of about 2–5% carbon dioxide will counter the injurious effect of exposure to $0.2\,\mu l\,l^{-1}$ ethylene for two days and will also suppress the surge of ethylene which occurs from the flowers at the end of their natural senescence (Nichols, 1968). It follows that the environmental conditions predisposing flowers to produce enhanced levels of ethylene such as might occur in densely packed market boxes kept at high temperatures will also favour accumulation of respiratory carbon dioxide. The high temperature will hasten deterioration of the produce but not necessarily as a result of accumulation of ethylene provided the carbon dioxide remains high. Under normal circumstances ethylene and carbon dioxide remain in phase. Danger arises when they become out of phase as a result of ethylene exposure originating from an external source.

Sources of ethylene

External sources could include ripening fruits, e.g. as in transit with tomatoes, or an industrial source of pollution. Another common source of ethylene is that produced as a result of infection of flowers or their vegetative parts with micro-organisms (e.g. fungi). Decay of the tissues produces enhanced levels of ethylene inside a container which may then affect other flowers. Furthermore, some flowers produce increased levels of ethylene at the end of their natural life as a result of an autocatalytic production and if these are included with fresher flowers, the latter may be affected and in turn be stimulated into wilting and further ethylene production.

It has been mentioned earlier that the rate of production of ethylene is increased as a result of pollination. The style appears to initiate the ethylene response which stimulates further production from petals. This phenomenon may take as long as two days or more, depending on ambient temperature, from pollination to petal wilt, and therefore flowers which are not showing symptoms at the time of harvest may be potential sources of ethylene somewhat later. The importance of this clearly depends on whether or not the species is a significant producer of ethylene on the one hand and the restrictions placed on dispersal of the gas on the other, assuming that viable pollen and suitable vectors are available.

CONCENTRATION AND EXPOSURE TIME

It may be inferred from the considerations of the influence of the environment on the response to ethylene that both concentration and time of exposure (dose) are important. Sublethal doses of ethylene may cause symptoms which are reversible. For instance, partial in-rolling of carnation petals may be caused by exposure to $0.1\,\mu l\,l^{-1}$ ethylene for 16 h but on return to ethylene-free air turgor of the petals is restored. The difficulty from the practical standpoint is that, without sophisticated analytical equipment, there is no way of assessing whether such symptoms are transient or indicative of irreversible, accelerated senescence. It seems likely that changes in colour are generally indicative of ageing in common commercial flowers whereas petal flaccidity may not be. However, the latter may be indicative of exposure of the flowers to high temperature resulting in water deficits and metabolic changes which hasten flower deterioration.

Symptoms of ethylene damage

It is predictable that the effects of ethylene on flowers are complicated by the nature of the flower itself since it is composed of reproductive and vegetative tissues. Not only does the metabolism of these tissues differ at harvest but they also change with respect to each other as the flower ages. Nonetheless, it is possible to identify symptoms which seem fairly typical for a given genus (*Table 28.1*). These

Table 28.1 SYMPTOMS OF ETHYLENE DAMAGE

Symptom	Genus
Dry sepal	Orchids (Davidson, 1949)
Abscission	Antirrhinum
	Calceolaria (Fischer, 1949)
	Zygocactus (Cameron and Reid, 1981)
Floret abscission	Digitalis (Stead and Moore, 1983)
In-rolling of petals 'sleepiness'	Carnations (Nichols, 1968)
In-rolling of corollas	Ipomoea (Kende and Baumgartner, 1974)
Wilting	Kalanchoe (Marousky and Harbaugh, 1979)
	Alstroemeria (Harkema and Woltering, 1981)
Colour changes	Orchids (Arditti, Hogan and Chadwick, 1973)

are visible expressions of the morphological effects of exogenous ethylene on the exposed flower; in some species the developing shoot is affected and this may occur during storage of the perennating organ as in some bulbs. As mentioned earlier, sensitivities of genera differ. Inflorescences of some flowers such as *Gerbera* fade but only at very high concentrations of ethylene (Nowak and Plich, 1981); it is suggested that this species is resistant to ethylene.

Numerous articles have appeared concerned with the effects of ethylene on flowers (Hasek, James and Sciaroni, 1969; Beyer, 1980) and these are dealt with comprehensively in review papers concerned with the post-harvest physiology of cut flowers (Halevy and Mayak, 1979; 1981).

Control of ethylene effects

It seems very probable that the pathway of ethylene biosynthesis in flowers is the same or very similar to that described by Adams and Yang (1979) for other plant tissues, namely:

methionine → S-adenosyl methionine → ACC → ethylene

The correlative changes between ACC and ethylene during senescence of carnation (Bufler *et al.*, 1980) suggest that regulation of this pathway should lead to control of senescence of this flower species. Thus removal of ethylene, blocking its synthesis or sites of activity should alter the course of senescence and result in increased longevity of the flower.

REMOVAL OF ETHYLENE

The simplest approach is to avoid exposure of the produce to known sources of ethylene. These include exhausts from combustion engines, fruit and cool stores in which fruit may have been kept, rotting vegetation, fumes from gas-filled boilers and plastic materials which may be undergoing chemical deterioration, and this can be achieved by adequate ventilation with non-polluted air, ensuring, however, that dehydration of the tissue does not occur. This may not be practicable but certain measures can be taken, such as the ventilation of contaminated stores prior to usage. Hypobaric storage would prevent exposure to external ethylene and lower the concentrations of endogenous ethylene (and oxygen), which probably accounts for its reported beneficial effects, but the construction of a hypobaric store is expensive and the results with flowers have been variable and dependent on the cultivar (Halevy and Mayak, 1981).

Chemical absorption of ethylene is practicable and can be achieved by dispersing potassium permanganate on an inert carrier material. Commercial sachets, 'blankets' and so on are available, formulated for inclusion inside boxes of produce and packing sheds.

There are a few reports specifically concerned with these materials in association with flower marketing (Peles et al., 1977; cited in Halevy and Mayak, 1981; Tompsett, 1979). It appears that air movement particularly within closed containers may be insufficient to permit effective ethylene absorption. An alternative approach would be to pretreat flowers with an antagonist of ethylene action if ethylene absorption or dispersal proved impracticable.

PREVENTION OF ETHYLENE SYNTHESIS

Synthesis of ACC requires a pyridoxyl phosphate-mediated enzyme and inhibitors of this type of enzyme have been shown to increase longevity of flowers. Amino-oxyacetic acid (AOA) and aminoethoxyvinylglycine (AVG) have been used experimentally in this context (Baker et al., 1977; Broun and Mayak, 1981). Polyamines have also been shown to inhibit ethylene production (Suttle, 1981) and increase longevity of carnations (Wang and Baker, 1980). AVG or AOA will delay senescence accelerated by stigma wounding of *Petunia* corollas (Nichols and Frost, unpublished data) which implicates participation of ethylene in the wounding response. At present, these compounds provide valuable tools for research concerned with those aspects of flower maturation, fruit ripening and growth phenomena in which ethylene is involved. The effects of plant growth regulators, cytokinins, inhibitors of protein synthesis, metal ions and many other chemicals in the general context of flower senescence have been extensively reviewed by Halevy and Mayak (1981).

PREVENTION OF ETHYLENE ACTIVITY

Silver salts block ethylene action (Beyer, 1976). The discovery of the formulation, silver thiosulphate (STS), as a mobile silver salt within plant tissues (Veen and Van de Geijn, 1978), has led to promising treatments not only for preventing ethylene injury but also for increasing the longevity of certain flowers. Veen (1983) has reviewed the literature on this topic. The longevity of certain genera (*Dianthus*, *Lathyrus odoratus*, *Dendrobium* and *Delphinium*) is increased by treatment with STS; for others, a pretreatment with STS has a smaller effect on longevity but it does confer resistance to exogenous ethylene. Reid et al. (1980) have estimated that

Figure 28.2 Effect of silver thiosulphate on pollen-induced wilting of carnation flowers. Flowers were photographed on day 6 after the following treatments: control (left); pollinated on day 0 (centre); pollinated and treated with STS (2 mM) for 1 h on day 0 (right)

(a)　　　　(b)　　　　(c)　　　　(d)　　　　(e)

Figure 28.3 Reversal of ethephon inhibition of tulip stem extension by silver thiosulphate. Flowers were photographed on day 5 after the following treatments: (a) water, (b) ethephon, (c) 0.1 mM STS → ethephon; (d) 0.5 mM STS → ethephon; (e) 0.5 mM STS. All stems were stood in water after treatment (detail in Nichols and Kofranek, 1982)

about 0.5 µmol of silver is sufficient to increase longevity of carnations. We find that an STS-pretreatment will prevent the accelerated senescence of carnations caused by pollination (*Figure 28.2*). STS substantially increases the life of these flowers in water as reported by other workers. Silver thiosulphate also delays pollination accelerated senescence of petunia flowers (Whitehead, Halevy and Reid, 1984).

An example of prevention of an ethylene effect by silver is shown in *Figure 28.3*. Cut tulip stems were pretreated with a 10 min pulse of STS and then stood in 48 mg l^{-1} ethephon (an ethylene generating solution) as described by Nichols and Kofranek (1982). The pretreatment with STS allowed normal or near-normal extension growth of the stem to occur compared with the partial elongation of the untreated stems. In this instance, the ethephon appears to have an effect on vegetative cells, although the flowers tend to remain closed, since the elongation of the stem is caused by extension of cells in the apical internode, in the region immediately adjacent to the receptacle.

Treatments with STS have been reported to have beneficial effects on bulb crops such as *Lilium*, apparently preventing flower bud abscission in plants grown in low light (van Meeteren and de Proft, 1982) and improving flower quality. Although the precise action of the STS is not fully understood, the potent anti-ethylene effects of silver indicate the involvement of ethylene. A further beneficial effect of silver is to protect flowers from damage by exogenous ethylene which can be particularly deleterious in dry storage (Swart, 1981).

In this context it is pertinent that treatment with ethylene may have some desirable commercial effects. It is possible that one benefit from 'burning-over' of crops of narcissus and iris to promote earlier flowering might be attributed to ethylene in the smoke as for freesias (Imanishi and Fortanier, 1982/1983; Uyemura and Imanishi, 1983), although other hydrocarbons, carbon monoxide and carbon dioxide may add to or modify the response to ethylene (Imanishi and Fortanier, 1982/1983).

Conclusion

It is evident that ethylene has many diverse effects on flowers. However, in considering specific problems encountered with their post-harvest behaviour, ethylene should be viewed as perhaps only one of a number of possible causes, others may also lead to symptoms which resemble those induced by ethylene. For instance prolonged cool storage may result in failure of buds of some species to open; excessively high temperatures during transport may lead to desiccation and accelerated senescence; unsuitable handling procedures may result in failure of stems to take up water resulting in premature wilting of flowers and leaves. Ethylene may be secondarily involved in some of these phenomena but the primary cause should be recognized.

References

ABELES, F.B. (1973). *Ethylene in plant biology*. Academic Press, New York and London
ADAMS, D.O. and YANG, S.F. (1979). *Proceedings of the National Academy of Sciences of the United States of America*, **76**, 170–174
ARDITTI, J., HOGAN, N.M. and CHADWICK, A.V. (1973). *American Journal of Botany*, **60**, 883–888
BAKER, J.E., WANG, C.Y., LIEBERMAN, M. and HARDENBURG, R.E. (1977). *HortScience*, **12**, 38–39
BARDEN, L.E. and HANAN, J.J. (1972). *Journal of the American Society for Horticultural Science*, **97**, 785–788
BEYER, E.M. (1976). *Plant Physiology*, **58**, 268–271
BEYER, E.M. (1980). *Florists' Review*, **165**, 26–28
BROUN, R. and MAYAK, S. (1981). *Scientia Horticulturae*, **15**, 272–282
BUFLER, G., MOR, Y., REID, M.S. and YANG, S.F. (1980). *Planta*, **150**, 439–442

BURG, S.P. and DIJKMAN, M.J. (1967). *Plant Physiology*, **42**, 1648–1650
CAMERON, A.C. and REID, M.S. (1981). *HortScience*, **16**, 761–762
CAMPRUBI, P. and NICHOLS, R. (1978). *Journal of Horticultural Science*, **53**, 17–22
DAVIDSON, O.W. (1949). *Proceedings of the American Society for Horticultural Science*, **53**, 440–446
FISCHER, C.W. (1949). *New York State Flower Growers Bulletin*, **52**, 5–8
HALEVY, A.H. and MAYAK, S. (1979). *Horticultural Reviews*, **1**, 204–236
HALEVY, A.H. and MAYAK, S. (1981). *Horticultural Reviews*, **3**, 59–143
HALL, I.V. and FORSYTH, F.R. (1967). *Canadian Journal of Botany*, **45**, 1163–1166
HARKEMA, H. and WOLTERING, E.J. (1981). *Vakblad voor de Bloemisterij*, **22**, 40–42
HASEK, R.F., JAMES, H.A. and SCIARONI, R.H. (1969). *Florists' Review*, **144**, 3721: 21, 65–68, 79–82; 3722: 16–17, 53–56
IMANISHI, H. and FORTANIER, E.J. (1982/83). *Scientia Horticulturae*, **18**, 381–389
KENDE, H. and BAUMGARTNER, B. (1974). *Planta*, **116**, 279–289
LÜRSSEN, K., NAUMANN, K. and SCHROEDER, R. (1979). *Zeitschrift für Pflanzenphysiologie*, **92**, 285–294
MAROUSKY, F.J. and HARBAUGH, B.K. (1979). *HortScience*, **14**, 505–507
NICHOLS, R. (1968). *Journal of Horticultural Science*, **43**, 335–349
NICHOLS, R., BUFLER, G., MOR, Y., FUJINO, D.W. and REID, M.S. (1983). *Journal of Plant Growth Regulation*, **2**, 1–8
NICHOLS, R. and KOFRANEK, A.M. (1982). *Scientia Horticulturae*, **17**, 71–79
NOWAK, J. and PLICH, H. (1981). *Rósliny Ozdobne, Prace Ínstytutu Sadownictwa i Kwiaciarstwa*, seria B, tom 6, 89–97
PELES, A., VALIS, G., KIRSCHOLTZ, Y., COHEN, B. and MAYAK, S. (1977). *Agrexco and Flower Market Board Bulletin, Tel Aviv, Israel*
REID, M.S., PAUL, J.L., FARHOOMAND, M.B., KOFRANEK, A.M. and STABY, G.L. (1980). *Journal of the American Society for Horticultural Science*, **105**, 25–27
SMITH, W.H., MEIGH, D.F. and PARKER, J.C. (1964). *Nature*, **204**, 92–93
SMITH, W.H. and PARKER, J.C. (1966). *Nature*, **211**, 100–101
SMITH, W.H., PARKER, J.C. and FREEMAN, W.W. (1966). *Nature*, **211**, 99–100
STEAD, A.D. and MOORE, K.G. (1983). *Planta*, **157**, 15–21
STRAUSS, M.S. and ARDITTI, J. (1982). *Botanical Gazette*, **143**, 286–293
SUTTLE, J.C. (1981). *Phytochemistry*, **20**, 1477–1480
SWART, A. (1981). *Acta Horticulturae*, **113**, 45–49
TOMPSETT, A.A. (1979). *Annual Report 1978 Rosewarne and Isles of Scilly Experimental Horticulture Stations*, 39–41, 46
UYEMURA, S. and IMANISHI, H. (1983). *Scientia Horticulturae*, **20**, 91–99
VAN MEETEREN, U. and DE PROFT, M. (1982). *Physiologia Plantarum*, **56**, 236–240
VEEN, H. (1983). *Scientia Horticulturae*, **20**, 211–224
VEEN, H. and VAN DE GEIJN, S.C. (1978). *Planta*, **140**, 93–96
WANG, C.Y. and BAKER, J.E. (1980). *HortScience*, **15**, 805–806
WHITEHEAD, C.S., FUJINO, D.W. and REID, M.S. (1983). *Scientia Horticulturae*, **21**, 291–297
WHITEHEAD, C.S., HALEVY, A.H. and REID, M.S. (1984). *Physiologia Plantarum*, **61**, 643–648

29
SIGNIFICANCE OF ETHYLENE IN POST-HARVEST HANDLING OF VEGETABLES

S.P. SCHOUTEN
Sprenger Instituut, Wageningen, The Netherlands

Introduction

Vegetables may originate from many plant parts. For instance stem sprout (asparagus), main bud (lettuce), leaf blade (spinach), petioles (celery), swollen leaf base (leek), stem tuber (potato), swollen tap root (carrot), swollen hypocotyl (beetroot), bulb (onion), axillary bud (Brussels sprouts), flower bud (artichoke) and swollen inflorescence (cauliflower) are all examples of vegetables. Furthermore several commodities which are botanically classified as fruit (e.g. tomato, cucumber, egg plant, bean) may be considered vegetables by the consumer (Wills *et al.*, 1983).

During their storage and distribution, vegetables are sometimes exposed to ethylene. This gas may originate from several sources, for instance ripening fruit, exhaust fumes from trucks and cars and air pollution (Reid, Chapter 22; Abeles, Chapter 23). Ethylene can have a profound effect on the post-harvest physiology of vegetables. Sometimes it may improve the quality of a particular commodity, but more commonly its effects are detrimental. Research at the Sprenger Institute is primarily concerned with the latter effects of ethylene and in particular is centred on ways to prevent ethylene damage or decrease the influence of the gas.

Sensitivity of vegetables to ethylene

The majority of vegetables produce less than $1.0 \, \mu l \, kg^{-1} \, h^{-1}$ of ethylene. Only those vegetables which are botanically classified as fruits (such as ripening tomatoes) may produce higher levels. However, the sensitivity of vegetables to applied ethylene

Table 29.1 SENSITIVITY OF VEGETABLES TO ETHYLENE

Sensitivity	Commodities
Low	Artichoke, beetroot, carrot, celeriac, egg plant, kohlrabi, dry onion, pepper, raddish, rhubarb, turnip
Moderate	Asparagus, beans, celery, escarole, kale, peas, potatoes, leeks
High	Broccoli, Brussels sprouts, chinese cabbage, green cabbage, red cabbage, savoy cabbage, cucumber, cauliflower, endive, sweet corn, lettuce, spinach, tomato

Table 29.2 EFFECTS OF EXPOSURE TO ETHYLENE

Effect	Commodity
Undesirable	
Accelerated senescence and degreening	Spinach, cucumber
Accelerated ripening	Egg plant
Russet spotting	Lettuce
Souring	Carrots
Sprouting	Potatoes
Leaf abscission	Cauliflower, cabbage
Toughening	Asparagus
Desirable	
Acceleration of abscission to aid harvesting	Fruit
Stimulation of sprouting	Seed potato
Acceleration and synchronization of ripening	Fruit

varies considerably and three categories can be distinguished, low, moderate or high (*Table 29.1*). Exposure to ethylene may lead to the appearance of many undesirable effects (*Table 29.2a*) and these are particularly apparent in those commodities which are highly sensitive to the gas. Occasionally the effects of ethylene may be desirable (*Table 29.2b*) but these are in general limited to fruit.

Commercial application of ethylene

Commercial application of ethylene is presently restricted to tomatoes (Lürssen, Chapter 30). Many tomatoes are harvested at the mature-green stage, when they are less sensitive to damage by bruising, and subsequent exposure to ethylene ensures synchronized ripening (Kader, 1979). Ethylene in the store originates from gas cylinders or is derived from ethylene generators, one can also use ethylene-releasing chemicals such as ethephon (2-chloroethanephosphoric acid) prior to harvest. This latter treatment is employed in the Netherlands to ripen the final trusses of glasshouse grown tomatoes which otherwise would remain unripe due to limiting growth circumstances. However, ethephon treatment may impair the quality of the final product and reduce its shelf-life (Buitelaar, 1978; Boon, 1980).

The occurrence of ethylene

The results of research carried out by Morris *et al.* (1978) in California to monitor ethylene levels at various stages between harvest and consumption of lettuce are shown in *Table 29.3*. It was apparent that a high concentration of the gas can occur in the field as a result of pollution and in holding areas as a consequence of exhaust fumes from trucks. Ethylene also accumulated during storage and in distribution centres. The highest concentrations were found during retailing, where the most important likely cause was the presence of other commodities. Normal duration of transit for the lettuce was five to eight days and exposure to 0.1 ppm ethylene during this time was sufficient to cause commercially important damage.

Concentrations of ethylene up to 2 ppm have been reported from auctions in the Netherlands (Boerrigter and Molenaar, 1984; Uffelen, private communication). The gas was found to originate from the exhausts of forklifts and trucks. These high

Table 29.3 LEVELS OF ETHYLENE FOUND IN THE EXTERNAL ATMOSPHERE AND IN PACKED CARTONS WITH LETTUCE AT VARIOUS LOCATIONS BETWEEN FIELD AND CONSUMPTION

Sample locations		Ethylene concentration (ppm)		No. of samples analysed	Potential sources
		Range	Mean		
Field	A	<0.10–0.12	—	21	Air pollution
Field to cooler	B	0.03–0.11	0.07	3	Mechanically injured lettuce; Exhaust from truck; pollution
Holding areas prior to vacuum cooling	A	0.01–0.61	0.05	47	Exhaust from trucks and forklifts
	B	0.01–0.80	0.16	12	
Immediately following cooling	B	0.01–0.29	0.12	11	(vacuum cooling removes much of the C_2H_4 inside cartons)
Cold storage rooms at vacuum coolers	A	0.01–2.78	0.33	144	Exhaust from forklifts other commodities
	B	0.01–1.56	0.22	73	
Inside rail cars at destination	A	0.01–0.19	0.06	14	Decay, other pollution sources
	B	0.01–0.02	0.01	3	
Inside truck units at destination	A	0.04–0.22	0.08	9	Decay, other pollution sources
	B	0.08–0.11	0.09	4	
Distribution centres warehouses	A	0.03–2.49	0.25	22	Exhaust, other commodities
	B	0.01–0.78	0.08	43	
Retail storage areas	A	0.02–2.95	0.36	19	Other commodities
	B	0.06–2.88	0.41	18	
Home refrigerator	A	0.02–1.58	0.25	33	Other commodities

A = atmosphere external to carton, B = inside carton
From Morris *et al.* (1978)

levels of ethylene may decline in future as the use of trucks in auction rooms decreases as a result of dockyards being built for them. The use of electric forklifts or at least the provision of a catalyst (platinum) scrubbing exhaust system would further improve the situation (Boerrigter and Molenaar, 1984).

Measurement of ethylene concentrations gives no more than an indication of the potential danger to the vegetable crop. Other conditions such as temperature and duration of exposure, are also significant and these together with concentration give a more complete picture.

Ethylene in storage rooms

ORIGIN OF ETHYLENE

Ethylene levels were determined in cold rooms used at a Dutch auction to store tomatoes and cucumbers. Levels of ethylene increased tenfold in the cucumber store after an adjacent room was filled with tomatoes. It is unlikely that this increase in ethylene concentration can be explained by the transient burst in ethylene production which is known to occur after cucumber harvest (Saltvict and McFeeters, 1980). It is more likely to have been caused by the diffusion of the gas through the walls from the adjacent tomato store, since ripening tomatoes produce ten or even 100 times more ethylene than cucumbers. A similar observation has been reported where ethylene from apple storerooms diffused into an adjacent coldroom ($-1.5\,°C$) used to store savoy cabbage (Bouman, personal communication). In this instance levels of ethylene reached 20–70 ppm in the cabbage store. Although this concentration is known to cause leaf abscission and loss of colour in cabbages held at 1 °C (Ilker and Morris, 1980) the effects at $-1.5\,°C$ are unknown. Indeed it seems that for some commodities ethylene damage is reduced at low temperatures. For instance chicory roots exhibited no detectable damage after exposure to 25 ppm ethylene for three months at 0 °C (Pelleboer, private communication). Further research is required to determine the efficacy of low temperature during the storage of ethylene sensitive vegetables.

EFFECTS OF TEMPERATURE

The effect of temperature on ethylene production in mixed stores containing both vegetables and fruit has been investigated (Boerrigter and Damen, 1983). Two storage rooms were set up, the first at 5 °C contained fruit, cauliflower, lettuce and tomatoes, the second at 10 °C contained cucumber, tomatoes and fruit.

Ethylene concentrations were monitored in both rooms over a period of 63 h using a mobile gas chromatography system able to detect >0.01 ppm of the gas (Boerrigter, 1980). In the 5 °C room ethylene levels reached 2.8 ppm within 16 h and attained a maximum of 5.2 ppm. In the 10 °C room ethylene levels reached 19 ppm within 16 h and attained a maximum of 30 ppm. Even at these temperatures levels of ethylene present are sufficient to cause detrimental effects to the stored commodities.

Ethylene accumulation during transport

It is common practice to transport mixed loads of fruit and vegetables. Such conditions are often only protected by the sail of the sailing car which may allow ethylene to accumulate. This phenomenon was investigated using pallets containing either tomatoes, or tomatoes and cucumbers (Damen and Boerrigter, 1982). Two pallets of each were set up and one of each pair enclosed with shrinkfilm. Ethylene levels were determined over 16 h and the results are shown in *Figure 29.1*. Although levels of ethylene reached a maximum of 2 ppm in the tomato and cucumber pallet enclosed in shrinkfilm this did not cause any detectable damage to the cucumbers. This may be due to the short exposure time and/or the accumulation of CO_2 within the pallet.

Another common practice is the long-term transport of mixed loads in closed refrigeration trucks. In order to determine ethylene accumulation under these conditions the mobile gas chromatograph was employed. A truck combination loaded with several commodities, maintained at 10–12 °C, was followed and ethylene measured several times on a trip from the Netherlands to Sweden. The results of this study are shown in *Figure 29.2*. It is clear that in both the closed truck and ventilated trailer the ethylene level rose both during loading and transport on the ferry boat. There was also a detectable increase after the arrival in Gothenburg. During transport ethylene escaped from both the closed truck and the trailer although in the truck the level remained constant (Damen, Boerrigter and Schouten, 1983). The maximum level of CO_2 in the closed truck was 6%. One of the commodities transported was cucumbers. Samples of this vegetable were taken

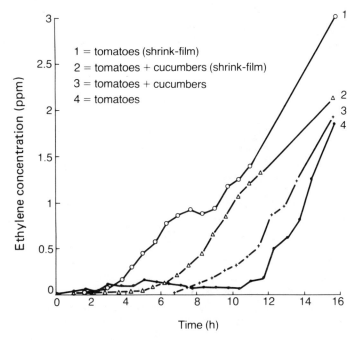

Figure 29.1 Time course of ethylene accumulation in pallets with tomatoes and cucumbers placed in a standing truck

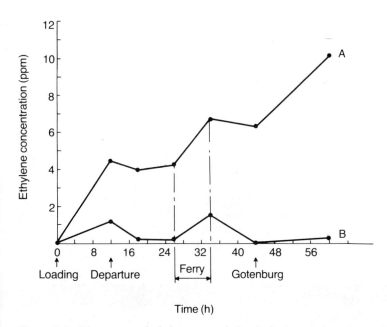

Figure 29.2 Time course of ethylene accumulation during long-term transport of commodities from Holland to Sweden. A, Closed truck; B, ventilated trailer

Table 29.4 COLOUR RATINGS OF CUCUMBERS TRANSPORTED TO SWEDEN

Treatment of cucumbers	Colour rating[a] after	
	6 days	10 days
Transport to Sweden, sealed	5.9	4.8
Transport to Sweden, not sealed	5.6	4.6
Stayed in Holland, sealed	6.2	5.8
Stayed in Holland, not sealed	5.9	5.3

[a]Colour rating in a scale from 1 to 9 (9 = completely green, 4 = 50% yellow)

prior to the journey and held in store at the Sprenger Institute. Further samples were taken at the end of the journey and returned to the Sprenger Institute. The extent of cucumber degreening was determined in both sets of samples after six and ten days at 15 °C (*Table 29.4*). Those cucumbers maintained at the Institute showed less degreening than those transported to Sweden. However, the difference was only slight, perhaps as a result of the high CO_2 concentration in the closed truck. Further research is necessary to determine the levels of CO_2 required to prevent damage. One conclusion that can be drawn from this experiment is that during transport of ethylene sensitive products levels of the gas should not be allowed to accumulate. One approach to overcome this problem is to increase ventilation. However, during the winter months such treatment may result in chilling injury to the transported commodity. This may be prevented by using a bulkhead in the car which allows air exchange to take place independently of driving speed (Nieuwenhuizen, 1983).

Threshold values for ethylene effects

The concentration of ethylene required for ripening varies between commodities but in the majority of instances is between 0.1 and 1 ppm. The duration of exposure necessary also differs but normally periods of 12 h or longer are most effective (Reid, 1981). The concentration of ethylene which causes damage to vegetables has a similar threshold. For instance 0.1 ppm and 0.5 ppm are sufficient to impair the quality of lettuce and egg plant respectively (Morris *et al.*, 1978; Schouten and Stork, 1977).

In order to study the interaction between ethylene concentration and duration of exposure on cucumber degreening, the following experiment was performed. Fruit were exposed to various concentrations of ethylene for a period of three days at 20 °C (*Figure 29.3a*) or to a fixed level of ethylene for differing periods of time (*Figure 29.3b*). The extent of degreening was assessed over the following nine-day period. It is clear that both the concentration of, and duration of exposure to ethylene influenced the rate of degreening. During the experiment described in *Figure 29.3b* it was noted that the number of cucumber fruits with decay rose with increasing exposure time to ethylene (*Table 29.5*). From these experiments it can be concluded that the threshold values for ethylene induced damage in cucumbers is 1 ppm for two days, 1–5 ppm for one day or >5 ppm for 12 hours (Schouten and Stork, 1981). Recent work has suggested that these values may be too high and that 0.1 ppm for 24 h may be sufficient to hasten degreening.

A further parameter currently under investigation is the effect of temperature on ethylene threshold. The preliminary results of this study suggest that temperature is

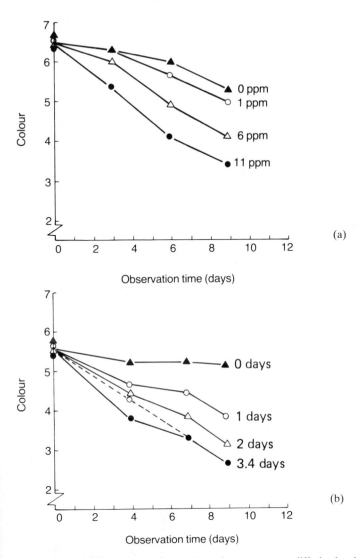

Figure 29.3 (a) Degreening of cucumbers after exposure to differing levels of ethylene for three days (T = 20 °C). (b) Degreening of cucumbers exposed to 4–5 ppm ethylene for several exposure times (T = 20 °C)

Table 29.5 PERCENTAGE CUCUMBER FRUIT WITH DECAY AFTER EXPOSURE TO ETHYLENE (4–5 ppm) AT 20 °C FOR DIFFERENT TIME PERIODS

Exposure time (days)	Fruits with decay (%)	
	After exposure for 7 days	After exposure for 9 days
0	11	14
1	17	22
2	39	58
3	39	58
4	42	67

very important in determining the extent of degreening and incidence of decay during exposure to ethylene.

Methods of controlling undesirable effects

Several methods exist to reduce undesirable effects of ethylene. The simplest way is to remove sources of the gas during storage and distribution. However, since fruit and vegetables are frequently transported and retailed together it may be impractical to use separate rooms and displays for them. Thus more complex methods must be employed; these include scrubbers, controlled atmosphere (CA) storage and various packaging techniques.

SCRUBBING

The simplest and cheapest way to remove ethylene is to ensure adequate ventilation is maintained. Alternatively there are chemical means employing the use of agents such as $KMnO_4$ as ethylene absorbants. Ethylene absorbing materials based on $KMnO_4$ as the active agent are efficient scrubbers if all the air is passed through them, however their activity declines sharply at 90–95% relative humidity (Rudolphij and Boerrigter, 1981). A further disadvantage is that this material cannot be regenerated and so may only be used once.

Those vegetables with low sensitivity and production of ethylene (*Table 29.1*) do not require the use of a scrubber during their storage. Those vegetables with high sensitivity to ethylene only require the use of a scrubber if stored along with commodities which produce high levels of the gas.

As well as ventilation and scrubbing, maintenance of the store at a low temperature may also reduce ethylene-induced damage. This is particularly useful for the storage of vegetables since in general they have the capacity to withstand very low temperatures. However, when vegetables are stored along with chilling sensitive commodities, such as tomatoes, much higher temperatures have to be utilized.

CONTROLLED-ATMOSPHERE STORAGE

The effect of low storage temperatures in combination with CA-storage is to inhibit ethylene production (Staden, 1981). The effects of high CO_2 and low O_2 concentrations on the degreening of cucumbers are shown in *Table 29.6*. The cucumbers were

Table 29.6 DEGREENING OF CUCUMBERS AFTER MAINTENANCE UNDER DIFFERENT STORAGE REGIMES AT 20 °C

Treatment	Loss of colour after 9 days[a]	
	Without ethylene	With ethylene[b]
Normal O_2 + KOH	2.0	7.9
Normal O_2 − KOH	0.5	3.7
N_2 flush + KOH	1.0	3.1
N_2 flush − KOH	0.2	0.9

[a]Loss of colour was quantified as the difference between day 0 and day 9
[b]The ethylene concentration varied from 8 to 25 ppm; CO_2 concentration (−KOH) varied from 8 to 11%

stored along with tomatoes. Degreening was reduced if either CO_2 was allowed to rise, O_2 to decline, or both (Uffelen, 1975). CA-storage can also be employed for cabbage and Brussels sprouts (Pelleboer, 1982).

PACKAGING

Many consumer packages employ plastic films which allow the development of high CO_2 and low O_2 atmospheres around the commodities. If the film is impermeable to these gases then very high CO_2 or very low O_2 levels may result in damage to the produce. This problem may be overcome by the use of films which allow restricted passage of these gases and the modified atmosphere produced with these films undoubtedly limits ethylene damage (Geeson and Browne, 1983). Advantages of this technique have been established for cauliflower, leeks and sweetcorn (De Maaker, 1983).

Control of ethylene in wholesale markets and retail shops

During their distribution and retailing vegetables are often mixed with ripening fruits and thus exposed to high ethylene levels. As a consequence of this and other factors, quality is often poor. Several measures may be taken to avoid ethylene damage and keep vegetables in a good condition. These measures include:

(1) use of low temperatures during storage and display;
(2) avoidance of unnecessary mixing of vegetables with fruit;
(3) maintenance of adequate ventilation;
(4) use of appropriate packaging material.

Conclusions

With the exception of tomatoes, all vegetables suffer undesirable effects from ethylene exposure. This chapter has outlined a number of measures which may be taken to prevent these undesirable effects from occurring. These include (pre-)cooling the commodities, scrubbing, ventilation, CA-storage and appropriate packaging during handling. It is clearly important to bring this information to the knowledge of people working in the distribution of fruit and vegetables.

References

BOERRIGTER, H.A.M. (1980). *Rapport no. 2140*. Sprenger Instituut, Wageningen, Netherlands
BOERRIGTER, H.A.M. and DAMEN, P.M.M. (1983). *Interimrapport no. 21*. Sprenger Instituut, Wageningen, Netherlands
BOERRIGTER, H.A.M. and MOLENAAR, W.H. (1984). *Vakblad voor de Bloemisterij*, **4**, 114–117
BOON, H.Th.M. (1980). *Groenten en Fruit*, **36** (9), 32–35
BUITELAAR, K. (1978). *Groenten en Fruit*, **33** (50), 37

DAMEN, P.M.M. and BOERRIGTER, H.A.M. (1982). *Groenten en Fruit*, **37** (32), 18–19
DAMEN, P.M.M., BOERRIGTER, H.A.M. and SCHOUTEN, S.P. (1983). *Vakblad A.G.F.*, **37**, 23–25
GEESON, J.D. and BROWNE, K.M. (1983). *Grower*, **100**(2), 35–37
ILKER, J. and MORRIS, L.L. (1980). *Survey of Ethylene Production and Responses by Fruit and Vegetables at Various Temperatures*. University of California, Davis, California and Sea-Land Services Inc., New Jersey
KADER, A.A. (1979). *Perishables handling*, **44**, 2–5
MAAKER, J. DE (1983). *Vakblad Handel A.G.F.*, **37** (37), 16–17
MORRIS, L.L., KADER, A.A., KLAUSTERMEYER and CHENEY, C.C. (1978). *California Agriculture*, **32** (6), 14–15
NIEUWENHUIZEN, G.H. VAN (1983). *Vakblad Handel AGF*, **37** (49), 115–121
PELLEBOER, H. (1982). *Groenten en Fruit*, **38** (6), 60–61
REID, M.S. (1981). In *Plant Science 196 Syllabus. Postharvest Technology of Horticultural Crops*. Kader, A.A., Kasmire, R.F., Mitchell, F.G., Reid, M.S., Sommer, N.F. and Thompson, I. University of California, Davis
RUDOLPHIJ, J.W. and BOERRIGTER, H.A.M. (1981). *Bedrijfsontwikkeling*, **12** (3), 307–312
SALTVEIT, M.E. Jr. and McFEETERS, R. (1980). *Plant Physiology*, **66**, 1019–1023
SCHOUTEN, S.P. and STORK, H.W. (1977). *De Tuinderij*, **17** (9), 48
SCHOUTEN, S.P. and STORK, H.W. (1981). *Groenten en Fruit*, **36**, (37), 42–43
STADEN, O.L. (1981). *Bedrijfsontwikkeling*, **12** (1), 79–81
UFFELEN, J.A.M. VAN (1975). *Landbouwkundig Tijdschrift/pt.*, **87** (11), 295–299
WILLS, R.H.H., LEE, T.H., GRAHAM, D., McGLASSON, W.B. and HALL, E.G. (1983). *Post harvest: An introduction to the physiology and handling of fruit and vegetables.* Granada Publishing, London

30

RELATIONSHIP BETWEEN ETHYLENE PRODUCTION AND PLANT GROWTH AFTER APPLICATION OF ETHYLENE RELEASING PLANT GROWTH REGULATORS

K. LÜRSSEN and J. KONZE
Bayer AG, Pflanzenschutz, Anwendungstechnik, Biologische Forschung, Leverkusen, FRG

Ethylene related plant growth regulators in agriculture

As early as the last century, before the concept of ethylene as a plant hormone was formulated, the gas was used in agriculture as a plant growth regulator (PGR). A pineapple farmer on the Azores experimenting with fumes in the greenhouse to kill insects observed that plants flowered earlier after fumigation. As a result of this observation the use of fumes to induce pineapple flowering in greenhouses became common practice on the Azores. Since the turn of the century our knowledge of the effects of ethylene on plant growth and development has grown rapidly and has been extensively documented (Zimmer, 1968; Abeles, 1973; Burg, 1973).

Due to its gaseous nature, the only commercial application of ethylene until recently has been to ripen bananas held in store and it has taken more than half a century until ethylene-releasing PGRs have been developed (for references and further details *see* Lürssen, 1984). Today there are many possible applications for ethylene-related plant growth regulators in agriculture and horticulture (*Table 30.1*). However, as yet practical applications are limited to only a few of these possibilities.

Table 30.1 POSSIBLE APPLICATION FOR ETHYLENE RELATED PLANT GROWTH REGULATORS

Physiological effect	*Possible use in agriculture or horticulture*
Promotion of abscission	Loosening of fruits for harvest aid purposes
	Fruit thinning
	Defoliation (harvest aid in grape; use in tree nurseries)
Promotion of ripening	Yield increase and harvest aid in tomato
	Harvest aid and quality improvement in tobacco
	Improvement of quality and colour in fruits and vegetables
Promotion of flowering	Improved management in pineapple farms and in horticulture (ornamental Bromeliaceae)
Induction of female flowers in Cucurbitaceae	Yield increase and earlier harvest in Cucurbitaceae
Growth inhibition	Prevention of lodging, mainly in cereals
Stimulation of latex flow	Yield increase and harvest aid in Hevea
Stimulation of germination	Improvement of weed control in some weed species with long lasting germination periods

A number of problems are associated with the use of ethylene-related PGRs for fruit loosening. Often leaves are affected as well as fruit. Physiologically older leaves are particularly susceptible in autumn, when the application of such PGRs would be of greatest practical significance. Furthermore the efficacy of PGRs varies from year to year, depending on the physiological status of the plant tissue, which in turn depends on environmental factors. The application of ethylene related PGRs for fruit thinning also suffers from lack of reproducibility. The use of ethylene related PGRs for defoliation of grape vines as an aid to harvest suffers from the same problems: either the performance may be insufficient or berry drop may occur. 1-Aminocyclopropane-1-carboxylic acid (ACC) is the only PGR with sufficient defoliation activity for use in tree nurseries. As a result of these problems no suitable ethylene related PGR is commercially available for use as a promoter for abscission except perhaps for the use of Etacelasil for olive fruit loosening.

Ethephon is used to synchronize ripening of tomato fruit particularly in the USA, where fruits are mechanically harvested and further processed. The advantage of this procedure is to increase yield since more fruit are ripe at the time of harvest. For other vegetables and fruit such as red pepper and apple, the advantages of ripening stimulation could result in an earlier harvest and better coloration of the produce. However, practical application in this field seems to be limited. Senescence in tobacco leaves can be accelerated by ethephon. However, this does not allow a single harvest, but reduces the number from about six to three.

Ethylene induces flowering in Bromeliaceae and a few other closely related plants. Pineapple growers use ethephon to achieve uniform flowering in their fields, which without treatment is erratic. Normally more than 95% flower induction can be achieved. Ethephon is also used for flower induction of Bromeliaceae in the ornamental plant market. Other ethylene related PGRs are also very effective in flower induction, but they are unavailable commercially.

In the Cucurbitaceae family, sex expression may be regulated by ethylene and gibberellins. Ethylene induces female flower formation while gibberellins induce male flower formation. Ethephon can be used in the Cucurbitaceae to achieve an earlier harvest, as female flower set occurs earlier during plant development.

Recently ethephon has been employed to prevent lodging in cereals (e.g. barley and wheat). The mechanism by which it works is via the influence of ethylene upon cell enlargement. In the presence of ethylene longitudinal expansion is inhibited and cells expand laterally.

The most interesting commercial use of ethylene related PGRs is probably the stimulation of latex flow in *Hevea* trees. Such treatment leads to both an improvement of harvest and an increase in yield. Until now ethylene-related PGRs have not been used to stimulate the germination of weeds as an aid to their eradication. Although the germination of some weed species is stimulated by ethylene, the application to the seeds in the soil may be difficult in practice.

Chemistry and mode of action of ethylene related PGRs

Commercially the most important ethylene releasing chemical is 2-chloroethylphosphonic acid (ethephon). Also available on the market is 2-chloroethyl-tris-(2-methoxyethoxy)silan (Etacelasil). Both compounds spontaneously release ethylene upon contact with water. Ethylene is the compound responsible for the physiological effects. The experimental compounds 2-chloroethylsulphonic acid (HOL 1302) and (2-chloroethyl)-sulphonyl-methanol (HOL 1274) show a similar mode of action (for details and references *see* Lürssen, 1982 and 1984).

The discovery of ACC (Schröder and Lürssen, 1978; Lürssen, Naumann and Schröder, 1979; Adams and Yang, 1979) has led to a new practical class of ethylene-related PGRs. The pathway of ethylene biosynthesis is now well established: methionine—S-adenosyl-methionine-(ACC)—ethylene (Yang *et al.*, Chapter 2, Kende, Acaster and Guy, Chapter 3). As well as the natural intermediate ACC, the synthetic derivative N-formyl-ACC (NF-ACC) will be discussed in this article. NF-ACC is metabolized by the plant to ACC which is then further converted enzymatically to ethylene. There is no evidence for the direct conversion of NF-ACC to ethylene nor can ACC be converted nonenzymatically in the plant to ethylene.

Performance of ethylene related PGRs

Though there can be no doubt that ethylene is the compound that is responsible for the effect of the above mentioned PGRs, they differ markedly in their efficacy in practice. A comparison of their performance is given in *Table 30.2*. Possible

Table 30.2 COMPARISON OF THE PERFORMANCE OF ETHYLENE RELATED PGRS

Effect	Plant species	Plant growth regulator					
		Hol 1274	Hol 1302	Ethephon	Etacelasil	ACC	NF–ACC
Stimulation of ripening	Tomato,	+	+	+++		+	+++
	pineapple,	−	−	+++		−	
	red pepper,			+++		+	+++
	tobacco	+	+	+++		+	+++
Growth inhibition	Barley	+	−	++		+	+++
Flower induction	Pineapple	+++	++	++		++	−
Defoliation	Grape,	+	+	+		+++	+
	nursery trees	+	+	+		+++	++
Fruit loosening	Olive,	+++	+++	++	+++	+++	++
	cherry,	++	++	+++		+++	++
	apple	−		+		+++	+
Stimulation of latex flow	Rubber tree	+++	++	+++	++	−	−

− no effect; + to +++ increasing effect

reasons for these differences have been discussed previously (Lürssen, 1982). It is apparent that uptake, transport, rate and duration of ethylene evolution can all play a role in the kind of effects which ethylene related PGRs will have on plants. The absolute amount of ethylene evolved may be of less importance. However, the performance of NF-ACC especially in the promotion of fruit ripening cannot be satisfactorily explained, as there is a discrepancy between the performance of NF-ACC and ACC and the amount of ethylene produced after application of these compounds.

Plant responses in relation to ethylene evolution

In *Table 30.3* the amount of ethylene produced one day and between two and eight days after application of the PGRs is compared. Application of HOL 1274 and Etacelasil resulted in the release of most ethylene during the first day. In contrast ethephon and NF-ACC released more ethylene two to eight days after application than during the first day. ACC behaved in an intermediate fashion releasing about equal amounts during both the first day and the period from day 2 to 8. Comparing these data with the performance of the PGRs in the field (*Table 30.2*), it was concluded, that for the promotion of ripening, a long-lasting ethylene release is necessary while for the promotion of abscission only a short pulse of ethylene is required (Lürssen, 1982).

Table 30.3 TIME COURSE OF ETHYLENE EVOLUTION[a]

Compound	Ethylene produced per incubation vial in nmol		Ratio day 2–8/day 1
	Day 1	Day 2–8	
HOL 1274	11.55	0.14	0.012
Etacelasil	3.35	0.36	0.107
ACC	1.38	1.72	1.246
Ethephon	2.11	4.25	2.014
NF-ACC	0.06	0.13	2.167
Control (water)	0.05	0.08	1.600

[a]Leaf discs 10 mm in diameter were cut from primary leaves of young soybean plants. The discs were placed in 12.8 ml vials floated on 1 ml solution containing 10 nmol of the PGR under test. The vials were incubated at 25°C in the dark. Evaporated water was replaced during the experiment. Alternate vials were sealed for 24 h and the ethylene content of the gas space determined by gas chromatography. Ethylene evolution for days 2–8 is additive

However, the very low amounts of ethylene evolved after application of NF-ACC and the excellent performance of this compound in the acceleration of ripening were still puzzling. Therefore further experiments were performed using NF-ACC.

The longer duration of the action of NF-ACC compared with ACC was demonstrated in an experiment with young tomato plants. The compounds were applied at a concentration of 250 ppm. One day after application both compounds caused the same degree of epinasty. Over the following days the epinastic effect induced by ACC disappeared while in the NF-ACC treated plants epinasty was maintained in the young growing leaves for more than two weeks.

In a further experiment the effect of ACC and NF-ACC on ethylene production during the stimulation of epinasty in tomato plants was investigated (*Figure 30.1*). The experiment shows that NF-ACC caused a greater epinastic response although it stimulated ethylene production less than the ACC treatment. Mixtures of NF-ACC and ACC caused an intermediate response.

In *Table 30.4* the effects of NF-ACC and ACC on tomato ripening are summarized. Ethylene evolution was measured at intervals during the experiment. In untreated plants ethylene evolution stayed approximately constant throughout the experiment. Both NF-ACC and ACC enhanced ethylene production considerably on day 1. Over the next three days the amount of ethylene evolved dropped in the ACC-treated plants. NF-ACC caused a lower stimulation of ethylene evolution

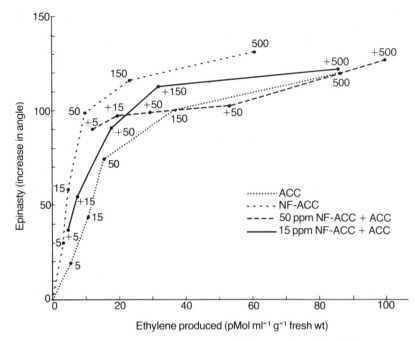

Figure 30.1 Young tomato plants were sprayed with water (control) or different concentrations of ACC or NF-ACC or mixtures thereof. The applied concentrations are indicated in the figure. The concentrations marked '+' indicate that additional amounts of ACC were applied to either 15 or 50 ppm NF-ACC. The plants were put into an airtight glass jar. Eighteen hours later the ethylene concentration in the jar was measured by gas chromatography. The epinastic response was measured as the increase in angle between plant stem and leaf axis over the corresponding angle of the control plants

Table 30.4 TIME COURSE OF ETHYLENE PRODUCTION DURING TOMATO RIPENING[a]

Treatment	Ethylene production 1 to 5 days after treatment					Ripe fruit at day 5
	1	2	3	4	5	
Water	27.8	23.0	27.1	23.5	18.8	37%
0.15% ACC	73.8	60.2	47.1	28.7	32.0	50%
0.15% NF-ACC	53.6	48.1	45.8	47.9	62.7	95%

[a]Tomato plants were grown in the greenhouse. When 15% of the fruit were ripe, four plants per treatment were sprayed with water, 0.15% solution of ACC or 0.15% solution of NF-ACC. Each day the four plants of each treatment were sealed in an 86 litre transparent plastic bag for 4 h. An aliquot of the air was then taken and analysed for ethylene content by gas chromatography. Ethylene production was calculated as $\mu l \, l^{-1} \, h^{-1}$ for the four plants in the 86 litre bag

than ACC from day 1 to day 3. On the fourth and fifth days the ethylene production of the NF-ACC-treated plants rose considerably to double the value of ACC-treated plants. ACC caused only a slight increase in ripeness of the tomato fruit, while NF-ACC caused nearly all fruit to be ripe by the fifth day after treatment. The increase in ethylene production on days 4 and 5 by the NF-ACC-treated plants can be explained by the endogenous ethylene climacteric of the ripening fruit. Thus during the critical phase of ripening induction the ethylene production is lower in

Table 30.5 EFFECT OF ACC AND NF-ACC ON PEA SEEDLING ELONGATION GROWTH AND ETHYLENE PRODUCTION[a]

Treatment	% of growth compared to water control	Ethylene production (pmol h^{-1} g^{-1} fresh wt)
Water	100	10
0.1 mM ACC	68	55
0.4 mM ACC	50	900
1.0 mM ACC	24	1350
0.1 mM NF-ACC	73	50
0.4 mM NF-ACC	50	290
1.0 mM NF-ACC	22	460

[a]The peas were germinated in open 50 ml syringes in the presence of water, ACC-solution or NF-ACC-solution. Ethylene production was determined by closing the syringes six days after treatment and measuring the amount of trapped ethylene by gas chromatography

NF-ACC-treated plants than in ACC-treated plants, yet NF-ACC has the greater effect on the ripening process.

The response of pea plants to ACC and NF-ACC compared to the ethylene produced are summarized in *Table 30.5*. Both compounds have the capacity to inhibit elongation growth and promote ethylene production by the tissue. The growth inhibition achieved by the application of equimolar concentrations of ACC and NF-ACC was approximately the same. However, as in the other experiments, the amount of ethylene evolved in the ACC treatments was much higher than the amount evolved in the NF-ACC treatments. Thus the ethylene produced in the NF-ACC treatment seems to be more effective than that produced by the ACC treatment.

Opening of the plumular hook of mungbean seedlings in the light is inhibited by ethylene. *Table 30.6* shows the results of an experiment in which hook opening was inhibited by ACC and NF-ACC. Once again the reaction of the plant parts to equimolar concentrations of ACC and NF-ACC is similar but the amount of ethylene produced is much higher in the ACC treatment than in the NF-ACC treatment. Ethylene, 0.03 nmol, produced in the NF-ACC treatment caused the same response as 0.85 nmol ethylene produced in the ACC treatment.

Table 30.6 EFFECT OF ACC AND NF-ACC ON PLUMULAR HOOK OPENING AND ETHYLENE PRODUCTION OF ETIOLATED MUNGBEANS[a]

Treatment	Hook opening	Amount of ethylene produced by ten hooks (nMol)
Water	0.17	0.02
0.1 mM ACC	2.50	0.85
0.4 mM ACC	3.00	5.00
1.0 mM ACC	3.00	28.00
0.1 mM NF-ACC	2.60	0.03
0.4 mM NF-ACC	3.00	0.07
1.0 mM NF-ACC	3.00	0.18

[a]For each treatment ten isolated hooks were placed in a closed glass vial in the test solution in the light. Hook opening after 4 h is expressed as the mean value for the ten replications of a linear scale from 0 (no inhibition of opening) to 3 (complete inhibition of opening). The amount of ethylene produced in 4 h by the ten hooks was measured by gas chromatography

Mode of action of NF–ACC

The discrepancy between the capacity of NF-ACC and ACC to promote ethylene production and their ability to effect ethylene related developmental events raises the question of the mode of action of these PGRs. Does NF-ACC really act as an ethylene precursor? Does the molecule itself induce the typical ethylene effects? Does NF-ACC and perhaps ACC act by inducing ethylene biosynthesis?

In *Table 30.7* the results of experiments confirming the role of NF-ACC and ACC as ethylene precursors are shown. Pieces of etiolated pea internodes were incubated in U-^{14}C-L-methionine to label the ethylene precursor pool. The internodes were then incubated in water/IAA to induce biosynthesis of ^{14}C-ethylene, and in mixtures of ACC/IAA or NF-ACC/IAA. As expected, the ethylene production by pea internode tissue was highest in the ACC/IAA treatment, intermediate in the NF-ACC/IAA treatment, and lowest in the IAA treatment. Compared to the IAA treatment the specific radioactivity of the

Table 30.7 SPECIFIC RADIOACTIVITY OF ETHYLENE[a]

Incubation medium	Amount of ethylene after 6 h (nMol)	Specific radioactivity of ethylene (Ci mol^{-1})
0.1 mM IAA + water	1.345	0.958
0.1 mM IAA + 3 mM NF-ACC	3.515	0.270
0.1 mM IAA + 3 mM ACC	6.904	0.063

[a]Pieces of etiolated pea seedlings were incubated in U-^{14}C-L-methionine followed by an incubation in IAA-containing media to stimulate ethylene biosynthesis

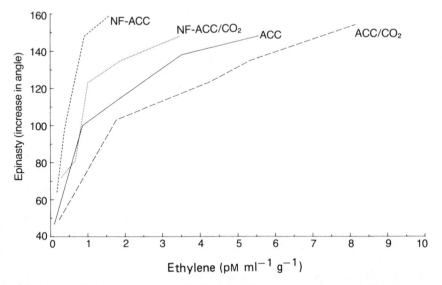

Figure 30.2 Young tomato plants were sprayed with water or concentrations of 15, 50, 150 and 500 ppm ACC or NF-ACC. After 1 h they were put into 2.2 litre glass jars containing air or air plus 1% CO_2. Five hours later the ethylene content of the gas phase was measured. After 6 h of incubation epinasty was determined and qualified as the increase in angle between plant stem and leaf axis over the corresponding angle of the control plants

ethylene produced dropped in both mixtures, the dilution being greater in the ACC treatment than in the NF-ACC treatment. This experiment clearly shows that both ACC and NF-ACC are converted to ethylene by feeding into the natural biosynthetic pathway with ACC being closer to the ethylene forming reaction than NF-ACC.

Carbon dioxide is a well known inhibitor of ethylene action which operates by interfering with the binding site for ethylene. We investigated the epinastic response of tomato plants treated with ACC and NF-ACC in the presence or absence of CO_2 to determine whether ACC or NF-ACC act directly or by their ability to promote ethylene production. The results of this experiment are illustrated in *Figure 30.2*. CO_2 inhibited the action of both ACC and NF-ACC, indicating that the action of these compounds is mediated by ethylene.

The compounds—aminovinylglycine (AVG) and amino-oxyacetic acid (AOA), inhibit the conversion of S-adenosylmethionine (SAM) to ACC. They therefore should influence neither the conversion of ACC to ethylene nor the conversion of NF-ACC to ethylene. However, in some experiments ethylene production from NF-ACC was partly inhibited by AVG or AOA while in others it was not. Also the action of NF-ACC and of ACC on plants or plant parts could occasionally be inhibited by these compounds. As yet we cannot explain the lack of reproducibility of these experiments. We cannot exclude the possibility that NF-ACC and perhaps ACC may act by inducing ethylene biosynthesis.

Discussion

The results of our investigation show that NF-ACC has a potent effect in ethylene biotests and in typical ethylene responses of whole plants although it has little ability to stimulate ethylene production.

The hypothesis that NF-ACC might increase the sensitivity of the plant to ethylene is not supported by the experiments detailed here. For instance applications of mixtures of ACC and NF-ACC reveal an intermediate response of the plant compared with ACC or NF-ACC treatments (*Figure 30.1*). Similar results were achieved in greenhouse and field trials in which growth inhibition and defoliation were investigated (unpublished data). If NF-ACC increases the sensitivity of plant tissue to ethylene it would be expected that an increased response would be noted when ACC and NF-ACC are applied in combination.

Since CO_2 acts as an inhibitor of NF-ACC action it would appear that the effects of NF-ACC are mediated via ethylene. The hypothesis that different pools of ethylene and ethylene precursor may exist in the plant cell (Konze and Lürssen, 1983) might be used to explain these results, regardless of what the mechanism of action of NF-ACC may be—conversion to ethylene or induction of ethylene biosynthesis. According to this theory of compartmentation it is suggested that ACC and NF-ACC are transported with differing efficiencies to different compartments within the plant cell (*Figure 30.3*). This assumption may be justified, since both compounds differ in their physicochemical data. It is further suggested that ACC—and perhaps also NF-ACC—can be converted to ethylene in at least two compartments (Pool A and Pool B in *Figure 30.3*) within the cell. There must also be an additional compartment, in which ACC is converted to N-malonyl-ACC (Pool C in *Figure 30.3*), an inactive conjugation product of ACC (Amrhein *et al.*, 1981). Additionally it is known, that ACC can accumulate in cells without being

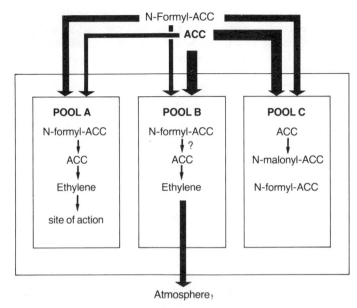

Figure 30.3 Proposed compartmentation of ethylene, ACC and NF-ACC in relation to the primary site of action of ethylene (after Konze and Lürssen, 1983)

converted immediately to ethylene (Burroughs, 1957; Vähätalo and Virtanen, 1957), thus there may even be a fourth pool of ACC in cells. In this compartmentation hypothesis as suggested NF-ACC is preferentially transported to Pool A, which is adjacent to the site of primary action of ethylene (*Figure 30.3*). There it is converted to ACC and further to ethylene. The ethylene developed near its site of action should of course be highly effective. Pool A cannot easily be reached by ethylene produced outside the pool. ACC on the other hand is not as effectively transported to Pool A as NF-ACC. It is mainly converted to ethylene in Pool B. Ethylene produced in Pool B is released to the atmosphere. This is the ethylene mainly measured in the experiments, but it is rather ineffective in stimulating plant responses. If NF-ACC acts to induce ethylene biosynthesis a compartmentation similar to that described above and in *Figure 30.3* can be considered. The only difference would be, that NF-ACC and ACC stimulate ethylene biosynthesis in Pool A instead of serving as ethylene precursors.

In contrast to all other effects ACC is more active in inducing abscission than is NF-ACC. In this case it must be assumed that either the site of action is different or the accessibility to the primary site of action is changed when plants become responsive to ethylene with regard to abscission.

Our theory implies that very small amounts of ethylene reaching the primary site of action may cause the full plant response. Thus amounts of ethylene measured with the usual techniques after application of ethylene related plant growth regulators are meaningless in relation to their effectiveness.

The very low rate of ethylene metabolism which seems to be correlated with ethylene action (Beyer, 1979a, 1979b; Beyer, Chapter 12) could support our theory. A possible way to test our theory of compartmentation would be to study the metabolism of ethylene after application of NF-ACC and ACC labelled in position 3 or 4. If NF-ACC serves as an ethylene precursor nearer the site of primary action of ethylene than ACC we would expect a higher specific activity of the ethylene metabolites after treatment with NF-ACC than after treatment with ACC. However, if NF-ACC and ACC act by inducing ethylene biosynthesis, the outcome of such an experiment would be difficult to interpret. In any case NF-ACC may be an appropriate tool for further investigations into the primary site of action of ethylene.

References

ABELES, F.B. (1973). *Ethylene in Plant Biology*. Academic Press, New York
ADAMS, D.O. and YANG, S.F. (1979). *Proceedings of the National Academy of Sciences USA*, **76**, 170–174
AMRHEIN, N., SCHNEEBECK, D., SKORUPKA, H., TOPHOFF, S. and STÖCKIGT, J. (1981). *Naturwissenschaften*, **68**, 619–620
BEYER, E.M. (1979a). *Plant Physiology*, **63**, 169–173
BEYER, E.M. (1979b). *Plant Physiology*, **64**, 971–974
BURROUGHS, L.F. (1957). *Nature*, **179**, 360–361
BURG, S.P. (1973). *Proceedings of the National Academy of Sciences USA*, **70**, 591–597
KONZE, J. and LÜRSSEN, K. (1983). In *Hohenheimer Arbeiten*, Vol. 129, Regulation des Phytohormongehaltes, pp. 145–166. Ed. by F. Bangerth. Verlag Eugen Ulmer, Stuttgart
LÜRSSEN, K., NAUMANN, K. and SCHRÖDER, R. (1979). *Zeitschrift für Pflanzenphysiologie*, **92**, 285–294
LÜRSSEN, K. (1982). In *Chemical Manipulation of Crop Growth and Development*, pp. 67–78. Ed. by J.S. McLaren. Butterworths, London
LÜRSSEN, K. (1984). In *Physiology of Plant Growth Substances*, Ed. by S.S. Purohit, Bikaner, Agro Botanical Publishers, India, in press
SCHRÖDER, R. and LÜRSSEN, K. (1978). *Deutsche Offenlegungsschrift*, **28** (24), 517
VÄHÄTALO, M.-L. and VIRTANEN, A.I. (1957). *Acta Chemica Scandinavica*, **11**, 741–756
ZIMMER, K. (1968). *Die Gartenbauwissenschaft*, **33**, 415–462

31

COMMERCIAL SCALE CATALYTIC OXIDATION OF ETHYLENE AS APPLIED TO FRUIT STORES

C.J. DOVER
East Malling Research Station, East Malling, Maidstone, Kent, UK

Introduction

To develop an effective ethylene removal system for the commercial storage of fresh produce it is important to define the requirements adequately. When a beneficial effect of low ethylene storage has been determined its realization is dependent on the quantification of three main factors.

(1) How low is it necessary to keep the ethylene concentration and for how long?
(2) What ethylene production rate can be expected at these concentrations?
(3) Are there likely to be any external sources of ethylene that may appreciably increase the effective production rate?

In addition there are other aspects to consider; for example, will ethylene removal alone be sufficient to maintain a low fruit production rate; are additional chemical or other treatments necessary; will harvesting requirements be consistent with the quality of the final product? Much of this information can be obtained from laboratory scale storage experiments and it should then be possible to establish whether ethylene removal on a commercial scale is likely to be feasible for the produce concerned using a particular ethylene removal technique. An example of one such approach designed to maintain very low levels of ethylene in apple stores is given here.

The catalytic converter

An ethylene scrubbing system using a catalytic converter has been developed specifically to control superficial scald on Bramley's seedling apples without the post-harvest application of an antioxidant. Knee and Hatfield (1981) showed that to achieve scald control it is necessary to maintain an ethylene concentration of below $1\,\mu l\,l^{-1}$ for most of the storage period. The maximum ethylene production rate associated with these 'low ethylene' Bramley apples was approximately $1\,\mu l\,kg^{-1}\,h^{-1}$. Systems were thus designed to maintain $1\,\mu l\,l^{-1}$ in the storage atmosphere of fruit producing $1\,\mu l\,kg^{-1}\,h^{-1}$. This meant that early in storage, when the production rate is $0.1\,\mu l\,kg^{-1}\,h^{-1}$ or less, a store concentration of approximately

$0.1\,\mu l\,l^{-1}$ could be expected. In developing the catalytic converter system the performance of catalyst material was assessed in laboratory scale units described by Dover and Sharp (1982).

The conversion efficiency of a catalyst is defined as:

$$\text{Conversion efficiency} = \left[1 - \frac{E_o}{E_i}\right] \times 100$$

where in this case E_o = ethylene concentration leaving catalyst, E_i = ethylene concentration entering catalyst and is the percentage of ethylene removed by the catalyst.

Typical performance curves for a platinum catalyst are shown in *Figure 31.1* from which it can be seen that by operating at 200 °C small changes in temperature will have little effect on the efficiency of ethylene removal by the system. Conversion efficiency is, however, slightly reduced and is more sensitive to temperature change at the lower ethylene concentration.

Space velocity is the number of catalyst bed volumes passed in 1 h; it thus relates flow to catalyst volume and is a parameter that is kept constant when scaling a system. The effect of gas flow rate on conversion efficiency is shown in *Figure 31.2*. These data suggest that a space velocity of 30 000–35 000 bed volumes/h represents

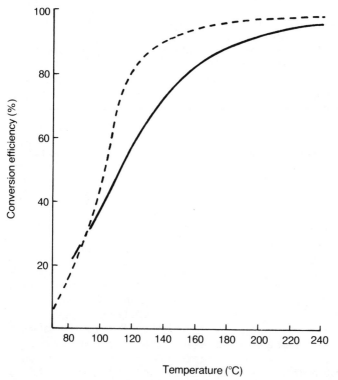

Figure 31.1 Effect of catalyst temperature on the C_2H_4 conversion efficiency of a platinum catalyst: (———) C_2H_4 1 $\mu l\,l^{-1}$; (- - -) C_2H_4 20 $\mu l\,l^{-1}$

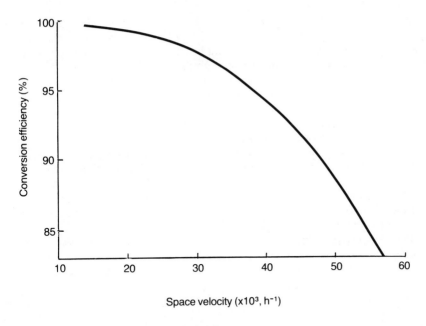

Figure 31.2 Effect of space velocity through a platinum catalyst on its C_2H_4 conversion efficiency

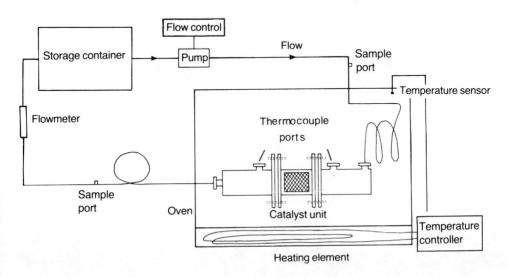

Figure 31.3 Catalytic converter system for the removal of ethylene from experimental storage cabinets

a reasonable compromise between high conversion efficiency and a small catalyst volume.

A schematic diagram of the catalyst system that has been produced to control the ethylene concentration in experimental storage cabinets containing up to 360 kg of apples is shown in *Figure 31.3*. The storage atmosphere is pumped from the cabinet, into an oven, through a coil of copper pipe to achieve the oven temperature and into the catalyst unit which is a tube with a flanged centre section to hold catalyst samples. The outer sections of the unit have, in addition to the inlet and outlet connections, thermocouple ports to measure the temperature onto and off the catalyst. The gas is then returned to the storage cabinet through a sufficient length of tubing inside the cold storage room for it to reach a temperature close to that of the cabinet.

The catalytic converter system designed to operate on a 20 tonne semicommercial fruit store has already been described (Dover, 1983). It differs from the cabinet system in that the storage atmosphere is heated by an element inside the pipework before passing over the catalyst and returns to the store through two heat exchangers. *Figure 31.4* shows conversion efficiency against temperature for the semicommercial system; more than 90% of the ethylene passing into the unit is removed when operating at 200 °C.

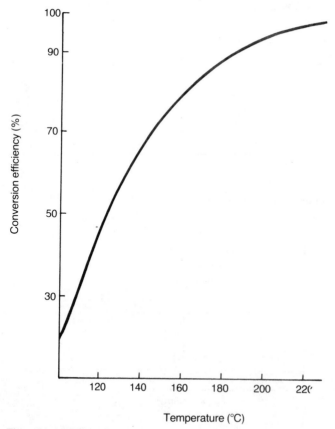

Figure 31.4 Effect of catalyst temperature on C_2H_4 conversion efficiency of the 20 tonne store catalytic converter system

Ethylene removal during storage of Bramley's Seedling apples

Bramley's Seedling apples that were not treated with an antioxidant at harvest were stored at 4 °C in 9% CO_2 12% O_2 in the semicommercial 20 tonne store and in experimental storage cabinets during the 1981/82 and 1982/83 storage seasons. The 20 tonne store and one cabinet containing 360 kg of fruit were fitted with catalytic converter systems. In 1981/82 a further 360 kg of untreated fruit were kept in a cabinet without an ethylene scrubbing system. In 1982/83 an additional unscrubbed cabinet of fruit treated with the antioxidant ethoxyquin (0.25% solution) was stored. Results obtained during 1981/82 have already been reported (Dover, 1983); here the results for 1982/83 are presented and compared with those for 1981/82.

To ensure that fruit in the first year would be at risk from superficial scald it was picked early (Fidler, 1959) approximately two weeks before the recommended start of commercial harvesting. However, early picking may be expected to retard subsequent ethylene production by the fruit, making it easier for the scrubbing system to maintain low ethylene levels during storage. Thus, in 1982, fruit was picked close to the recommended start of commercial harvesting. That this is the case is demonstrated in *Figure 31.5* which shows that for the first 100 days the ethylene concentration in the control cabinet in 1982/83 was 25–30 days ahead of the 1981/82 levels. In 1981/82 ethylene was maintained at a low level (0.01–0.02 µl l^{-1}) in the scrubbed store for 250 days followed by a rapid rise (*Figure 31.6*).

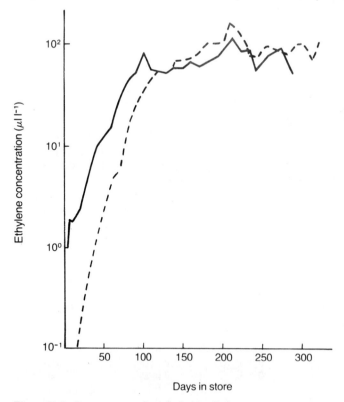

Figure 31.5 Log concentration of ethylene in the storage atmosphere of the unscrubbed (control) cabinet: (---) 1981/82; (——) 1982/83

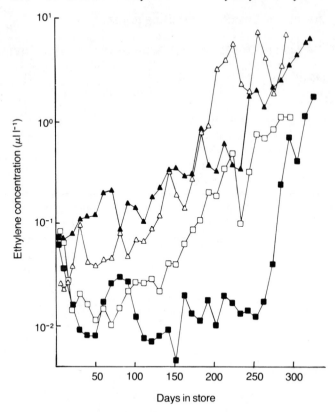

Figure 31.6 Log concentration of ethylene in the storage atmosphere: ethylene scrubbed store ■ 1981/82, □ 1982/83; ethylene scrubbed cabinet ▲ 1981/82, △ 1982/83

However, in 1982/83, this low level was maintained for less than 100 days but was followed by a much slower rate of increase. Technical problems delayed attainment of full scrubbing capacity in the low ethylene cabinet in 1981. These difficulties resulted in an initially higher ethylene concentration in the cabinet in 1981 compared to 1982 (*Figure 31.6*). This was maintained for approximately 170 days after which the production rate of the 1982/83 (later-picked) fruit began to exceed that of the 1981/82 (earlier picked) fruit. This can also be seen in *Table 31.1* where, comparing the scrubbed cabinet for the two years, a reduced scrubbing capacity for

Table 31.1 THE PRODUCTION OF ETHYLENE BY FRUIT STORED IN A LOW ETHYLENE ATMOSPHERE

Days in store	Ethylene production rate ($\mu l\ kg^{-1}\ h^{-1}$)			
	Scrubbed cabinet		Scrubbed store	
	1981/82	1982/83	1981/82	1982/83
10	0.01	0.038	0.049	0.11
20	0.023	0.023	0.021	0.024
50	0.072	0.020	0.011	0.011
100	0.068	0.028	0.007	0.023
150	0.13	0.18	0.008	0.028
200	0.16	0.9	0.007	0.10
280	1.8	2.0	0.21	1.9

Table 31.2 CONCENTRATION OF ETHYLENE IN THE CORE CAVITY OF THE FRUIT AND IN THE STORAGE ATMOSPHERE

Days in store		Ethylene concentration ($\mu l\ l^{-1}$)			
		1981/82		1982/83	
		220	320	170	287
Scrubbed store	Internal	0.012	5.1	0.06	3.8
	External	0.011	1.7	0.05	0.8
Scrubbed cabinet	Internal	0.025	12.3	0.21	9.8
	External	0.020	6.2	0.20	5.5

the first 30 days storage in 1981 allowed the ethylene production to increase. It is interesting that, although the production rate in 1981/82 remained steady and at a higher rate during the earlier part of the storage period, this did not result in an earlier rapid rise in production rate than in 1982/83. In the 20 tonne store, ethylene production declined from a relatively high rate, remained steady and then increased towards the end of the storage period. In 1981/82 (Dover, 1983) internal ethylene concentrations in fruit from scrubbed cabinets were close to those in the storage atmosphere until late in the season when they diverged. This was also seen in 1982/83. *Table 31.2* shows internal and storage atmosphere concentrations at the middle and end of the storage period. Early in storage it was observed that there were a few fruits that had considerably higher internal ethylene concentrations than the rest of the fruit in the sample and had thus not responded to ethylene removal. Although the numbers of ripening fruits tended to increase during storage, a stage was eventually reached when internal ethylene concentrations generally exceeded that of the store atmosphere indicating that ethylene production rates were increasing in the fruit population as a whole.

Effect of ethylene removal on superficial scald

The incidence of superficial scald was assessed as an index of severity on a five point scale based on the percentage area of the skin affected (Johnson, Allen and Warman, 1980). An index maximum of 100 was recorded where more than 50% of the skin of all fruits in a sample were affected by scald. The index of scald for unscrubbed control fruit not treated with an antioxidant increased in a similar way during both 1981/82 and 1982/83 (*Figure 31.7*) reaching 60 after 250 days when all fruit were severely affected. Scald was evident, in both the scrubbed cabinet and store, earlier in 1982/83 than in 1981/82. As the rate of development of scald in the untreated fruit was similar in both seasons, the earlier occurrence of scald symptoms on fruit from the scrubbed systems in 1982/83 is probably the result of the poorer control of ethylene in that season. Consistent with this is the association between improved scald control in the store compared with the scrubbed cabinet and the lower ethylene concentrations in the store.

Effect of ethylene removal on fruit quality

The effect of ethylene removal on fruit firmness is shown in *Figure 31.8*. Fruit was softer when it went into store in 1982 than in 1981; however, firmness loss was

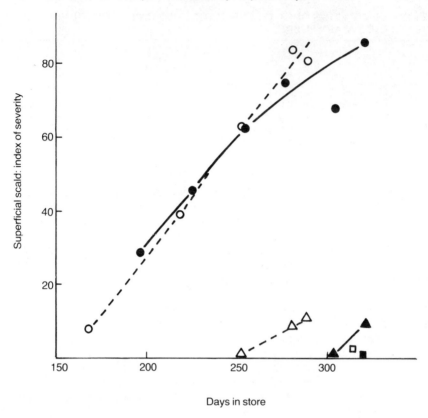

Figure 31.7 Effect of ethylene removal on the incidence of superficial scald as measured by an index of severity: Untreated control ● 1981/82, ○ 1982/83; ethylene scrubbed cabinet ▲ 1981/82, △ 1982/83; ethylene scrubbed store ■ 1981/82, □ 1982/83

smaller in 1982. In both storage seasons low ethylene fruit was consistently firmer than unscrubbed fruit when removed from storage. Even after 12 days at 18 °C a firmness advantage was still evident in the low ethylene fruit. Soluble pectin increased more in control fruit than in low ethylene fruit (*Figure 31.9*). This is in contrast to small changes in both scrubbed and unscrubbed Bramley's seedling reported by Knee (1975). Acid levels dropped from 13 to 8 mg g^{-1} and sugar increased from 64 to 90 mg g^{-1} in all treatments. Colour measurements showed no difference in green colour between ethylene scrubbed and ethoxyquin treated fruit; chlorophyll loss was also similar. The severity of scald on untreated controls precluded meaningful colour measurements.

Core flush was assessed using an index of severity, slight (1), moderate (2), severe (3), giving an index maximum of 60 for a 20 fruit sample. Removing ethylene from the storage atmosphere reduced core flush in both seasons (*Table 31.3*). As core flush in Bramley is not normally seen until late in storage in 9% carbon dioxide at 4 °C, the observed effect could result from ethylene removal slowing fruit maturation in store and delaying the appearance of the disorder.

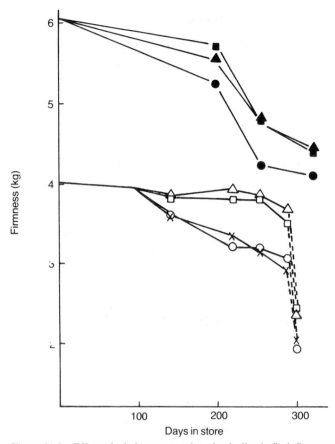

Figure 31.8 Effect of ethylene removal on the decline in flesh firmness in store: ethylene scrubbed store ■ 1981/82, □ 1982/83; ethylene scrubbed cabinet ▲ 1981/82, △ 1982/83; untreated control ● 1981/82, ○ 1982/83; treated control × 1982/83. ——, firmness on removal from store; ---, decline in firmness when fruit held at 18°C (12 days). Flesh firmness measured using 8 mm plunger on a semi-automatic penetrometer (Topping, 1981)

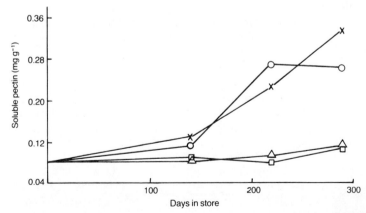

Figure 31.9 Effect of ethylene removal on increase in soluble pectin, 1982/83; ethylene scrubbed store □; ethylene scrubbed cabinet △; untreated control cabinet ○; treated control cabinet ×

Table 31.3 THE EFFECT OF ETHYLENE REMOVAL ON CORE FLUSH AT THE END OF STORAGE

	Core flush index			
	1981/82		1982/83	
	On removal	+14 days at 18°C	On removal	+12 days at 18°C
Scrubbed store	6.7	14	0.5	5
Scrubbed cabinet	4.8	7.9	0.6	3
Untreated control	21	38	4.9	29
Treated control			2.9	29

Conclusions

The small scale catalyst system described here provides an effective method for controlling ethylene concentrations in experimental storage containers so that the effects of low external ethylene on various types of fresh produce can be investigated. In this way operating parameters can be established to enable any benefits to be applied on a commercial scale. The control of ethylene in stored Bramley apples in the second season, when fruit was picked later, was not as good as in the first season when the fruit was picked two weeks early. However, it was still adequate to control superficial scald and markedly retard softening throughout ten months storage. It should be noted that Bramley's seedling apples are harvested well before their respiratory climacteric, and in 1982/83 the harvested fruit still took 10–13 days at 12°C to reach the preclimacteric minimum. These findings are not inconsistent with results for McIntosh apples (Liu, 1978; Chapter 32) where ethylene removal was found to be less effective in retarding softening if the fruit were not harvested early (i.e. about eight days prior to the onset of the climacteric). Unlike most dessert varieties of apple, the normal commercial picking date for the processing apple Bramley is 10–14 days preclimacteric and hence this represents a distinct advantage in the application of ethylene removal techniques.

Once the internal ethylene concentration of the fruit begins to increase above that of the storage atmosphere there may be no benefit in continuing to remove ethylene, since Dover (1984) has shown that the development of superficial scald and decline in flesh firmness in Bramley were largely unaffected by ceasing ethylene removal towards the end of storage. If this can be established generally, an ethylene removal system might not be required for the latter storage period for apples.

The 20 tonne store system has been developed into a commercially available unit for Bramley apple stores of 100 tonnes capacity (a typical store size in the UK). The unit's capacity for other produce is dependent on the answers to the questions posed in the introduction, particularly the ethylene concentration that must be maintained and the associated production rate.

Acknowledgements

Thanks are due to Johnson Matthey Chemicals Ltd for supplying catalyst material and for discussion and advice on its use. The author is also indebted to Dr M. Knee for his interest in this work and to Miss J. Coppins for technical assistance.

References

DOVER, C.J. (1983). *Preprints of 16th International Congress of Refrigeration Commission C2*, pp. 165–170

DOVER, C.J. (1984). *East Malling Research Station Report for 1983* (in press)

DOVER, C.J. and SHARP, L.E. (1982). *East Malling Research Station Report for 1981*, 138

FIDLER, J.C. (1959). *Proceedings of Xth International Congress of Refrigeration, Copenhagen*, **111**, 181–185

JOHNSON, D.S., ALLEN, J.G. and WARMAN, T.M. (1980). *Journal of the Science of Food and Agriculture*, **31**, 1189–1194

KNEE, M. (1975). *Journal of Horticultural Science*, **50**, 113–120

KNEE, M. and HATFIELD, S.G.S. (1981). *Annals of Applied Biology*, **98**, 157–165

LIU, F.W. (1978). *Journal of the American Society for Horticultural Science*, **103**, 388–392

TOPPING, A.J. (1981). *Journal of Agricultural Engineering Research*, **26**, 179–183

LOW ETHYLENE CONTROLLED-ATMOSPHERE STORAGE OF MCINTOSH APPLES

F.W. LIU
Cornell University, Ithaca, New York, USA

Introduction

Despite contradictory results from many early studies on the benefit of ethylene removal during apple storage (Liu, 1977b), low ethylene controlled-atmosphere (CA) storage of apples may have some commercial advantages. For instance, ethylene removal during CA storage is beneficial for the retention of fruit firmness in McIntosh apples (Forsyth, Eaves and Lightfoot, 1969; Lougheed et al., 1973; Liu, 1977a, 1978, 1979; Lidster, Lightfoot and McRae, 1983), in Cox's Orange Pippen apples (Knee, Cockburn and Watt, 1975; Knee, Hatfield and Cockburn, 1982) and in Bramley's seedling apples (Knee and Hatfield, 1981). This report summarizes some of the important findings obtained from research carried out at Cornell University between 1975 and 1983 on the effects of low ethylene CA storage on McIntosh apples.

Storage conditions

EXPERIMENTAL

McIntosh apples were harvested at various dates from mature trees grown in the orchard of Cornell University in Ithaca, New York, USA. On the day of harvest, 40–50 apples were placed into each of many 19-litre glass jars, which were left unstoppered at 3.3 °C overnight and then stoppered the next day. In experiments to test the effects of daminozide concentration, ethylene concentration, harvest date, etc. on storage life, three jars were used as three replicates each containing apples from one tree. The effect of daminozide was tested by spraying trees with aqueous solutions of either $1000\,\mathrm{mg\,l^{-1}}$ or $2000\,\mathrm{mg\,l^{-1}}$ about two months prior to harvest. Throughout the storage period the humidified gas mixture (2.5–3% O_2; 3–5% CO_2; 92.0–94.5% N_2 and 90–95% RH) flowed through each storage jar at a rate of $200\,\mathrm{ml\,min^{-1}}$, with the aid of a flow control board. The oxygen and carbon dioxide concentrations were monitored using an Orsat gas analyser and controlled to within ± 10% error. To test the effect of ethylene concentration on storage life, apples were exposed in the storage jars to three levels of ethylene, low, medium and high. Medium and high ethylene conditions were created by metering ethylene (2% in nitrogen) into the gas mixture to obtain a constant ethylene level of 10 and 500 µl l^{-1} respectively. These treatments commenced one month after the start of each experiment. Low ethylene conditions were those encountered in storage jars to which no ethylene was added. When levels of ethylene in these 'low' jars exceeded 1 µl l^{-1}, about 50 g of alumina beads impregnated with KMnO$_4$ (commercially sold as Purafil) in a mesh bag were placed into the jar. This procedure commonly maintained the ethylene concentration in the low treatments at <1 µl l^{-1}. Since commercial CA storage normally contains ≥500 µl l^{-1} ethylene, the 'high' ethylene CA treatment was actually a simulation of 'normal' CA. Ethylene concentration was monitored with a gas chromatograph fitted with an activated alumina column. At the end of the storage period, the apples in each jar were divided into two groups, each containing 20 to 25 apples, which were kept in paper bags at 21 °C in the air. The apples from one bag were used for quality evaluation after one day out of storage and from the other bag seven days out of storage.

COMMERCIAL

During the 1982–83 storage season, about 30 metric tons of McIntosh apples were stored in a semicommercial sized (inside dimension: 8 m × 4.6 m × 4.3 m) CA room. The apples were harvested on September 15th–16th, 1982 from trees which had been sprayed with 2000 mg l^{-1} daminozide in mid-July. The CA room was sealed on September 17th and then purged with nitrogen. The oxygen concentration in the room was reduced to 3% by September 20th and carbon dioxide increased to 3% by September 23rd. The room was operated at 2.2–3.3 °C with 2.5–3.0% O$_2$ and 3% CO$_2$ from September 23rd to October 22nd and with 2.5–3.0% O$_2$ and 5% CO$_2$ thereafter. An ethylene scrubber with outside dimension of 46 cm × 61 cm × 183 cm, containing 45 kg of 'Purafil', and a 22.3 m^3 min^{-1} blower was placed inside the room. The ethylene concentration in the room reached 0.23 µl l^{-1} on September 22nd, before the scrubber was turned on, but was maintained at <0.04 µl l^{-1} after September 23rd with the scrubber operating. The storage test was terminated on May 2nd, 1983, after 7.5 months of storage. Six samples each containing 20 apples were randomly selected from six storage bins for quality evaluation one day after removal from storage. A further six samples were taken from the same six bins and kept at 21 °C for quality evaluation one week after removal from storage.

In the quality evaluations, fruit firmness was measured with an Effegi penetrometer equipped with an 11.1 mm diameter plunger tip. The measurements were expressed in Newtons (N; 1 N = 0.102 kg force). The soluble solid content in the fruit juice was measured with a hand refractometer. The acidity of the juice was measured by titration against 0.1 N NaOH and expressed in grams of malate per litre of juice.

Effect of ethylene concentration on storage

From 1975 to 1978, a total of 14 paired comparisons were made of apples stored in CA with either 10 µl l^{-1} or 500 µl l^{-1} ethylene. Early and late harvested apples, with or without daminozide tree sprays, were included in the comparisons (*Table 32.1*). In only two of the 14 paired comparisons were apples stored in 10 µl l^{-1} ethylene significantly firmer than those stored in 500 µl l^{-1} ethylene; and the largest

Table 32.1 EFFECT OF ETHYLENE CONCENTRATION IN CA STORAGE ON THE FIRMNESS OF McINTOSH APPLES

Year	Date of harvest	Daminozide applied (mg l^{-1})	Duration of storage (month)	Firmness (N) after storage*		
				(<1 µl l^{-1} ethylene)	(10 µl l^{-1} ethylene)	(500 µl l^{-1} ethylene)
1975	Sept. 19	0	5.5	36.3 a†	31.9 b	30.5 b
			7.5	35.7 a	31.8 b	30.0 c
	Sept. 26	0	5.5	40.2 a	38.2 ab	36.7 b
			7.5	34.9 a	33.0 ab	31.5 b
1976	Sept. 11	0	5	56.9 a	53.9 a	55.9 a
	Sept. 17	0	5	54.9 a	52.0 ab	51.0 b
	Sept. 22	0	5	48.1 ab	50.0 a	46.1 b
1977	Sept. 15	0	7.5	53.9 a	49.0 b	46.1 b
	Sept. 21	0	7.5	47.1 a	46.1 a	44.1 a
	Sept. 22	1000	7.5	67.7 a	50.0 b	48.1 b
	Sept. 28	1000	7.5	59.8 a	46.1 b	44.1 b
1978	Sept. 28	0	7.5	35.2 a	35.5 a	36.5 a
		1000	7.5	40.6 a	38.2 ab	36.5 b
		2000	7.5	45.1 a	38.7 b	37.0 b
1979	Sept. 18	0	7.5	40.0 a	—	38.3 b
		1000	7.5	46.7 a	—	39.1 b
		2000	7.5	48.0 a	—	39.1 b
	Sept. 26	1000	7.5	42.3 a	—	35.1 b
		2000	7.5	43.6 a	—	35.6 b
1981	Sept. 16	0	5	52.8 a	—	54.9 a
		0	7.5	48.6 a	—	47.0 a
		2000	5	70.9 a	—	61.9 b
		2000	7.5	67.3 a	—	52.9 b
	Sept. 21	0	5	44.9 a	—	44.3 a
		0	7.5	41.7 a	—	39.8 a
		2000	5	65.4 a	—	54.8 b
		2000	7.5	63.4 a	—	44.9 b
	Sept. 26	2000	5	64.9 a	—	50.5 b
		2000	7.5	55.2 a	—	38.7 b

*Values are means of triplicate samples each containing 20 to 25 apples
†Mean separation within rows by Duncan's multiple range test, 5% level. Values with the same letter are not significantly different

difference in firmness was <4 N. These comparisons indicate that ethylene concentrations of 10 µl l^{-1} and 500 µl l^{-1} in CA would not make a commercially significant difference to McIntosh apple firmness.

From 1975 to 1981, a total of 15 paired comparisons were made of control apples (not sprayed with daminozide) stored in CA with <1 µl l^{-1} and 500 µl l^{-1} ethylene. In seven of the 15 paired comparisons, apples stored in <1 µl l^{-1} ethylene were significantly firmer than those stored in 500 µl l^{-1} ethylene (*Table 32.1*). The differences, which were statistically significant at $P = 0.05$ level, ranged from 1.7 to 6.8 N. The significant differences were found more often in early harvested apples than in late harvested apples. These results are in agreement with those of Lougheed *et al*. (1973): who found that 'Lowered ethylene levels were often but not always coincident with firmer fruit of early harvests after extended storage'.

In all 14 paired comparisons made between 1977 and 1981 to compare daminozide sprayed apples stored in CA with <1 µl l^{-1} and 500 µl l^{-1} ethylene, the apples stored in low ethylene CA were always firmer than those stored in high ethylene CA (*Table 32.1*). The firmness differences ranged from 4.1 to 19.6 N.

Effect of daminozide

In four years of observations, daminozide sprayed McIntosh apples were always firmer than control apples after 5 to 7.5 months of storage in low ethylene ($<1\,\mu l\,l^{-1}$) CA (*Table 32.2*). The firmness differences between daminozide sprayed and control apples ranged from 5.4–21.7 N. However, there was little difference between daminozide concentrations of 1000 and 2000 mg l^{-1}. In only one of six paired comparisons were apples sprayed with 2000 mg l^{-1} daminozide significantly firmer ($P = 0.05$) than those sprayed with 1000 mg l^{-1} daminozide; and the firmness difference was only 4.5 N. Thus a spray concentration of 1000 mg l^{-1} seems to be adequate for McIntosh apples. Daminozide sprayed McIntosh apples kept very firm for 7.5 months in low ethylene CA, but control apples did not. The best lot of

Table 32.2 EFFECT OF DAMINOZIDE ON THE FIRMNESS OF McINTOSH APPLES STORED IN LOW ETHYLENE CA

Year	Date of harvest	Duration of storage (month)	Firmness (N) after storage*		
			(0 mg l^{-1} daminozide)	(1000 mg l^{-1} daminozide)	(2000 mg l^{-1} daminozide)
1977	Sept. 21–22	7.5	47.1 b†	67.7 a	—
1978	Sept. 28	7.5	35.2 c	40.6 b	45.1 a
1979	Sept. 18	7.5	40.1 b	46.7 a	48.0 a
	Sept. 26	7.5	—	42.3 a	42.3 a
1981	Sept. 16	5	52.8 b	—	70.9 a
		7.5	48.6 b	—	67.3 a
	Sept. 21	5	44.9 b	—	65.4 a
		7.5	41.7 b	—	63.4 a
1982	Sept. 14	7.5	—	61.6 a	60.2 a
	Sept. 19	7.5	—	59.1 a	59.7 a
	Sept. 24	7.5	—	57.5 a	58.4 a

*Values are means of triplicate samples each containing 20 to 25 apples
†Mean separation within rows by Duncan's multiple range test, 5% level
Values with the same letter are not significantly different

Table 32.3 EFFECT OF DAMINOZIDE ON THE FIRMNESS OF McINTOSH APPLES STORED IN CA WITH 500 $\mu l\,l^{-1}$ ETHYLENE

Year	Date of harvest	Duration of storage (month)	Firmness (N) after storage*		
			(0 mg l^{-1} daminozide)	(1000 mg l^{-1} daminozide)	(2000 mg l^{-1} daminozide)
1977	Sept. 21–22	7.5	44.1 a†	48.1 a	—
	Sept. 28	7.5	—	44.1	—
1978	Sept. 28	7.5	36.5 a	36.5 a	37.0 a
1979	Sept. 18	7.5	38.3 a	39.1 a	39.1 a
	Sept. 26	7.5	—	35.1 a	35.6 a
1981	Sept. 16	5	54.9 b	—	61.9 a
		7.5	47.0 b	—	52.9 a
	Sept. 21	5	44.3 b	—	54.8 a
		7.5	39.8 b	—	44.9 a
	Sept. 26	5	—	—	50.5
		7.5	—	—	38.7

*Values are means of triplicate samples each containing 20 to 25 apples
†Mean separation within rows by Duncan's multiple range test, 5% level
Values with the same letter are not significantly different

Table 32.4 EFFECT OF DAMINOZIDE ON THE SENESCENT BREAKDOWN OF McINTOSH APPLES AFTER 7.5 MONTHS OF STORAGE IN CA WITH 500 µl l^{-1} ETHYLENE

Year	Date of harvest	Senescent breakdown apples (%)*		
		(0 mg l^{-1} daminozide)	(1000 mg l^{-1} daminozide)	(2000 mg l^{-1} daminozide)
1977	Sept. 21–22	2.2	7.8	—
	Sept. 28	—	13.3	—
1978	Sept. 28	0.0	6.7	22.3
1979	Sept. 18	2.5	7.1	7.8
	Sept. 26	—	37.8	74.4
1981	Sept. 16	6.4	—	0.0
	Sept. 19	15.2	—	0.0
	Sept. 26	—	—	10.0

*Values are sums of senescent breakdown that occurred in storage and in the seven-day holding period at 21 °C after storage to the total of 60 to 75 apples in triplicated samples

daminozide sprayed apples had an average firmness of 67.7 N after 7.5 months of storage, but the best control apples had an average firmness of only 48.6 N.

Daminozide sprayed apples were firmer than control apples after 7.5 months of storage in high ethylene (500 µl l^{-1}) CA in only one out of four years of observation (*Table 32.3*). The interaction between daminozide treatment and high ethylene storage also resulted in a high percentage of senescent breakdown compared to control fruit (*Table 32.4*). The greatest incidence of senescent breakdown occurred in apples which were sprayed with a high concentration of daminozide, harvested late and stored in high ethylene CA. Thus daminozide sprays not only fail to prevent McIntosh apples from softening in high ethylene CA but also have a tendency to aggravate McIntosh senescent breakdown in high ethylene CA. In comparison, McIntosh apples stored in low ethylene CA for up to 7.5 months never developed a significant incidence of senescent breakdown. Apples were classified as having senescent breakdown when the flesh firmness was <31 N. Most of the breakdown apples which had been sprayed with daminozide had brown cortex tissues, but the breakdown apples which were not sprayed with daminozide had little tissue browning.

Effect of harvest date

For McIntosh apples which were not sprayed with daminozide (control), a significant benefit of low ethylene (<1 µl l^{-1}) over high ethylene (500 µl l^{-1}) CA was found only when the apples were harvested early, i.e. at the preclimacteric stage (*Table 32.1*). Low ethylene CA had little benefit for late harvested control apples.

Since the date of the onset of the climacteric (DOC) differs from one fruit to another on the same tree and from one tree to another in the same orchard, it was rather difficult to define the DOC for an orchard. Before 1981, DOC was judged from the respiration and ethylene production rates of multiple five-apple samples collected at intervals. After 1981, DOC was determined by the internal ethylene content of fruit samples. Three representative trees were selected from an orchard. Ten apples were picked from each tree at three-day intervals beginning in early September; and the internal ethylene of each fruit was measured within 8 h after picking. Of the total of 30 apples tested each time, if three or more apples representing at least two trees were found to have $\geq 0.2\,\mu l\,l^{-1}$ internal ethylene, that date was defined as the DOC of that orchard. If one of the three trees had three or more apples with $\geq 0.2\,\mu l\,l^{-1}$ internal ethylene but neither of the other two trees had any apples containing that high level of ethylene, five more apples were picked from each of these two trees. If any fruit of the supplementary picking contained $\geq 0.2\,\mu l\,l^{-1}$ ethylene, that date was defined as the DOC of the orchard. The validity of this procedure was supported by two observations. Firstly, after the DOC was defined for an orchard, subsequent 30-apple samples frequently had three or more apples representing at least two trees containing $\geq 0.2\,\mu l\,l^{-1}$ of ethylene. Secondly, significant fruit abscission was always found within five days after the DOC. Judged by this procedure, the DOC of control apples in the Cornell Orchard was September 11th in 1981 and was September 9th in 1982.

Table 32.5 VARIATION IN FIRMNESS OF LOW ETHYLENE CA McINTOSH APPLES HARVESTED ON DIFFERENT DATES IN DIFFERENT YEARS

Year	Daminozide applied $(mg\,l^{-1})$	Date of harvest	Firmness (N)*		Loss of firmness in storage (%)
			Before storage	After 7.5 months of storage	
1977	1000	Sept. 22	64.7	67.7	−4.6
		Sept. 28	59.8	59.8	0.0
1978	1000	Sept. 28	—	40.6	—
	2000	Sept. 28	—	45.1	—
1979	1000	Sept. 18	70.4	46.7	33.7
		Sept. 26	68.6	42.3	38.3
	2000	Sept. 18	71.8	48.0	33.1
		Sept. 26	68.9	43.6	36.7
1981	2000	Sept. 16	70.7	67.3	4.8
		Sept. 21	65.4	63.4	3.1
		Sept. 26	61.4	55.2	10.1
1982	1000	Sept. 14	65.4	61.6	5.8
		Sept. 19	64.5	59.1	8.4
		Sept. 24	65.4	57.5	12.1
	2000	Sept. 14	65.8	60.2	8.5
		Sept. 19	65.8	59.7	9.3
		Sept. 24	63.6	58.4	8.2

*Values are means of triplicate samples each containing 20 to 25 apples

Each year the DOC of daminozide sprayed McIntosh apples was found to be at least ten days later than that of control apples. Normally daminozide sprayed apples harvested within ten days from the DOC of control apples had a good storage life in low ethylene CA. Although there was a tendency for later picked daminozide sprayed apples to be somewhat softer than earlier picked ones, the differences were small if the picking dates were within ten days from the DOC of control apples (*Table 32.5*).

For practical applications of this procedure, three representative trees in each orchard should be set aside as control trees while other trees are sprayed with daminozide. The DOC of control apples can be identified by the above mentioned procedure. The daminozide sprayed apples should be harvested for low ethylene CA storage within ten days from the DOC.

Variation among storage seasons

The 1977, 1981 and 1982 apple crops had excellent firmness retention in low ethylene CA, but the 1978 and 1979 apple crops did not (*Table 32.5*). In 'good' storage seasons such as 1977–78, 1981–82 and 1982–83, daminozide sprayed apples which were harvested within ten days from the DOC of control apples had lost <10% of their original firmness after 7.5 months of low ethylene CA storage, but in 'bad' storage seasons such as 1978–79 and 1979–80, comparable apples lost >30% of firmness in storage (*Table 32.5*). The cause(s) of this big variation has not been identified. It might be due to experimental error, such as error in daminozide application or in storage gas composition. It might alternatively be due to weather and cultural conditions. Even in 'bad' storage seasons, however, the daminozide sprayed low ethylene CA apples were significantly firmer than 'high' ethylene CA apples (*Table 32.1*). Therefore, low ethylene CA storage was an effective procedure for apple firmness retention, despite the existence of seasonal variations.

Storing McIntosh apples in a low ethylene CA room

In the 1982–83 trial storage of McIntosh apples in a semi-commercial sized low ethylene CA room, daminozide sprayed apples were harvested six to seven days after DOC of the control apples. When measured one day after harvest, these apples had an average firmness of 64.9 N. After 7.5 months in low ethylene CA, the average firmness was 62.8 N (*Table 32.6*), a loss of only 3.2% of the original firmness. Among the 120 apples randomly selected from six storage bins after storage, the softest single apple had a firmness of 57.8 N, a desirable firmness for consumers (Liu and King, 1978). After seven days in air at 21 °C, the average firmness was 52.8 N (*Table 32.6*), an acceptable firmness to most consumers. No senescent breakdown was found in any of the apples.

The soluble solids and acidity of these apples were not measured before storage. Previously published results (Liu, 1979) and unpublished data indicated that McIntosh apples did not change their soluble solid composition significantly but did

Table 32.6 FIRMNESS, SOLUBLE SOLID CONTENT AND ACIDITY OF McINTOSH APPLES STORED IN LOW ETHYLENE CA IN 1982–83*

	Firmness (N)	Soluble solids (%)	Acidity (g l^{-1} juice)
One day after storage			
Mean	62.8	12.1	4.93
Mean minimum	61.6	11.7	4.49
Mean maximum	63.4	12.5	5.66
Absolute minimum	57.8	—	—
Absolute maximum	68.9	—	—
Seven days after storage			
Mean	52.8	12.0	4.37
Mean minimum	48.3	11.4	3.72
Mean maximum	55.5	12.5	5.03
Absolute minimum	40.0	—	—
Absolute maximum	60.0	—	—

*Six samples each containing 20 apples randomly selected from a storage bin were evaluated at each time. The mean firmness of similar apples before storage was 64.9 N

lose about 26% of the original acidity during 7.5 months of low ethylene CA storage. The mean soluble solids of 12.1% for these apples (*Table 32.6*) was slightly above normal and the mean acidity of 4.93 g malate per litre juice for these apples (*Table 32.6*) was comparable to most low ethylene CA apples stored in jars in previous years. The $KMnO_4$ impregnated alumina beads in the ethylene scrubbing chamber were examined after the trial storage. Less than 10% of the beads had lost their pink colour. This indicated that <4.5 kg of the 45 kg of beads originally placed in the scrubber were actually used up in absorbing ethylene generated by the 30 metric tons of apples stored in the room.

The low ethylene CA stored apples from the 1982–83 season were sold in the retail store at Cornell Orchard. According to the manager's estimate, three out of four customers responded very favourably to the product. The fruit had normal sweetness and acidity. The fruit was praised for having excellent firmness and freshness. However, the fruit was criticized by some customers for lack of full ripe McIntosh aroma, green colour and poor flavour when cooked.

Conclusions

These results show that daminozide sprayed McIntosh apples retained their firmness, freshness and consumer appeal for up to 7.5 months whilst held in a semi-commercial low ethylene CA-storage room. The apples produced small amounts of ethylene but this was easily removed using Purafil, as an ethylene purger. Thus it seems that low ethylene CA storage of McIntosh apples may prove to be a commercially viable proposition.

References

FORSYTH, F.R., EAVES, C.A. and LIGHTFOOT, H.J. (1969). *Canadian Journal of Plant Science*, **49**, 567–572

KNEE, M., COCKBURN, J.T. and WATT, J.B. (1975). *East Malling Research Station Report for 1974*, p. 79.

KNEE, M. and HATFIELD, S.G.S. (1981). *Annals of Applied Biology*, **98**, 157–165

KNEE, M., HATFIELD, S.G.S. and COCKBURN, J.T. (1982). *East Malling Research Station Report for 1981*, p. 132

LIDSTER, P.D., LIGHTFOOT, H.J. and McRAE, K.B. (1983). *Scientia Horticulturae*, **20**, 71–83

LIU, F.W. (1977a). *Journal of the American Society for Horticultural Science*, **102**, 93–95

LIU, F.W. (1977b). In *Horticultural Report No. 28, Controlled Atmospheres for the Storage and Transport of Perishable Agricultural Commodities*, pp. 86–96. Edited by D.H. Dewey. Department of Horticulture, Michigan State University, East Lansing, Michigan, USA

LIU, F.W. (1978). *Journal of the American Society for Horticultural Science*, **103**, 388–392

LIU, F.W. (1979). *Journal of the American Society for Horticultural Science*, **104**, 599–601

LIU, F.W. and KING, M.M. (1978). *HortScience*, **13**, 162–163

LOUGHEED, E.C., FRANKLIN, E.W., MILLER, S.R. and PROCTOR, J.T.A. (1973). *Canadian Journal of Plant Science*, **53**, 317–322

33

A COMMERCIAL DEVELOPMENT PROGRAMME FOR LOW ETHYLENE CONTROLLED-ATMOSPHERE STORAGE OF APPLES

G.D. BLANPIED, JAMES A. BARTSCH and J.R. TURK
Cornell University, Ithaca, New York, USA

Introduction

The New York programme for the commercialization of low ethylene controlled-atmosphere (CA) storage for apples was divided into two phases, firstly the estimation of the ethylene climacteric in orchard blocks that were scheduled for low ethylene CA, and secondly the evaluation of ethylene scrubbers that were operating under commercial CA conditions. The programme was put into practice during the 1983–84 season, except where otherwise stated.

Estimations of the ethylene climacteric

MCINTOSH

In 1983 apples from 14 orchard blocks were scheduled for low ethylene CA at two western New York storage facilities. All blocks were within a 50 km radius of a

Table 33.1 DATES OF THE ETHYLENE CLIMACTERIC OF ATTACHED McINTOSH APPLES IN WESTERN NEW YORK

Climacteric date, 1983	Location	Strain	Understock	Year planted
Sept. 9	Wolcott	Spurmac	26	1977
20	Geneva	Macspur	9	1976
	Wolcott	Macspur	9	1977
23	Geneva	Rogers	VII	1965
26	Geneva	Macspur	9	1976
	Wolcott	Rogers	9/106	1975
		Spurmac	9/111	1978
28	Lafayette	Cornell	106	1971
		Cornell	VII	1967
	Mexico	Rogers	VII	1971
	Wolcott	Macspur	9/106	1976
		Rogers	VII	1967
29	Wolcott	Standard	sdlg	1935
Oct 1	Wolcott	Cornell	106	1969

location midway between these two CA facilities. The 14 blocks were sampled at three or four-day intervals between the 24th August and the 1st October. Each orchard block sample was 21 apples: seven apples picked from each of three trees. Individual apple internal ethylene analyses were performed at the university post-harvest laboratory within a period of no more than 6 h after picking. The date of the ethylene climacteric for an orchard block was defined as the date on which at least two of the three trees had at least one apple with an internal ethylene concentration in excess of $1.0\,\mu l\,l^{-1}$. All ethylene climacteric dates listed in *Table 33.1* were verified by ethylene analyses on the next sample date. The ethylene climacteric occurred very early in one block. In the remaining 13 blocks it occurred during the 12 day period from 20th September to 1st October. There was no apparent relationship between timing of the ethylene climacteric and orchard location, strain of McIntosh, understock or age of the trees (*Table 33.1*).

EMPIRE

Observations of the 1981 crop indicated that delayed harvest diminished the beneficial effects of low ethylene CA (*Table 33.2*). Analyses of fruit internal ethylene during the 1980 harvest period suggested the effect of harvest date on the

Table 33.2 EFFECT OF HARVEST DATE ON THE RESPONSE OF EMPIRE APPLES TO LOW ETHYLENE CA

Crop year	Measured variable	Harvest date		
1981		29th September	6th October	10th October
	Firmness[a] (kg)	1.0	0.7	0.5
	% Breakdown[b]	12.1	6.0	5.4
1980		2nd October	8th October	15th October
	Firmness[a]	1.7	1.8	0.5
	Harvest ethylene[c]	0.08	0.21	0.31
	May ethylene[d]	0.18	0.35	7.05

[a] Firmness (low ethylene CA − regular CA)
[b] Percentage senescent breakdown (regular CA − low ethylene CA)
[c] Average fruit internal ethylene concentration ($\mu l\,l^{-1}$)
[d] Average ethylene production ($\mu l\,l^{-1}\,h^{-1}$) in air at 0 °C by low ethylene CA Empire apples

benefits of low ethylene CA might be related to the development of the ethylene climacteric (*Table 33.2*). Liu (1978; Chapter 32) had shown that a beneficial response of McIntosh apples to low ethylene CA was dependent upon their harvest before the ethylene climacteric was initiated in the crop. Internal ethylene concentrations are plotted (*Figure 33.1a*) for individual apples harvested on 2nd, 8th and 15th October 1980, and are compared with the firmness responses of comparable apples to the removal of ethylene from the CA environment. Similar data are shown in *Figure 33.1b* for apples picked on 28th September 1982 from three different blocks. Internal ethylene analyses at harvest indicated that lots Y and Z were considerably more advanced on the ethylene climacteric curve than were apples from the 15th October harvest in 1980 (12.5 and 31.3% *versus* none of the apples with more than $0.5\,\mu l\,l^{-1}$). These observations suggested that firstly late picked Empire apples showed little or no benefit from low ethylene CA, and secondly Empire apples picked at an early calendar date benefited from low

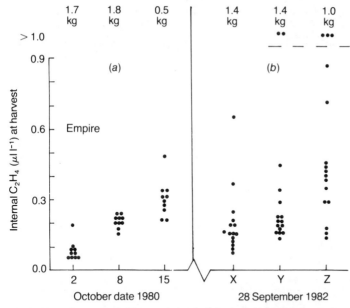

Figure 33.1 Internal ethylene concentrations for individual Empire apples sampled from one block on three dates in 1980 (a) and from three blocks on one date in 1982 (b). Numbers at the top of the figure indicate differences in poststorage flesh firmness attributable to low ethylene CA in comparison with normal CA

ethylene CA even though the ethylene climacteric had been partially initiated in the crop.

The procedures used to observe the ethylene climacteric in the 1983 Empire crop were similar to those described for McIntosh. Seven Empire blocks (A-1 to A-7) were located in one large orchard and four additional blocks (B-E) were located on separate farms that were within 16 km of the 'A' blocks and at approximately the same elevation. The internal ethylene concentration for the 11 blocks indicated the ethylene climacteric date for the area was on or about 27th September (*Table 33.3*), if the climacteric date was defined as the earliest date on which 10% or more of the apples had internal ethylene concentrations in excess of $0.5\,\mu l\,l^{-1}$. However, there appeared to be significant between-block variations in ethylene climacteric dates, even when comparisons were made between adjacent 'A' blocks (*Table 33.4*). The analyses indicated ethylene climacteric dates ranged from 21st September (A-1, A-2, A-4) to 27th September (A-3, A-6) and 30th September or later for the other six blocks.

Table 33.3 INTERNAL ETHYLENE ANALYSES FOR EMPIRE APPLES FROM 11 ORCHARD BLOCKS IN ULSTER COUNTY (1983)

September sample date	No. of fruits	Average internal ethylene for fruits with $<0.5\,\mu l\,l^{-1}$ ($\mu l\,l^{-1}$)	Fruits with $>0.5\,\mu l\,l^{-1}$ ethylene (% of sample)
7	165	0.04	1.1
14	165	0.11	2.4
21	231	0.10	6.5
27	220	0.20	14.1
30	220	0.19	11.8

Table 33.4 PERCENTAGES OF EMPIRE APPLE SAMPLES WITH >0.5 µl l^{-1} INTERNAL ETHYLENE. ULSTER COUNTY (1983)

Block	Rootstock	Year planted	September sample date				
			7 (%)	14 (%)	21 (%)	27 (%)	30 (%)
A-1	VII	1977			10	30	25
A-2	VII	1977			20	30	25
A-3	106	1974				20	10
A-4	VII	1977			10	20	25
A-5	106	1972				15	
A-6	III	1975				10	10
A-7	106	1972					10
B	106	1972			10		10
C	106	1977				10	
D	106	1972					
E	VII	1975				10	

Note: No entry indicates 5% or less of the sample with >0.5 µl l^{-1}

A Hudson Valley CA operator received samples from six neighbours who wanted to store Empire apples in his low ethylene CA store. Internal ethylene analyses of the samples indicated a wide range in physiological maturity, i.e. from 100% of the apples below 0.25 to 35% above 0.50 µl l^{-1} (*Figure 33.2*). Based on observations in 1982 (*Figure 33.1b*), the CA operator was advised to accept lots M to Q and reject lot R.

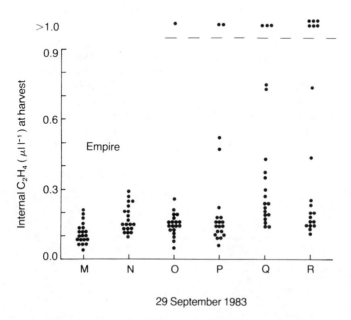

Figure 33.2 Internal ethylene concentrations for individual Empire apples sampled from six neighbouring orchard blocks on 27th September 1983

INFERENCES FOR THE 1984 CROP

Observations on several McIntosh and Empire blocks indicated that the ethylene climacteric for a given cultivar in a limited geographical area may occur during a seven- to ten-day period, but a few blocks may reach this stage of physiological maturity either earlier or later in the season. It is the earlier, not the later maturing blocks that limit the usefulness of marker or index blocks to establish the ethylene climacteric dates for the harvest of low ethylene CA apples. If all blocks are not sampled and analysed before low ethylene CA, one or two early maturing blocks may be placed into a low ethylene CA store and subsequently jeopardize the ability of the ethylene scrubber to maintain the concentration of ethylene below $1.0\,\mu l\,l^{-1}$ in the storage atmosphere.

The necessity for a sequence of internal ethylene samples is illustrated in *Table 33.5*. The percentages of apples with more than $1.0\,\mu l\,l^{-1}$ ethylene were never very

Table 33.5 PERCENTAGES OF McINTOSH APPLES IN 21-FRUIT SAMPLES WITH $>1.0\,\mu l\,l^{-1}$ INTERNAL ETHYLENE

Sample date 1983	% of sample
29 August	0.0
2 September	0.0
6 September	0.0
9 September	19.0
12 September	9.5
15 September	19.0
20 September	33.0
23 September	14.3
26 September	10% drop of crop

high because apples of this cultivar (McIntosh) tend to drop soon after the initiation of the ethylene climacteric. If the block in *Table 33.5* had been sampled once on 23rd September, the 14.3% of the crop with more than $1.0\,\mu l\,l^{-1}$ might not have been considered to be a serious risk for low ethylene CA. If the block was then harvested and placed into low ethylene CA, it probably would have jeopardized the successful out-turn from the low ethylene CA store, because the ethylene climacteric date for that block was two weeks earlier, indicating the block was much more mature than the 14.3% figure indicated.

Use of ethylene scrubbers in CA storage

PRELIMINARY LABORATORY TESTS WITH KMnO$_4$/ALUMINA BEAD ABSORBANTS

In preliminary laboratory tests we found a positive relationship between the rate of ethylene absorption by KMnO$_4$/alumina beads and a number of factors. These included the concentration of ethylene in the atmosphere, the length of residence time of atmosphere in the absorbant bed, the oxygen concentration of the atmosphere, the temperature of the atmosphere/absorbant, the dryness of the atmosphere/absorbant, the smallness in diameter of the absorbant beads and the shortness of time the absorbant was in use. We also found up to threefold

398 Commercial CA storage of apples

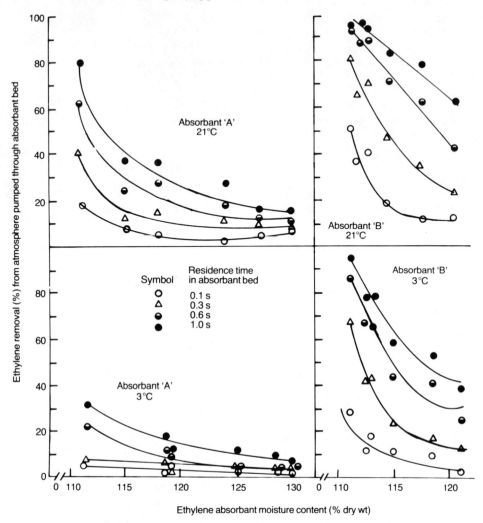

Figure 33.3 Effect of moisture content, flow rate, and temperature on ethylene removal efficiency of ethylene absorbants 'A' and 'B'

differences in the rates of ethylene absorption by beads provided by different manufacturers.

Data in *Figure 33.3* show the effects of four of these variables on the removal of ethylene from a CA apple store atmosphere containing 0.75–$1.50\,\mu l\,l^{-1}$ ethylene. Absorbant B was more efficient than absorbant A under the same conditions of temperature, atmosphere residence time in the absorbant bed, and moisture content of the beads. Under the same conditions of atmosphere residence time and moisture content, both absorbants were more efficient at 21 than at 3 °C. For any set of four curves, an increase in residence time of atmosphere in the absorbant bed was associated with an increase in ethylene removal from the atmosphere. For any

Table 33.6 THE EFFECTS OF ATMOSPHERE HUMIDIFICATION AND DRYING ON ABSORBANT WEIGHT AND ETHYLENE ABSORPTION BY KMnO$_4$/ALUMINA BEADS AT 1 °C[a]

Elapsed hours	Atmosphere treatment between CA store and ethylene absorption column	Ethylene absorbant wt (% of original)	Ethylene removal by absorbant (%)
0.0	Humidifier	100	100
14.2	Humidifier	108	100
23.8	Humidifier	114	96
26.6	Humidifier	116	93
29.6	Humidifier	118	90
38.6	Humidifier	120	58
47.6	Humidifier	122	51
48.1	Desiccant	—	—
48.9	Desiccant	122	54
52.6	Desiccant	120	73
56.6	Desiccant	119	88
73.0	Desiccant	116	94
100.5	Desiccant	101	100

[a] Atmosphere from CA apple store pumped at 250 ml min^{-1} through 8 mm diameter column containing 25 g dry absorbant

set of four curves, the higher the moisture content of the beads the lower the ethylene removal rate. The loss of ethylene removal capacity with increased moisture content was found to be reversible (*Table 33.6*). Not shown in *Figure 33.3* was the observation that under the same conditions of relative humidity, absorbant A had a higher equilibrium moisture content than absorbant B. Thus, under operating CA conditions, where the storage atmosphere is close to 3 °C and the relative humidity usually exceeds 90%, the differences in ethylene removal by absorbants A and B would be even greater than the differences shown at the bottom (left versus right) of *Figure 33.3*.

1982–83 COMMERCIAL CA TEST

The 1982–83 test was conducted in a commercial CA store which held 80 tons of Empire and Delicious apples. Ethylene absorbant beads (140 kg) were placed 1 cm deep on shelves in a plywood box, through which the store atmosphere was blown by a centrifugal fan. The ethylene scrubber was located inside the CA store to eliminate the need for making it airtight. Ethylene levels in the store were maintained at below 2 µl l^{-1} from the late September date of sealing the store until mid-January when the test was abandoned because the store was too leaky. This test demonstrated that ethylene absorbing beads could be used for low ethylene CA under commercial conditions. We also learned that it is preferable to locate an airtight ethylene scrubber outside the CA store, where the beads can be easily inspected and replaced when they are spent.

1983–84 COMMERCIAL CA TESTS

Orchard climacteric dates were measured in several western New York McIntosh blocks and in several Hudson Valley Empire blocks for low ethylene CA operators

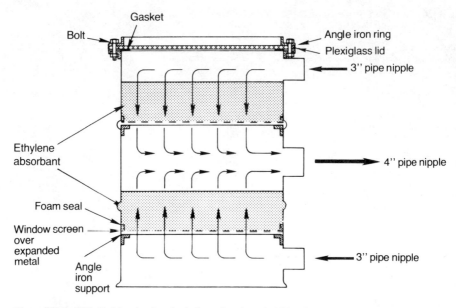

Figure 33.4 KMnO$_4$/alumina bead ethylene absorbant held in 55-gallon drum outside low ethylene CA store. Blower fan located inside store circulated store atmosphere through ethylene absorbant bed

Table 33.7 180 TON LOW ETHYLENE CA McINTOSH TEST (1983)

Days after sealing store	Scrubber ethylene (μl l^{-1})	
	Intake from store	Effluent to store
−12	Start to fill store	
− 1	16.0	—
0	Flush sealed store with N$_2$	
	Start C$_2$H$_4$ scrubber[a]	
4	4.9	4.3
5	Replace absorbant beads	
9	3.9	—
12	3.6	3.0
23	1.15	0.82
	Replace absorbant beads	
30	0.32	0.27
39	0.60	0.57
45	2.56	1.82
51	16.6	16.6
52	Replace absorbant beads	
58	5.48	4.80
63	37.4	27.5
70	140.0	140.0

[a]Two parallel 27-kg Purafil beds in 55-gallon drum, 630 m^3 atmosphere h^{-1}

in each area. One McIntosh store was discontinued in December for reasons discussed below. The remaining McIntosh and Empire stores are presently operating with low ethylene atmospheres. The ethylene scrubbers used for these two stores will be discussed later in this section.

Five factors may have contributed to the failure of the ethylene scrubber (*Figure 33.4*) to maintain low ethylene concentrations in the 180 ton McIntosh store (*Table 33.7*). Firstly although picking of the apples for this store was completed before the ethylene climacteric dates in the blocks scheduled for low ethylene CA, the operator took 12 instead of the recommended three to four days to fill the room. Secondly the operator overestimated the crop in the blocks scheduled for low ethylene CA. Consequently, apples, whose ethylene climacteric date had not been determined were used to finish filling the store. Apples from this block may have had an earlier climacteric date. Thirdly ethylene in the store during the long loading period, $16\,\mu l\,l^{-1}$ on the last day of loading, may have stimulated the ethylene production by the apples. This high concentration of ethylene shows that ethylene scrubbers should be in operation during the store loading period. Fourthly the ethylene scrubber was off for several periods during the early storage season because it was not wired to its own circuit breaker. Finally the required scrubber capacity may have been underestimated.

All apples in the 200 ton McIntosh store (*Table 33.8*, top) were picked preclimacteric from blocks whose ethylene climacteric dates had been established by several twice-weekly internal ethylene measurements. The oxygen was below

Table 33.8 PERFORMANCE DATA FOR TWO ETHYLENE SCRUBBERS IN 1983–84 COMMERCIAL TEST OF LOW ETHYLENE CA FOR APPLES

Test run start	Elapsed days	Scrubber ethylene ($\mu l\,l^{-1}$)		Ethylene removal	
		Intake from store	Effluent to store	(%)	(ml h^{-1})
		Heated catalyst — 200 tons McIntosh			
15 Dec	1	3.08	0.53	83	1275
	14	2.96	0.22	93	1370
	28	1.80	0.20	89	800
	43	1.60	0.27	83	665
	57	1.25	0.20	84	525
	80	5.60	1.20	79	2200
	84	4.32	0.32	93	2000
	91	2.12	0.30	87	910
	$r = 0.987$[a]				
		Ethylene absorbant beads — 200 tons Empire			
21 Dec	1	0.48	0.20	58	280
	13	0.90	0.75	17	150
	27	1.16	1.00	16	160
	34	0.44	0.39	11	50
	$R^2 = 0.84$[b]				
19 Feb	1	0.42	0.18	57	240
	5	0.23	0.16	30	70
	9	0.25	0.19	24	60
	16	0.57	0.32	44	250
	23	0.34	0.26	24	80
	30	0.35	0.27	20	70
	$R^2 = 0.98$[b]				

[a] For ethylene removal (ml h^{-1}), store ethylene ($\mu l\,l^{-1}$)
[b] For ethylene removal (ml h^{-1}), elapsed days + store ethylene ($\mu l\,l^{-1}$)

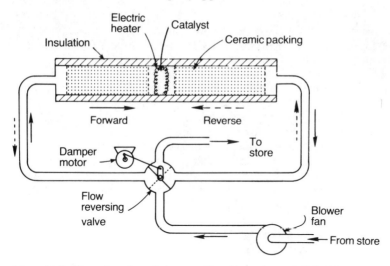

Figure 33.5 Heated catalyst ethylene scrubber (Swingtherm E-500) with ceramic packing heat exchangers and flow reversing valve

3% four days after the earliest picked fruit was loaded into the store. A drum ethylene scrubber (*Figure 33.4*) controlled the ethylene until two months after sealing the store. Then ethylene scrubbing was diverted to the heated catalyst scrubber (*Figure 33.5*), because the drum scrubber required very frequent renewal of the absorbant beads to maintain low ethylene concentrations in the store. The ethylene scrubbing performance of the heated catalyst scrubber is summarized in *Table 33.8*, top. Quantitative removal (ml h^{-1}) of ethylene from the store atmosphere was highly correlated ($r = 0.987$) with the concentration of ethylene in the store. The percentage of ethylene removed from the atmosphere blown through the heated catalyst remained almost constant (87±6%) over the range of observed ethylene concentrations. The heated catalyst scrubber was operated with one Fugi 801A ring compressor and a motorized valve. Atmosphere temperature rise was negligible across the heated catalyst scrubber. The 28 °C temperature rise across the Fugi ring compressor could have been removed by a simple heat exchanger located down stream from the heated catalyst scrubber.

The 200 ton Empire store was sealed six days after the earliest picked apples were loaded into the store. When the drum scrubber (*Figure 33.4*) failed to remove all the ethylene from the effluent of the propane burner used to reduce the oxygen in the store, the room was flushed with fresh air and then with nitrogen gas to establish the low oxygen atmosphere. Two drums in series were then used to maintain the ethylene below 1 µl l^{-1} in the store. The blower was located inside the CA store and the scrubber drums were located outside the CA store. Every two weeks the drum lids and the top absorbant beds were removed to permit stirring of the beads to minimize channelling of atmosphere in the absorbant beds. At this time the beads were visually inspected and replaced when the centres of the beads turned from purple to brown. The vertical bed scrubber (*Figure 33.6*), which replaced the drum scrubber in mid-December, was designed to reduce the time required for the two-weekly maintenance. The ethylene scrubbing performances of the drum and vertical bed scrubbers were similar, but the latter was much easier to

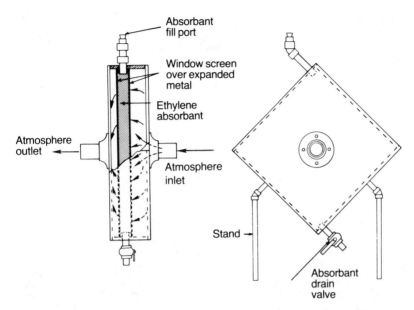

Figure 33.6 KMnO$_4$/alumina bead ethylene absorbant held in vertical bed outside low ethylene CA store. Blower fan located inside CA store circulated store atmosphere through ethylene absorbant bed

Table 33.9 SOME CHARACTERISTICS OF TWO ETHYLENE SCRUBBERS TESTED WITH 200 TON LOW ETHYLENE CA STORES FILLED WITH McINTOSH (HEATED CATALYST) OR EMPIRE (ABSORBANT BEADS) APPLES (1983–84 STORAGE SEASON)

	Heated catalyst[a]	*Ethylene absorbant beads*[b]
Capacity		
Atmosphere (m^3 h^{-1})	500	1000
Ethylene removal (ml h^{-1})		
0.1 µl l^{-1} in store atmosphere	43[c]	20[d]
0.5	213	100
1.0	425	200
Electrical usage (kW)	8.0	1.2
Estimated costs ($ US)		
Capital	10 800	300
Annual operating (270 days)		
Electricity ($.08 kw^{-1})		
Fan	3162	622
Heater, etc.	985	0
Chemical	0	2200
Total	4147	2822
5-year total	31 535	14 410
$ bushel^{-1} season^{-1}	0.63	0.29

[a] Swingtherm E-500
[b] 68 kg Purafil in 0.9 × 0.9 × 0.1 m bed
[c] Assuming 85% reduction in ethylene across the scrubber (*Table 33.8*)
[d] Assuming 20% reduction in ethylene across the scrubber (*Table 33.8*)

maintain. Each week 25% of the 68 kg charge of absorbant beads were drained from the bottom of the scrubber, visually examined for colour, and then reintroduced into the top of the scrubber if the insides of the beads were purple. The movement of the beads in the bed minimized the loss of efficiency associated with channelling of the atmosphere in the absorbant bed.

The quantitative (ml h^{-1}) removal of ethylene by the vertical bed scrubber was influenced by the concentration of ethylene in the store atmosphere and the number of days the beads were in use (*Table 33.8*, middle and bottom). The scrubber operated at about 20% ethylene removal efficiency during more than half of the ethylene absorbing life of the beads.

Comparative data (*Table 33.9*) for the two types of ethylene scrubbers indicated that the ethylene removal capacity of the heated catalyst scrubber was 2.1 times the ethylene removal capacity of the ethylene absorbant beads: the comparative five-year cost per unit of apples stored was 2.2 times greater. If we assume the costs for smaller heated catalyst scrubbers are proportionately less and the costs for larger scrubbers with absorbant beads are proportionately more, the choice between these two scrubber types may not be based on economics.

The profitability of low ethylene CA will ultimately depend upon the cost/benefit ratio. Therefore, the monetary premiums (benefit) received for low ethylene CA of rapid ethylene producers (McIntosh) must be greater than the monetary premiums received for slow ethylene producers (Empire). We intend to obtain some monetary premium data when the rooms are opened in June.

Reference

LIU, F.S. (1978). *Journal of the American Society for Horticultural Science*, **103**, 388–392

APPENDIX

Units and conversion factors

Concentrations of ethylene are commonly expressed on a volume/volume basis and authors have been encouraged to use $\mu l\ l^{-1}$ in their chapters. This is not an SI unit and there are several alternative forms by which concentration can be expressed. The conversion between the various units is that:

$$1\ \mu l\ l^{-1} = 1\ ppm = 1\ mm^3\ dm^{-3} = 10^{-6} m^3 m^{-3} = 0.1\ Pa$$

Rates of ethylene production are expressed either as volume or moles, per unit weight and time. One mole of ethylene at STP occupies a volume of 22.4 litres. Thus at NTP (25°C) 1 nl = 0.041 nmoles and 1 nmole = 24.5 nl of ethylene.

LIST OF PARTICIPANTS

Abeles, Dr F.B.	Appalachian Fruit Research Station, Box 45, Kearneysville WV 25430, USA
Acaster, Dr M.A.	School of Biological Sciences, Biochemistry Department - 4 West, University of Bath, Claverton Down, Bath BA2 7AY, UK
Achilea, Mr O.	Agricultural Research Organization, The Volcani Center, PO Box 6, Bet-Dagan 50-250, Israel
Alderson, Dr P.G.	University of Nottingham, Department of Agriculture & Horticulture, Sutton Bonington, Loughborough, LE12 5RD, UK
Aldrick, Miss S.	University of Nottingham, Department of Applied Biochemistry & Food Science, Sutton Bonington, Loughborough, LE12 5RD, UK
Allen, Professor M.	University of Nottingham, Department of Physiology & Environmental Science, Sutton Bonington, Loughborough, LE12 5RD, UK
Almutaw, Mr M.	Botany Department, University College of Wales Aberystwyth, Penglais, Dyfed SY23 3DA, UK
Alwan, Mr T.	Botany Department, University College of Wales Aberystwyth, Penglais, Dyfed SY23 3DA, UK
Anderson-Taylor, Dr G.	FBC Ltd., Chesterfield Park Research Station, Saffron Walden, Essex CB11 3DJ, UK
Atherton, Dr J.G.	University of Nottingham, Department of Agriculture & Horticulture, Sutton Bonington, Loughborough, LE12 5RD, UK
Baker, Miss A.J.	Plant Science Department, Lincoln College, Canterbury, New Zealand
Banks, Dr N.H.	Postharvest Unit, Department of Applied Biology, University of Cambridge, New Museum Site, Pembroke Street, Cambridge CB2 3DX, UK
Bartley, Dr I.M.	East Malling Research Station, East Malling, Maidstone, Kent ME19 6BJ, UK
Bathgate, Mr B.	University of Nottingham, Department of Physiology & Environmental Science, Sutton Bonington, Loughborough, LE12 5RD, UK
Beyer, Dr E.M.	E.I. du Pont de Nemours & Co. Inc., Agricultural Chemicals Department, Building 402, Experimental Station, Wilmington, DE, USA 19898
Black, Dr C.R.	University of Nottingham, Department of Physiology & Environmental Science, Sutton Bonington, Loughborough, LE12 5RD, UK
Blanpied, Dr G.D.	Pomology Department, Cornell University, Ithaca, NY, 14853, USA

List of participants

Bradbury, Mr G.K.	Hunter Produce, Saphir South, The Old Malt House, Minnis Road, Birchington, Kent CT7 9SG, UK
Bramlage, Dr W.J.	Department of Plant and Soil Sciences, University of Massachusetts, Amherst, MA, 01003, USA
Brennan, Ms A.M.	Department of Botany and Microbiology, University College of Wales Aberystwyth, Penglais, Dyfed SY23 3DA, UK
Browne, Mr T.L.	Waveney Apple Growers, Aldeby, Beccles, Suffolk, UK
Bruinsma, Professor J.	Department of Plant Physiology, Agricultural University, Arboretumlaan 4, 6703 BD Wageningen, The Netherlands
Bufler, Dr G.	Institut für Obstbau, Universitat Hohenheim, 7000 Stuttgart 70, Federal Republic of Germany
Burdon Mr J.	Department of Biological Science, University of Stirling, Stirling, UK
Caygill, Dr J.C.	Tropical Development & Research Institute, 56/62, Gray's Inn Road, London WC1X 8LU, UK
Clarke, Mr F.	Banana-Rite Ltd, 35, Greenes Road, Whiston, Merseyside L35 3RE, UK
Crookes, Mr P.	University of Nottingham, Department of Physiology & Environmental Science, Sutton Bonington, Loughborough LE12 5RD, UK
Dalziel, Dr J.	I.C.I. Plc., Plant Protection Division, Jealott's Hill Research Station, Bracknell, Berks RG12 6EY, UK
Deeley, Ms S.M.	Butterworths Publishers, Borough Green, Sevenoaks, Kent TN15 8PH, UK
Dover, Mr C.J.	East Malling Research Station, East Malling, Maidstone, Kent ME19 6BJ, UK
Elyatem, Mr S.	Applied Biology Department, University of Cambridge, Pembroke Street, Cambridge, CB2 3DX, UK
Field, Dr R.J.	Plant Science Department, Lincoln College, Canterbury, New Zealand
Finn, Dr G.A.	American Cyanamid, PO Box 400, Princeton NJ 08540, USA
Frazier, Dr H.W.	Monsanto Europe, Rue Laid Burniat, Louvin-la-Neuve, B-1348, Belgium
Fritsch, Dr H.	BASF AG, Landw. Versuchsstation, D-6703 Limburgerhof, W. Germany
Frost, Mrs C.E.	Glasshouse Crops Research Institute, Worthing Road, Littlehampton, W. Sussex BN16 3PU, UK
Gomez Lim, Mr M.A.	Botany Department, Edinburgh University, Kings Building, West Mains Road, Edinburgh EH9 3JH, UK
Gower, Mr D.J.	Johnson Matthey Chemicals Ltd, Orchard Road, Royston, Herts SG8 5HE, UK
Graham, Mr K.W.	Kirdford Growers Ltd, Kirdford, Billingshurst, W. Sussex RH14 0NO, UK
Grierson, Dr D.	University of Nottingham, Department of Physiology & Environmental Science, Sutton Bonington, Loughborough LE12 5RD, UK
Grimwade, Miss J.A.	University of Nottingham, Department of Physiology & Environmental Science, Sutton Bonington, Loughborough LE12 5RD, UK
Halder-Doll, Dr H.	Institut fur Obstban, Universitat Hohenheim, 7000 Stuttgart 70, Federal Republic of Germany
Hall, Professor M.A.	Department of Botany and Microbiology, University College of Wales, Aberystwyth, Penglais, Dyfed, SY23 3DA, UK

Harman, Miss J.	Glasshouse Crops Research Institute, Worthing Road, Littlehampton, Sussex BH16 3PU, UK
Hatfield, Mr S.G.S.	East Malling Research Station, East Malling, Maidstone, Kent ME19 6BJ, UK
Heap, Mr R.D.	Shipowners Refrigerated Cargo Research Association, 140, Newmarket Road, Cambridge CB5 8HE, UK
Hill, Miss S.E.	Botany Department, Royal Holloway College, University of London, Egham Hill, Egham, Surrey TW20 0EX, UK
Hillman, Professor J.R.	Department of Botany, The University, Glasgow, G12 8QQ, UK
Hoad, Dr G.V.	Long Ashton Research Station, Long Ashton, Bristol BS18 9AF, UK
Hobson, Dr G.E.	Glasshouse Crops Research Institute, Worthing Road, Rustington, Littlehampton, W. Sussex BN16 3PU, UK
Horton, Dr R.F.	Department of Botany and Genetics, University of Guelph, Guelph, Ontario, Canada
Howarth, Miss C.J.	Department of Botany and Microbiology, University College of Wales, Aberystwyth, Penglais, Dyfed, SY23 3DA, UK
Jackson, Dr M.B.	Agricultural & Food Research Council, Letcombe Laboratory, Wantage, Oxfordshire OX12 9JT, UK
Jarvis, Mr R.	Geest Industries Ltd, Spalding, Lincs, UK
Johnson, Mr D.S.	East Malling Research Station, East Malling, Maidstone, Kent ME19 6BJ, UK
Justin, Mr S.H.F.W.	Department of Plant Biology and Genetics, University of Hull, Hull, HU6 7RX, UK
Khalifa, Mr M.M.	National Academy for Science Research, PO Box 8004, Tripoli, Libya
Knapp, Miss J.	University of Nottingham, Department of Physiology & Environmental Science, Sutton Bonington, Loughborough LE12 5RD, UK
Knee, Dr M.	East Malling Research Station, East Malling, Maidstone, Kent ME19 6BJ, UK
Knight, Dr J.N.	East Malling Research Station, East Malling, Maidstone, Kent ME19 6BJ, UK
Liu, Dr F.	Department of Pomology, Cornell University, Ithaca, NY 14853, USA
Love, Mr J.	Produce Technical & Development Department, Scientific Services Division, J. Sainsbury Plc, Stamford Street, London SE1 9LL, UK
Lürssen, Dr K.	Pflanzenshutz, Auwendungstecknik, Biologische Forschung, Bayer AG, D 5090, Leverhusen, Federal Republic of Germany
Manning, Mr K.	Glasshouse Crops Research Institute, Worthing Road, Rustington, Littlehampton, W. Sussex BN16 3PU, UK
Marston, Mr G.G.	Tesco Group of Companies, Tesco House, Delemere Road, Cheshunt, Herts, UK
Maunders, Dr M.	University of Nottingham, Department of Physiology & Environmental Science, Sutton Bonington, Loughborough LE12 5RD, UK
McLaren, Dr J.S.	Monsanto Europe, Rue Laid Burniat, Louvain-la-Neuve, B-1348, Belgium
Medlicott, Mr A.P.	Department of Biological Science, The Polytechnic, Wolverhampton, UK
Menzies, Mr M.H.	Produce Department, Marks and Spencer plc, 47, Baker Street, London W1, UK

List of participants

Miah, Mr M.A.S.	Department of Botany, University College of Wales Aberystwyth, Penglais, Dyfed SY23 3DA, UK
Mitchell, Mr B.B.	Stay Fresh Ltd, Glenavon House, 39, Common Road, Claygate, Esher, Surrey KT10 0HG, UK
Murfitt, Mr R.F.A.	Silsoe College, Silsoe, Bedford MK45 4DT, UK
Nicol, Miss F.B.	Department of Botany and Microbiology, University College of Wales, Aberystwyth, Penglais, Dyfed SY23 3DA, UK
Nichols, Dr R.	Glasshouse Crops Research Institute, Worthing Road, Rustington, Littlehampton, W. Sussex BN16 3PU, UK
Norton, Dr G.	University of Nottingham, Department of Applied Biochemistry & Food Science, Sutton Bonington, Loughborough LE12 5RD, UK
Orchard, Dr B.	Horticultural Advisory Service, St. Martins, Guernsey
Osborne, Dr D.J.	Developmental Botany, AFRC Weed Research Organisation, Begbroke Hill, Oxford OX5 1PF, UK
Partis, Mr J.P.	Luddington Experimental Horticulture Station, Stratford-Upon-Avon, Warwicks, CV37 9SJ, UK
Pearson, Mr J.	Marks and Spencer Plc, 47, Baker Street, London W1A 1ND, UK
Plant, Miss A.L.	University of Nottingham, Department of Physiology & Environmental Science, Sutton Bonington, Loughborough LE12 5RD, UK
Proctor, Ms F.J.	Tropical Development & Research Institute, 56–62 Gray's Inn Road, London WC18 XLU, UK
Purton, Miss M.E.	University of Nottingham, Department of Physiology & Environmental Science, Sutton Bonington, Loughborough LE12 5RD, UK
Reid, Dr D.M.	Department of Biology, Plant Physiology Research Group, University of Calgary, Alberta, Canada
Ridge, Dr I.	Biology Department, The Open University, Walton Hall, Milton Keynes, MK7 6AA, UK
Robbie, Miss F.A.	East Malling Research Station, East Malling, Maidstone, Kent ME19 6BT, UK
Roberts, Dr I.	Department of Cell Biology, John Innes Institute, Colney Lane, Norwich NR4 7UH, UK
Roberts, Dr J.A.	University of Nottingham, Department of Physiology & Environmental Science, Sutton Bonington, Loughborough, LE12 5RD, UK
Rossall, Dr S.	University of Nottingham, Department of Physiology & Environmental Science, Sutton Bonington, Loughborough, LE12 5RD, UK
Sanders, Mr I.O.	Department of Botany and Microbiology, University College of Wales, Aberystwyth, Penglais, Dyfed, SY23 3DA, UK
Schindler, Miss C.B.	University of Nottingham, Department of Physiology & Environmental Science, Sutton Bonington, Loughborough LE12 5RD, UK
Schouten, Dr S.P.	Sprenger Instituut, Haagsteeg 6, 6700 AA, Wageningen, The Netherlands
Schuch, Dr W.	I.C.I. Corp. Bio Science Group, PO Box 11, The Heath, Runcorn, Cheshire, UK
Schulz, Dr G.	BASF, ZHP/FB-A30, 6700 Ludwigshafen am Rhein, Federal Republic of Germany
Scrine, Mr G.R.	Shipowners Refrigerated Cargo Research Association, 140, Newmarket Road, Cambridge CB5 8HE, UK
Sexton, Dr R.	Department of Biological Science, Stirling University, Stirling, Scotland, UK

List of participants   411

Seymour, Mr G.B.	Tropical Development & Research Institute, 56–62, Gray's Inn Road, London WC1X 8LU, UK
Sharples, Dr R.O.	East Malling Research Station, Fruit Storage Division, East Malling, Maidstone, Kent ME19 6BT, UK
Shipway, Dr M.R.	Kirton Experimental Horticulture Station, Kirton, Boston, Lincs, UK
Skorupka, Mr H.	Ruhr-Universitat Bochum, Lehrstuhl fur Pflanzenphysiologie, Gebaude ND, Postfach 10 21 48, 4630 Bochum 1, Federal Republic of Germany
Slater, Dr A.	University of Nottingham, Department of Physiology & Environmental Science, Sutton Bonington, Loughborough LE12 5RD, UK
Smith, Dr A.R.	Department of Botany and Microbiology, University College of Wales, Aberystwyth, Penglais, Dyfed, SY23 3DA, UK
Smith, Dr B.G.	Unilever Research, Colworth House, Sharnbrook, Bedford MK44 1, UK
Smith, Mr C.	University of Nottingham, Department of Physiology & Environmental Science, Sutton Bonington, Loughborough LE12 5RD, UK
Smith, Dr S.M.	East Malling Research Station, East Malling, Maidstone, Kent ME19 6BJ, UK
Sobeilt, Mr W.	University of Nottingham, Department of Agriculture & Horticulture, Sutton Bonington, Loughborough LE12 5RD, UK
Stead, Dr A.D.	Department of Botany, Royal Holloway and Bedford Colleges, Callow Hill, Virginia Water, Surrey GU25 42H, UK
Stow, Dr J.R.	East Malling Research Station, East Malling, Maidstone, Kent ME19 6BJ, UK
Taeb, Mr A.G.	University of Nottingham, Department of Agriculture & Horticulture, Sutton Bonington, Loughborough LE12 5RD, UK
Taylor, Dr I.B.	University of Nottingham, Department of Physiology & Environmental Science, Sutton Bonington, Loughborough LE12 5RD, UK
Taylor, Ms J.E.	Department of Botany and Microbiology, University College of Wales, Aberystwyth, Penglais, Dyfed, SY23 3DA, UK
Treharne, Professor K.J.	Long Ashton Research Station, Plant Sciences Division, Long Ashton, Bristol, UK
Tucker, Dr G.A.	University of Nottingham, Department of Applied Biochemistry & Food Science, Sutton Bonington, Loughborough LE12 5RD, UK
Tucker, Dr M.L.	Department of Molecular Plant Biology, University of California, Berkeley, USA
Venis, Dr M.A.	Sittingbourne Research Centre, Sittingbourne, Kent ME9 8AE, UK
Willümsen, Mr K.	Department of Vegetable Crops, Agricultural University of Norway, PO Box 22, 1432 AAS-NLH, Norway
Wilson, Dr J.M.	School of Plant Biology, University College of North Wales Bangor, Gwynedd, N. Wales, UK
Withers, Dr L.	University of Nottingham, Department of Agriculture & Horticulture, Sutton Bonington, Loughborough LE12 5RD, UK
Whittington, Professor W.J.	University of Nottingham, Department of Physiology & Environmental Science, Sutton Bonington, Loughborough LE12 5RD, UK
Woods, Miss S.L.	Long Ashton Research Station, Long Ashton, Bristol BS18 9AF, UK
Yang, Professor S.	Department of Vegetable Crops, University of California Davis, CA 95616, USA

INDEX

Abscisic acid,
  effect on mesocotyl elongation, 248
  role in abscission, 186–187, 192
  role in leaf senescence, 269–270
  signal in growth, 213
Abscission, 2–4, 64–66, 173–192, 216, 273
  effectors of, 136, 177–179
  ethylene sensitivity and, 179–183
  flower bud, 179–183, 199–200, 350
  fruit, 363–364
  leaf, 110, 134–135, 200–203
  role of ABA, 186–187
  ultrastructural changes, 207–211
Abscission zone, 134, 173–175, 199–207
ACC, 9–19, 23–26, 93–97, 364–372
  discovery of, 9
  effects of temperature on, 58–64
  effects of waterlogging on, 14, 244, 259
  in flowers, 74–75, 83, 86
  in fruit, 34–35, 94–96, 148
  in leaf, 271
  in pollen, 79–80, 344
ACC synthase,
  activity of, 9–10, 23–24, 63–64, 156, 159, 271
  effects of temperature on, 58
  effects of IAA on, 44, 63
  inhibition by rhizobitoxin, 310
  properties of, 23
  purification of, 24
Accommodation response, 229, 231, 234, 251
Acetylene, 107–108, 179, 321
S-Adenosylmethionine, 23
Adventitious rooting, 258
Aerenchyma, 2, 249, 253–255
Alfalfa, 127–128, 131
D-Amino acids, 16–18
Aminocyclopropane-1-carboxylic acid (*see* ACC)
Amino-2-ethylcyclopropane-1-carboxylic acid,
  conversion to 1-butene, 11, 26, 90
  effect on malonyltransferase, 18
  stereoisomers of, 11, 19, 25, 90

Aminoethoxyvinylglycine (AVG),
  and apple storage, 310
  as an inhibitor of ACC synthase, 2–3, 23, 62, 373
  effect on abscission, 75, 77, 185, 187, 189
  effect on flowers, 89–90, 348
  effect on growth, 217–218, 220, 248
  effect on roots, 255, 259
  effect on senescence, 271–272
Aminoxyacetic acid, 2, 3, 23, 248, 259, 348, 370
Amphibious plants, 44–45, 198–199, 229–239
  effect of CO_2 on, 251
  effect of silver on, 246
Annonaceous fruit (*see* Soursop fruit)
Anthocyanins, 148
Anthracnose, 327
Apical dominance (*see* Correlative inhibition)
Apple, 14, 47, 60–61, 267
  ripening of, 54, 307–309, 379–382
  storage of, 52, 308, 310–311, 377–382, 385–392, 393–404
Arnot–Schulz law, 288
Autocatalysis, 79, 148, 333–338, 342, 346
Auxin, 1–2
  binding sites, 101
  correlative inhibition and, 219–220
  regulation of ACC synthase, 24, 43–44, 186
  role in abscission, 177, 180–183, 186, 204–211,
  role in growth, 198–199, 213, 219–220, 229, 247, 250
  role in wilting, 88, 344
  transport, 181, 190, 221–225
Avena sativa, 41–42, 267–268, 278
Avocado, 65
  cellulase synthesis in, 163–171
  post-harvest handling of, 304, 324–326

Banana,
  ethylene forming enzyme in, 29–35
  post-harvest handling of, 319–322
Bean, 47, 52, 58

413

Bud dormancy, 4
Burg's action hypothesis, 136
1-Butene, 11, 18, 90–91, 107–108, 179

C_3 plants, 39
C_4 plants, 39
Callitriche, 249–251
L-Canaline, 2
Canola, 278, 281–282
Carbon dioxide,
    and ethylene metabolism, 128–130, 132–134, 136, 139–141
    effects on abscission, 177
    effects on ethylene binding, 107, 112
    effects on growth, 246, 249, 251, 307–308
    effects on leaf, 10, 37–45
    effects on wilting, 83, 345
Carbon monoxide, 107, 109, 128, 179
Carnation, 11, 64
    ethylene forming enzyme, 9, 84–86
    response to ethylene, 343–345
    senescence, 9, 127, 348–350
Cellulase, 2, 4
    and abscission, 174–176, 188–190, 199–200, 203–206, 209–210
    and ripening, 152, 163–171
Cell wall,
    acidification, 246, 251
    degradation, 152
Chilling injury,
    in fruit, 319–320, 325
    resistance, 49–53
Chitinase, 2–3, 190
Chloroethylsulphonic acid (HOL 1302), 365–366
2-(Chloroethyl)-sulphonyl-methanol (HOL 1274), 365–366
Chlorophyll, 147, 268–270, 272, 322–323, 327–328, 380
Chlorophyllase, 2
Chloroplast, 111, 151
Cholodny–Went theory, 223, 225
Chromoplast, 151
Cinnamate-4-hydroxylase, 2
β-Citraurin, 322
Citrus fruit, 93
    post-harvest handling of, 322–323
Cloning, 157–159, 167–170
Coiled sprout disorder, 221
Coleoptiles, 245–247
Controlled atmosphere storage, 29, 36, 307–308, 312, 360–361, 377, 382, 385–392, 393–404
Core flush, 380, 382
Corolla,
    abscission, 73–74, 79
    wilting, 74, 78, 344, 346
Correlative inhibition, 2, 213, 214, 216, 219
Cotton, 236
Cucumber, 49, 57–58, 127, 357–361
Cyanide-resistant respiration, 336
Cyanoalanine synthase, 13
Cycloalkenes, 107–109, 118–120

Cycloheximide, 5, 24, 90, 126, 188, 269
Cytochrome oxidase, 307
Cytochrome pathway, 336
Cytokinins, 66, 213, 311
    effect on abscission, 182, 269

Daminozide, 310–311, 385, 387–391
Defoliants, 177
Degreening, 148, 318, 321, 323, 357–361
Dictyosome ratio, 207–211
Digitalis purpurea, 71–73, 75–79
Diiodo-hydroxybenzoic acid (DIHB), 257, 259
Drought, 283–284
Dwarf bean, 50–51, 56–57, 63

Empire apples, 394–397, 399–404
Endoplasmic reticulum, 111–112
Epinasty, 2, 4, 198, 241, 244, 259, 300, 366, 369–370
Etacelasil, 364–366
Ethane, 52–54, 215, 219, 293
Ethephon, 66, 217–220, 249, 255, 323, 326, 354, 364–366
Ethoxyquin, 377, 380
Ethrel, 5
Ethylene action, 3–5, 120, 130–134
    and abscission, 187–190
    effects of inhibitors, 44–45, 133
    fruit ripening and, 159, 305
Ethylene binding sites, 54, 102–114, 117–123, 301, 304
    effects of inhibitors, 105–110, 118
    properties of, 102–113
    purification of, 102–105, 112–113, 117, 120, 122
    subcellular localization of, 110–112
Ethylene climacteric,
    during ripening, 393–397
    during senescence, 270–271
Ethylene forming enzyme, 16, 83–86
    induction of, 10, 86, 88
    localization of, 3, 11–12, 25–26, 90, 99, 306
    oxygen affinity of, 29–34, 307
    properties of, 9, 11, 25
    reaction mechanism, 12, 90
Ethylene glycol, 125–126, 129, 136, 139
Ethylene metabolism, 5, 41, 139–145
    during abscission, 183
    effectors of, 126, 128, 132, 139–145
    kinetics of, 129
    rates of, 139–144
    role of, 114, 127, 130, 371
    products of, 126
    working model of, 125
Ethylene oxide, 103, 125–126, 129–130, 134–136, 139, 144
Ethylene production,
    and pollination, 74–77
    effect of auxin on, 43–44
    effect of CO_2 on 39–41
    effect of light on, 37–39
    effect of stomatal opening on, 41–42

Ethylene production (cont.)
    effect of temperature on, 47–68
    effect of wounding on, 52, 55, 57
    role of membrane in, 3, 25–26, 49–52, 58–60
Ethylene receptors, 101
Ethylene scrubbers, 293–294, 305, 360–361, 373–378, 397–404
Explant abscission, 174, 180–182, 186

Farnesene, 309
Festuca pratensis, 269
Fick's law, 299
Flowers, 2, 71–80, 85, 88–90, 127, 343–350
Flower senescence, 71–74, 83, 272, 343–344, 348–350
N-Formyl-ACC, 66, 365–372
Freezing damage, 52
Fruit ripening, 1–5, 297–301, 304–307, 317–319
    and senescence, 272–273
    avocado, 163–171
    gene expression during, 147–160, 163–171
    temperature effect on, 65–66
    tomato, 10, 127, 147–160
Fruit storage, 297–312, 317–330, 373–382, 385–392, 393–404
Fungal infection, 93–96
Fusicoccin, 101

Gas chromatography, 1, 95, 140, 177, 277, 356
Germination, 14, 66–67, 113
Gibberellin, 67, 186, 213, 247, 249–250, 310, 319, 321
β1-3-Glucanase, 2–3, 190
Glucosamine, 95, 97
Gomphrena globosa, 38–39
Grapefruit, 65, 93–99
Gravitropism, 223–225, 255
Growth, 368
    effects of ethylene, 198–199
    inhibition, 133, 213–214
    of lateral buds, 213–220

Heterophylly, 250
Hydroxycinnamate CoA ligase, 2
Hypobaric,
    effects on abscission, 185, 188–189
    storage of fruit, 85, 298, 305, 348

IAA oxidase, 11, 90–91
Illuminating gas, 1, 176, 187, 287, 343
Immunoelectrophoresis, 203
Indole acetic acid (*see* Auxins)
Invertase, 2

Laccase, 2
Lateral bud growth, 213–220
Latex flow, 2, 364
Leaf,
    effects of wounding, 215
    senescence, 267–273
Light, 10, 37–39, 267, 269

Lycopene, 155–156

McIntosh apples, 385–392, 395–404
Malate dehydrogenase, 2
N-Malonyl-ACC, 9–17, 64, 370
Malonyltransferase, 16–19
Mango, 326–327
Membranes,
    and ethylene production, 11, 25–26, 271
    effects of temperature on, 49–52, 55–60
    ethylene binding sites, 117
    integrity during ripening, 148, 333
Mesocotyl, 247–248
Metal ions, 341
Micro-organisms, 256–257, 288, 291–293, 346
Mitochondria, 11, 111, 268
Modified atmosphere stores, 297, 305, 308
Monocarpic plants, 267
Mono-oxygenase, 136, 144
Mungbean, 11–12, 17, 25, 54, 61, 368
Mutants, 148, 151, 153, 269, 283, 311

NADH, 141–144
NADPH, 141–144
Naphthalene acetic acid, 183
Naptalam, 250
Nicotiana tabacum, 73, 104–105, 108, 269
Nodulation, 255
Non-climacteric fruit, 93, 148, 301, 304, 318
Norbornadiene, 107
Nymphoides peltata, 234–238, 250–251

Octyl glucoside, 112, 118
Ouchterlony double diffusion, 202–203
Ovary swelling, 344
Oxygen, 29–34, 179, 243–246, 249, 253–254, 307–308, 345

Panama disease, 320
Papaya, 327–328
Pear, 300, 303
Pectin, 380–381
Penicillium digitatum, 93–99
Peroxidase, 2, 11, 91, 190
Petals, 87, 89
Petiole development, 229–239
Petunia hybrida, 73–74, 79–80
Phase transition, 49–51
Phaseolus aureus, 104–105, 109–110, 112
Phaseolus vulgaris, 47, 49, 214–215
    abscission in, 199, 204–207, 209–210, 216
    bud growth in, 213–214, 217–220
    ethylene binding sites, 102–114, 117
Phenylalanine ammonia lyase, 2
Photosynthesis, 38–39, 268
Phytochrome, 39
Pineapple, 328
Pisum sativum, 11–12, 26, 43, 48–54, 127–141, 308, 368
Plant growth regulators, 363–366
Pollen, 72–75, 79–80, 86, 344
Pollination, 71–77, 83, 86, 184, 344

416 Index

Pollution,
  automobile exhaust, 289–290, 347, 354–355
  by ethylene, 277–285, 287–289, 293, 343, 346, 354–358
  control of, 293–294, 360–361
  industrial, 291
  vegetation, 290–291, 348, 355
Polyamine, 270, 348
Polygalacturonase, 2, 4, 148, 150–156, 176, 189–190
Positional differentiation, 198, 201
Propylene, 107, 132, 140, 144–145, 288, 344
Protein synthesis, 348
  during abscission, 174, 188, 199, 268
  during ripening, 148, 160, 163
Proteolysis, 268
Pyridoxal phosphate, 2, 23, 90

Ranunculus repens, 230–234, 238, 250–251
Ranunculus sceleratus, 38, 40–45
Refrigerated stores, 297, 305, 307
Respiration, 174, 268, 333–338, 340
Respiratory climacteric,
  as a harvest indicator in apples, 389–391, 393–394
  during ripening, 148, 154–159, 163, 167, 297–301, 318–319
  in soursop fruit, 333–338
Rhizobitoxin, 2, 3, 310
Ribonuclease, 2
Rice, 135–136, 245–249, 251–252, 255
RNA,
  synthesis during abscission, 174–176, 188–190
  synthesis during ripening, 149–157, 160, 163, 166–168
  synthesis during senescence, 268, 273
Roots, 2, 243, 251–256

Seed germination, 4, 66–67, 364
Senescence,
  of flowers, 5, 71–80, 83, 85, 91, 272, 343–344, 348–350
  of leaves, 267–274
Separation layer, 173–174, 176
Sigatoka disease, 320
Silver, 348
  effect on abscission, 179, 186
  effect on aerenchyma, 253
  effect on ethylene metabolism, 132, 140–141
  effect on growth, 133, 217–218, 246, 249, 255
  effect on ripening, 34, 310, 339–342
  effect on senescence, 83, 271–272, 348–350

Soil ethylene, 256–258, 288, 291–292
Solanum tuberosum, 221, 223
Soursop fruit, 333–337
Stolon, 259
Stomata, 41–42, 259, 284
Stomatal resistance, 284
Stress, 13–14, 96–97, 215–216
Style, 86
Sugar beet, 53
Sunflower, 278, 283–284
Superficial scald, 309, 373, 377, 379, 382

Target cells, 183, 197–212
Temperature, 243, 345
  effect on ethylene metabolism, 127
  effect on ethylene production, 47–68, 298, 305–306, 324, 356
  effect on ripening, 305, 307, 321
TH6241, 3
Tillering, 281–282
Tissue sensitivity, 259
  and abscission, 179–183, 185
  and flowers, 344–346
  and ripening, 155, 298, 300–303, 318
  and senescence, 271
  effect of ethylene oxide, 132, 134
  in vegetables, 358–360
Tomato, 10, 15–17, 23–24, 34, 49, 54, 127, 147–160, 300–305, 311, 354, 357, 367–369
Triple response, 67, 135, 221
Tropic curvature, 221

Uronic acid oxidase, 176, 190

Vacuoles, 12, 25–26
Vacuum extraction, 215
Vegetables,
  ethylene production, 355–358
  ethylene sensitivity, 353–354
Vicia faba, 26, 102–103, 126–127, 130–132, 139, 141–144
Vicia sativa, 12

Waterlogging, 14, 130, 241–260, 292
Water stress, 10, 13–15, 215, 216
Waxing, 325
Wheat, 14, 15, 64, 127
Wilting, 13–15, 73, 84, 88, 283–284, 344, 347
Wounding, 23–24, 34, 47–57, 62, 74, 259, 335, 341, 348